PHYSICS OF HOT PLASMAS

PHYSICS OF HOT PLASMAS

Scottish Universities' Summer School
1968

Edited by

B. J. RYE, B.Sc., Ph.D.

and

J. C. TAYLOR, B.Sc., Ph.D.

OLIVER & BOYD

EDINBURGH

OLIVER AND BOYD

Tweeddale Court
Edinburgh 1

(a division of Longman Group Ltd)

FIRST PUBLISHED   1970

05 002126 5

ISBN 978-1-4615-8641-8   ISBN 978-1-4615-8639-5 (eBook)
DOI 10.1007/978-1-4615-8639-5

PREFACE

THE ninth Scottish Universities' Summer School in Physics, sponsored jointly by the Scottish Universities and NATO was held at Newbattle Abbey from 28th July to 16th August 1968. This was the first Scottish Summer School to be devoted to plasma physics, the exact title for the School being the Physics of Hot Plasmas. Forty-three students were accepted, fourteen of these being resident in the United Kingdom. In addition there were eleven lecturers and seven other participants.

The choice of lecturers, particularly in experimental plasma physics, was limited to some extent by the fact that an international conference on controlled fusion was held at Novosibirsk during the first week in August. Notwithstanding this, it was possible to arrange a programme of lectures reasonably well balanced between theoretical and experimental plasma physics. The topics chosen included kinetic theory, waves and oscillations, instabilities, turbulence, collisionless shocks, computational methods, laser scattering and laser generated plasmas, plasma production and containment. Several seminars on special topics were given by invited speakers and by students.

One of the purposes of the Summer School was to show how theory and experiment were linked in the physics of hot plasmas. It was also our aim to highlight some of the current developments in plasma physics. To be more specific, the programme of lectures clearly showed the importance of kinetic theory (Professors C. Oberman and W. B. Thompson), and the theory of waves and oscillations (Professor R. J. Tayler) for the understanding of micro-instabilities (one of the two classes of instability considered by Professor E. G. Harris) that arise in physical experiments. A topic of great current interest is that of collisionless shocks. Dr H. Völk dealt with the theory of this subject, discussing the earth's bow shock, and also introducing the effects of turbulence, a topic that Professor M. G. Rusbridge lectured on. Dr J. W. M. Paul of Culham Laboratory discussed his experimental investigations on collisionless shocks created by an axial discharge. There is no doubt that collisionless shocks will receive much attention both theoretically and experimentally in the next few years.

The series of lectures on computational methods by Dr J. Killeen showed how useful numerical results could be obtained for many practical plasma physics experiments.

The creation of plasma by focused laser beams was a new topic for most of the students, and very few were familiar with the use of the high-power Q-switched laser as a diagnostic tool that yielded plasma temperature and density in a light scattering experiment. The theoretical and experimental aspects of these topics were admirably covered by Professors W. B. Thompson and S. A. Ramsden respectively.

A thorough treatment of optical diagnostics for the study of plasmas was given by Professor U. Ascoli-Bartoli, while Dr G. B. F. Niblett dealt with experiments on the production and containment of hot plasmas.

The Director is particularly dependent on his Executive Committee for the organization and running of the School, and I wish to express my thanks to Drs E. W. Laing (Secretary), D. E. Kidd (Treasurer), J. C. Taylor and B. J. Rye (Joint Editors), and J. Rae and R. S. Stewart (Stewards) for their efforts to make the School a success. They have been most ably assisted in their clerical work by Miss A. Currie of the University of Glasgow and Miss M. Smith of the University of Strathclyde, who was also involved with the registration of students and conference office duties during the first week. I am sure that all the participants in the School will join me in thanking all these helpers and also the members of the Staff of Newbattle Abbey, who contributed so essentially to the smooth and pleasant running of the School.

J. IRVING

Professor of Natural Philosophy,
University of Strathclyde

EDITORS' NOTE

The lecture notes were prepared by the lecturers themselves before the Summer School and in some cases later revised. As the page proofs were not checked by the lecturers, any misprints and similar errors are our responsibility.

We would like to express our appreciation to the printers for work which has demanded the highest performance (see Chapter 2!). Our thanks go also to the publishers, Oliver & Boyd.

B. J. RYE
J. C. TAYLOR

Professor J. Irving, Strathclyde, *Chairman*
Dr E. W. Laing, Glasgow, *Secretary*
Dr D. E. Kidd, Strathclyde, *Treasurer*
Dr J. Rae, Glasgow
Dr R. S. Stewart, Strathclyde
Dr B. J. Rye, Strathclyde, *Joint Editor*
Dr J. C. Taylor, Glasgow, *Joint Editor*

LECTURERS

Professor U. Ascoli-Bartoli, L.G.I., Frascati
Professor E. G. Harris, Tennessee
Dr J. Killeen, L.R.L., Livermore
Dr G. B. F. Niblett, Culham Laboratory
Professor C. Oberman, Princeton
Dr J. W. M. Paul, Culham Laboratory
Professor S. A. Ramsden, Hull
Professor M. G. Rusbridge, Manchester
Professor R. J. Tayler, Sussex
Professor W. B. Thompson, San Diego, California
Dr H. Völk, Max Planck Institut, Garching

PARTICIPANTS

Mr T. O. Aro, Zaria, Nigeria
Mr J. P. Baconnet, Villeneuve Saint Georges, France
Mr F. T. Barclay, Glasgow
Mr R. A. H. Bardet, Gif-sur-Yvette, France
Mr P. Bartholomew, Brighton
Mr A. Blanc, Villeneuve Saint Georges, France
Mr A. E. Blaugrund, Rehovot, Israel
Mr G. Bouligand, Fontenay-aux-Roses, France
Mr M. Brambilla, Gif-sur-Yvette, France
Mr R. A. Cairns, Glasgow
Mr F. Cipolla, Roma, Italy
Mr M. Dobrowolny, Rome, Italy
Mr M. Espedal, Bergen, Norway
Mr H. Gerhauser, Julich, West Germany
Mr J. Godart, Orsay, France
Mr C. Gormezano, Gif-sur-Yvette, France
Mr I. Grabec, Ljubljana, Yugoslavia
Mr G. Granata, Fontenay-aux-Roses, France
Mr H. Gustafsson, Stockholm, Sweden
Mr Andrew F. Hogan, Glasgow
Mr N. Hopfgarten, Stockholm, Sweden

Mr N. Hornqvist, Stockholm, Sweden
Mr J. H. Jackson, Newcastle-upon-Tyne
Mr L. Janicke, Julich, West Germany
Mr A. Lamont, Glasgow
Mr F. A. McFarlane, Glasgow
Mr D. P. Mason, Oxford
Mr C. J. Myerscough, Cambridge
Mr R. Nakach, Gif-sur-Yvette, France
Dr J. P. Nicholson, Glasgow
Mr P. Nielsen, Roskilde, Denmark
Mr A. H. Øien, Bergen, Norway
Mr L. Pieroni, Rome, Italy
Mr A. A. H. Poquerusse, Paris, France
Mr E. R. Priest, Leeds
Mr C. Salter, Brighton
Mr R. Shuker, Rochester, N.Y.
Mr Soubbaramayer, Fontenay-aux-Roses, France
Dr P. Stewart, Manchester
Mr G. Tonon, Villeneuve-Saint-Georges, France
Mr K. P. Tritton, Sussex
Dr F. Verheest, Prince on, N.J.
Mr J. Waldes, New York

CONTENTS

12. PLASMA DIAGNOSTICS BASED ON REFRACTIVITY *U. ASCOLI-BARTOLI*

1

INTRODUCTION TO KINETIC THEORY OF PLASMA

W. B. THOMPSON

University of California, La Jolla

1

INTRODUCTION

1.1. *Object*

The main object of kinetic theory is to produce a description of the macroscopic behaviour of large systems by considering the microscopic behaviour of their atomic and molecular parts. This should lead to a logical simplification in our understanding of matter, although perhaps at the expense of technical or formal complications. While the macroscopic behaviour of fluids is expressed in terms of variables such as velocity, density, pressure, temperature, electric current etc., with equations of state which describe "the properties of matter", the behaviour of the molecular components follows simple mechanical laws.

For plasmas in particular, kinetic theory is important; for, whereas for most fluids macroscopic equations of state are easily determined but the microscopic behaviour is only vaguely understood—the intermolecular forces being inferred from the kinetic theory calculations—for plasmas the situation is reversed: macroscopic properties are measured only with difficulty, and often basic parameters must be indirectly inferred, while the microscopic interparticle forces are well known and fairly simple.

For gaseous plasmas, a further simplification is available. It is usually enough to give the simple classical description of particle motion. For electrons the de Broglie wavelength $\lambda \simeq h/(mv) \simeq 10^{-7}\mathscr{E}^{-\frac{1}{2}}$ cm (where $\mathscr{E}$ is in eV), while the mean interparticle distance $n^{-\frac{1}{3}} \simeq 10^{-5}$ cm at $n \simeq 10^{15}$ cm^{-3}, which suggests that the overlap between wave packets is small so that a classical description is adequate. This is not completely true—close encounters should perhaps have a quantum description, and in the partition function, quantum effects alone prevent electrons from falling into ions.

On the other hand, the electron rest energy is 0·5 MeV, and for plasma temperatures less than a few keV relativistic effects are probably unimportant. Again, this argument must be viewed with some caution. The self-magnetic

B

interaction which is unscreened and may be important, is, in fact, a relativistic process. In what follows, however, we shall neglect relativistic processes as well as quantum effects.

1.2. *Levels of description*

As has been observed, the gross behaviour of a simple field is characterized by variables such as the density ρ, the velocity V, the temperature T, and the pressure, or stress tensor, p. Usually there exists an equation of state $\Phi(\rho, p, T) = 0$ relating p or some part of it to ρ, T, while the velocity and density are related to the stress tensor and any applied force density $\mathscr{F}$ by hydrodynamic equations; all quantities being determined as functions of position x and time t.

On the other hand, the microscopic description requires the specification of a set of $3N$ coordinates q_i, and momenta p_i, as functions of time, N being the (large) number of particles in the system.

These variables satisfy the equations of mechanics most usefully written in Hamiltonian form:

$$\mathscr{H} = \mathscr{H}(q_1, \ldots, q_N, p_1, \ldots, p_N)$$

$$\dot{q}_i = \frac{\partial \mathscr{H}}{\partial p_i}, \quad \dot{p}_i = -\frac{\partial \mathscr{H}}{\partial q_i}.$$

Note further that for function f of time t and the dynamical variables p and q, $f(t, \ldots, q_i, \ldots, p_i, \ldots)$

$$\frac{Df}{Dt} = \frac{\partial f}{\partial t} + \dot{q}_i \cdot \frac{\partial f}{\partial q_i} + \dot{p}_i \cdot \frac{\partial f}{\partial p_i} = \frac{\partial f}{\partial t} + \left[\frac{\partial \mathscr{H}}{\partial p_i} \cdot \frac{\partial f}{\partial q_i} - \frac{\partial \mathscr{H}}{\partial q_i} \cdot \frac{\partial f}{\partial p_i} \right].$$

The procedure used to connect microscopic and macroscopic phenomena is the introduction of a distribution function. The mechanical description can clearly be effected by introducing a function $D_N(q_1, q_2, \ldots, q_N, p_1, p_2, \ldots, p_N)$ such that the probability of finding the system in the hypervolume $d\tau_N = d^3q_1 \cdots d^3q_N d^3p_1 \cdots d^3p_N$ in the $6N$ dimensional phase space, near $q_1, \ldots, q_N, p_1, \ldots, p_N$ is $P = D_N(q_1, \ldots, q_N, p_1, \ldots, p_N)d\tau_N$. If the initial conditions for the motion are known, as say $q_1^0, \ldots, q_N^0, p_1^0, \ldots, p_N^0$ at $t = 0$ then the distribution at any subsequent time $D_N = K_N(q_1, \ldots, q_N, p_1, \ldots, p_N, t; q_1^0, \ldots, q_N^0, p_1^0, \ldots, p_N^0)$ is clearly

$$K = \prod^N \delta(q_1 - Q_1)\delta(q_2 - Q_2) \cdots \delta(q_N - Q_N)\delta(p_1 - P_1)\delta(p_2 - P_2) \cdots$$
$$\delta(p_N - P_N) \qquad \ldots[1.1]$$

where the quantities Q_i, P_i are solutions of the equations of motion, $\dot{Q}_i = \partial \mathscr{H}/\partial P_i, \dot{P}_i = -\partial \mathscr{H}/\partial Q_i; Q_i(0) = q_i^0, P_i(0) = p_i^0$, for K must clearly vanish for any $d\tau_N$ not containing Q_N, P_N and must yield unity when integrated over any volume $d\tau_N$ containing that $6N$ point. We can also construct the differential equation in the $(6N+1)$ space $(q_1, \ldots, q_N, p_1, \ldots, p_N, t)$

satisfied by K. Since K depends on t only through the Q, P which makes its dependence on q, p explicit,

$$\frac{\partial K}{\partial t} = \left[\frac{\partial K}{\partial Q_i} \cdot \dot{Q}_i + \frac{\partial K}{\partial P_i} \cdot \dot{P}_i\right] = -\left[\dot{Q}_i \cdot \frac{\partial K}{\partial q_i} + \dot{P}_i \cdot \frac{\partial K}{\partial p_i}\right].$$

The $\dot{Q}_i$, $\dot{P}_i$ are given in terms of the Hamiltonian as explicit functions of Q, P, t thus $\dot{Q}_i = (\partial/\partial P_i)[\mathscr{H}(Q, \ldots, P, \ldots)]$, $\dot{P}_i = -\partial\mathscr{H}/\partial Q_i$. Because of the δ-functions in K, the (explicit) Q, P here may be replaced by q, p. Hence $(\partial K/\partial t) + [\mathscr{H}, K] = 0$ and K is constant on trajectories.

If on the other hand, the probability of finding q_i^0, p_i^0 is given by $P(q_1^0, \ldots, q_N^0, p_1^0, \ldots, p_N^0)$ the subsequent probability

$$P(q_1, \ldots, q_N, p_1, \ldots, p_N, t) =$$
$$\int K(q_1, \ldots, q_N, p_1, \ldots, p_N, t, q_1^0, \ldots, q_N^0, p_1^0, \ldots, p_N^0)P(q^C, p^0)dt.$$

I can, however, identify $P(t)$ with $D_N(q, p, t)$; $P(q^0, p^0)$ with $D_N(q_1, \ldots, q_N, p, 0)$; and since P depends on time and the q's and p's only through the dependence of K, we obtain the differential equation satisfied by D_N: the Liouville equation

$$\frac{\partial D}{\partial t} + \sum_N \left(\dot{q}_i \cdot \frac{\partial D}{\partial q_i} + \dot{p}_i \cdot \frac{\partial D}{\partial p_i}\right) = \frac{\partial D_N}{\partial t} + [\mathscr{H}, D_N] = 0. \qquad \ldots[1.2]$$

This distribution function is that most closely related to the mechanical description, but is remote from the fluid dynamic description. To approach this, we can introduce a set of reduced distribution functions defined as follows

$$f_S = V^S \int D_N d^3q_{S+1} d^3q_{S+2} \cdots d^3q_N d^3p_{S+1} \cdots d^3p_N. \qquad \ldots[1.3]$$

Usually we assume that all N particles are equivalent. (If they are of different types, the analysis can be carried through for each *type* separately.) Thus it does not matter how the S particles are selected. V here is the total volume of the system, and represents a normalization of f. The relation to hydrodynamics is most clearly seen in the case of non-interacting particles, when the Hamiltonian $\mathscr{H}(q_1, \ldots, q_N, p_1, \ldots, p_N)$ may be written as a sum of terms each depending on a single q, p, thus $\mathscr{H} = \sum_i \mathscr{H}_i(q_i, p_i)$. If, moreover, there is initially no correlation between the particles so that the initial distribution can be factored

$$D_N(q_1, \ldots, q_N, p_1, \ldots, p_N, t) = \prod f_i(q_i, p_i)$$

then the f satisfy identical equations, provided all particles are equivalent:

$$\partial f_i/\partial t + [\mathscr{H}_i, f] = 0$$

or, in terms of the velocity v and the acceleration a,

$$\frac{\partial f}{\partial t} + v \cdot \frac{\partial f}{\partial x} + a \cdot \frac{\partial f}{\partial v} = 0 = Lf. \qquad \ldots[1.4]$$

Now, if we normalize f not to 1, but to N (the number of particles in V), then the hydrodynamic variables ρ and V can be formed thus:

$$\rho(x, t) = \sum m = \int mf(x, v, t)d^3v$$
$$\rho V(x, t) = \int mvfd^3v.$$

To get the hydrodynamic equations we construct the moments of $[1.4]$ thus:

$$\int mLfd^3v = \frac{\partial \rho}{\partial t} + \frac{\partial}{\partial x} \cdot \rho V + \int ma \cdot \frac{\partial f}{\partial v} d^3v.$$

The last term here can be handled by a partial integration, and, if $(\partial/\partial v) \cdot a = 0$, then the last term vanishes, and

$$\partial \rho/\partial t + (\partial/\partial x) \cdot \rho V = 0. \qquad \qquad ...[1.5]$$

To construct $\partial V/\partial t$, consider

$$\int mvLfd^3v = (\partial/\partial t) \int mvfd^3v + (\partial/\partial x) \cdot \int mvvfd^3v + \int mva : (\partial f/\partial v) d^3v = 0.$$

The first term here is just $(\partial/\partial t)\rho V$, while on introducing $v = V+c$ with c as a new independent variable, and $\int f(c)cd^3c = 0$, the second term becomes $(\partial/\partial x) \cdot (\rho VV) + (\partial/\partial x) \cdot (\rho\langle cc \rangle)$; while on a partial integration the last term becomes $-\int mafd^3v = -\mathscr{F}$ the force density; hence, with $\mathsf{p} = \rho\langle cc \rangle$, $(\partial/\partial t)\rho V + (\partial/\partial x) \cdot (\rho VV) + (\partial/\partial x) \cdot \mathsf{p} - \mathscr{F} = 0$. The first two terms may be written $[\partial \rho/\partial t + (\partial/\partial x) \cdot (\rho V)]V + \rho[\partial V/\partial t + V \cdot (\partial V/\partial x)]$, the first bracket vanishing from the equation of continuity; hence,

$$\rho(DV/Dt) + (\partial/\partial x) \cdot \mathsf{p} - \mathscr{F} = 0. \qquad \qquad ...[1.6]$$

(This, of course, involves p, a second moment of f.) To get at the components of p, we can, for example, form the moment $\int mv^2Lfd^3v$, which yields, introducing $(V+c)$,

$$(\partial/\partial t) \int \tfrac{1}{2}m(V^2+c^2+2V \cdot c)fd^3v + (\partial/\partial x) \cdot \int m(V+c)\tfrac{1}{2}(V^2+c^2+2V \cdot c)d^3v$$
$$+ \int \tfrac{1}{2}m(V+c)^2a \cdot (\partial f/\partial v)d^3v$$
$$= (\partial/\partial t)(\tfrac{1}{2}\rho V^2 + \tfrac{1}{2}\rho\langle c^2 \rangle) + (\partial/\partial x) \cdot (\tfrac{1}{2}\rho V^2 V + V\rho\langle \tfrac{1}{2}c^2 \rangle + \rho V \cdot \langle cc \rangle) - V \cdot \mathscr{F}$$

(where we have assumed $\langle c \cdot a \rangle = 0$, an inessential assumption). Some simplification can be made here; for we may write the equation as

$$\tfrac{1}{2}V^2[(\partial \rho/\partial t) + (\partial/\partial x) \cdot \rho V] + V \cdot \{\rho[(\partial V/\partial t) + V \cdot (\partial V/\partial x)] + (\partial/\partial x) \cdot \mathsf{p} - \mathscr{F}\}$$
$$+ (\partial/\partial t)(\tfrac{1}{2}\rho\langle c^2 \rangle) + V \cdot (\partial/\partial x)(\tfrac{1}{2}\rho\langle c^2 \rangle)$$
$$+ \rho\langle c^2 \rangle \operatorname{div} V + \mathsf{p} \cdot V + (\partial/\partial x)(\langle \tfrac{1}{2}\rho cc^2 \rangle) = 0. \qquad ...[1.7]$$

Note that the first two terms vanish, and that the last involves a third moment of f. Note that at each stage some information is required about the distribution function; so that the hydrodynamic set does not close. To produce closure, information about f is required, and the object of kinetic theory is to provide that information.

If f is known to be isotropic in c, then

$$p = \tfrac{1}{3}\rho\langle c^2\rangle = \tfrac{2}{3}C_V\kappa T = n\kappa T$$

(where n is the particle number density $= \rho/m$, and C_V and $n\kappa T$ define the specific heat and internal energy) and

$$
\left.
\begin{aligned}
\frac{D}{Dt}(\tfrac{3}{2}n\kappa T)+\tfrac{5}{2}n\kappa T \operatorname{div} V &= \frac{1}{n\kappa T}\frac{D}{Dt}(n\kappa T)-\frac{5}{3n}\frac{Dn}{Dt} \\
&= \frac{D}{Dt}(p\rho^{-\frac{5}{3}}) \\
&= 0.
\end{aligned}
\right\}
\quad \ldots[1.8]
$$

1.3. The B.B.G.K.Y. (*Bogoliubov, Born, Green, Kirkwood, Yvon*) hierarchy

If, for interacting particles, we could find f_1, then we could "easily" construct the hydrodynamic equations. D_N, however, is still a long way from f_1; but f_1 is a reduced distribution function; and hence, from the Liouville equation we can construct an equation for it. Let us first integrate [1.2] over all q except a particular q_i. If, for simplicity, we use Cartesian coordinates, and if the force acting on any particle consists of some external "applied" force F and a force derivable from the interparticle potential ϕ, [1.2] becomes

$$\frac{\partial D}{\partial t}+\sum_0^N\left[v_i\cdot\frac{\partial D}{\partial x_i}+\frac{F}{m}\cdot\frac{\partial D}{\partial v_i}\right]-\sum_{i,j}^{N,\,N-1}\frac{1}{m}\frac{\partial\phi(x_i,x_j)}{\partial x_i}\cdot\frac{\partial D}{\partial v_i}=0.$$

On performing the integral over all but x_i, v_i, we obtain (multiplying through by V):

$$\frac{\partial f_1}{\partial t}+v_i\cdot\frac{\partial f_1}{\partial x_i}+\frac{F}{m}\cdot\frac{\partial f_1}{\partial v_i}-\frac{1}{V}\sum_j\frac{1}{m}\int\frac{\partial\phi(x_i,x_j)}{\partial x_i}\cdot\frac{\partial f_2(i,j)}{\partial v_i}\,dq_j\,dp_j=0;$$

and, if all particles are equivalent, the last term may be written

$$\frac{N-1}{V}\int\frac{\partial\phi(x,x')}{\partial x}\cdot\frac{\partial f_2(x,x',v,v',t)}{\partial v}\,d^3x'\,d^3v';\quad\text{and}\quad\frac{N-1}{V}=n$$

the number density; hence, we may write an equation for f_1:

$$\frac{\partial f_1}{\partial t}+v\cdot\frac{\partial}{\partial x}f_1+\frac{F}{m}\cdot\frac{\partial}{\partial v}f_1-\frac{n}{m}\int\frac{\partial\phi(x,x')}{\partial x}\cdot\frac{\partial f_2(x,x',v,v')}{\partial v}\,d^3x'\,d^3v'=0.$$

$$\ldots[1.9]$$

Of course, to determine f_1, we now need the equation for f_2. By exactly similar arguments, it becomes

$$\frac{\partial f_2}{\partial t} + \mathbf{v} \cdot \frac{\partial}{\partial \mathbf{x}} f_2 + \mathbf{v}' \cdot \frac{\partial}{\partial \mathbf{x}'} f_2 + \frac{\mathbf{F}}{m} \cdot \frac{\partial}{\partial \mathbf{v}} f_2 + \frac{\mathbf{F}'}{m} \cdot \frac{\partial}{\partial \mathbf{v}'} f_2$$

$$- \frac{1}{m} \int \left[\frac{\partial \phi(\mathbf{x}, \mathbf{x}')}{\partial \mathbf{x}} \cdot \frac{\partial}{\partial \mathbf{v}} + \frac{\partial \phi(\mathbf{x}, \mathbf{x}')}{\partial \mathbf{x}'} \cdot \frac{\partial}{\partial \mathbf{v}'} \right] f_2 \, d^3 x' \, d^3 v' \qquad \ldots [1.10]$$

$$- \frac{n}{m} \int \left[\frac{\partial \phi(\mathbf{x}, \mathbf{x}'')}{\partial \mathbf{x}} \cdot \frac{\partial}{\partial \mathbf{v}} + \frac{\partial \phi(\mathbf{x}', \mathbf{x}'')}{\partial \mathbf{x}'} \cdot \frac{\partial}{\partial \mathbf{v}'} \right] f_3(\mathbf{x}, \mathbf{x}', \mathbf{x}'', \mathbf{v}, \mathbf{v}', \mathbf{v}'') \, d^3 x'' \, d^3 v''$$

$$= 0,$$

which clearly requires a knowledge of f_3 for its solution. Repetitions of this operation lead to equations which express the f_S and eventually f_1 in terms of the D_N; a clearly intractable system. A major activity in kinetic theory is the search for plausible methods of truncating this hierarchy, and for arguments to justify that truncation.

1.4. *The Boltzmann equation*

A particularly simple case arises, if the gas is dilute and the forces of interaction are of short range. Then, almost everywhere in the gas, $f_2 = f_1(1) f_2(2)$; and only where the interaction is actually occurring is it necessary to include the correlation of particle motions. Moreover, at most two particles can be expected to interact; hence it is reasonable to introduce the center of mass and the separation as coordinates, thus:

$$\mathbf{V} = (\mathbf{v}_1 + \mathbf{v}_2)/2; \quad \mathbf{g} = \mathbf{v}_1 - \mathbf{v}_2; \quad \mathbf{X} = (\mathbf{x}_1 + \mathbf{x}_2)/2; \quad \mathbf{x} = \mathbf{x}_1 - \mathbf{x}_2;$$

and the last (awkward) term in [1.9] may be written, if $\phi = \phi(x_1 - x_2)$,

$$- (1/m) \int (\partial/\partial \mathbf{x}) \phi(\mathbf{x}) \cdot (\partial/\partial \mathbf{g}) f_2(\mathbf{X}, \mathbf{V}, \mathbf{x}, \mathbf{g}) d^3 g d^3 x.$$

Now f_2 has the following properties: at the ingoing side of the "sphere of influence" (region within which ϕ matters), $f_2 = f_1(x) f_1(x')$; while on the outgoing side, f_2 is again $f_1(x) f_1(x')$. Between these points, the dependence on $\mathbf{x}$, $\mathbf{g}$ is given by $\delta(\mathbf{x} - \mathbf{X}(x_0, g_0, t)) \delta(\mathbf{g} - \mathbf{G}(x_0, g_0, t))$, where the δ-functions define the (relative) trajectory. Using the property of these δ-functions the integral may be written $- \int \dot{\mathbf{g}} \cdot (\partial/\partial \mathbf{g}) f_2 d^3 g d^3 x$. Furthermore, the volume integral can be carried out as follows. For any $\mathbf{g}$, we replace the sphere of influence by a cylinder with axes parallel to $\mathbf{g}$, and then introduce polar coordinates on the surface b, φ. For the third coordinate, we integrate along the trajectory of a particle $x = \int g dt$. This yields

$$\int \dot{\mathbf{g}} \cdot (\partial/\partial \mathbf{g}) f d^3 g b db d\varphi g dt = \int [f_2(\mathbf{g} + \Delta \mathbf{g}) - f_2(\mathbf{g})] g b db d\varphi.$$

(Use $x_f \simeq x_i$; energy conservation ensures that $g = \text{constant}$.) But the final result depends on f_2 on the surface and hence on $f_1(v) f_1(v')$. So the integral becomes

$$\int [f_1(V+\tfrac{1}{2}(g+\Delta g), X) f_1(V-\tfrac{1}{2}(g+\Delta g), X)$$
$$-f_1(V+\tfrac{1}{2}g, X) f_1(V-\tfrac{1}{2}g, X)] \, gb\,db\,d\varphi$$
$$= \int g[f(\bar{v}) f(\bar{v}') - f(v) f(v')] \, b\,db\,d\varphi. \qquad \ldots[1.11]$$

Note that the $\bar{v}$, $\bar{v}'$ are functions of v, v' and b. Note further that on a collision g can only change direction, for

$$\begin{aligned}
\tfrac{1}{2}m\bar{v}^2 + \tfrac{1}{2}m(\bar{v}')^2 &= \tfrac{1}{2}m\{[V+\tfrac{1}{2}(g+\Delta g)]^2 + [V-\tfrac{1}{2}(g+\Delta g)]^2\} \\
&= \tfrac{1}{2}m[2V^2 + \tfrac{1}{2}(g+\Delta g)^2] \\
&= \tfrac{1}{2}m[2V^2 + \tfrac{1}{2}g^2].
\end{aligned}$$

Hence, Δg is characterized by two angles, θ, φ and for central forces, the single angle θ. Since θ, $\varphi = \theta(b)$, $\varphi(b)$: $b = b(\theta, \varphi)$ we could go from b as a variable to θ, φ or the solid angle Ω. Then $2\pi b\,db = \sigma(\theta, \varphi)\,d\Omega$ and the right-hand side of the Boltzmann equation becomes $\int g[f_1(\bar{v}) f_1(\bar{v}') - f_1(v) f_1(v')]$ $\sigma d\Omega d^3 v$. Finally, replacing f_1 by $f = (N/V)f_1$, we may write the Boltzmann equation

$$\frac{\partial f}{\partial t} + v \cdot \nabla f + \frac{F}{m} \cdot \frac{\partial}{\partial v} f = \int g[f(\bar{v}) f(\bar{v}') - f(v) f(v')] \, \sigma d\Omega d^3 v'. \quad \ldots[1.12]$$

2

APPLICATION OF THE BOLTZMANN EQUATION TO PLASMAS

The Boltzmann equation was derived on the assumption that the interaction potential was of short range; thus we should not expect it to be applicable to a plasma. On the other hand, it seems (intuitively) a reasonably satisfactory way to describe collisions. In attempting to apply the Boltzmann equation to plasmas there is an immediate difficulty, which arises from the form of the cross section. This appears as

$$\sigma(g, \theta)d\Omega = \tfrac{1}{4}[e^2/(mg^2)]^2 \sin^{-4}(\theta/2)\,d\Omega \qquad \ldots[2.1]$$

and hence is singular for small scattering angles θ. This singularity, however, is not quite as bad as it looks, since $\bar{v}$, $\bar{v}'$ differ from v, v' by terms of order θ. Indeed, we can expand the integral $\mathscr{I}$ in $[1.12]$ for small values of Δg, noting that

$$\bar{v} = v + \tfrac{1}{2}\Delta g; \qquad \bar{v}' = v' - \tfrac{1}{2}\Delta g;$$

hence

$$[f(\bar{v}) f(\bar{v}') - f(v) f(v')] = \tfrac{1}{2}\Delta g \cdot \left[\frac{\partial f(v)}{\partial v} f(v') - f(v) \frac{\partial f(v')}{\partial v'} \right] + \tfrac{1}{8}\Delta g \Delta g :$$

$$\left[\frac{\partial^2 f(v)}{\partial v \partial v} f(v') + \frac{\partial^2 f(v')}{\partial v' \partial v'} f(v) - 2 \frac{\partial f(v)}{\partial v} \cdot \frac{\partial f(v')}{\partial v'} \right]. \qquad \ldots[2.2]$$

We now represent the components of Δg on an axis $\hat{g} \parallel g$, and two (arbitrary!) perpendicular axes m, n, whereupon (because of the conservation of $|g|$) $\Delta g = g[-\hat{g}(1-\cos\theta), m\sin\theta\cos\varphi, n\sin\theta\sin\varphi]$ and $\mathscr{I}$ becomes

$$
\begin{aligned}
\mathscr{I} &= \tfrac{1}{8}\left(\frac{e^2}{m}\right)^2 \int \frac{1}{g^3}\left\{\Delta g \cdot J + \tfrac{1}{4}\Delta g \Delta g : \left(\frac{\partial}{\partial v}-\frac{\partial}{\partial v'}\right)J\right\}\sin^{-4}(\theta/2)\,d\Omega \\
&= \frac{2\pi}{8}\left(\frac{e^2}{m}\right)^2 \int \frac{1}{g^3}\left\{-(1-\cos\theta)g\hat{g}\cdot J + \frac{g^2}{4}\left[(1-\cos\theta)\hat{g}\hat{g}\right.\right. \\
&\qquad \left.\left. + \tfrac{1}{2}\sin^2\theta(mm+nn)\right]:\left(\frac{\partial}{\partial v}-\frac{\partial}{\partial v'}\right)J\right\}\frac{\sin\theta}{\sin^4(\theta/2)}\,d\theta
\end{aligned}
$$

where $J = J(v, v') = f(v')(\partial f/\partial v) - f(v)(\partial f/\partial v')$.

To carry out the integral over θ:

$$
\begin{aligned}
\sin\theta\,d\theta &= 4\left[\cos(\theta/2)\sin(\theta/2)\,d(\theta/2)\right] \\
1-\cos\theta &= 2\sin^2(\theta/2), \\
\sin^2\theta &= 4\left[\sin^2(\theta/2)-\sin^4(\theta/2)\right]
\end{aligned}
$$

and the terms in the integral are of two types:

$$
\int_0^{\pi/2}\left[\cos(\theta/2)/\sin(\theta/2)\right]d(\theta/2) \quad\text{and}\quad \int_0^{\pi/2}\sin(\theta/2)\cos(\theta/2)\,d(\theta/2).
$$

Of these, the first diverges at $\theta = 0$, while the second remains bounded.

In order to use this expression, we must develop some argument that will justify limiting the scattering angle to $\theta > \theta_0$. A rough argument is as follows.

At small scattering angles, the impact parameter b, which is about equal to the minimum separation of the scatterers, is large, indeed

$$
b = [2e^2/(mg^2)]\cot(\theta/2).
$$

However, if the particle separation is large, the screening by the particles between them becomes important. To get at the range, note that the plasma will only transmit waves at frequencies $\omega > \omega_p$ (ω_p = plasma frequency). The potential acting on a particle as another passes at distance b is approximately $\phi(t) = e(b^2 + V^2 t^2)^{-\frac{1}{2}}$ and its Fourier components

$$
\phi(\omega) = \int_{-\infty}^{\infty} dt\, e^{-i\omega t}(b^2 + V^2 t^2)^{-\frac{1}{2}} = (1/V)K_0(\omega b/v),
$$

where K_0 is a Bessel function. K_0 is exponentially small if $\omega \gg (v/b)$; but if $\omega < \omega_0$, the field is screened by the plasma; hence if $(v/b) < \omega_0$, the particle interaction is weak: i.e. if $b > (v/\omega_0)$, and if we use an approximate average value for $v \simeq v_\theta$, we discover that the particle interaction is effectively cut off at separations $b \simeq \lambda_D \simeq (v_\theta/\omega_p)$ and at angles

$$\frac{\theta_0}{2} \simeq \frac{2e^2}{mv_\theta^2 b} = \frac{1}{4\pi}\left[\frac{1}{n\lambda_D^3}\right] = \Lambda^{-1}, \qquad \ldots[2.3]$$

where $4\pi n\lambda_D^3 \equiv$ number of particles in a Debye sphere, which is usually large. Indeed

$$4\pi n\lambda_D^3 = 4\pi n\left[mv_\theta^2/(8\pi ne^2)\right]^{\frac{3}{2}} = 1/(8\pi)^{\frac{1}{2}}\left[mv_\theta^2/(2e^2n^{\frac{1}{3}})\right]^{\frac{3}{2}}.$$

But $mv_\theta^2/(2e^2n^{\frac{1}{3}})$ represents the ratio of the mean particle kinetic energy to its potential energy, and must be large if no recombination occurs.

If we cut off the integration over θ at $\theta = \theta_0$, the otherwise singular term becomes $\ln[1/\sin(\theta_0/2)] = \ln(4\pi n\lambda_D^3) = \ln \Lambda$, and the large terms in $\mathscr{I}$ become:

$$\mathscr{I} = 2\pi(e^2/m)^2 \ln \Lambda \int d^3v' \{-(\hat{g}/g^2)\cdot J + [1/(4g)](1-\hat{g}\hat{g}):(\partial/\partial v - \partial/\partial v')J\}$$

where **1** is the unit tensor. This can be written more elegantly by using the properties of the tensor $w = (1/g)[1-\hat{g}\hat{g}]$. Note

$$\frac{\partial^2 g}{\partial g \partial g} = \left(\frac{\partial}{\partial g}\right)\frac{g}{g} = \frac{1}{g} - \frac{gg}{g^3} = w$$

$$\left(\frac{\partial}{\partial g}\right)\cdot w = \sum\left(\frac{\partial}{\partial g}\right)_i\left[\frac{\delta_{ij}}{g} - \frac{g_ig_j}{g^3}\right] = -\frac{g}{g^2}\cdot w - \left(\frac{1}{g}\right)\left[\frac{3g+g}{g^2} - \frac{2g}{g^2}\right] = -\frac{2\hat{g}}{g^2};$$

hence

$$\mathscr{I} = 2\pi(e^2/m)^2 \ln \Lambda \int d^3v'\{\tfrac{1}{2}(\partial/\partial g)\cdot w \cdot J + \tfrac{1}{4}w\cdot[(\partial/\partial v)-(\partial/\partial v')]\cdot J\}.$$

The last term may be integrated by parts yielding $\tfrac{1}{4}(\partial/\partial v')\cdot w\cdot J$. But $g = v - v'$; hence $(\partial/\partial v')\cdot w = -(\partial/\partial g)\cdot w = -(\partial/\partial v)\cdot w$ and

$$\mathscr{I} = \tfrac{1}{2}\pi(e^2/m)^2 \ln \Lambda (\partial/\partial v)\cdot \int d^3v'(w\cdot J) \qquad \ldots[2.4]$$

and the Boltzmann equation for an electron gas becomes

$$(\partial f/\partial t)+v\cdot \nabla f+(e/m)(E+v\times B)\cdot(\partial f/\partial v) = \tfrac{1}{2}\pi(e^2/m)^2 \ln \Lambda (\partial/\partial v)\cdot \int a^3v'(w\cdot J),$$

which is Landau's form of the Fokker-Planck equation. For a gas composed of singly charged ions as well as electrons, we note that $M_1v_1 = M_1V+m_rg$, $M_2v_2 = M_2V-m_rg$, where m_1 is the reduced mass, abbreviated m above, and again $|g+\Delta g| = |g|$; hence the results hold, provided that the velocity derivatives are replaced by momentum derivatives, and

$$\frac{\partial f^\pm}{\partial t}+\frac{p}{m^\pm}\cdot\frac{\partial f^\pm}{\partial x}\pm e\left[E+\left(\frac{1}{m^\pm}\right)p\times B\right]\cdot\frac{\partial f^\pm}{\partial p}$$

$$= \tfrac{1}{2}\pi e^4 \ln \Lambda \left(\frac{\partial}{\partial p^\pm}\right)\cdot\sum_{+,-}\int d^3p'\left[f'\frac{\partial f^\pm}{\partial p}-f^\pm\frac{\partial f'}{\partial p'}\right]:w.$$

$$\ldots[2.5]$$

3

TRANSPORT COEFFICIENTS FOR SIMPLE GASES

3.1. *Simple kinetic theory*

We will leave until later a "critical" derivation of the kinetic equation for the plasma, and will in the meantime consider the relationship between the Boltzmann equation and hydrodynamics. We will first discuss this for the case of a simple gas, showing the procedures used to obtain solutions, and also showing a useful, simple approximation, the relaxation approximation. We shall then apply this approximation to a derivation of the plasma transport coefficients, then compare these with the results of an accurate calculation (which we will not reproduce).

First let us consider some of the symmetry properties of the Boltzmann equation (which, of course, also hold for the Landau form).

Consider, then, the integral:

$$G(v) = \int \mathscr{I}(v)h(v)d^3v = \int d^3v \int d^3v' \int d\Omega \sigma g[f(\bar{v})f(\bar{v}')-f(v)f(v')]h(v).$$

Note, first, that, if v and v' are interchanged, the integral is unaltered ($g \to -g$; $\bar{v} \to \bar{v}'$); hence

$$G(v) = \int d^3v \int d^3v' \int d\Omega \sigma g \left[f(\bar{v})f(\bar{v}')-f(v)f(v') \right] h(v').$$

Further, note that, if v and v' are exchanged for $\bar{v}$ and $\bar{v}'$ in the integral, then $d^3\bar{v}\, d^3\bar{v}' = d^3v\, d^3v'$ (Liouville); so

$$\begin{aligned}
G(v) &= -\int d^3v \int d^3v' \int d\Omega \sigma g \left[f(\bar{v})f(\bar{v}')-f(v)f(v') \right] h(\bar{v}) \\
&= -\int d^3v \int d^3v' \int d\Omega \sigma g \left[f(\bar{v})f(\bar{v}')-f(v)f(v') \right] h(\bar{v}') \\
&= \tfrac{1}{4} \int d^3v \int d^3v' \int d\Omega \sigma g \left[f(\bar{v})f(\bar{v}')-f(v)f(v') \right] [h(v)+h(v')-h(\bar{v})-h(\bar{v}')].
\end{aligned}$$

$$...[3.1]$$

Using this relation, we can immediately prove the Boltzmann H theorem. Consider $S = -\int f \ln f d^3v d^3x$. Then $(\partial S/\partial t) \geqslant 0$, if f satisfies the Boltzmann equation, for

$$\begin{aligned}
(\partial S/\partial t) &= -\int (\ln f + 1)(\partial f/\partial t) d^3v d^3x \\
&= \int (\ln f + 1)\{[v \cdot (\partial f/\partial x)+(F/m) \cdot (\partial f/\partial v)] - \mathscr{I}(v)\} d^3v d^3x.
\end{aligned}$$

To deal with the first two terms, integrate the first by parts, thus:

$$\begin{aligned}
\int (\ln f)[v \cdot (\partial f/\partial x)+(F/m) \cdot (\partial f/\partial v)] d^3v d^3x \\
= -\int f [v \cdot (\partial/\partial x)+(F/m) \cdot (\partial/\partial v)](\ln f) d^3v d^3x \\
= -\int [v \cdot (\partial f/\partial x)+(F/m) \cdot (\partial f/\partial v)] d^3v d^3x
\end{aligned}$$

$$\therefore \quad \int (\ln f + 1) [v \cdot (\partial f/\partial x)+(F/m) \cdot (\partial f/\partial v)] d^3v d^3x = 0.$$

The term in $\mathscr{I}$ survives, and

$$\begin{aligned}
\partial S/\partial t = -\tfrac{1}{4} \int d^3x \int d^3v \int d^3v' d\Omega \sigma g \left[f(\bar{v})f(\bar{v}')-f(v)f(v') \right] [\ln f(v) \\
+ \ln f(v') - \ln f(\bar{v}) - \ln f(\bar{v}')]
\end{aligned}$$

$$= -\tfrac{1}{4} \int d^3x \int d^3v \int d^3v' \, d\Omega\sigma g \, [f(\bar{v})f(\bar{v}')$$
$$-f(v)f(v')] \ln \{f(v)f(v')/[f(\bar{v})f(\bar{v}')]\}.$$

The integrand has the form $(a-b) \log b/a \leqslant 0$; hence $\partial S/\partial t \geqslant 0$.

Boltzmann gave a solution to $\mathcal{I} = 0$ for any collision cross section, by noting that a function $f(\mathbf{r})$ could be constructed such that

$$f(\bar{v})f(\bar{v}')-f(v)f(v') = 0.$$

Consider first the problem of finding solutions to the equation $\int \mathcal{I}(v)h(v)d^3v = 0$. Clearly $h = \text{const.}$ is a solution and, since

$$\left.\begin{array}{c} mv+mv' = m\bar{v}+m\bar{v}' \\ \tfrac{1}{2}mv^2+\tfrac{1}{2}m(v')^2 = \tfrac{1}{2}m(\bar{v})^2+\tfrac{1}{2}m(\bar{v}')^2 \end{array}\right\}, \quad \dots[3.2]$$

mv and $\tfrac{1}{2}mv^2$ are also solutions. These are the collision invariants.

Suppose f is a function of the collision invariants which has the property $f(x)f(y) = f(x+y)$; then

$$f(\bar{v})f(\bar{v}')-f(v)f(v') = f(\bar{\mathscr{E}}+\bar{\mathscr{E}}', -+\bar{p}')-f(\mathscr{E}+\mathscr{E}', p+p'),$$

and the left-hand side vanishes since

$$\mathscr{E}+\mathscr{E}' = \bar{\mathscr{E}}+\bar{\mathscr{E}}', \quad p+p' = \bar{p}+\bar{p}'.$$

$\text{Exp}\,(\alpha x)$ is such a function; hence $f \propto \exp\left[-\alpha(\tfrac{1}{2}mv^2)+\gamma \cdot m\mathbf{r}\right]$ and a normalized solution is

$$\left.\begin{array}{c} f_{\mathrm{M}} = n[m/(2\pi\kappa T)]^{\frac{3}{2}} \exp\left[-\tfrac{1}{2}m(v-u)^2/(\kappa T)\right] \\ \int f_{\mathrm{M}}d^3v = n \\ \int f_{\mathrm{M}}(v)vd^3v = nu \\ \int f_{\mathrm{M}}(v)\tfrac{1}{2}m(v-u)^2d^3v = \tfrac{3}{2}n\kappa T. \end{array}\right\} \quad \dots[3.3]$$

This is, of course, the thermal equilibrium distribution.

3.2. *Normal solution*

To describe hydrodynamic behaviour in gases, we seek solutions that are close to thermal equilibrium. The distribution function might be expected to have this property for motions that are slowly varying in space and time on the molecular scale, and it is exactly this property that is exploited in the *normal solution*. Let us consider the approximate magnitude of the terms on the two sides of the Boltzmann equation. The left-hand side

$$Df = \frac{\partial f}{\partial t}+v \cdot \frac{\partial f}{\partial x}+\frac{F}{m} \cdot \frac{\partial f}{\partial v} \simeq \left[\frac{1}{\mathrm{T}}+\frac{v}{\mathrm{L}}+\frac{F}{mv}\right]f$$

while

$$\begin{aligned} \mathscr{I}(f,f) &= \int d^3v'd\Omega\sigma g[f(v)f(v')-f(\bar{v})f(\bar{v}')] \\ &\simeq \int d^3v'd\Omega\sigma g f(v')f(v) \\ &\simeq \langle n\sigma g\rangle f \\ &= (1/\tau)f \end{aligned}$$

where $(1/\tau)$ is the mean collision frequency, and T, L and F are respectively the time needed for a macroscopic change and the macroscopic scale length and force per particle. Now, if the time $T \gg \tau$, while $L \gg \lambda = \tau v_\theta$, the mean free path, and $(\dot{v}/v_\theta) \ll (1/\tau)$, then, in scale $\mathscr{I}(f,f) \gg Df$, and we could write $\mathscr{I}(f,f) = \epsilon Df$ and seek a solution expanded in powers of ϵ. To lowest order, then, $\mathscr{I}(f_0,f_0) = 0$. But, this is satisfied by f_M, with n, T and $\boldsymbol{u}$ arbitrary functions of position and time, that is, by a locally Maxwellian distribution function.

To first order, we could write $f = f_0(1+\epsilon\psi)$, and the linearized equation becomes

$$
\begin{aligned}
Df_0 &= \mathscr{I}(f_0, \psi) \\
&= \int d^3v' d\Omega \sigma g f_0(v) f_0(v')[\psi(\bar{v})+\psi(\bar{v}')-\psi(v)-\psi(v')]
\end{aligned} \qquad ...[3.4]
$$

(where we have used $f_0(v)f_0(v') = f_0(\bar{v})f_0(\bar{v}')$). The left-hand side is clearly symmetric in v, $v'(v \leftrightarrow v'$, $\bar{v} \leftrightarrow \bar{v}'$, $\sigma g = \sigma g')$, and $[3.4]$ has the form $Df_0 = \int K(v,v')\psi(v')d^3v'$, an inhomogeneous integral equation with a symmetric kernel. Moreover, we know the solutions to the homogeneous equations, $h(v)$, which are the collision invariants, 1, mv, $\frac{1}{2}mv^2$. Consider now

$$
\int h(v)\,Df_0 d^3v = \int d^3v \int h(v)K(v,v')\psi(v')d^3v'
$$

where

$$
\int K(v,v')h(v')d^3v' = 0 = \int h(v)K(v,v')d^3v
$$

since K is symmetric in v, v'). Then

$$
\int h(v)\,Df_0(v)d^3v = 0.
$$

Hence

$$
\int Df_0 \begin{Bmatrix} 1 \\ mv \\ \frac{1}{2}mv^2 \end{Bmatrix} d^3v = 0.
$$

These equations, however, are exactly the adiabatic equations of motion. Hence

$$
\begin{aligned}
(\partial n/\partial t)+(\partial/\partial x) \cdot \boldsymbol{n}\boldsymbol{u} &= 0, \\
\rho[(\partial u/\partial t)+\boldsymbol{u} \cdot (\partial u/\partial x)]+(\partial/\partial x)(n\kappa T)-\mathscr{F} &= 0, \\
(\partial/\partial t)(n^{-\frac{2}{3}}\kappa T)+\boldsymbol{u} \cdot (\partial/\partial x)(n^{-\frac{2}{3}}\kappa T) &= 0.
\end{aligned}
$$

Now, in order that $[3.4]$ should be integrable, the quantities n, $\boldsymbol{u}$, T must satisfy these equations. However, if they are, in fact, the true density, velocity and temperature, they will satisfy, not these equations, but the hydrodynamic equations determined by the solution to the Boltzmann equation. To avoid this dilemma, let us consider the form of Df_0.

$$
\begin{aligned}
D &\equiv (\partial/\partial t)+\boldsymbol{v} \cdot (\partial/\partial x)+(F/m) \cdot (\partial/\partial v) \\
&= (\partial/\partial t)+\boldsymbol{u} \cdot (\partial/\partial x)+\boldsymbol{c} \cdot (\partial/\partial x)+(F/m) \cdot (\partial/\partial v),
\end{aligned}
$$

where $c = v-u$; since f_0 depends on x, t only through n, $\boldsymbol{u}$, T,

$$Df_0 = f_0\left[\left(\frac{\partial}{\partial t}+\boldsymbol{v}\cdot\frac{\partial}{\partial\boldsymbol{x}}\right)\left(\ln n-\tfrac{3}{2}\ln T-\tfrac{1}{2}\frac{mc^2}{\kappa T}\right)-\boldsymbol{F}\cdot\frac{\boldsymbol{c}}{\kappa T}\right]$$

$$= f_0\left\{\left(\frac{\partial}{\partial t}+\boldsymbol{u}\cdot\frac{\partial}{\partial\boldsymbol{x}}\right)\left[\tfrac{3}{2}\ln (n^{-\frac{2}{3}}T)-\tfrac{1}{2}\frac{mc^2}{\kappa T}\right]\right.$$

$$\left.+\boldsymbol{c}\cdot\frac{\partial}{\partial\boldsymbol{x}}\left[\ln (nT)-\tfrac{5}{2}\ln T-\tfrac{1}{2}\frac{mc^2}{\kappa T}\right]-\boldsymbol{F}\cdot\frac{\boldsymbol{c}}{\kappa T}\right\}.$$

Now observe that $[(\partial/\partial t)+\boldsymbol{u}\cdot(\partial/\partial\boldsymbol{x})]\tfrac{1}{2}c^2 = -\boldsymbol{c}\cdot[(\partial/\partial t)+\boldsymbol{u}\cdot(\partial/\partial\boldsymbol{x})]\boldsymbol{u}$ since $\boldsymbol{c} = \boldsymbol{v}-\boldsymbol{u}$, and

$$\frac{1}{T}\left(\frac{\partial}{\partial t}+\boldsymbol{u}\cdot\frac{\partial}{\partial\boldsymbol{x}}\right)T = \frac{2}{3n}\left(\frac{\partial n}{\partial t}+\boldsymbol{u}\cdot\frac{\partial n}{\partial\boldsymbol{x}}\right) = -\tfrac{2}{3}\frac{\partial\boldsymbol{u}}{\partial\boldsymbol{x}};$$

hence

$$Df_0 = f_0\left\{\left(\frac{\partial}{\partial t}+\boldsymbol{u}\cdot\frac{\partial}{\partial\boldsymbol{x}}\right)[-\tfrac{3}{2}\ln (n^{-\frac{2}{3}}T)]\right.$$

$$+\frac{m\boldsymbol{c}}{\kappa T}\cdot\left[\frac{\partial\boldsymbol{u}}{\partial t}+\boldsymbol{u}\cdot\frac{\partial\boldsymbol{u}}{\partial\boldsymbol{x}}+\frac{1}{nm}\left(\frac{\partial}{\partial\boldsymbol{x}}\right)(n\kappa T)-\frac{\boldsymbol{F}}{m}\right]$$

$$\left.+\left(\tfrac{1}{2}\frac{mc^2}{\kappa T}-\tfrac{5}{2}\right)\frac{\boldsymbol{c}}{T}\cdot\frac{\partial T}{\partial\boldsymbol{x}}+\left(\tfrac{1}{3}\frac{mc^2}{\kappa T}\operatorname{div}\boldsymbol{u}-\frac{m}{\kappa T}\boldsymbol{cc}:\frac{\partial\boldsymbol{u}}{\partial\boldsymbol{x}}\right)\right\}.$$

Now, if in this expression we use the zero order equations of motion to eliminate the time derivatives, then the conditions required for the solution are satisfied, and we need make no further comment on the time variation of n, T, $\boldsymbol{u}$ etc. Indeed, we assume that the local time derivatives differ by some higher order quantity $O(\tau/T)$ from their adiabatic values, although, of course, as functions of time, the solutions could finally differ by a larger amount. In effect, we are expanding the time derivatives $(\partial/\partial t)+\epsilon(\partial/\partial\tau)$ by introducing a variety of time scales. The (consistent) integral equation then becomes

$$\int K(\boldsymbol{v},\boldsymbol{v}')\psi(\boldsymbol{v}')d^3v' = f_0\left[\tfrac{1}{3}\frac{mc^2}{\kappa T}\operatorname{div}\boldsymbol{u}-\frac{m}{\kappa T}\boldsymbol{cc}:\nabla\boldsymbol{u}\right]$$

$$+\left[\tfrac{5}{2}-\tfrac{1}{2}\frac{mc^2}{\kappa T}\right]\frac{1}{T}\boldsymbol{c}\cdot\nabla T. \qquad\qquad ...[3.5]$$

To handle this, we first note that K and ψ could be written as functions of the dimensionless w, where $w^2 = \tfrac{1}{2}[mc^2/(\kappa T)]$, and the solution clearly has the form

$$[A(w)/T](2\kappa T/m)^{\frac{1}{2}}\boldsymbol{w}\cdot\nabla T+B(w)[\tfrac{1}{3}w^2\operatorname{div}\boldsymbol{u}-\boldsymbol{ww}:\nabla\boldsymbol{u}],$$

i.e. $\qquad A(w)\boldsymbol{w}\cdot\boldsymbol{V}(\boldsymbol{x},t)+B(w)\boldsymbol{ww}:\boldsymbol{T}(\boldsymbol{x},t);$

and the equation can be separated into a vector equation for A and a tensor equation for B. The general form is $\int KT d^3v' = R$. Approximate solutions can be obtained by using the following variational principle: consider the functions $t(v)$ which satisfy the integral constraint

$$\int Kt(v)t(v')d^3vd^3v' = \int R(v)t(v)d^3v,$$

which need *not* be solutions. (They may be required to satisfy further constraints, which are applied to the solutions T, in order that n, u, T be correctly given:

$$\int f_0 \begin{Bmatrix} T \\ t \end{Bmatrix} \begin{Bmatrix} 1 \\ mc \\ \tfrac{1}{2}mc^2 \end{Bmatrix} d^3v = 0,$$

while of course $\int Kt(v)T(v')d^3vd^3v' = \int R(v)t(v)d^3v$.) Now, let us look at the properties of the integrals on the left. Write

$$[\chi, \psi] = -\int \chi(v)K(v, v')\psi(v')d^3vd^3v'$$
$$= -\int d^3v'd^3vd\Omega\sigma g\chi(v)f_0(v)f_0(v')[\psi(\bar{v})+\psi(\bar{v}')-\psi(v)-\psi(v')]$$
$$= \tfrac{1}{4}\int d^3v'd^3vd\Omega\sigma g f_0(v)f_0(v')[\chi(\bar{v})+\chi(\bar{v}')-\chi(v)-\chi(v')][\psi(\bar{v})+\psi(\bar{v}')$$
$$-\psi(v)-\psi(v')]$$

using the symmetry property [3.1]. In particular $[\psi, \psi] > 0$. We now have the constraint $[t, t] = -\int Rt d^3v$ and $[t, t] = [T, t] = [t, T]$ and the solution $[T, T] = -\int RT d^3v$. Consider now

$$[t-T, t-T] = [t, t]-2[t, T]+[T, T] = -[t, T]+[T, T] \geqslant 0.$$

Hence $[t, t] = [t, T] < [T, T]$. Thus, if we can find a trial function t satisfying the required constraints and having available some variational parameters, then that choice which maximizes $[t, t]$ comes closest to the true solution. Application of this method is simplified by the occurrence of certain polynomials—Sonine's polynomials—S_m^n, which satisfy

$$\int \exp(-w^2)w^{2n+1}S_m^n(w)S_{m'}^n(w)dw = \delta_{m, m'}\frac{(n+m)!}{2m!}$$

$$S_m^n = \sum_{j=0}^m (-1)^j \frac{(n+m)!}{(m+j)!\,(n+j)!\,j!}\,w^{2j}.$$

For example $S_0^{\frac{3}{2}} = 1; \quad S_1^{\frac{3}{2}} = (\tfrac{5}{2}-w^2)$
                            $S_0^{\frac{5}{2}} = 1; \quad S_1^{\frac{5}{2}} = (\tfrac{7}{2}-w^2)$

Consider, for example, the problem of calculating an approximation to the thermal conductivity. For this we must evaluate A. Let us introduce an approximation $t = A(w)w$. Then

$$\int Rt d^3v = (4/\pi^{\frac{1}{2}}) \int (\tfrac{5}{2}-w^2)Aw^2 \exp(-w^2)w^2dw.$$

If we write $t = aS_1^{\frac{3}{2}}w = [\tfrac{5}{2}-w^2]w,$

then $\int R t d^3 v = a(4/\pi^{\frac{1}{2}}) \int dw \exp(-w^2) w^4 (S_1^{\frac{3}{2}})^2 = a(4/\pi^{\frac{1}{2}})(\frac{5}{2}!)/2! = (15/4)a$.

The constraint on t, $\int R t d^3 v = -[t, t]$ then becomes

$$\frac{15}{4}a = \int d^3 v' d^3 v \int d\Omega \sigma g f(v) f(v') [t(\bar{v}) + t(\bar{v}') - t(v) - t(v')]$$

$$= \frac{a^2}{16}\left(\frac{e^4}{m^2}\right)\left(\frac{m}{2\kappa T}\right)^{\frac{3}{2}} \int d^3 w d^3 w' \int d\Omega \frac{\exp\{-[w^2 + (w')^2]\}}{g^3 \sin^4(\theta/2)}$$

$$[S(\bar{w})\bar{w} + S(\bar{w}')\bar{w}' - S(w)w - S(w')w']$$

$$= a^2(1/\tau)(1/16)\{S^{\frac{3}{2}}, S^{\frac{3}{2}}\}$$

hence

$$a = \frac{15}{4}\frac{16\tau'}{\{S^{\frac{3}{2}}, S^{\frac{3}{2}}\}} \quad \text{where} \quad \tau' = \left(\frac{2\kappa T}{m}\right)^{\frac{3}{2}}\left(\frac{m}{e^2}\right)^2.$$

To get the thermal conductivity we need

$$q = -\lambda \nabla T$$

$$= \kappa T \left(\frac{2\kappa T}{m}\right)^{\frac{1}{2}} \frac{a}{\pi^{\frac{3}{2}}} \int w^2 w(\tfrac{5}{2} - w^2) w \cdot \frac{\nabla T}{T} \exp(-w^2) d^3 w$$

$$= \frac{60\tau'}{\{S, S\}} \frac{\kappa T}{T}\left(\frac{2\kappa T}{m}\right)^{\frac{1}{2}} \frac{2}{\pi^{\frac{1}{2}}} \int_0^\infty \exp(-w^2) w^6 (\tfrac{5}{2} - w^2) dw \nabla T$$

$$= \frac{(15)^2}{2} \frac{\tau'}{\{S, S\}} \kappa \left(\frac{2\kappa T}{m}\right)^{\frac{1}{2}} \nabla T.$$

There remains the problem of evaluating $\{S, S\}$. All velocities are normalized to thermal speeds, so we can introduce $V = (w + w')/2$; $g = w - w'$. The square bracket then becomes

$$[(V + \tfrac{1}{2}(g + \Delta g))^2(V + \tfrac{1}{2}(g + \Delta g)) + (V - \tfrac{1}{2}(g + \Delta g))^2(V - \tfrac{1}{2}(g + \Delta g))$$
$$- (V + g)^2(V + g) - (V - g)^2(V - g)]$$
$$= (V \cdot g)(V \cdot (g + \Delta g))(\Delta g)^2 + (V \cdot \Delta g)^2 g^2 + (V \cdot g)(V \cdot \Delta g)(g \cdot \Delta g).$$

We need

$$\{S^{\frac{3}{2}}, S^{\frac{3}{2}}\} = \frac{2\pi}{\pi^3} \int d^3 V \exp(-2V^2) \int \frac{d^3 g}{g^3} \exp(-g^2/2) \int d\theta \frac{\sin\theta}{\sin^4(\theta/2)}$$

$$\{(V \cdot g)^2 g^2[(1 - \cos\theta)^2 + \sin^2\theta] + (V \cdot g)^2 g^2(1 - \cos\theta)^2$$
$$+ \tfrac{1}{2}V_\perp^2 g^4 \sin^2\theta + (V \cdot g)^2 g^2(1 - \cos\theta)^2\}$$

where the integral over the "ignorable" scattering angle φ, has been carried out. Now, to carry out the integrations over the scattering angle θ, we use

$$\int d\theta[\sin\theta/\sin^4(\theta/2)](1 - \cos\theta)^2 = \tfrac{1}{2}\sin^2(\theta/2)$$
$$\int d\theta[\sin\theta/\sin^4(\theta/2)]\sin^2\theta = 16 \ln \Lambda,$$

and only the second term need be retained;

$$\{S^{\frac{3}{2}}, S^{\frac{3}{2}}\} = \frac{32}{\pi^2} \ln\Lambda \int d^3 V \exp(-2V^2) \int \frac{d^3 g}{g^3} g^4 \exp(-\tfrac{1}{2}g^2)[(V \cdot \hat{g})^2 + \tfrac{1}{2}V_\perp^2]$$

$$= (32/\pi) \ln\Lambda \int d^3 V \exp(-2V^2) \int d^3 g g^3 \exp(-\tfrac{1}{2}g^2) V^2$$

$$= (32/\pi) \ln\Lambda \cdot 2^3 \cdot 4\pi \cdot (\pi^{\frac{1}{2}}/2^5) \cdot (3/8)$$

$$= 12\pi^{\frac{1}{2}} \ln\Lambda$$

$$q = \frac{75}{4\pi^{\frac{1}{4}}}\left(\frac{m}{e^2}\right)^2 \frac{1}{\ln\Lambda}\left(\frac{\kappa T}{m}\right)^2 \kappa\nabla T.$$

It is clear that this operation can be carried out for a simple gas; it can, moreover, be shown that for a two-component gas a similar operation can be performed, although, if the forces acting on the two gases differ, the components do not satisfy the adiabatic equations in zero order. Instead, the method requires that the difference between the two gases also be small; and a term arises from the difference between them. This leads to a first order current. For the gas mixture again a variational principle can be set up, in a form similar to that used here.

If the gas is in a magnetic field there are serious complications. In the first place, the scaling law $\tau \ll T$ becomes uncertain, as for a diffuse plasma in a magnetic field $\omega\tau > 1$; hence the magnetic term must be taken seriously, and considered as of the same order as the collision term. This unfortunately invalidates the simple form of the variational principle. It is, however, possible to generate a new variational principle—firstly a stationary principle, and secondly a mini-max principle have been found—and to estimate the conductivities etc. in terms of [,] integrals, which then must be evaluated. It is also possible to get a useful approximation to the transport coefficients by using a much simpler and cruder method. To understand this, let us first consider the problem of the decay of an isotropic, time dependent, velocity-dependent disturbance $\delta f = f_0\psi$ applied to a Maxwellian f_0. Then, to lowest order

$$f_0(\partial\psi/\partial t) = \int K(v, v')\psi(v', t)d^3v'.$$

Suppose now, we can factor ψ as $\tilde\psi(v)a(t)$. Then

$$(\dot a/a)f_0\tilde\psi = \int K(v, v')\tilde\psi(v')d^3v'.$$

Now, if we multiply this by $\tilde\psi(v)$ and integrate over all v:

$$(\dot a/a)\int f_0\tilde\psi\tilde\psi d^3v = \int \tilde\psi(v)K(v, v')\tilde\psi(v')d^3vd^3v' = -[\tilde\psi, \tilde\psi]$$
$$(\dot a/a) = -[\tilde\psi, \tilde\psi]/\int f_0\tilde\psi^2 d^3v \leqslant 0,$$

i.e. $(\dot a/a) = -1/\tau$, $a = \exp(-t/\tau)$. Thus, the effect of this collision term is to cause a disturbance δf to relax back toward equilibrium. This must, of course, be done in such a way that the integral constants of motion are preserved, and furthermore, the relaxation time τ will be a functional of $\tilde\psi$. Indeed, we do not expect that an arbitrary $\psi(v)$ will be a normal mode of the system. Nevertheless, neglecting these complications still leaves useful results.

4

The Relaxation Approximation

The simplest application of the relaxation method is to the simple gas. In $[3.5]$ we replace the left-hand side by $\int K(v, v')\psi(v')d^3v' = -[\psi(v)/\tau]f_0$.

Hence

$$f_0\psi = -\tau f_0\left[\frac{mc^2}{3\kappa T}\operatorname{div}\boldsymbol{u} - \frac{m}{\kappa T}\boldsymbol{cc}:\boldsymbol{\nabla u}\right] + \left[\frac{5}{2} - \frac{mc^2}{2\kappa T}\right]\boldsymbol{c}\cdot\frac{\boldsymbol{\nabla}T}{T} \qquad \dots[4.1]$$

is the appropriate form of δf. If we now form the moments, note first that $\delta n = 0$ and $\delta(mc^2) = 0$, since

$$\int \exp\left[-\tfrac{1}{2}mc^2/(\kappa T)\right]\boldsymbol{cc}:\boldsymbol{\nabla u}d\Omega = (4\pi/3)mc^2(\boldsymbol{\nabla}\cdot\boldsymbol{u})\exp\left[-\tfrac{1}{2}mc^2/(\kappa T)\right]$$
$$= \int\tfrac{1}{3}mc^2(\operatorname{div}\boldsymbol{u})\exp\left[-\tfrac{1}{2}mc^2/(\kappa T)\right]d\Omega,$$

while $\langle\boldsymbol{c}\rangle = 0$ since

$$\int\left\{\tfrac{5}{2} - \tfrac{1}{2}[mc^2/(\kappa T)]\right\}c^2\exp\left[-\tfrac{1}{2}mc^2/(\kappa T)\right]d^3c \simeq \int(\tfrac{5}{2} - w^2)w^4\exp(-w^2)dw$$
$$= (\tfrac{5}{2}\cdot\tfrac{3}{2}\cdot\tfrac{1}{2} - \tfrac{5}{2}\cdot\tfrac{3}{2}\cdot\tfrac{1}{2})$$
$$= 0.$$

The components of the stress tensor are

$$\begin{aligned}
\mathsf{p} &= \langle nm\boldsymbol{cc}\rangle = 2n\kappa T\langle\boldsymbol{ww}\rangle \\
&= n\kappa T(\tau/\pi^{\frac{3}{2}})\int d^3w\exp(-w^2)\boldsymbol{ww}(\tfrac{1}{3}w^2\operatorname{div}\boldsymbol{u} - \boldsymbol{ww}:\boldsymbol{\nabla u}) \\
&= n(\kappa T/2)\tau[\tfrac{1}{3}(\operatorname{div}\boldsymbol{u})\mathbf{1} - \tfrac{1}{2}\boldsymbol{Du}] \qquad\qquad \dots[4.2]
\end{aligned}$$

where
$$(\boldsymbol{Du})_{ij} = \frac{\partial u_j}{\partial x_i} + \frac{\partial u_i}{\partial x_j}$$

while the heat flux vector is

$$\boldsymbol{q} = (\boldsymbol{\nabla}T/T)\tau\kappa T[2\kappa T/(m\pi)]^{\frac{1}{2}}(2/3)\int(\tfrac{5}{2} - w^2)w^6\exp(-w^2)dw$$

$$= \tfrac{5}{4}\tau\left(\frac{2\kappa T}{m}\right)^{\frac{1}{2}}\kappa\boldsymbol{\nabla}T. \qquad\qquad \dots[4.3]$$

Preliminary to an application of the relaxation approximation, let us consider some elementary consequences of the Boltzmann equation as applied to an ionized gas. This may be written

$$\left.\begin{aligned}
(\partial f_+/\partial t) + \boldsymbol{v}\cdot\boldsymbol{\nabla}f_+ + (e/m)_+(\boldsymbol{E} + \boldsymbol{v}\times\boldsymbol{B})\cdot(\partial f_+/\partial\boldsymbol{v}) &= \mathscr{I}_{++} + \mathscr{I}_{+-} \\
(\partial f_-/\partial t) + \boldsymbol{v}\cdot\boldsymbol{\nabla}f_- + (e/m)_-(\boldsymbol{E} + \boldsymbol{v}\times\boldsymbol{B})\cdot(\partial f_-/\partial\boldsymbol{v}) &= \mathscr{I}_{--} + \mathscr{I}_{-+}
\end{aligned}\right\} \quad \dots[4.4]$$

If we construct the moment equations for these two equations separately, we obtain, as before, $(\partial\rho_{+,-}/\partial t) + \boldsymbol{\nabla}\cdot(\rho_{+,-}\boldsymbol{u}_{+,-}) = 0$, the r.h.s. vanishing since, if we omit ionization processes, particle number is conserved on collision, and

$$\rho_{+,-}[(\partial\boldsymbol{u}_{+,-}/\partial t) + \boldsymbol{u}_{+,-}\cdot\boldsymbol{\nabla u}_{+,-}] + \boldsymbol{\nabla}\cdot\mathsf{p}_{+,-} + (ne)_{+,-}(\boldsymbol{E} + \boldsymbol{u}_{+,-}\times\boldsymbol{B})$$
$$= \Delta\boldsymbol{p}_{+-,-+} \qquad\qquad \dots[4.5]$$

where $\Delta\boldsymbol{p}_{+-,-+}$ represents the transfer of momentum between ions and electrons by collision and has the property that $\Delta\boldsymbol{p}_{+-} + \Delta\boldsymbol{p}_{-+} = 0$ (conservation of total momentum).

If we add the equations for ions and electrons, there results

$$(\partial/\partial t)(\rho_+ + \rho_-) + \nabla \cdot (\rho_+ u_+ + \rho_- u_-) = 0 = (\partial \rho/\partial t) + \nabla \cdot (\rho V),$$

defining the centre of mass motion V by $\rho V = (\rho_+ u_+ + \rho_- u_-)$, $\rho = \rho_+ + \rho_-$, and

$$(\partial/\partial t)(\rho_+ u_+ + \rho_- u_-) + \nabla \cdot (\rho_+ u_+ u_+ + \rho_- u_- u_-) + \nabla \cdot (p_+ + p_-)$$
$$- [(ne)_+ + (ne)_-]E - [(neu)_+ + (neu)_-] \times B = 0;$$

and, if $u_{+,-} = V + v_{+,-}$; $\rho_+ v_+ + \rho_- v_- = 0$, then

$$(\partial/\partial t)\rho V + \nabla \cdot \rho V V + \nabla \cdot [(\rho_- v_+ v_+ + \rho_- v_- v_-) + p_+ + p_-]$$
$$- Q[E + V \times B] - j \times B = 0,$$

where $j = ne_+ v_+ + ne_- v_-$. This could be written

$$\rho[(\partial V/\partial t) + V \cdot \nabla V] + \nabla \cdot p - Q(E + V \times B) - j \times B = 0 \quad \ldots[4.6]$$

where p is defined as $(p_+ + p_- + \rho_+ v_+ v_- + \rho_- v_- v_+)$. This is a generalization of the familiar MHD equation. If the assumption of quasi-neutrality is made $Q = 0$ and

$$\rho[(\partial V/\partial t) + V \cdot \nabla V] + \nabla \cdot p + j \times B = 0. \quad \ldots[4.7]$$

To obtain an equation for the current, the equations should be multiplied by the charge to mass ratio, and then added. If we use the condition of charge neutrality, and observe then that the large ratio of ion to electron mass means $u_+ \simeq 0$, we obtain

$$(Dj/Dt) + (e/m)_- \nabla \cdot p_- + (ne^2/m)\{E + V \times B + [1/(ne)]j \times B\} = (e/m)(\Delta p_{-+}).$$

This is most conveniently solved for $(E + V \times B)$:

$$E + V \times B = -[m/(ne^2)](Dj/Dt) - [1/(ne)][\nabla \cdot p_- + j \times B]$$
$$+ (e/m)[m/(ne^2)](\Delta p_{-+}).$$

If we make the plausible assumption that $\Delta p_{-+} \simeq (mv)_-$, then we may write $[m/(ne^2)](e/m)\Delta p_{-+} = \eta j$ and introduce the plasma frequency $\omega_p = 4\pi ne^2/m$ so the generalized Ohm's law becomes

$$E + V \times B = -\frac{4\pi}{\omega_p^2} \frac{Dj}{Dt} + \eta j - \frac{1}{(ne)_-} (\nabla \cdot p_- + j \times B). \quad \ldots[4.8]$$

To proceed with the normal solutions, we note first that

$$\mathscr{I}_{++} + \mathscr{I}_{+-} = \int d^3 v' \int d\Omega \{\sigma_{++}[f_+(\bar{v})f_+(\bar{v}') - f_+(v)f_+(v')]$$
$$+ \sigma_{+-}[f_+(\bar{v})f_-(\bar{v}') - f_+(v)f_-(v')]\}.$$

The equilibrium solutions are easily seen to be

$$(f_0)_{+,-} = n[m/(2\kappa T)]^{\frac{3}{2}} \exp\{-[\tfrac{1}{2}m_{+,-}(v-u)^2/(\kappa T)]\},$$

where the functions n, T and u are the *same* for ions and electrons. The collision invariants, however, are now *not* solutions of the linearized equations. We do, however, know some vanishing moments. If we take $\psi_{+,-} = m_{+,-}$, $m_{+,-}v$, $\frac{1}{2}m_{+,-}v^2$, then the sum of the two moments will vanish, from conservation of mass, momentum and energy for the gas as a whole. The constraints on n, u, T then become

$$\left.\begin{array}{c} (\partial n/\partial t)+\text{div}\,(nu) = 0 \\ n(m_{+}+m_{-})[(\partial u/\partial t)+u\cdot\nabla u]+(\partial/\partial x)2n\kappa T = 0 \\ (\partial/\partial t)(Tn^{-\frac{2}{3}})+u\cdot(\partial/\partial x)(Tn^{-\frac{2}{3}}) = 0. \end{array}\right\} \quad ...[4.9]$$

The left-hand sides of the two components of the Boltzmann equation then become

$$(f_0)_{+}\{[(\partial/\partial t)+v\cdot(\partial/\partial x)][\ln n-\tfrac{3}{2}\ln T-\tfrac{1}{2}Mc^2/(\kappa T)]+[1/(\kappa T)]\,E\cdot c\}.$$

In evaluating this, the constraining equation on u, instead of vanishing, yields $c\cdot[(m_{-}-m_{+})/(m_{-}+m_{+})(1/p)(\partial p/\partial x)+eE/(\kappa T)]$,

$$\left.\begin{array}{l} Df_{+} = (f_0)_{+}\left\{\left[\dfrac{m_{+}}{\kappa T}\,cc:\nabla u-\dfrac{1}{3}\dfrac{m_{+}c^2}{\kappa T}\,\text{div}\,u\right]\right. \\[3mm] \qquad +c\cdot\left[\tfrac{1}{2}\dfrac{m_{+}c^2}{\kappa T}-\tfrac{5}{2}\right]\dfrac{1}{T}\dfrac{\partial T}{\partial x} \\[3mm] \qquad \left.+c\cdot\left[\left(\dfrac{m_{-}-m_{+}}{m_{-}+m_{+}}\right)\dfrac{1}{p}\dfrac{\partial p}{\partial x}+\dfrac{eE}{\kappa T}\right]\right\} \\[5mm] \text{while} \\[3mm] Df_{-} = (f_0)_{-}\left\{\left[\dfrac{m_{-}}{\kappa T}\,cc:\nabla u-\dfrac{1}{3}\dfrac{m_{-}c^2}{\kappa T}\,\text{div}\,u\right]\right. \\[3mm] \qquad +c\cdot\left[\tfrac{1}{2}\dfrac{m_{-}c^2}{\kappa T}-\tfrac{5}{2}\right]\dfrac{1}{T}\dfrac{\partial T}{\partial x} \\[3mm] \qquad \left.+c\cdot\left[\left(\dfrac{m_{+}-m_{-}}{m_{+}+m_{-}}\right)\dfrac{1}{p}\dfrac{\partial p}{\partial x}-\dfrac{eE}{\kappa T}\right]\right\}. \end{array}\right\} \quad ...[4.10]$$

In seeking a solution we may use a modified form of the variational principle, whereupon results become somewhat complex. It is *somewhat* simpler to use the relaxation model, replacing the collision integral by $-(1/\tau)f_0\psi$, and the results are quantitatively correct. We write, then

$$\begin{aligned} -\frac{1}{\tau_{+,-}}\,\delta f_{-,+} = (f_0)_{-,-}\Bigg\{&\left[\frac{m_{+,-}}{\kappa T}\,cc:\nabla u-\frac{1}{3}\frac{m_{+,-}c^2}{\kappa T}\,\text{div}\,u\right] \\ &+c\cdot\left[\tfrac{1}{2}\frac{m_{+,-}c^2}{\kappa T}-\tfrac{5}{2}\right]\frac{1}{T}\frac{\partial T}{\partial x} \\ &\mp c\cdot\left[\frac{eE}{\kappa T}+\left(\frac{m_{-}-m_{+}}{m_{-}+m_{+}}\right)\frac{1}{p}\frac{\partial p}{\partial x}\right]\Bigg\}. \end{aligned} \qquad ...[4.11]$$

If we now form the moments, we reproduce the previous results with two alterations. The term in $eE/(\kappa T) + (m_- - m_+)/(m_- + m_+)(1/p)(\partial p/\partial x)$ leads to a charge-dependent change in velocity, hence to an electric current:

$$j = ne\langle c_+ \rangle - ne\langle c_- \rangle \simeq (ne^2/m)\tau_-\{E + [1/(ne)]\nabla p\}$$
$$= (ne^2/m)\tau d. \qquad \qquad ...[4.12]$$

Note that this result agrees with the one obtained above, by rather simpler arguments, provided that $B = 0$ and $Dj/Dt = 0$. When q is formed

$$q \simeq \tfrac{5}{4}\tau(2\kappa T/m)^{\frac{1}{2}}\kappa\nabla T + (\kappa T/e)j. \qquad ...[4.13]$$

These results suffer from a serious flaw; they do not satisfy the Onsager relations.‡

In spite of this serious failing, we will apply this method to the magnetized plasma, in which case we must remember that the gyro-frequency $\omega = eB/(mc)$ may easily be of the same order as the collision frequency $1/\tau$, and write the Boltzmann equation as

$$\mathscr{I}(f,f) - \frac{e}{m}(c \times B) \cdot \frac{\partial f}{\partial c} = \frac{\partial f}{\partial t} + v \cdot \frac{\partial f}{\partial x} + \frac{e}{m}(E + u \times B) \cdot \frac{\partial f}{\partial v}. \quad ...[4.14]$$

The normal solution procedure now leads to exactly the same expression as above, with E replaced by $E + u \times B$. In the relaxation approximation $\mathscr{I} \to -(1/\tau)\delta f$, and in cylindrical polars $(e/m)(c \times B) \cdot (\partial f/\partial c) = -\omega(\partial f/\partial \varphi)$ so the relaxation model yields

$$(\partial/\partial\varphi)\delta f - [1/(\omega\tau)]\delta f = (1/\omega)P \qquad ...[4.15]$$

$$P = 2w_i w_j[(\partial u_i/\partial x_j) - \tfrac{1}{3}(\text{div } u)\delta_{ij}] - [(\tfrac{5}{2} - w^2)\nabla T/T - d][m/(2\kappa T)]^{\frac{1}{2}} \cdot w$$
$$\delta f = (1/\omega) \int^\varphi d\varphi' \exp[(\varphi - \varphi')/(\omega\tau)]P(\varphi')d\varphi'.$$

P may be Fourier expanded in φ, and use made of the following result:

$$\int \{\exp[(\varphi - \varphi')/(\omega\tau)]\}[\cos n\varphi' + i\sin n\varphi']d\varphi' = [\exp(in\varphi)]/[1/(\omega\tau) + in]$$

$$= \frac{\omega\tau}{1 + (n\omega\tau)^2}[(\cos n\varphi + n\omega\tau \sin n\varphi) + i(\sin n\varphi - n\omega\tau \cos n\varphi)].$$

The vector part of $P = [(\tfrac{5}{2} - w^2)\nabla T/T - d][m/(2\kappa T)]^{\frac{1}{2}} \cdot w = W \cdot w$ is easily handled and yields

$$\delta f_w = f_0\tau\{W_\parallel w_\parallel + [1/(1 + \omega^2\tau^2)][W_\perp \cdot w_\perp + \omega\tau b \cdot (W \times w)]\}$$
$$W_\parallel = (W \cdot b)b; \quad W_\perp = W_\perp - W_\parallel; \qquad \qquad \bigg\} \quad ...[4.16]$$

‡ *Note on the Onsager relations.* Departures from thermal equilibrium are characterized by forces, F_i, and fluxes, J_i; the fluxes carry the system back towards equilibrium, are linearly related to F_j, $J_i = \alpha_{ij}F_j$, and generate entropy at a rate $(dS/dt) = \Sigma(J \cdot F)$ $(J, F$ are conjugates!). Then $(dS/dt) = \Sigma \alpha_{ij}F_iF_j = \Sigma \alpha_{ji}F_jF_i$; $\alpha_{ij} = \alpha_{ji}$ since the $F_{i,j}$ are given and arbitrary. Therefore the α are symmetric. Since q contains a term $\propto d$, j must contain the conjugate term $\propto \nabla T$

where b is the unit vector along B. From this follow the electric current and the heat flux. Writing

$$d = [E + u \times B + (1/ne)\nabla p]$$
$$j = \sum (ne^2/m)\tau[d_\parallel + (d_\perp + \omega\tau d \times b)/(1 + \omega^2\tau^2)]$$
$$= \sigma_0\{[1/(1 + \omega_-^2\tau_-^2)][d + \omega_-\tau_- d \times b + (\omega_-\tau_-)^2(d \cdot b)b]$$
$$- (\tau_+/\tau_-)(m/M)[1/(1 + \omega_+^2\tau_+^2)][d + \omega_+\tau_+ d \times b + (\omega_+\tau_+)^2(d \cdot b)b]\}$$
$$= \sigma_0 T \cdot d. \qquad\qquad\qquad ...[4.17]$$

Usually only the electron component is relevant here, and we may solve for d; or $E + u \times B$,

$$E + u \times B = \eta j - [1/(ne)](\nabla p + j \times B) \qquad ...[4.18]$$

which again agrees with our simple results except for the term Dj/Dt. This has been thrown out by our time scaling arguments.

If
$$\nabla\Theta = \nabla T + (4/5)(\kappa T/e)[m/(2\kappa T)]^{\frac{1}{2}}j,$$
$$q = \tfrac{5}{4}\tau_-(2\kappa T/m)^{\frac{1}{2}}\kappa T : \nabla\Theta = L \cdot \nabla\Theta \qquad ...[4.19]$$

where
$$L = \lambda_0\{T_- \cdot \nabla\Theta + (\tau_+/\tau_-)(m/M)^{\frac{1}{2}}T_+ \cdot \nabla\Theta\}$$

and
$$\lambda_0 = \tfrac{5}{4}\tau_-(2\kappa T/m)\kappa.$$

The tensor part may be resolved thus (with Oz along b):

$$\tfrac{1}{2}P_T = \left(\frac{\partial u_x}{\partial x} + \frac{\partial u_y}{\partial y}\right)\frac{w_\perp^2}{2} + \frac{\partial u_z}{\partial z}w_\parallel^2 - \tfrac{1}{3}(\text{div } u)w^2$$
$$+ w_\parallel w_\perp\left[\left(\frac{\partial u_z}{\partial x} + \frac{\partial u_x}{\partial z}\right)\cos\varphi + \left(\frac{\partial u_z}{\partial y} + \frac{\partial u_y}{\partial z}\right)\sin\varphi\right]$$
$$+ \frac{w_\perp^2}{2}\left[\left(\frac{\partial u_x}{\partial x} + \frac{\partial u_y}{\partial y}\right)\sin 2\varphi + \left(\frac{\partial u_x}{\partial x} - \frac{\partial u_y}{\partial y}\right)\cos 2\varphi\right]$$

Hence $\quad\delta f_T = 2\tau f_0\left\{\left[(\text{div } u_\perp^+)\frac{w_\perp^2}{2} + (\text{div } u\parallel)w_\parallel^2 - \tfrac{1}{3}(\text{div } u)w_2^2\right]\right.$
$$+ \frac{w_\parallel w_\perp}{1 + \omega^2\tau^2} \cdot \left[\left(\frac{\partial u_\parallel}{\partial x_\perp} + \frac{\partial u_\perp}{\partial z}\right) + \left(\frac{\partial u_\parallel}{\partial x_\perp} + \frac{\partial u_\perp}{\partial z}\right) \times b\omega\tau\right]$$
$$+ \tfrac{1}{2}\frac{w_\perp^2}{1 + 4\omega^2\tau^2}\left[\left(\frac{\partial u_x}{\partial y} + \frac{\partial u_y}{\partial x}\right)(\sin 2\varphi - 2\omega\tau\cos 2\varphi)\right.$$
$$\left.\left. + \left(\frac{\partial u_x}{\partial x} - \frac{\partial u_y}{\partial y}\right)(\cos 2\varphi + 2\omega\tau\sin 2\varphi)\right]\right\}.$$

From this we form the perturbed part of the stress tensor, which may be written

$$p = p_0 \mathbf{1} - \tau p[(2 \operatorname{div} \mathbf{u}_\perp + \operatorname{div} \mathbf{u}_\parallel - \tfrac{5}{3} \operatorname{div} \mathbf{u})\mathbf{1} + (2 \operatorname{div} \mathbf{u}_\parallel - \operatorname{div} \mathbf{u}_\perp)\mathbf{bb}]$$
$$+ \frac{\tau p}{1 + \omega^2 \tau^2} \left[\left(\frac{\partial \mathbf{u}_\perp}{\partial x_\parallel} + \frac{\partial \mathbf{u}_\parallel}{\partial x_\perp} \right) - \omega\tau \left(\frac{\partial \mathbf{u}_\perp}{\partial x_\parallel} + \frac{\partial \mathbf{u}_\parallel}{\partial x_\perp} \right) \times \mathbf{b} \right]$$
$$+ \frac{\tau p}{(1 + 4\omega^2\tau^2)} [(\nabla_\perp \mathbf{u}_\perp + \mathbf{b} \times \nabla \mathbf{u} \times \mathbf{b}) + 2\omega\tau(\nabla_\perp \mathbf{u}_\perp \times \mathbf{b} - \mathbf{b} \times \nabla \mathbf{u}_\perp)]$$
$$...[4.20]$$

Note that at large $\omega\tau$ the surviving terms are

$$p = p_0\mathbf{1} - \tau p[(2 \operatorname{div} \mathbf{u}_\perp + \operatorname{div} \mathbf{u}_\parallel - \tfrac{5}{3} \operatorname{div} \mathbf{u})\mathbf{1} + (2 \operatorname{div} \mathbf{u}_\parallel - \operatorname{div} \mathbf{u}_\perp)\mathbf{bb}]$$
$$+ \frac{p}{\omega} \left(\frac{\partial \mathbf{u}_\perp}{\partial x_\parallel} + \frac{\partial \mathbf{u}_\parallel}{\partial x_\perp} \right) \times \mathbf{b} + \frac{p}{2\omega} (\nabla_\perp \mathbf{u}_\perp \times \mathbf{b} - \mathbf{b} \times \nabla \mathbf{u}_\perp).$$

Results obtained from detailed theory differ from these in only modest ways; there is a current $j \propto \nabla T$ needed to satisfy the Onsager relations, and the functions $\omega\tau/(1 + \omega^2\tau^2)$ etc. of $\omega\tau$ become a good deal more complicated.

The term in p proportional to τ has an interesting property. Suppose that we allowed for a time variation of δf. This could be handled by a Laplace transform and would be replaced by $[s + 1/\tau]$, and in p the second term would become

$$[1/(s + 1/\tau)]p[(2 \operatorname{div} \mathbf{u}_\perp + \operatorname{div} \mathbf{u}_\parallel - \tfrac{5}{3} \operatorname{div} \mathbf{u})\mathbf{1} + (2 \operatorname{div} \mathbf{u}_\parallel - \operatorname{div} \mathbf{u}_\perp)\mathbf{bb}].$$

In the limit of small $1/\tau$ this may be expressed in terms of the displacement $\boldsymbol{\xi} = \int \mathbf{V}dt$ and, if we note that $p_0 = p(0)(1 - \tfrac{5}{3} \operatorname{div} \boldsymbol{\xi})$ (adiabatic compression), we obtain

$$p = p(0)\mathbf{1} - p[(2 \operatorname{div} \boldsymbol{\xi}_\perp + \operatorname{div} \boldsymbol{\xi}_\parallel)(\mathbf{1} - \mathbf{bb}) + (3 \operatorname{div} \boldsymbol{\xi}_\parallel + \operatorname{div} \boldsymbol{\xi}_\perp)\mathbf{bb}],$$
$$...[4.21]$$

the linearized double adiabatic result.

5

THE LORENTZ GAS (I)

A good deal can be learned about the behaviour of the ionized gas—beyond the normal solution—by considering a simple model, that of the Lorentz gas. In this model the ions are taken as infinitely heavy, and the electrons are considered to collide only with the ions. The Boltzmann equation then becomes

$$\frac{\partial f}{\partial t} + \mathbf{v} \cdot \frac{\partial f}{\partial x} + \frac{e}{m} (\mathbf{E} + \mathbf{v} \times \mathbf{B}) \cdot \frac{\partial f}{\partial v} = 2\pi n \left(\frac{e^2}{m} \right)^2 \ln \Lambda \frac{\partial}{\partial v} \cdot \mathbf{w}(v) \cdot \frac{\partial f}{\partial v}. \quad ...[5.1]$$

Solutions to the equilibrium problem are now much wider than just the Maxwellian distribution. Indeed, if f is any function of v^2, then $\partial f/\partial v = \mathbf{v}[\partial f/\partial(\tfrac{1}{2}v^2)]$, and since $\mathbf{w}(v) = (1 - \hat{\mathbf{v}}\hat{\mathbf{v}})/v$, $\mathbf{v} \cdot \mathbf{w} = 0$. Hence any function of

v^2 is a solution to the equilibrium problem. Indeed, if we consider the form of the differential operator, using

$$(\partial/\partial \boldsymbol{v}) \cdot \boldsymbol{v}/v^2 = 3/v^2 - 2\boldsymbol{v} \cdot \boldsymbol{v}/v^4 = 1/v^2 \quad \text{and} \quad \hat{\boldsymbol{v}} \cdot \partial/\partial \boldsymbol{v} \equiv \partial/\partial v,$$

then $\quad \dfrac{\partial}{\partial \boldsymbol{v}} \cdot \left[\dfrac{1}{v}(1 - \hat{\boldsymbol{v}}\hat{\boldsymbol{v}}) \right] \cdot \dfrac{\partial}{\partial v} f = \dfrac{1}{v} \dfrac{\partial}{\partial \boldsymbol{v}} \cdot \dfrac{\partial}{\partial \boldsymbol{v}} f - \dfrac{1}{v^2} \dfrac{\partial f}{\partial \boldsymbol{v}} - \dfrac{1}{v^2} \dfrac{\partial f}{\partial \boldsymbol{v}} - \dfrac{1}{v} \dfrac{\partial^2 f}{\partial v^2}$

$$= \frac{1}{v} \left[\frac{\partial}{\partial \boldsymbol{v}} \cdot \frac{\partial}{\partial \boldsymbol{v}} - \frac{1}{v^2} \frac{\partial}{\partial v} \left(v^2 \frac{\partial}{\partial v} \right) \right] f$$

$$= \frac{1}{v} \nabla_\theta^2 f$$

(where ∇_θ^2 is the azimuthal part of ∇^2), and in spherical polar coordinates

$$\frac{1}{v} \nabla_\theta^2 f = \frac{1}{v^3} \left[\frac{1}{\sin \theta} \frac{\partial}{\partial \theta} \left(\sin \theta \frac{\partial}{\partial \theta} \right) + \frac{1}{\sin^2 \theta} \frac{\partial^2}{\partial \varphi^2} \right] f. \qquad \ldots [5.2]$$

For the time-dependent problem, $f = f(t, \boldsymbol{v})$, we can then write down the eigenmodes $f = f_l^n P_l^n$, where P represents a spherical harmonic and

$$\frac{\partial f_l^n}{\partial t} = -\frac{2\pi n}{v^3} \left(\frac{e^2}{m} \right)^2 \ln \Lambda \, (l; l+1) f_l^n.$$

Hence

$$f_l^n(t) = f_l^n(0) \exp \left[-\frac{2\pi n}{v^3} \left(\frac{e^2}{m} \right)^2 \ln \Lambda \, (l; l+1) t \right], \qquad \ldots [5.3]$$

f_l^n being an arbitrary function of v^2. Note that this result means that the relaxation time varies as $1/v^3$, and hence that the relaxation model will have questionable validity.

Let us now consider the value of the electrical conductivity. Suppose a distribution f_0 is given, and an electric field is switched on; we can then inquire as to the change in the steady distribution function. We find, linearizing in E, that

$$\frac{e}{m} \boldsymbol{E} \cdot \frac{\partial f_0}{\partial \boldsymbol{v}} = e\boldsymbol{E} \cdot \boldsymbol{v} \frac{\partial f_0}{\partial \mathscr{E}}$$

$$= -2\pi n \left(\frac{e^2}{m} \right)^2 \frac{\ln \Lambda}{v^3} \nabla_\theta^2 (\delta f),$$

where $f_0 = f_0(\tfrac{1}{2}mv^2) = f_0(\mathscr{E})$. But, introducing a polar axis along E, the left-hand side becomes $eEv(\partial f_0/\partial \mathscr{E}) P_1(\theta, \varphi)$ and

$$\delta f = -1/(4\pi n)(m/e^2)^2 (1/\ln \Lambda) v^4 (\partial f_0/\partial \mathscr{E}) eE P_1.$$

The electric current:

$$j = \langle nev \cos\theta \rangle = \frac{ne^2}{2m} \frac{v_\theta^3 E}{4\pi n (e^2/m)^2 \ln\Lambda}$$

$$= \omega_p^2 \tau / (2\pi) E \qquad \qquad ...[5.4]$$

$$\left[\frac{1}{\tau} = \frac{4\pi n}{v_\theta^3} \left(\frac{e^2}{m}\right)^2 \ln\Lambda; \; v_\theta = \left(\frac{2\kappa T}{m}\right)^{\frac{1}{2}}; \; \omega_p^2 = \frac{4\pi n e^2}{m} \right].$$

We can ask for the relaxation time for this current; that is, supposing that the electric field is suddenly switched on at time $t = 0$, what is the time behaviour of the current? We have

$$\frac{\partial}{\partial t}(\delta f) + \frac{2\pi n}{v}\left(\frac{e^2}{m}\right)^2 \ln\Lambda \; \nabla_\theta^2(\delta f) = -eE \cdot v \frac{\partial f_0}{\partial \mathscr{E}}$$

and, as before,

$$\delta f = P_1 g(v, t)$$

$$\partial g/\partial t + (1/\tau)(v_\theta/v)^3 g = -(e/m) E(t) v \partial f_0/\partial \mathscr{E}.$$

We solve this with $g(0) = 0$,

$$g = -\int_0^t \exp\left[-\left(\frac{v_\theta}{v}\right)^3 \frac{t-t'}{\tau}\right] \frac{e}{m} E(\,') v \frac{\partial f_0}{\partial \mathscr{E}} \, dt'.$$

The current

$$j = -\frac{e^2}{m} \int d^3 vv \int d\Omega \cos^2\theta \int_0^t \exp\left[-\left(\frac{v_\theta}{v}\right)^3 \frac{t-t'}{\tau}\right] E(t') v \frac{\partial f_0}{\partial \mathscr{E}} \, dt'.$$

$$...[5.5]$$

For the switch-on problem

$$E = \begin{cases} \text{const.,} \; t > 0 \\ 0, \; t < 0 \end{cases}$$

so

$$j(t) = (e^2/m) E \int dv \int d\Omega v^4 \cos^2\theta (v/v_\theta)^3 \tau \{1 - \exp\left[-(v_\theta/v)^3 t/\tau\right]\}$$

$$= j_\infty \left\{ 1 - \frac{1}{3} \int_0^\infty \exp(-w^2) \exp\left[-t/(\tau w^3)\right] w^7 \, dw \right\}$$

$$= j_\infty \left[1 - F(t/\tau)\right], \qquad \qquad ...[5.6]$$

where j_∞ is the d.c. value of the current. $F(t/\tau)$ starts at 1 and decreases steadily. For small t/τ,

$$F \simeq 1 - \frac{1}{8}\pi^{\frac{1}{2}} t/\tau + \frac{1}{3}(t/\tau)^2,$$

while for large values of t/τ, a saddle point integration yields

$$F \simeq \frac{1}{3}(\pi/6)^{\frac{1}{2}}(\frac{3}{2}t/\tau)^{\frac{7}{5}} \exp\left[-\frac{5}{3}(\frac{3}{2}t/\tau)^{\frac{2}{5}}\right].$$

Hence j relaxes only slowly to equilibrium (as exp $[-(t/\tau)^{\frac{1}{2}}]$ not as exp $(-t/\tau)$).

If $E = E_0 e^{i\omega t}$ for $t \geqslant -\infty$,

$$j = -e^2 E_0 \int dv d\Omega v^4 \cos^2 \theta \{1/[i\omega+1/(\tau w^3)]\}(\partial f_0/\partial \mathscr{E}) \exp(-i\omega t)$$

$$= n_0 \frac{e^2}{m} \frac{E_0}{i\omega} \frac{8\pi}{3\pi^{\frac{1}{2}}} \int_0^\infty \frac{dw\, w^7 \exp(-w^2)}{w^3 - i/(\omega \tau)}. \qquad \ldots[5.7]$$

For large values of $1/(\omega\tau)$ (small values of $\omega\tau$), $j \simeq j_\infty$; and for small values of $1/(\omega\tau)$ (large values of $\omega\tau$),

$$j(t) = n_0 (e^2/m)[E_0/(i\omega)] \exp(i\omega t)\{1+[i/(\omega\tau)][4/(3\pi^{\frac{1}{2}})]\}.$$

If E varies in both space and time, $E = E_0\, e^{i\omega t}\, e^{ik \cdot x}$, we can get a causal solution by Laplace transforming, whereupon

$$(S+ik \cdot v)\delta f + [1/(\tau w^3)]\nabla_\theta^2 (\delta f) = -eE \cdot v(\partial f_0/\partial \mathscr{E}).$$

Now, if $(S+ik \cdot v) > \tau^{-1}$, then

$$\delta f \simeq -\frac{eE \cdot v}{S+ik \cdot v} \frac{\partial f_0}{\partial \mathscr{E}};$$

and if $(S+ik \cdot v) \ll \tau^{-1}$, $\delta f_2 \simeq -\tau(w^3/2)eE \cdot v(\partial f_0/\partial \mathscr{E})$; and to next order δf_3 has the form, for $(S+ik \cdot v) > \tau^{-1}$,

$$\delta f_3 \simeq \frac{1}{(S+ik \cdot v)} \frac{1}{\tau w^3} \nabla_\theta^2 \left(\frac{eE \cdot v}{S+ik \cdot v}\right) \frac{\partial f_0}{\partial \mathscr{E}}.$$

For small E and finite k, this yields (approximately)

$$\delta f_3 \simeq -[1/(S+ik \cdot v)][1/(\tau w^3)]eE(s) \cdot v(\partial f_0/\partial \mathscr{E})(k \cdot v)^2(\partial^2/\partial s^2)[1/(S+ik \cdot v)].$$

If $E = E_0 e^{i\omega t}$, $E(s) = 1/(S-ik \cdot v)$ and from δf_1, δf_3 we must select only the pole at $i\omega$:

$$\delta f_1 + \delta f_3 = -\mathscr{P}\left[\frac{1}{i(\omega+k \cdot v)} - \frac{1}{(\omega+k \cdot v)^2} \frac{(k \cdot vt)^2}{\tau w^3}\right] eE \cdot v \frac{\partial f_0}{\partial \mathscr{E}}.$$
$$\ldots[5.8]$$

Note that this yields a damping that increases as $(k \cdot vt)^2$—a result that holds only for large k and should lead to a rapid damping of the waves.

6

THE LORENTZ GAS (II)

It is possible to discuss the Lorentz gas in an alternative and instructive way. Instead of starting from the Landau form of the Fokker-Planck equation, we could begin from the "collisionless Boltzmann equation" and intro-

duce the field of the ions explicitly as a potential $\phi(x)$. The electron distribution function would then satisfy

$$\frac{\partial f}{\partial t} + v \cdot \frac{\partial f}{\partial x} - \frac{e}{m} \frac{\partial \phi}{\partial x} \cdot \frac{\partial f}{\partial v} = 0. \qquad \qquad \ldots [6.1]$$

The solution to $[6.1]$ will of course display the rapid spatial variation associated with ϕ, hence will scarcely define the smoothed Boltzmann function
we seek. Indeed, we could take $[6.1]$ and average it over some volume of
space, small compared with the scale on which the smoothed Boltzmann
function varies, but containing many ions, i.e. large on the scale of variation
of ϕ. Then

$$\frac{\partial \bar{f}}{\partial t} + v \cdot \frac{\partial \bar{f}}{\partial x} - \frac{e}{m} \overline{\frac{\partial \phi}{\partial x} \cdot \frac{\partial f}{\partial v}} = 0. \qquad \qquad \ldots [6.2]$$

To form the average of the last term, let us write

$$\bar{f} = f_0(v), \quad f = f_0 + \sum_k e^{ik \cdot x} f_k(v), \quad \text{and} \quad \phi = \sum_k e^{ik \cdot x} \phi_k.$$

Then the coefficient of $\exp(ik \cdot x)$ linearized in ϕ_k becomes

$$\partial f_k/\partial t + i(k \cdot v) f_k - (e/m) ik \cdot (\partial f_0/\partial v) \phi_k = 0,$$

which yields a lowest approximation for f_k:

$$f_k = (e/m) \int \exp[-i(k \cdot v)(t - t')] ik \phi_k \cdot (\partial f_0/\partial v) dt'. \qquad \ldots [6.3]$$

To form the average we may now substitute this result in $[6.2]$ replacing
sums by integrals:

$$\frac{e}{m} \overline{\frac{\partial \phi}{\partial x} \cdot \frac{\partial f}{\partial v}}$$

$$= \frac{1}{V} \int d^3x \int d^3k \int d^3k' \left(\frac{e}{m}\right)^2 e^{i(k+k') \cdot x} \phi_{k'} ik' \cdot \frac{\partial}{\partial v} \int e^{-i(k \cdot v)(t-t')} \phi_k ik \cdot \frac{\partial}{\partial v} f_0 dt'$$

$$= \left(\frac{e^2}{m}\right)^2 \frac{(2\pi)^3}{V} \frac{\partial}{\partial v} \int d^3k \, kk \, \phi_k \int e^{-ik \cdot v(t-t')} \phi_k \cdot \frac{\partial f_0}{\partial v} \, dt'. \qquad \ldots [6.4a]$$

Note that this could have been written

$$\left(\frac{e}{m}\right)^2 \frac{\partial}{\partial v} \cdot \langle E \int dt' E[t', x(t')] \rangle \cdot \frac{\partial}{\partial v} f_0, \qquad \ldots [6.4b]$$

where the inner integral is carried along a particle trajectory with $x = x_0 + vt$.
To evaluate this, we need the value of ϕ_k. If we include *only* the field of the
ions this is easily written down, for the ion charge density is

$$q(x) = e \sum_i \delta(x - X_i),$$

the X_i being the random positions of the ions; so

$$q(k) = e/(2\pi)^3 \int d^3x \exp(-ik \cdot x) q(x) = \sum_i 1/(2\pi)^3 \exp(-ik \cdot X_i)$$

and $\mathbf{V}^2\phi = -4\pi q$, hence

$$\phi_k = \frac{e}{2\pi^2}\frac{1}{k^2}\sum_i \exp\left(-ik \cdot X_i\right). \qquad \ldots[6.5]$$

Now, we seek a value of f which is steady on the $k \cdot v$ time scale, and the inner integral in [6.4] should be extended to $-\infty$: the assumption being that the correlation between $\phi(x)$ and $\phi(x+vt)$ will vanish for values of t during which f_0 changes negligibly. The value of the integral is then $\mathscr{P}[1/(ik \cdot v)]+\frac{1}{2}\pi\delta(k \cdot v)$, and on carrying the integral over all k, the first (odd) function disappears, and we are left with

$$\frac{e}{m}\overline{\frac{\partial\phi}{\partial x} \cdot \frac{\partial f}{\partial v}} = (2\pi)^3\left(\frac{e}{m}\right)^2\frac{\pi}{2V}\frac{\partial}{\partial v} \cdot \int d^3k\delta(k \cdot v)kk\,\phi_k\phi_{-k} \cdot \frac{\partial}{\partial v}f_0$$

and substituting [6.5] yields

$$\frac{e}{m}\overline{\frac{\partial\phi}{\partial x} \cdot \frac{\partial f}{\partial v}} = \left(\frac{e^2}{m}\right)^2\frac{1}{V}\frac{\partial}{\partial v} \cdot \int d^3k\frac{kk}{k^4}\delta(k \cdot v)\sum_{i,j}e^{ik \cdot (X_i-X_j)} \cdot \frac{\partial}{\partial v}f_0. \qquad \ldots[6.6]$$

In the sum here, the X_i and X_j are distributed at random, and the sum may be split into two parts

$$\sum_i e^{ik \cdot (X_i-X_j)}+\sum_i\sum_{j \neq i}e^{ik \cdot (X_i-X_j)} = N+\sum_{i,\,j \neq i}e^{ik \cdot (X_i-X_j)}$$

$$\simeq N.$$

The second term is the sum of a series of terms each of unit magnitude and random phase, and these add to zero; hence, with $n = N/V$,

$$\frac{e}{m}\overline{\frac{\partial\phi}{\partial x} \cdot \frac{\partial f}{\partial v}} = n\left(\frac{e^2}{m}\right)^2\frac{\partial}{\partial v}\int d^3k\frac{kk}{k^4}\delta(k \cdot v) \cdot \frac{\partial}{\partial v}f_0. \qquad \ldots[6.7]$$

To evaluate the inner integral here, we note that $kk = k^2\hat{k}\hat{k}$ and $\delta(k \cdot v) = [1/(kv)]\delta(\hat{k} \cdot \hat{v})$, $\hat{k}$ and $\hat{v}$ being unit vectors; hence

$$\mathscr{I} = \int d^3k\frac{kk}{k^4}\delta(k \cdot v) = \frac{1}{v}\int d\Omega\hat{k}\hat{k}\delta(\hat{k} \cdot \hat{v})\int\frac{dk}{k}. \qquad \ldots[6.8]$$

The inner integral (over k) must be cut off both at high and low k, (k_{max}, k_{min}) which represents a loss from the Boltzmann equation, where only the lower cut-off (b_{max}) was required. That analysis however, enables us to identify the upper cut-off $k_{max} = b_{min} = \frac{1}{2}\,(mv^2/e^2)$ unambiguously, and we may identify this integral with the $\ln\Lambda$ of the Boltzmann equation. We are left with the integral over solid angles. If we write $\mu = \hat{k} \cdot \hat{v}$ (choosing v as a polar axis), then the integral becomes

$$\int d\Omega\,\hat{k}\hat{k}\,\delta(\mu) = \int_{-1}^1 d\mu\int_0^{2\pi} d\varphi\,\delta(\mu)\,\hat{k}\hat{k}.$$

The tensor $\hat{k}\hat{k}$, which is symmetric, may be written

$$\hat{k}\hat{k} = \begin{bmatrix} \mu^2 & \mu(1-\mu^2)^{\frac{1}{2}}\cos\varphi & \mu(1-\mu^2)^{\frac{1}{2}}\sin\varphi \\ \mu(1-\mu^2)^{\frac{1}{2}}\cos\varphi & (1-\mu^2)\sin^2\varphi & (1-\mu^2)\sin\varphi\cos\varphi \\ \mu(1-\mu^2)^{\frac{1}{2}}\sin\varphi & (1-\mu^2)\sin\varphi\cos\varphi & (1-\mu^2)\cos^2\varphi \end{bmatrix}.$$

In the integral, terms off the main diagonal vanish on integration over φ, while $\delta(\mu)$ destroys the first term; hence

$$\int \hat{k}\hat{k}\delta(\mu)d\Omega = \pi(1-\hat{v}\hat{v}) \qquad \text{...[6.9]}$$

$$\mathscr{I} = 2\pi(e^2/m)^2 \ln \Lambda (\partial/\partial v)(1/v)(1-\hat{v}\hat{v})\cdot(\partial/\partial v)f_0$$
$$= 2\pi\left(\frac{e^2}{m}\right)^2 \ln \Lambda \frac{\partial}{\partial v}\cdot w\cdot\frac{\partial}{\partial v}f_0 \qquad \text{...[6.10]}$$

which is exactly the Landau result.

We can improve this result somewhat by including in the calculation of ϕ_k the effect of the electrons considered as a distributed charge. To get ϕ_k then, we note that the field due to the ions will alter the electron charge distribution. In the usual way we write

$$\delta f_k = (e/m)\int_{-\infty}^{t} e^{-i(k\cdot v)(t-t')}ik\cdot(\partial f_0/\partial v)\phi_k dt'$$
$$= -(e/m)k\cdot(\partial f_0/\partial v)\{\mathscr{P}[1/(k\cdot v)]+i\pi\delta(k\cdot v)\}\phi_k$$
$$q_{\text{ind}} = -(e^2/m)\int k\cdot(\partial f_0/\partial v)[1/(k\cdot v)+i\pi\delta(k\cdot v)]\phi_k d^3v$$
$$= \frac{ne^2}{\kappa T}\phi_k = \frac{1}{4\pi}k_D^2\phi_k$$

and

$$\nabla^2\phi = -4\pi q_{\text{ind}}-4\pi q_{\text{app}}$$

yielding

$$\phi_k = \frac{4\pi}{(2\pi)^3}\frac{\sum e^{ik\cdot x_i}}{k^2+k_D^2} \qquad \text{...[6.11]}$$

hence

$$\phi_k = \frac{4\pi}{(2\pi)^3}\frac{\sum e^{ik\cdot x_i}}{k^2+k_D^2}.$$

The only effect of this on our analysis is to replace the integral over k by

$$\int dk \frac{k^3}{(k^2+k_D^2)^2} = \frac{1}{2}\ln(k^2+k_D^2)\Big]_0^{k_{\max}} + \frac{k_D^2}{k^2+k_D^2}\Big]_0^{k_{\max}}.$$

We may now extend the lower limit of integration to zero, thus removing the long range cut-off. $k_{\max}$, the short range cut-off has already been identified with the minimum effective impact parameter $b_{\min} = e^2/(mv^2)$ produced by the Boltzmann integral. We have not correctly calculated this, since we have worked only to second order in the field strength, while the close collisions involve strong fields.

If now $$k_{max} \gg k_D$$

$$\int_0^{k_{max}} dk[k^3/(k^2+k_D^2)^2] \simeq \ln(k_{max}/k_D) = \ln \Lambda \qquad ...[6.12]$$

and we obtain the Landau result, *without* the arbitrary long range cut-off.

7

THE KINETIC EQUATION FOR THE PLASMA

Let us now attempt to apply these concepts to an actual plasma. For the real case, we must observe, first, that the ions do not remain at rest, but have thermal motions; and second, that the electrons are also point particles and act as sources of a field; and third, that the appropriate form of the averaged Vlasov equation is

$$\frac{\partial f}{\partial t} + v \cdot \frac{\partial f}{\partial x} - \frac{\partial}{\partial v} \cdot \left(\frac{e}{m} \frac{\partial \phi}{\partial x} f \right) = 0. \qquad ...[7.1]$$

Why is this? If we go back to the Liouville equation, and the B.B.G.K.Y. hierarchy, we recognize the last term here as

$$n(e/m) \int d^3v'd^3x' (\partial/\partial x) \phi(x, x') \cdot (\partial/\partial v) f_2(x, x', v, v'),$$

and note that this could be written

$$(\partial/\partial v) n(e/m)[\int d^3v' (\partial \phi/\partial x) \Psi(x', v'; x, v)] f_1(x_1, v_1),$$

where Ψ represents the conditional probability of finding particles at x_2, given x_1. It is the bracketed expression here that gives our $\partial \phi/\partial x$, the actual field acting on a particle at x_1.

Now the fluctuating charge due to the ions has the form

$$q_+(k, \omega) = e_+/(2\pi)^3 \sum_i e^{-ik \cdot X_i} \delta(\omega + k \cdot V_i) \qquad ...[7.2]$$

since $q_+(x, t) = e_+ \sum \delta[x - (X_i + V_i t)]$, X_i being the ith position at $t = 0$. The spectral density of q_+ is then

$$\langle q_+ q_+^*(\omega, k) \rangle = \frac{e_+^2}{(2\pi)^6} \frac{1}{V} \int d^3x dt \int d^3k' d\omega' e^{i(k+k') \cdot x} \sum_i e^{-ik \cdot X_i} \delta(\omega + k \cdot V_i)$$
$$\sum_j e^{-ik' \cdot X_j} \delta(\omega' + k' \cdot V_j)$$

$$= e_+^2/(2\pi)^3 \int d^3k' d\omega' \sum_{i,j} e^{-ik \cdot X_i} e^{-ik' \cdot X_j} \delta(\omega + k \cdot V_i)$$
$$\delta(\omega' + k' \cdot V_j) \delta(k+k')$$

$$= e_+^2/(2\pi)^3 \int d\omega' \sum_{i,j} e^{ik \cdot (X_i - X_j)} \delta(\omega + k \cdot V_i) \delta(\omega' - k \cdot V_j)$$
$$...[7.3]$$

and, using the random phase approximation (r.p.a.),

$$\sum e^{i\mathbf{k}\cdot(\mathbf{X}_i-\mathbf{X}_j)}\delta(\omega+\mathbf{k}\cdot\mathbf{V}_i) = \int d^3v f(v)\delta(\omega+\mathbf{k}\cdot\mathbf{v});$$

so $\qquad \langle qq^*(\omega,\mathbf{k})\rangle = e_+^2/(2\pi)^3 \int d^3v f_+(v)\delta(\omega+\mathbf{k}\cdot\mathbf{v}).$ ...[7.4]

The potential due to the fluctuations is now

$$\phi = 4\pi/(2\pi)^3\{e_-\sum e^{i\mathbf{k}\cdot\mathbf{X}_i}\delta(\omega+\mathbf{k}\cdot\mathbf{v}_i)+e_+\sum e^{i\mathbf{k}\cdot\mathbf{X}_i}\delta(\omega+\mathbf{k}\cdot\mathbf{V}_i)\}/(k^2\epsilon),$$

where

$$\epsilon = 1-4\pi\sum_{+,-}\frac{e^2}{m}\frac{1}{k^2}\int d^3v\,\frac{\mathbf{k}\cdot(\partial f/\partial\mathbf{v})}{\omega+\mathbf{k}\cdot\mathbf{v}}. \qquad ...[7.5]$$

This may sometimes conveniently be written as

$$\epsilon = 1-\frac{k_D^2}{k^2}\left[G_-\left(\frac{\omega}{kv_\theta^-}\right)+\frac{T_-}{T_+}G_+\left(\frac{\omega}{kv_\theta^+}\right)\right], \qquad ...[7.6]$$

where $f(v)$ has been written $v_\theta^{-3}f(v/v_\vartheta)$; $k_D^2 = 4\pi ne^2/(mv_\theta^2) = 8\pi ne^2/(\kappa T)$;

$$G = \int dt\,g(t)\{\mathscr{P}[1/(u+t)]+i\pi\delta(u+t)\};$$

$$g(v/v_\theta) = (1/n)(1/v_\theta)\int d^2(\mathbf{v}_\perp/v_\theta)\,v_\theta\hat{\mathbf{k}}_\perp\cdot(\partial f/\partial\mathbf{v}). \qquad ...[7.7]$$

Now a final remark. Since the distribution f in [7.1] describes the motion of particles which are themselves sources of the field, the mean value of ϕ_k does not vanish at the particle. Indeed, the field at the particle is given by

$$\mathbf{E}(\mathbf{x}+\mathbf{v}t) = \frac{4\pi e}{(2\pi)^3}\int\delta(\omega+\mathbf{k}\cdot\mathbf{v})\frac{i\mathbf{k}}{k^2\epsilon(\omega,\mathbf{k})}\,d^3kd\omega.$$

Now, on integration over ω, $\mathbf{k}$, only the real part of this will survive, and this is

$$\mathbf{E} = (4\pi)^2/(2\pi)^3e^2\sum\pi\int d^3v'[\delta(\omega+\mathbf{k}\cdot\mathbf{v})/(k^4\,|\,\epsilon\,|^2)]\mathbf{k}\mathbf{k}\cdot(\partial f/\partial\mathbf{v}')$$

$$\delta(\omega+\mathbf{k}\cdot\mathbf{v}')d^3kd\omega. \qquad ...[7.8]$$

We can now use the arguments advanced for the Lorentz gas case, to write the last term in the kinetic equation as

$$\int d^3kd\omega\{(\partial/\partial\mathbf{v})\cdot[(e/m)\mathbf{E}_k f]-(e/m)(\partial/\partial\mathbf{v})\cdot i\mathbf{k}\phi\delta f\},$$

where

$$\delta f = \left[\mathscr{P}\frac{1}{i(\omega+\mathbf{k}\cdot\mathbf{v})}+\pi\delta(\omega+\mathbf{k}\cdot\mathbf{v})\right]i\mathbf{k}\cdot\frac{e}{m}\frac{\partial f}{\partial\mathbf{v}}\phi_k, \qquad ...[7.9]$$

and keeping the real term

$$(\partial/\partial\mathbf{v})\cdot\int d^3kd\omega(e/m)i\mathbf{k}\phi\delta f$$

$$= -\tfrac{1}{2}\pi(\partial/\partial\mathbf{v})\cdot(e/m)^2\int d^3kd\omega\,\delta(\omega+\mathbf{k}\cdot\mathbf{v})\mathbf{k}\mathbf{k}\cdot(\partial f/\partial\mathbf{v})\phi_k\phi_{-k}$$

and $\phi_k\phi_{-k} = (2/\pi) \sum e^2 \int d^3v' f(v')\delta(\omega+k\cdot v')/|k^2\epsilon|^2.$ ...[7.10]

We obtain for the total term

$$\frac{2e^2}{m}\frac{\partial}{\partial v}\cdot\int d^3k d\omega \sum_{+,-}\int d^3v' \frac{\delta(\omega+k\cdot v')}{[k^4|\epsilon|^2]}\delta(\omega+k\cdot v)kk\cdot\left(\frac{e^2}{m}\right)J(v',v)$$

...[7.11]

where J has been defined in the discussion following [2.2], and on integrating over ω, the kinetic equation for the plasma becomes

$$\frac{\partial f}{\partial t}+v\cdot\frac{\partial f}{\partial x}+2e^4\frac{\partial}{\partial p}\cdot\sum_{+,-}\int d^3k\int d^3v' \frac{kk}{k^4|\epsilon(-k\cdot v,k)|^2}\delta[k\cdot(v-v')]$$
$$\cdot J(p',p)=0 \qquad ...[7.12]$$

an expression sometimes called the Lenard-Balescu equation.

Let us consider the dominant term in this by first integrating the tensor coefficient over the magnitude of k. Thus,

$$T = 2e^4\int d^3k\, kk\delta[k\cdot(v-v')]/[k^4|\epsilon(-k\cdot v,k)|^2];$$

and writing $g = v-v'$; $G_+ + G_- = -X+iY$; $\delta(k\cdot g) = \delta(\hat{k}\cdot\hat{g})/(kg)$; where X and Y are functions of $k\cdot v/(kv_\theta) = \hat{k}\cdot\hat{v}v/v_\theta$,

$$T = \frac{2e^4}{g}\int d\Omega \hat{k}\hat{k}\delta(\hat{k}\cdot\hat{g})\int dk \frac{k^3}{(k^2+k_D^2X)^2+k_D^4Y^2} \qquad ...[7.13]$$

The inner integral becomes

$$\int_0^\infty dk \frac{k(k^2+k_D^2X-k_D^2X)}{(k^2+k_D^2X)^2+k_D^4Y^2} = \tfrac{1}{4}\ln(k^2+k_D^2X)^2-\tfrac{1}{2}\frac{X}{Y}\tan^{-1}\left[\frac{k^2+k_D^2X}{k_D^2Y}\right].$$

...[7.14]

This again diverges at large values of k, reflecting our neglect of the strong field region. However, we may again invoke the Boltzmann cut-off here, so $k_{max} = mv_\theta^2/e^2$ using the mean value in a logarithm. If $(k_{max})^2 \gg k_D^2$ while X, Y remain bounded, the inner integral may be written

$$\ln\left(\frac{k_{max}}{k_D}\right)-\tfrac{1}{2}\ln\left(\frac{X}{1+(k_D^2/k_{max}^2)X}\right)+\frac{X}{Y}\left[\frac{\pi}{2}-\tan^{-1}\left(\frac{X}{Y}\right)\right]\simeq\ln\Lambda,$$

provided Y does not vanish in the range of integration: i.e. provided $\int d^3v(\partial f/\partial v)\delta[k\cdot(v-v')]$ does not vanish, or become too small. For very superthermal particles or for unstable distributions these conditions are *not* satisfied; but, when the first term dominates great simplification results; the integral over k reducing to $\ln\Lambda$. The integral over solid angles becomes, as before, $\pi(1/g)(1-\hat{g}\hat{g}) = \pi w$; $T = 2\pi e^4 w\ln\Lambda$ so that the Lenard-Balescu equation reduces to the Landau form.

8

SOME PROPERTIES OF THE KINETIC EQUATION

(1) If we write the kinetic equation as

$$\frac{\partial f}{\partial t} + v \cdot \frac{\partial f}{\partial x} + \frac{F}{m} \cdot \frac{\partial f}{\partial v} = \mathscr{I}, \left.\vphantom{\begin{array}{c}a\\b\end{array}}\right\} \quad \ldots[8.1]$$

$$\mathscr{I} = -2e^4 (\partial/\partial p) \cdot \int d^3k \int d^3v' [\delta(k \cdot g)/(k^4 \epsilon^2)] kk \cdot J(p', p),$$

we note first that $\mathscr{I} = 0$ for f Maxwellian, for

$$J(p', p) = \frac{\partial f(p')}{\partial p'} f(p) - f(p') \frac{\partial f(p)}{\partial p} = -\kappa T \left[\frac{p'}{m'} - \frac{p}{m}\right] f(p')f(p)$$

$$= +\kappa T f(p')f(p) g$$

and $\delta(k \cdot g) k \cdot g = 0$; hence the integrand vanishes.

(2) The Boltzmann H theorem holds.

$$\partial S/\partial t = -(\partial/\partial t) \int d^3 p f \ln f = -\int d^3 v \mathscr{I}(f)(1 + \ln f)$$

$$= -2e^4 \int d^3 p \int d^3 p' \int d^3 k \frac{\partial}{\partial p} \cdot \frac{kk\delta(k \cdot g)}{k^4 |\epsilon^2|} \cdot J(p', p)(1 + \ln f) \ldots[8.2]$$

and on a partial integration

$$\frac{\partial S}{\partial t} = 2e^4 \int d^3 p \int d^3 p' \int d^3 k \, kk \frac{\delta(k \cdot g)}{k^4 |\epsilon(k, -k \cdot v)|^2} : J(p', p) \frac{1}{f} \frac{\partial f}{\partial p}.$$

We could also have used p' as a variable, whereupon

$$\frac{\partial S}{\partial t} = 2e^4 \int d^3 p \int d^3 p' \int d^3 k \, kk \frac{\delta(k \cdot g)}{k^4 |\epsilon(k, -k \cdot v')|^2} : J(p, p') \frac{1}{f} \frac{\partial f}{\partial p'}$$

and since the δ-function is even in its argument

$$\frac{\delta[k \cdot (v' - v)]}{k^4 |\epsilon(k, -k \cdot v)|^2} = \frac{\delta[k \cdot (v - v')]}{k^4 |\epsilon(k, -k \cdot v')|^2}$$

and on adding

$$\frac{\partial S}{\partial t} = 2e^4 \int d^3 p d^3 p' f(p) f(p') \frac{\delta[k \cdot (v - v')]}{k^4 |\epsilon(k, -k \cdot v)|^2} kk : J(p', p) J(p', p). \quad \ldots[8.3]$$

But the r.h.s. $= \int d^3 p d^3 p' \frac{f(p)f(p')}{k^4 |\epsilon|^2} \delta[k \cdot (v - v')][k \cdot J(p', p)]^2$

which is positive semi-definite and S cannot decrease.

(3) The form of $\mathscr{I}$ for a Maxwellian is sometimes important. Consider this in the dominant approximation,

$$\mathscr{I} = 2\pi e^4 \ln \Lambda \sum (\partial/\partial p) \cdot \int d^3p'w \cdot J(p', p).$$

Consider the term that apparently contains $f(p)$, i.e.

$$\{(\partial/\partial p) \cdot \int d^3p'w \cdot [\partial f(p')/\partial p']\}f(p).$$

This could be written $1/(mm')\{(\partial/\partial v) \cdot \int d^3v'w \cdot [\partial f(v')/\partial v']\}f(v)$ and, on integrating by parts, $-1/(mm')[\int d^3v'f(v')(\partial/\partial v)(\partial/\partial v'):w]f(v)$. But, since $w = w(g)$ this is $(\partial/\partial v)(\partial/\partial v'):w = 0$. $\mathscr{I}$ may therefore be written

$$-2\pi e^4 \ln \Lambda \left[\frac{1}{mm'} \frac{\partial}{\partial v} \cdot \int d^3v'wf(v') \cdot \frac{\partial}{\partial v} f(v) \right.$$

$$\left. - \frac{1}{m^2} \frac{\partial}{\partial v} \cdot \int d^3v'wf(v') \cdot \frac{\partial}{\partial v} f(v) - \frac{1}{m^2} \int d^3v'wf(v') : \frac{\partial^2}{\partial v \partial v} f(v) \right]$$

$$= -2\pi \left(\frac{e^2}{m}\right)^2 \ln \Lambda \left(\frac{m}{m'} - 1\right) \left[\frac{\partial}{\partial v} \cdot \int d^3v'wf(v') \cdot \frac{\partial}{\partial v} f(v) \right.$$

$$\left. - \sum \int d^3v'wf(v') : \frac{\partial^2}{\partial v \partial v} f(v) \right]. \qquad \ldots[3.4]$$

Note that this makes the dynamic friction term depend only on interspecies interaction.

We may now ask the form of $\int d^3v'wf(v')$ for a Maxwellian $f(v')$. We need

$$T = \pi^{-\frac{3}{2}} \int d^3(v'/v_\theta) \exp\left[-(v'/v_\theta)^2\right](1/g)(1-\hat{g}\hat{g})$$

$$= (\pi^{-\frac{3}{2}}/v_\theta) \int d^3g \exp\left[-(v-g)^2\right](1/g)(1-\hat{g}\hat{g}),$$

where v and g have been normalized to $v_\theta = (2\kappa T/m)^{\frac{1}{2}}$, so

$$T = (\pi^{-\frac{3}{2}}/v_\theta) \int_0^{2\pi} d\varphi \int_{-1}^1 d\mu \int_0^\infty dgg \exp\left[-(v^2+g^2-2vgu)\right](1-\hat{g}\hat{g}).$$

The vector $\hat{g}$ can be written as $[\hat{v}\mu + t_1(1-\mu^2)^{\frac{1}{2}} \cos \varphi + t_2(1-\mu^2)^{\frac{1}{2}} \sin \varphi]$, introducing unit vectors $\hat{v}$ along v and t_1 and $t_2 \perp v$. Integration over φ then yields

$$T = [2/(\pi^{\frac{1}{2}}v_\theta)] \int_{-1}^1 d\mu \int_0^\infty dg\{\exp\left[-(v^2+g^2-2vgu)\right][\tfrac{1}{2}(1-\hat{v}\hat{v})(1+\mu^2)$$

$$- \hat{v}\hat{v}(1-\mu^2)]\}$$

$$= (1-\hat{v}\hat{v})\mathscr{I}_1 - \hat{v}\hat{v}\mathscr{I}_2 \qquad \ldots[3.5]$$

$$\mathscr{I}_1 = [2/(\pi^{\frac{1}{2}}v_\theta)] \int_{-1}^1 d\mu \int_0^\infty dg \exp\left[-(v^2+g^2-2vgu)\right][\tfrac{1}{2}(1+\mu^2)]$$

$$\mathscr{I}_2 = [2/(\pi^{\frac{1}{2}}v_\theta)] \int_{-1}^1 d\mu \int_0^\infty dg \exp\left[-(v^2+g^2-2vgu)\right](1-\mu^2).$$

The two scalars may be written

$$\mathscr{I}_1 = 2\pi^{\frac{1}{2}}(v_\theta/v^2)\exp\left[-(v/v_\theta)^2\right]+2\pi^{\frac{1}{2}}\left[(2v^2-v_\theta^2)/v^3\right]\int^{v/v_\theta}\exp\left(-t^2\right)dt$$

$$\mathscr{I}_2 = -4\pi^{\frac{1}{2}}(v_\theta/v^2)\exp\left[-(v/v_\theta)^2\right]+4\pi^{\frac{1}{2}}(v_\theta^2/v^3)\int^{v/v_\theta}\exp\left(-t^2\right)dt.$$

$$...[8.6]$$

For small values of v, $\mathscr{I}_1 \to 4\pi^{\frac{1}{2}}$, $\mathscr{I}_2 \to 4\pi^{\frac{1}{2}}$; while for large v, $\mathscr{I}_1 \to 4\pi/v$, $\mathscr{I}_2 \to 4\pi^{\frac{1}{2}}(v_\theta^2/v^3)$.

8.1. *The drag coefficient*

Differentiating $\mathscr{I}_2$, we obtain the dynamical friction or drag,

$$D = 2\pi(e^2/m)^2\ln\Lambda\left[(m/m'-1)/\pi^{\frac{1}{2}}\right]\int d^3(v/v_\theta)(g/g^3)\exp\left[-(v/v_\theta)^2\right]$$

$$= 4\pi^{\frac{1}{2}}\left(\frac{e^2}{m}\right)^2\ln\Lambda\left(\frac{m}{m'}-1\right)\frac{1}{v_\theta^2}\left(\frac{v_\theta}{v}\right)^2\left\{\left[\int^{v/v_\theta}\exp\left(-t^2\right)dt\right.\right.$$

$$\left.\left.-\left(\frac{v}{v_\theta}\right)\exp\left[-\left(\frac{v}{v_\theta}\right)^2\right]\right\}. \quad ...[8.7]$$

For small v,

$$D \to -\tfrac{16}{3}\pi^{\frac{1}{2}}(e^2/m)^2\ln\Lambda\left[(m-m')/(\kappa T)\right]v/v_\theta$$

and for large v,

$$D \to -4\pi^{\frac{1}{2}}(e^2/m)^2\ln\Lambda\left[(m-m')/(\kappa T)\right](v_\theta/v)^3.$$

It is the maximum in D that leads to the so-called runaway electron process. Very roughly, the current carried in the presence of an electric field is given by that value of the velocity for which $D = (e/m)E$. If, however, D decreases with velocity, the current is unstable against small increases in velocity $(\partial D/\partial v < 0)$, hence on a small fluctuation increasing the velocity, drag is reduced, and the velocity of the electrons increases still further. This occurs at

$$\frac{eE}{m} = 4\pi^{\frac{1}{2}}\left(\frac{e^2}{m}\right)^2\ln\Lambda\left(1-\frac{m}{m'}\right)\frac{1}{v_\theta^2}$$

or $\qquad\qquad eE\lambda = \kappa T.$

$$...[8.8]$$

This is the critical field above which an electron beam forms and becomes essentially free of the ions.

Since D is a function of v we might expect, just as in the Lorentz gas, to find a suprathermal electron tail developing in a much more modest electric field (for any field and some v, electrons can run away). An examination of this, however, would call for a more careful treatment of solutions to the kinetic equation than we have presented.

9

The Electron Correlation Function and Radiation Scatter

In the process of deriving the kinetic equation for the plasma, we have incidentally developed a program for calculating the electron correlation function. This quantity, however, is susceptible to direct measurement, since it determines the effective cross section for the inelastic scattering of radiation.

To see this, consider first the process of radiation and recall a few electromagnetic formulae. Radiation is conveniently described by the Hertz vector $\boldsymbol{\Pi}(r, t)$, from which the Lorentz gauge potentials may be derived, thus

$$A = (1/c)\partial\boldsymbol{\Pi}/\partial t; \quad \phi = \operatorname{div} \boldsymbol{\Pi}$$

hence

$$E = \frac{1}{c}\frac{\partial A}{\partial t} - \nabla\phi = \frac{1}{c^2}\frac{\partial^2 \boldsymbol{\Pi}}{\partial t^2} - \nabla(\nabla\cdot\boldsymbol{\Pi}). \qquad \text{...[9.1]}$$

The Hertz vector, $\boldsymbol{\Pi}$ is useful because of its simple relation to the polarization, P, $\square^2\boldsymbol{\Pi} = 4\pi P$; hence, if the polarization is confined to some small volume V, then the Hertz vector at great distance r is

$$\boldsymbol{\Pi}(r, t) = \int_V d^3r' P(r', t - R/c)/R.$$

If we select an origin in the scattering volume, then $R = |r - r'|$ and $r' \ll r$.

We are usually interested in the harmonic composition of $\boldsymbol{\Pi}$, $\boldsymbol{\Pi}(t) = \int d\omega\, e^{i\omega t}\, \boldsymbol{\Pi}(\omega)$, and are often given the harmonic composition of P. Then

$$\boldsymbol{\Pi}(t) = \int d\omega\, e^{i\omega t}\, \boldsymbol{\Pi}(\omega) = \int d\omega \int d^3r' P(r', \omega)\, e^{i\omega(t - R/c)}/R.$$

However,

$$\begin{aligned} R &= [(r - r')^2]^{\frac{1}{2}} = [r^2 - 2r\cdot r' + (r')^2]^{\frac{1}{2}} \\ &\simeq r - \hat{r}\cdot r' + \tfrac{1}{2}[(r')^2/r][1 - (\hat{r}\cdot\hat{r}')^2], \end{aligned}$$

where $\hat{r}$, $\hat{r}'$ are unit vectors along r, r'. Now, if the phase shift due to the last term $(r')^2/(\lambda r)$ is small (or random), we can write

$$i\omega(t - R/c) = i\omega(t - r/c) + ir'\cdot\hat{r}\omega/c,$$

and

$$\boldsymbol{\Pi}(r) = \int d^3r' P(r')e^{i(\omega/c)\hat{r}\cdot r'} e^{i\omega(t - r/c)}/r. \qquad \text{...[9.2]}$$

Asymptotically, the coefficient of the wave propagating along $\hat{r}$ is then $\int d^3r' P(r')\exp[i(\omega/c)\hat{r}\cdot r']$. To get the radiated spectral intensity, we must first form

$$E = 1/c^2(\partial^2/\partial t^2)\boldsymbol{\Pi} - \nabla(\nabla\cdot\boldsymbol{\Pi}) = -\omega^2/c^2[\boldsymbol{\Pi} - \hat{r}(\hat{r}\cdot\boldsymbol{\Pi})],$$

and then the intensity

$$
\begin{aligned}
I &= c/(4\pi)\langle E \cdot E \rangle \\
&= c/(4\pi) \int d\omega\, E(\omega) \cdot E^*(\omega) \\
&= \int d\omega[\omega^4/(4\pi c^3)][\Pi - \hat{r}(\hat{r} \cdot \Pi)]^2 \\
&= \int d\omega[\omega^4/(4\pi c^3)][\Pi \cdot \Pi^* - (\hat{r} \cdot \Pi)(\hat{r} \cdot \Pi^*)] \\
&= \int d\omega[\omega^4/(4\pi c^3 r^2)] \int d^3r' \int d^3r'' \{P(r') \cdot P(r'') - [\hat{r} \cdot P(r')][\hat{r} \cdot P(r'')]\} \\
&\qquad \exp[i(\omega/c)\hat{r}_0 \cdot (r' - r'')]. \qquad\qquad ...[9.3]
\end{aligned}
$$

For the scattering problem, the polarization P is, in the simplest case, the product of a polarizability p and the field strength E. The field strength E is the sum of an incident field E_0 with some frequency Ω_i and some wave vector $K = (\Omega_i/c),\hat{K}$ and the scattered wave, which is (for the plasma case) small by comparison with E_0, even at the scattering particle. Hence the polarization $P \simeq pE_0$. Scattering arises because of rapid fluctuations in p, and $P(r') = p(r', t)E_0$; $P(\Omega_s) = p(r', \Omega_s - \Omega_i) \exp[i(\Omega_i t - K \cdot r')]$ and $\Pi(\omega) = \int d^3r' p(r', \Omega_s - \Omega_i) \exp\{i[(\Omega_s/c)\hat{r} - K] \cdot r'\} \exp\{i\Omega_s[t - (r/c)]\}/r$. Now the quantity $\Omega_s - \Omega_i = \omega$, which is the frequency shift on scattering; while $(\Omega_s/c)\hat{r} - K = k$, the shift in wave number. To a good approximation, k may be expressed in terms of the scattering angle θ, as $k = 2K \sin(\theta/2)$.

Moreover, the scalar product $\hat{r} \cdot P = p\hat{r} \cdot E = (K + k) \cdot E$. Since, however, the input is electromagnetic $K \cdot E = 0$, and again in terms of the scattering angle $k \cdot E = kE \sin\theta \cos\varphi$, where φ is the angle between the plane of polarization of the incident wave and the scattering plane (see Fig. 1)

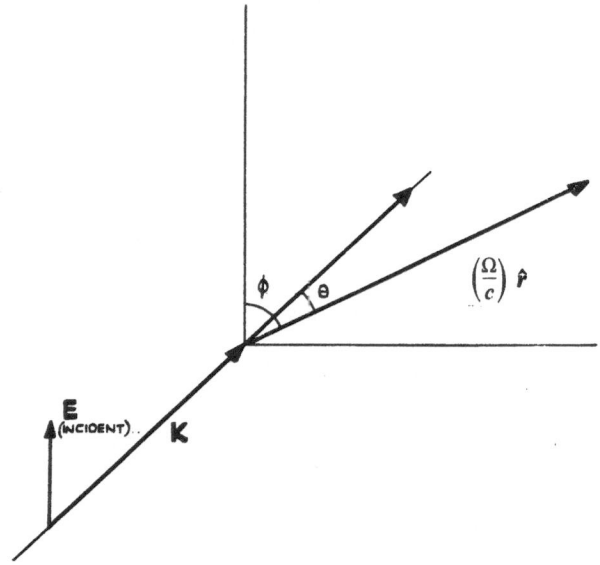

FIG. 1. Scattering geometry.

$$I d\omega = d\omega [\omega^4/(4\pi c^3 r^2)] \int d^3r' \int d^3r'' p(\Omega_s - \Omega_i, r') p^*(\Omega_s - \Omega_i, r'')$$
$$e^{i k \cdot (r' - r'')} E_0 \cdot E_0 (1 - \sin^2 \theta \cos^2 \varphi)$$

or

$$\frac{I}{I_0} = \frac{\omega^4}{c^4} \frac{1}{r^2} \int d^3r' \int d^3r'' p(\varpi, r') p^*(\varpi, r'') e^{i k \cdot (r' - r'')} (1 - \sin^2 \theta \cos^2 \varphi).$$
$$...[9.4]$$

For an unmagnetized plasma, $4\pi p = -\omega_p^2/\omega^2$ and

$$\frac{I(\omega)}{I_0} = \left(\frac{e^2}{mc^2}\right)^2 \frac{1}{r^2} \int d^3r' \int d^3r'' n(\omega, r') n^*(\omega, r'') e^{i k \cdot (r' - r'')}$$
$$(1 - \sin^2 \theta \cos^2 \varphi). \qquad ...[9.5]$$

If the density, on the relevant scale, is a random variable, then

$$\int d^3r' n n^* e^{i k \cdot (r' - r'')} \sim n V = N.$$

For electrons at rest,

$$\int d^3r' \int d^3r'' n n^* e^{i k \cdot (r' - r'')} = \sum_{ij} \int d^3r\, e^{i k \cdot (x_i - x_j)}$$
$$= N$$

(using the random phase approximation). Hence

$$\frac{I}{I_0} = \left(\frac{e^2}{mc^2}\right)^2 \frac{1}{r^2} N(1 - \sin^2 \theta \cos^2 \varphi) \delta(\omega),$$

and the total scattered intensity $= \frac{8}{3}\pi [e^2/(mc^2)]^2 N/r^2.$ $\qquad ...[9.6]$

The quantity $\frac{8}{3}\pi [e^2/(mc^2)]^2$ is the classical Thomson cross-section $\simeq 10^{-25}$ cm^2.

If the electrons have a Maxwellian distribution, but are otherwise uncorrelated, $n n^* = N \int d^3v f(v) \delta(\omega + k \cdot v)$. The scattered intensity is then

$$\frac{I}{I_0} = N \left(\frac{e^2}{mc^2}\right)^2 \frac{1}{r^2} \frac{1}{\pi^{\frac{1}{2}} v_\theta} \exp \left[-\frac{c^2(\omega/\Omega_s)^2}{4v_\theta^2 \sin^2(\theta/2)}\right] (1 - \sin^2 \theta \cos^2 \varphi) \quad ...[9.7]$$

a Gaussian line with half-width determined by the electron temperature.

Let us now calculate the total electron density fluctuation, allowing for correlations.

$$\delta f = \delta f_{\text{free}} + \delta f_2; \quad \delta f_2 = \frac{e}{m} \frac{\phi k \cdot (\partial f_0/\partial v)}{\omega + k \cdot v}.$$

But

$$\phi_k = \frac{4\pi}{k^2 \epsilon} \sum [e_- e^{-i k \cdot x_i} \delta(\omega + k \cdot v_i) + e_+ e^{-i k \cdot x_i} \delta(\omega + k \cdot V_i)]$$

and

$$\delta f = \sum_{-} e^{-ik \cdot x_i} \delta(\omega + k \cdot v_i) \left[1 + \frac{4\pi(e^2/m)}{k^2\epsilon} \frac{k \cdot (\partial f_-/\partial v)}{\omega + k \cdot v} \right]$$
$$+ \sum_{+} e^{-ik \cdot x_i} \delta(\omega + k \cdot V_i) \frac{4\pi(e_- e_+/m)}{k^2\epsilon} \frac{k \cdot (\partial f_-/\partial v)}{\omega + k \cdot v},$$

$\delta n(\omega, k)$ being of course the integral of this over v:

$$\delta n = \sum_{-} e^{-ik \cdot x_i} \delta(\omega + k \cdot v_i) \left[1 + \frac{4\pi n(e^2/m)}{k^2\epsilon} \int d^3v \frac{k \cdot (\partial f_-/\partial v)}{\omega + k \cdot v} \right]$$
$$+ \sum_{+} e^{-ik \cdot x_i} \delta(\omega + k \cdot V_i)(e_+/e_-) \frac{4\pi n(e^2/m)}{k^2\epsilon} \int d^3v \frac{k \cdot (\partial f_-/\partial v)}{\omega + k \cdot v}.$$

$$...[9.8]$$

ϵ, however, may be written $1 - (k_D^2/k^2)(G_- + G_+)$ and

$$\delta n = \sum_{-} e^{-ik \cdot x_i} \delta(\omega + k \cdot v_i) \left[1 + \frac{(k_D^2/k^2) G_-}{\epsilon} \right]$$
$$+ \frac{e_+}{e_-} \sum_{+} e^{-ik \cdot x_i} \delta(\omega + k \cdot V_i) \frac{(k_D^2/k^2) G_-}{\epsilon}$$
$$= \sum_{-} e^{-ik \cdot x_i} \delta(\omega + k \cdot v_i) \frac{1 + (k_D^2/k^2) G_+}{1 - (k_D^2/k^2)(G_+ + G_-)}$$
$$+ \frac{e_+}{e_-} \sum_{+} e^{ik \cdot x_i} \delta(\omega + k \cdot V_i) \frac{(k_D^2/k^2) G_-}{1 - (k_D^2/k^2)(G_+ + G_-)}. \qquad ...[9.9]$$

Now, to form the correlation function $\langle \delta n \delta n^* \rangle$ we use the r.p.a. on the sums.

$$\langle \delta n \delta n^* \rangle = N \int d^3v$$
$$\frac{\tilde{f}_-(v)\delta(\omega + k \cdot v) \left| 1 - (k_D^2/k^2) G_+ \right|^2 + \tilde{f}_+(v)(e_+/e_-)\delta(\omega + k \cdot v)(k_D^2/k^2) \left| G_- \right|^2}{\left| 1 - (k_D^2/k^2)(G_+ + G_-) \right|^2}.$$

$$...[9.10]$$

Note that this depends only on the variables k_D^2/k^2 and $\omega/(kv_\theta)$, where $k^2 = 4K^2 \sin^2(\theta/2)$ and ω is the frequency shift on scattering. (It depends, of course, also on the form of the Boltzmann distribution function f_0.)

For small values of k_D/k (large scattering angles, etc.) it reduces to the Doppler broadened Thomson scattering result. For large values of k_D/k, however, the terms involving G, which represent collective effects, become dominant. For small values of its argument $G \to 2$; but, since f_+ is much narrower than f_-, it is the second term that dominates, and near the line center, there is a peak whose width is given by the *ion* thermal speed. That is, the scattering is produced by the Debye screening clouds surrounding the ions. For larger values of ω/k, this term of course disappears, and the small electron effect is left. However, whenever the denominator ϵ is small, i.e. wherever an oscillation is self-sustaining, we can expect a peak in the scattering.

One such peak occurs at the Langmuir frequency, where for small k and $\omega = \omega_p$ a resonance occurs, and the scattering exhibits a plasma-produced Raman line.

If the system is not in thermal equilibrium, ϵ can have other zeros, and new phenomena are to be expected. In particular, if $T_- \gg T_+$, ion waves can grow, and we expect to find a line at $\omega/k = (\kappa T_-/m_+)^{\frac{1}{2}}$ corresponding to ion waves.

It is this scattering function whose properties are discussed in the lectures of Ramsden. Experiments with the scattering of laser beams have demonstrated that the main features of the scattering are rather well predicted by our expression. However, this scattering function leads to unsatisfactory results at the electron plasma frequency; and the difficulties in the theory at this point are concealed by the difficulty of making measurements. Nevertheless, an examination of the magnitude of the spectrum at ω_p and large k, is, instructive.

If $\omega/k \gg v_\theta^-$; the dielectric function takes the form

$$\epsilon = 1 - \frac{\omega_p^2}{\omega^2} + 2i\pi \frac{k_D^2}{k^2} \frac{\omega}{kv_\theta} e^{-[\omega/(kv_\theta)]^2} \qquad \ldots[9.11]$$

and at $\omega = \omega_p$

$$\epsilon = 2i\pi(k_D/k)^3 e^{-(k_D/k)^2}$$

The potential field has the form

$$\begin{aligned}
\phi\phi^* d\omega &= 1/(\pi^{\frac{1}{2}}kv_\theta) e^{-(k_D/k)^2}/[k^2\epsilon]^2 d\omega \\
&= 1/(\pi^{\frac{1}{2}}kv_\theta) e^{-(k_D/k)^2}/[2\pi k_D^2(k_D/k) e^{-(k_D/k)^2}]^2 d\omega \\
&= ne^2 (d\omega/\omega_p)(k/k_D^5) e^{(k_D/k)^2} \qquad \ldots[9.12]
\end{aligned}$$

which is exponentially large. The line width is, of course, $\propto e^{-(k_D/k)^2}$ so that the energy in the line is fixed, but the existence of a near resonance in the spectral density indicates that a weakly damped wave exists in the plasma, and this is enough to invalidate the assumptions upon which our kinetic theory was based—the argument of time scale separation. We had to assume that the field correlation vanished on a short time scale. For the low phase velocity parts of the spectrum this is indeed true, the correlation time being of order $\tau \leqslant 1/(k_D v_\theta) = \omega_p^{-1}$; however, for the plasma waves, correlation times can be long, of order the collision period. This makes no contribution to the *dominant* term, but it casts some suspicion on the form of the correlation function at those points where $k^2\epsilon$ is small.

If the plasma is *electrostatically unstable*, then $k^2\epsilon = 0$ at some points, and the kinetic integral is not even defined. In addition, Poisson's equation permits exponentially growing homogeneous solutions, and our linear treatment would predict unbounded oscillations in ϕ_k. The discussion of the plasma spectrum in these regions forms the subject of plasma turbulence, of which only a particular branch—quasi-linear theory—has been developed. These subjects are considered in the lectures of Rusbridge and Harris; and their connection with kinetic theory is considered in the lectures of Oberman. For many purposes, and for stable plasmas, however, the treatment presented here gives adequate results.

REFERENCES

Section 1

N. Bogoliubov. 1946. *Problems of a Dynamical Theory in Statistical Physics*. Translated in *Statistical Physics*, edited by J. de Boer and G. E. Ulenbeck (North Holland), 1962.
M. Born and H. S. Green. 1949. *A General Kinetic Theory of Liquids*. Cambridge University Press.
J. G. Kirkwood. 1947. *J. Chem. Phys.*, **14**, 180.
D. C. Montgomery and D. A. Tidman. 1964. *Plasma Kinetic Theory*, McGraw-Hill, New York.
T. Y. Wu. 1966. *Kinetic Equations of Gases and Plasmas*. Addison Wesley, Reading, Mass.
J. Yvon. 1935. *La théorie des fluides et l'équation d'état*. Herman, Paris.

Section 2

L. Landau. 1937. *J.E.T.P.*, **7**, 203.

Section 3

S. Chapman and T. G. Cowling. 1951. *The Mathematical Theory of Non-Uniform Gases* (second edition). Oxford University Press.
D. Hilbert. 1912. *Math. Ann.*, **7**, 562.
R. Jancel and T. Kahan. 1963. *Electrodynamique des Plasmas*. Dunod, Paris.
R. Landshoff. 1943. *Phys. Rev.*, **82**, 442.

Section 4

A. N. Kaufman. 1959. *Proceedings of the Les Houches Summer School in Theoretical Physics*. Herman, Paris.
R. Landshoff. 1943. *Phys. Rev.*, **82**, 442.
W. Marshall. 1958. *A.E.R.E.*, *Harwell, Reports T/R* 2247, 2352, 2419.
B. B. Robinson and I. B. Bernstein. 1962. *Ann. Phys.*, **18**, 110.
W. B. Thompson. 1963. *Rendiconti della Scuola Internazionale de Fisica "E. Fermi" XXV Corso*, Chapter I. Academic Press, New York.

SECTIONS 6, 7 & 8

R. BALESCU. 1960. *Phys. Fluids*, **3**, 52.
S. CHANDRASEKHAR. 1943. *Rev. Mod. Phys.*, **15**, 1.
D. GABOR. 1952. *Proc. Roy. Soc.*, *A*, **213**, 73.
J. HUBBARD. 1961. *Proc. Roy. Soc.*, *A*, **260**, 114.
A. LENARD. 1960. *Ann. Phys.* **10**, 390.
N. ROSTOKER and M. N. ROSENBLUTH. 1960. *Phys. Fluids*, **3**, 1.
W. B. THOMPSON and J. HUBBARD. 1960. *Rev. Mod. Phys.*, **32**, 714.

SECTION 9

J. P. DOUGHERTY and D. T. FARLEY. 1960. *Proc. Roy. Soc.*, *A*, **259**, 79.
J. A. FEJER. 1960. *Can. J. Phys.*, **38**, 1114.
E. E. SALPETER. 1960. *Phys. Rev.*, **120**, 1528.
W. B. THOMPSON. 1961. *Princeton University Plasma Physics Laboratory Reports, MATT* 91.

2

ADVANCED KINETIC THEORY

CARL OBERMAN

Plasma Physics Laboratory, Princeton University

1

INTRODUCTION

IN the previous lectures of Professor Thompson we were presented with the important intuitive origins of plasma kinetic theory, largely by treating several simplified models. With such a basis, the development of kinetic theory has proceeded until recently in two diverging directions where either particle discreteness or collective behaviour plays the dominant role.

The theory of Balescu, Guernsey, and Lenard[1, 18, 19] (B.G.L.) for describing the time evolution of the one-particle distribution function for spatially homogeneous stable plasma systems in the classical limit rests on Bogoliubov's conjecture that after a short transient period a gas will reach a kinetic regime. That is, after a time longer than a collision duration period τ_1 ($\approx$ Debye length/thermal speed = inverse plasma frequency) but much shorter than the average 90° deflection time τ_2, all particle correlation functions relax to functionals of the one-particle distribution function. The existence of the small parameter τ_1/τ_2 permits a consistent truncation of the previously mentioned B.B.G.K.Y. hierarchy in which, in particular, the pair correlation function is solved for in terms of the local value (in time) of the one-particle distribution function. The result, when inserted in the equation for the one-particle distribution function, yields a Markoffian kinetic equation of Fokker-Planck type.

In a plasma, collective motions are made possible by the long range of the Coulomb interactions (compared with the interparticle distance), and the validity of Bogoliubov's conjecture rests upon the assumption of the rapid decay of these plasma oscillations. However, it has been recognized that the B.G.L. equation for the kinetic description of stable plasma (and *a fortiori* if unstable), is improper, because at least the small k portion of the fluctuation spectrum (plasmons) is long-lived, and hence the Bogoliubov Ansatz used in deriving the equation is invalid in this portion of the spectrum. This weakness is vividly reflected when the coefficient of spatial diffusion is computed using this kinetic description.[3] The coefficient approaches infinity if one considers a stable initial distribution function deformed toward instability. This impropriety of the Bogoliubov Ansatz represents at the same time the failure of Rostoker's superposition principle[4] (dressed, then uncorrelated particles).

Quasi-linear theories were devised to describe those situations in which the plasma is weakly unstable according to linearized theory. Here, the wave spectrum I_k is long-lived and there is no relaxation of the particle correlations. The system is no longer Markoffian in the one-particle distribution function $F^{(1)}$ alone, but is in the enlarged space $\{F^{(1)}, I_k\}$. There are two distinct approaches to the problem of weak turbulence. In the first [5-8] the starting-point is the set of non-linear Vlasov equations with *smooth* initial conditions. Effects associated with the discreteness of the plasma particles are systematically discarded with this procedure. In the second approach[9,10] the starting point is once again the B.B.G.K.Y. hierarchy of equations. With an ordering appropriate to weak turbulence (and which still permits closure of the hierarchy) these theories also emphasize the collective aspect of plasma behaviour. It is assumed that the Coulomb interaction energy in the stable portion of the fluctuation spectrum is small compared with the wave energy in the unstable portion of the fluctuation spectrum, which in turn is much smaller than the average kinetic energy. Stabilization results from the effect of diffusion on the particle distribution functions. The decay of the waves is dominated by nonlinear Landau damping and the emission by Cerenkov radiation.

A common feature of all the kinetic theories just reviewed is that they do not describe adequately the crossing from instability to stability. Furthermore, they are inappropriate to describe those regimes where the decay of the waves by collisional damping, and emission by longitudinal bremsstrahlung radiation are of importance.

It is our main purpose here to indicate some recent developments toward correcting these deficiencies.‡[11] We can proceed most simply using an adiabatic treatment within the Klimontovich formalism.[12,13] The plasma is described via a phase-space density N, which satisfies the non-linear Vlasov equations with *singular* initial conditions reflecting the discreteness of the plasma particles. A kinetic description is obtained by averaging over a statistical ensemble.

A basic idea of the present theory is that discreteness effects of plasma and collective behaviour are not generally separable. We are forced, therefore, to consider the evolution of the wave spectrum I_k and the one-particle distributions $F^{(1)}$ on a par. The kinetic description then embraces not only the Bogoliubov regime and the weakly unstable situation, but also those regimes which are dominated by collisional absorption and/or emission of the waves via wave-particle and particle-particle scattering (bremsstrah-

‡ Most of this material is derived from the Ph.D. thesis of André Rogister, Princeton University, 1967; see also Princeton University, *Plasma Physics Laboratory Report*, MATT-583, 1967.

lung). The equations also provide the continuous transition from instability to stability, so that an initially (weakly) unstable plasma can be followed all the way to thermal equilibrium. In the spirit of this recently developed kinetic theory, we generalize the test-particle problem of Rostoker to describe both unstable and stable plasma. We then compute the coefficient of spatial diffusion across a uniform magnetic field. In different limits we recover previous results,[3, 14] but obtain the proper transition between stable and unstable regimes.

We shall now proceed by first reviewing carefully the systematic derivation of the B.G.L. equation from the B.B.G.K.Y. hierarchy. We discuss its properties and point the limitations as imposed by Bogoliubov's conjecture.

Next, we introduce the Klimontovich description of plasma via a phase-space density N. The relation to the B.B.G.K.Y. hierarchy is given. The fundamental equations of the problem are the non-linear Vlasov equations (with *singular* initial conditions). We show how the conventional B.G.L. theory and Quasi-linear theory (in its simplest version) are arrived at by various assumptions regarding the solution of the linearized Vlasov equations. The shortcomings of these theories are pointed out, and we indicate how we shall systematically correct them.

From the solution of the linearized Vlasov equations we proceed to obtain a kinetic theory in the enlarged space $\{F^{(1)}, I_k\}$ which reduces either to the B.G.L. equation, or the Pines-Schrieffer equations[15, 16] in the appropriate regions of the (k,v) space. The necessity of including non-linear effects is discussed.

We then discuss the iterative solution of the nonlinear Vlasov equations which the phase-space density satisfies. We can then show how to improve in a systematic fashion the equation for the time evolution of the fluctuation spectrum of the waves. This equation now includes collisional damping[17] and non-linear Landau damping[8, 9] of the waves, as well as nonlinear processes of emission: via particle-particle scattering, wave-wave scattering, and wave-particle scattering. We show that in general it is not possible to separate the different mechanisms, since in the non-approximated form their expressions individually diverge; this is a manifestation of the fact that discreteness effects cannot be separated from collective behaviour. We demonstrate how the synchronization of Cerenkov emission comes about and how the Markoffian behaviour is preserved. We also discuss the appropriate modification of the equation for the time evolution of the one-particle distribution function to include non-linear effects.

Finally, we point out how to generalize the kinetic theory of a test particle, to include both stable and unstable regimes, according to the scheme just developed. The coefficient of spatial diffusion across a homogeneous magnetic field is then computed. In the proper limits, it reduces to known results, but does not show the characteristic divergence of Rostoker's formula.[3]

2

THE BALESCU-GUERNSEY-LENARD EQUATION AND ITS PROPERTIES

Since we shall be concerned with collections of large numbers of particles interacting with their self-created and/or externally imposed electromagnetic fields, and since it is a problem of prohibitive difficulty to follow the detailed motion of all the particles, we must rest content to describe the plasma in some average or statistical sense.

The starting-point for much contemporary work has been the Liouville equation, which describes the temporal evolution of an ensemble of systems identical to the one under consideration. (For good reasons a classical description of high temperature plasma is valid over a large range of wave numbers and frequencies; but as a matter of fact most of the techniques we shall discuss can, to a large extent, be carried over to the corresponding equation for the Density Matrix, now paying full respect to the exclusion and uncertainty principles.)

Each system (of N particles) in the ensemble is represented by a point of the $6N$-dimensional phase (Γ space).

Then

$$D_N(x_1, v_1, \ldots, x_N, v_N)d^3x_1 \cdots d^3v_N$$

can be taken to be the probability of finding a given system in the phase volume $d^3x_1 \cdots d^3v_N$. With this interpretation as a probability

$$\int \cdots \int d^3x_1 \cdots d^3v_N D_N = 1. \qquad \ldots[2.1]$$

How does D_N evolve in time? Consider an infinitesimal volume $\delta\Omega = \delta^3x_1 \cdots \delta^3v_N$ in the phase space surrounding a given system point at some time t, with the boundary of the volume formed by some surface of neighbouring system points. In the course of time the volume will move about and deform. It is clear that the number of systems within $\delta\Omega$, $D_N\delta\Omega$, cannot change with time; for if any system point were to cross the boundary, it would occupy at that time the same portion in phase space as one of the system points defining the boundary. Since the subsequent motion of a system is uniquely determined by its location in phase space at a particular time, the two points would travel together from there on. Hence no system can leave (or enter) the volume.

Hence following the motion

$$\frac{d}{dt}(D_N\delta\Omega) = 0, \qquad \ldots[2.2]$$

or

$$\frac{dD_N}{dt}\delta\Omega + D_N\frac{d(\delta\Omega)}{dt} = 0. \qquad \ldots[2.3]$$

Since we shall show immediately $d/dt(\delta\Omega) = 0$, then

$$\frac{dD_N}{dt} = \frac{\partial D_N}{\partial t} + \sum_{i=1}^{N} \dot{x}_i \cdot \frac{\partial D_N}{\partial x_i} + \sum_{i=1}^{N} \dot{v}_i \cdot \frac{\partial D_N}{\partial v_i} = 0, \qquad \text{...[2.4]}$$

or, using Newton's equations of motion,

$$\frac{\partial D_N}{\partial t} + \sum_{i=1}^{N} v_i \cdot \frac{\delta D_N}{\partial x_i} + \sum_{i=1}^{N} \frac{F_i}{m_i} \cdot \frac{\partial D_N}{\partial v_i} = 0. \qquad \text{...[2.5]}$$

Here F_i is the force on the ith particle due to all the other particles and any external fields. To show $\delta\Omega = 0$ (Law of Conservation of Extension in Phase), consider equations of motion in first-order form

$$\dot{z}_i = g_i(z_1, \ldots, z_{6N})$$

where z_i may be either a coordinate or velocity component. Consider the extension in phase $\delta\Omega$.

Then

$$\frac{\delta\dot{\Omega}}{\delta\Omega} = \sum_i \frac{\delta\dot{z}_i}{\delta z_i} \qquad \text{...[2.6]}$$

$$= \sum_i \frac{1}{\delta z_i} [\dot{z}_i(z_1, \ldots, z_i + \delta z_i, z_{i+1}, \ldots) - \dot{z}_i(z_1, \ldots, z_i, \ldots)] \text{ ...[2.7]}$$

$$= \sum_i \frac{\partial g_i}{\partial z_i} + O(\delta\Omega) \qquad \text{...[2.8]}$$

$$= \nabla_{6N} \cdot g + O(\delta\Omega). \qquad \text{...[2.9]}$$

Therefore

$$\delta\dot{\Omega} = \delta\Omega \nabla_{6N} \cdot g + O(\delta\Omega^2). \qquad \text{...[2.10]}$$

Now a sufficient condition for the vanishing of $\delta\dot{\Omega}$ is the vanishing of $\nabla_{6N} \cdot g$. A sufficient (but not necessary!) condition for this to be true is that the system be Hamiltonian, which we assume.

Now we have

$$\frac{\partial D_N}{\partial t} + \sum_i v_i \cdot \frac{\partial D_N}{\partial x_i} + \sum_i \left(\sum_{j \neq i} \frac{F_{ij}}{m_i} + \frac{F_i^{\text{ext}}}{m_i} \right) \cdot \frac{\partial D_N}{\partial v_i} = 0, \qquad \text{...[2.11]}$$

where we have resolved the force on the ith particle into that due to all other particles of the system, and the external force.

Next we specialize to plasma where the interaction between particles is coulombic

$$F_{ij} = \frac{Z_i Z_j e^2 (x_i - x_j)}{|x_i - x_j|^3}. \qquad \text{...[2.12]}$$

We emphasize that the Liouville equation is exact and contains all the information we need concerning the statistical temporal evolution of the system at one time.

In a sense, it contains more information than we want; for the expectation values of most observables of interest usually involve only the lowest members of the so-called B.B.G.K.Y. hierarchy.

Define the set of functions:

$$f_s = V^s \int d^6\xi_{s+1} \cdots d^6\xi_N D_N \qquad \dots [2.13]$$

where f_s/V^s is the probability of finding particles 1, through s, in the vicinity of ξ_1, (with $\xi_i = (x_i, v_i)$) through ξ_s, respectively, regardless of the location of the remaining particles. To find the equation of motion for f_s multiply the Liouville equation by V^s and integrate out the $6N-6s$ remaining co-ordinates to obtain

$$\frac{\partial f_s}{\partial t}(\xi_1, \ldots, \xi_s) + \sum_{i \in s} v_i \cdot \frac{\partial f_s}{\partial x_i} + \sum_{\substack{i, j \in s \\ i \neq j}} \frac{F_{ij}}{m_i} \cdot \frac{\partial f_s}{\partial v_i} + \sum_{i \in s} \frac{F_i^{ext}(\xi_i)}{m_i} \cdot \frac{\partial f_s}{\partial v_i}$$

$$+ \frac{1}{V} \sum_{\substack{i \in s \\ j \in N-s}} \int d^6\xi_j \frac{F_{ij}}{m_i} \cdot \frac{\partial f_{s+1}}{\partial v_i} = 0. \qquad \dots [2.14]$$

We shall now make the assumption that we have a plasma composed of electrons, with overall charge neutrality provided by a smooth uniform background of positive charge. The straightforward generalization to many species will be stated only if needed. Further, we shall assume that f_s is symmetric under interchange of particles, which remains true if initially true.

We then have

$$\frac{\partial f_s}{\partial t} + \sum_{i \in s} v_i \cdot \frac{\partial f_s}{\partial x_i} + \sum_{i \in s} \frac{F_i^{ext}}{m} \cdot \frac{\partial f_s}{\partial v_i} + \sum_{\substack{i, j \in s \\ i \neq j}} \frac{F_{ij}}{m} \cdot \frac{\partial f_s}{\partial v_i} + \frac{N-s}{mV} \sum_i \int d^6\xi_{s+1} F_{is+1} \cdot \frac{\partial f_{s+1}}{\partial v_i}$$

$$= 0, \qquad \dots [2.15]$$

It is now the standard procedure in statistical mechanics, in order to eliminate surface effects, to let $N, V \to \infty$ but in such a way that $N/V = n$ remains finite. Finally we write (discarding the external field for the time being)

$$\frac{\partial f_s}{\partial t} + \sum_{i=1}^{s} v_i \cdot \frac{\partial f_s}{\partial x_i} + \frac{1}{m} \sum_{\substack{i, j \in s \\ i \neq j}} F_{ij} \cdot \frac{\partial f_s}{\partial v_i} + \frac{n}{m} \sum_i \int d^6\xi_{s+1} F_{is+1} \cdot \frac{\partial f_{s+1}}{\partial v_i} = 0.$$

$$\dots [2.16]$$

There are several remarks now to be made.

(a) Some irreversible phenomena have been introduced into the system (e.g. infinite Poincaré recurrence time), but the equations themselves are time-reversible.

(b) Notice that each number of the hierarchy is connected to the next higher number. In general, the hierarchy does not terminate.

(c) Notice that the next-to-last term in the preceding equation describes the interaction of the s particles among themselves, while the last term relates to the average force on each of the s particles due to the remaining particles of the system. Our backlog of experience with sample problems tells us we should expect that the direct interaction of any handful of particles is much smaller than the interaction of any one particle with the average field of all the other $(N-1)$ particles of the system. Now can we see this? If we scale the dimensions of the system t, l, v to those typical of plasma $(1/\omega_p, \lambda_D, \bar{v})$ we find in the new dimensionless units that the direct interaction term carries the factor $\varepsilon(=1/(n\lambda_D^3))$. Another physical way is to view the plasma limit where each particle is continually halved but in such a way that $n \to \infty$, $e \to 0$, $m \to 0$, but e/m, ne remain infinite.

Thus we find that in plasma physics we have been blessed, in that we find we can truncate the hierarchy in a systematic way.

Thus, the dominant plasma approximation (0-order in ε) is to neglect the direct-interaction term from each member of the hierarchy.

It is then observed that the resultant hierarchy is satisfied by

$$f_s(\xi_1, \ldots, \xi_s) = \prod_{i=1}^{s} f_1(\xi_i) \qquad \ldots[2.17]$$

if $f_1(\xi)$ satisfies the collisionless (or correlationless) kinetic equation

$$\frac{\partial f_1}{\partial t} + v \cdot \frac{\partial f_1}{\partial x} + \frac{1}{m}(eE + F^{\text{ext}}) \cdot \frac{\partial f_1}{\partial v} = 0, \qquad \ldots[2.18]$$

where E is the self-consistent electric field determined by

$$\begin{aligned} \nabla \cdot E &= 4\pi e\left[\int f_1(\xi)d^3v - n_0\right] \\ \nabla \times E &= 0. \end{aligned} \qquad \ldots[2.19]$$

(Here the term n_0 arises from smooth ion background and has been neglected in our equations so far.)

It will be the concern of other lectures to examine the consequences of this system of equations. We note that, if the system is spatially homogeneous,

$$F^{\text{ext}} = 0, \quad \frac{\partial f_1}{\partial x} = 0, \quad E = 0,$$

then $\partial f_1/\partial t = 0$ and we must proceed to next order in ε to see, for example the time evolution on the now collisional time scale.

Let us now consider a stable spatially homogeneous system. Since correlations (collisions, fluctuations) are considered weak, we suppose[27]

$$\begin{aligned} f_1(1, t) &= f(1, t) \\ f_2(1, 2, t) &= f_1(1, t)f_1(2, t) + \varepsilon g(1, 2, t) \\ f_3(1, 2, 3, t) &= f_1(1)f_1(2)f_1(3) + \varepsilon[f_1(1)g(2, 3) + \ldots] + \varepsilon^2 h(1, 2, 3) \end{aligned}$$

Then it is observed that the whole hierarchy is satisfied if

$$\frac{\partial f}{\partial t}(\mathbf{v}_1, t) = \frac{n}{m} \int d^6\xi_2 \frac{\partial \phi}{\partial x_1} (|\, \mathbf{x}_1 - \mathbf{x}_2 \,|) \cdot \frac{\partial g}{\partial \mathbf{v}_1} (1, 2, t)$$

$$\frac{\partial g}{\partial t}(1, 2, t) + (\mathbf{v}_1 - \mathbf{v}_2) \cdot \frac{\partial g}{\partial x_1} = \frac{1}{m} \frac{\partial \phi}{\partial x_1} \cdot \left(\frac{\partial}{\partial \mathbf{v}_1} - \frac{\partial}{\partial \mathbf{v}_2}\right) f(1, t) f(2, t) \qquad ...[2.20]$$

$$+ \frac{n}{m} \int d^6\xi_3 \frac{\partial \phi}{\partial x_1} \cdot \frac{\partial f(1)}{\partial \mathbf{v}_1} g(2, 3) + \frac{n}{m} \int d^6\xi_3 \frac{\partial \phi}{\partial x_2} \cdot \frac{df(2)}{\partial \mathbf{v}_2} g(1, 3).$$

Here we have taken advantage of the fact that for spatially homogeneous systems $g(\mathbf{x}_1, \mathbf{x}_2, \mathbf{v}_1, \mathbf{v}_2, t) = g(\mathbf{x}_1 - \mathbf{x}_2, \mathbf{v}_1 \mathbf{v}_2, t)$.

We are going to make two assumptions (Bogoliubov Ansatz) based on our observation that f is to change on the much longer collisional time scale whereas g changes significantly on the time scale characterized by the plasma period (time for a typical thermal particle to cross the Debye sphere of another).

Therefore we assume:

(1) That $f(1, t)$, and $f(2, t)$ occurring in the equation for $g(1, 2, t)$ may be considered time-independent as far as the solution of that equation is concerned.

(2) Only the solution $g(1, 2, t)$ as $t \to \infty$ need be found. The function $g(1, 2, \infty)$—a functional of $f(1, t)$ and $f(2, t)$—substituted back into the equation for the one-particle distribution function, yields the kinetic equation describing the time evolution of f.

(We shall consider in our next lecture important situations and regimes where the Bogoliubov Ansatz is untenable.) It is quite clear that, *if the initial transients* (depending on the initial conditions) *die away sufficiently rapidly*, then we find $g(1, 2, t \to \infty) = \lim_{p \to 0} pg(1, 2, p)$ in terms of the time Laplace transform. We also Fourier transform in the variable $\mathbf{x}_1 - \mathbf{x}_2$. Hence we may write

$$\frac{\partial f(\mathbf{v}, t)}{\partial t} = -\frac{\partial}{\partial \mathbf{v}} \cdot \mathbf{J}(\mathbf{v}) \qquad ...[2.21]$$

where

$$\mathbf{J}(\mathbf{v}) = \frac{i}{(2\pi)^3} \int d^3k \, \mathbf{k} \tilde{\phi}(k) h(\mathbf{k}, \mathbf{v}) \qquad ...[2.22]$$

and

$$h(\mathbf{k}, \mathbf{v}) = \int d^3v' \tilde{g}(\mathbf{k}, \mathbf{v}, \mathbf{v}'). \qquad ...[2.23]$$

Here

$$\tilde{\phi}(k) = \frac{4\pi e^2}{mk^2}, \qquad \tilde{g}(\mathbf{k}, \mathbf{v}, \mathbf{v}') = \int d^3x \, g(\mathbf{x}, \mathbf{v}, \mathbf{v}') e^{-i\mathbf{k} \cdot \mathbf{x}} \qquad ...[2.24]$$

The equation for $\tilde{g}(k, v, v', \infty)$ now satisfies

$$\tilde{g}(k, v, v') = \frac{\phi}{u - u' - i\varepsilon} \left\{ \frac{\partial f(v)}{\partial u} f(v') - \frac{\partial f(v')}{\partial u'} f(v) + \frac{\partial f(v)}{\partial u} \int d^3\eta\, \tilde{g}^*(v', \eta) \right.$$
$$\left. - \frac{\partial f(v')}{\partial u'} \int d^3\eta\, \tilde{g}(v, \eta) \right\}. \qquad ...[2.25]$$

(Note that u is defined by $ku = k \cdot v$ and $\tilde{g}(-k, v, v') = \tilde{g}^*(k, \dot{v}, v')$ because $g(x_1 - x_2, v_1, v_2)$ is real. Thus the k-dependence is suppressed in the above equation.)

    We now observe that only $h(k, v)$ is needed for the kinetic equation, and, further, only the imaginary part of h. Therefore, integrating the last equation over v' yields

$$h(v) = \phi \int \frac{d^3v'}{u - u' - i\varepsilon} \left\{ \frac{\partial f(v)}{\partial u} f(v') - \frac{\partial f(v')}{\partial u'} f(v) + \frac{\partial f(v)}{\partial u} h^*(v') - \frac{\partial f(v')}{\partial u'} h(v) \right\}.$$
$$...[2.26]$$

Notice that the kernel of this last integral equation depends only on u, so that $h(v)$ is determined algebraically once we know

$$H(k, u) = \int d^3v\, h(k, v)\delta(u - \hat{k} \cdot v). \qquad ...[2.27]$$

Therefore we obtain an equation for $H(u)$ by integrating the equation for $h(v)$ over the two directions perpendicular to k to obtain

$$H(u) = \phi \int_{-\infty}^{\infty} \frac{du'}{u - u' - i\varepsilon} \left\{ \frac{\partial F(u)}{\partial u} F(u') - \frac{\partial F(u')}{\partial u'} F(u) + \frac{\partial F}{\partial u} H^*(u') \right.$$
$$\left. - \frac{\partial F(u')}{\partial u'} H(u) \right\}. \qquad ...[2.28]$$

Here

$$F(u) = \int d^3v\, \delta(u - \hat{k} \cdot v) f(v). \qquad ...[2.29]$$

We shall now rewrite the equation for $h(v)$ as

$$h(v) = \phi \int \frac{du'}{u - u' - i\varepsilon} \left\{ \frac{\partial f(v)}{\partial u} F)u') - \frac{\partial F(u')}{\partial u.} f(v) + \frac{\partial f(v)}{\partial u} H^*(u') - \frac{\partial F(u')}{\partial u'} h(v) \right\}.$$
$$...[2.30]$$

Now we make the important observation, that in [2.28] it is consistent to assume that the $H(u)$ is real, since the imaginary part of both sides of the equation vanishes. Since $H(u)$ is to be uniquely determined by $F(u)$, we conclude H is indeed real. (A direct proof of the reality of H can be given; but we do not give it here.)

    We can get a simpler form for the relation between $h(v)$ and $H(u)$ by

eliminating the term where the kernel operates on H. That is, we multiply
[2.30] by $[\partial F(u)]/\partial u$, [2.28] by $[\partial f(v)]/\partial u$, and subtract, to obtain

$$\frac{\partial F}{\partial u}\, h(v) - \frac{\partial f(v)}{\partial u}\, H(u) = -\left\{\frac{\partial F}{\partial u}\, f(v) - \frac{\partial f(v)}{\partial u}\, F(u)\right\}\frac{\tilde{\phi}\psi}{1+\tilde{\phi}\psi} \quad \ldots[2.31]$$

where

$$\psi(\hat{k}, u) = \int \frac{du'}{u-u'-i\varepsilon}\, \frac{\partial F(\hat{k}, u')}{\partial u'}. \qquad \ldots[2.32]$$

If we now take the imaginary part of [2.31], we find

$$\operatorname{Im} h(v) = \frac{-\pi\tilde{\phi}}{|1+\tilde{\phi}\psi|^{2}}\left\{\frac{\partial F(u)}{\partial u}\, f(v) - \frac{\partial f(v)}{\partial u}\, F(u)\right\}. \qquad \ldots[2.33]$$

Thus

$$J(v) = \int d^{3}v'\, Q(v, v') \cdot \left\{\frac{\partial f(v)}{\partial v}\, f(v') - \frac{\partial f(v')}{\partial v'}\, f(v)\right\}, \qquad \ldots[2.34]$$

where

$$Q(v, v') = -\frac{1}{8\pi^{2}}\int d^{3}k\,\delta(k\cdot v - k\cdot v')\,\frac{kk\tilde{\phi}^{2}(|\,k\,|)}{|\,1+\tilde{\phi}\psi(\hat{k}, \hat{k}\cdot v)\,|^{2}} \qquad \ldots[2.35]$$

and is a symmetric dyadic in v and v'.

We now make several remarks:

(1) The collision term represents the weak scattering of two particles dynamically shielded by the presence of all the other particles. (Note: $1+\tilde{\phi}\psi = \varepsilon(k, -k\cdot v)$ where ε is the usual dielectric or dispersion function.)

(2) The integral is convergent for small k, the shielding providing the necessary and natural fall-off.

(3) It remains divergent for large k. However, if we just cut off the integral at $k_{max} \approx (mv^{2})/2e^{2}$ the result is still true to dominant order in ε.

Now is our kinetic equation a bona fide one? Namely, does it possess the following properties:

(a) If $f > 0$ at $t = 0$ is it positive for $t > 0$?

(b), (c), (d) Are the mean particle density, mean velocity, and mean square velocity independent of time?

(e) Is a Maxwellian distribution a stationary solution?

(f) As $t \to \infty$, does any solution approach a Maxwellian solution?

To prove (a) let us assume that f is positive everywhere initially and becomes negative at some later time. Then there must be a time when its minimum first becomes negative. At this time there exists a v such that

$$(i) \ f(\mathbf{v}) = 0, \qquad (ii) \ \frac{\partial f(\mathbf{v})}{\partial \mathbf{v}} = 0, \qquad (iii) \ \frac{\partial^2 f(\mathbf{v})}{\partial \mathbf{v} \partial \mathbf{v}}$$

is a non-negative definite tensor, and (iv) $\partial f / \partial t < 0$.

We shall now show that these conditions are incompatible with the kinetic equation.

Using (i) and (ii) we may write the kinetic equations as

$$\frac{\partial f}{\partial t} = -\int d^3v' f(\mathbf{v}') Q(\mathbf{v}, \mathbf{v}') \frac{\partial^2 f(\mathbf{v})}{\partial \mathbf{v} \partial \mathbf{v}}. \qquad \ldots[2.36]$$

Now let us choose an appropriate coordinate system such that at the point considered

$$\frac{\partial^2 f}{\partial v_i \partial v_j} = \delta_{ij} \Lambda_i. \qquad \ldots[2.37]$$

Then using condition (iii)

$$\Lambda_i \geqslant 0. \qquad \ldots[2.38]$$

Then

$$\frac{\partial f(\mathbf{v})}{\partial t} = -\int d^3v' f(\mathbf{v}') \sum_i \Lambda_i Q_{ii}(\mathbf{v}, \mathbf{v}'); \qquad \ldots[2.39]$$

but

$$Q_{ii} = -\frac{1}{8\pi^2} \int d^3k \, \delta(\mathbf{k} \cdot \mathbf{v} - \mathbf{k} \cdot \mathbf{v}') \frac{k_i^2 \tilde{\phi}^2}{|\varepsilon(k, -\mathbf{k} \cdot \mathbf{v})|^2} < 0 \qquad \ldots[2.40]$$

Hence $\dfrac{\partial f(\mathbf{v})}{\partial t} > 0$ contradicting (iv). Thus (a) is satisfied.

Property (b) is automatically satisfied, because the kinetic equation is written as a divergence in velocity space.

To prove (c) we write

$$\frac{d}{dt} \int d^3v \, \mathbf{v} f = -\int d^3v \, \mathbf{v} \frac{\partial}{\partial \mathbf{v}} \cdot \mathbf{J}(\mathbf{v}) = \int d^3v \, \mathbf{J}(\mathbf{v})$$

$$= \int d^3v \int d^3v' Q(\mathbf{v}, \mathbf{v}') \cdot \left\{ \frac{\partial f(\mathbf{v})}{\partial \mathbf{v}} f(\mathbf{v}') - \frac{\partial f(\mathbf{v}')}{\partial \mathbf{v}'} f(\mathbf{v}) \right\}.$$

$$\ldots[2.41]$$

This is a double integral whose integrand changes sign upon interchanging $\mathbf{v}$ and $\mathbf{v}'$. Hence it is zero.

To prove (d) we write

$$\frac{d}{dt}\int d^3v\,\frac{v^2}{2}\,f(v) = -\int d^3v\,\frac{v^2}{2}\,\frac{\partial}{\partial v}\cdot J(v) = \int d^3v\,v\cdot J$$

$$= \int d^3v\int d^3v'\,v\cdot Q(v,v')\cdot\left\{\frac{\partial f(v)}{\partial v}f(v')-\frac{\partial f(v')}{\partial v'}f(v)\right\}$$

$$= \tfrac{1}{2}\int d^3v\int d^3v'\,(v-v')\cdot Q(v,v')\cdot\left\{\frac{\partial f(v)}{\partial v}f(v')-\frac{\partial f(v')}{\partial v'}f(v)\right\}.$$

$$...[2.42]$$

But the definition of Q shows that $(v-v')\cdot Q(v,v')$ is identically zero in v and v'. Hence (d) is true.

To prove property (e), let $f(v) = \exp(-\tfrac{1}{2}Av^2+B\cdot v+C)$ with $A > 0$, B and C constants. Then

$$\frac{\partial f}{\partial v} = (-Av+B)f(v) \qquad\qquad ..[2.43]$$

and

$$\frac{\partial f(v)}{\partial v}f(v')-\frac{\partial f(v')}{\partial v'}f(v) = -A(v-v')f(v)f(v'). \qquad ...[2.44]$$

Substituting into expression [2.34] for $J(v)$, and again using the fact that $(v-v'\cdot)Q(v,v') = 0$ for all v, v', we have the result.

The proof of property (f) (the H-Theorem) follows the usual derivation of kinetic theory, which is not modified by the shielding of the interaction.

3

KLIMONTOVICH EQUATIONS OF PLASMA

In this section, we introduce the Klimontovich-Dupree description of plasma via the phase-space density N, and indicate the correspondence with the B.B.G.K.Y. hierarchy of equations.

We consider a plasma with s species of particles, $r = 1, \ldots, s$, and define for every species the phase-space density:

$$N_r(X, t) = \frac{V}{\mathcal{N}_r}\sum_{i=1}^{\mathcal{N}_r}\delta(X-X_{ir}(t)) \qquad\qquad ...[3.1]$$

There are $\mathcal{N}_r$ particles of species r, and V is the volume of the system. $X \equiv [x, v]$ is a generic point in phase space, and $X_{ir}(t)$ describes the trajectory of the ith particle of species r in this space. If we note that

$$\frac{\partial}{\partial t}\delta(X-X_{ir}(t)) = -\frac{\partial}{\partial t}X_{ir}(t)\cdot\frac{\partial}{\partial X}\delta(X-X_{ir}(t)), \qquad ...[3.2]$$

then it is clear that the Klimontovich function satisfies a Liouville equation:

$$\frac{d}{dt} N_r(X, t) = \left[\frac{\partial}{\partial t} + \dot{x} \cdot \frac{\partial}{\partial x} + \dot{v} \cdot \frac{\partial}{\partial v}\right] N_r(X, t) = 0 \qquad \dots[3.3]$$

provided the commutator $[\dot{v} \cdot (\partial/\partial v)]$ is zero.

Let E_0 and B_0 be the externally produced electric and magnetic fields and E, B the self-consistent electromagnetic field due to the plasma. Then equation [3.3] can be explicitly written as follows:

$$\left\{\frac{\partial}{\partial t} + v \cdot \frac{\partial}{\partial x} + \frac{q_r}{m_r}(E + E_0) \cdot \frac{\partial}{\partial v} + \frac{q_r}{m_r}\frac{v}{c} \times (B + B_0) \cdot \frac{\partial}{\partial v}\right\} N_r(X, t) = 0 \qquad \dots[3.4]$$

where the self-consistent fields obey Maxwell's equations:

$$\nabla \times B = \frac{4\pi}{c} J + \frac{1}{c}\frac{\partial E}{\partial t} \qquad\qquad \nabla \times E = -\frac{1}{c}\frac{\partial B}{\partial t}$$

$$\nabla \cdot B = 0 \qquad\qquad \nabla \cdot E = 4\pi\rho$$

$$J = \sum_r q_r n_r \int d^3v N_r(X, t)\, v \qquad \rho = \sum_r q_r n_r \int d^3v N_r(X, t). \qquad \dots[3.5]$$

Here, q_r and m_r are the charge and mass, respectively, of species r. The initial conditions for equation [3.4] are:

$$N_r(X, 0) = \frac{V}{\mathcal{N}_r} \sum_{i \in r} \delta(X - X_{ir}(0)). \qquad \dots[3.6]$$

Although it is not explicitly stated, all particle self-interaction terms except for the radiation reaction are excluded in equation [3.4]. This question is discussed by Dupree. We shall simply state, as we go along, where self-field effects are left out.

Let $D_1(\{X_i\}, t)$ be the usual Liouville function of the total system, satisfying the Liouville equation in all the $6n$ particle variables ($n = \sum_r \mathcal{N}_r$):

$$\left(\frac{\partial}{\partial t} + \{\dot{X}_i\} \cdot \frac{\partial}{\partial\{X_i\}}\right) D_1(\{X_i\}, t) = 0 \qquad \dots[3.7]$$

with the smooth initial condition $D_1(\{X_i\}, 0)$ given. We take $D_1(\{X_i\}, 0)$ normalized to unity and symmetric under the interchange of like particles; these properties persist in time. $\{X_i\}$ is a point of the full phase-space Γ.

We define the phase-space average of any phase function $\Psi(X; \{X_i\})$ as follows:

$$\langle\Psi(X, t)\rangle \equiv \int d^3\{X_i\} D_1(\{X_i\}, t)\Psi(X; \{X_i\}). \qquad \dots[3.8]$$

The integration $d^3\{X_i\}$ is over Γ. We shall find it useful to write equation [3.8], using Liouville's Theorem and the law of conservation of extension in phase, as

$$\langle\Psi(X, t)\rangle \equiv \int d^3\{X_{io}\} D_1(\{X_{io}\}, 0)\Psi[X; \{X_i(\{X_{io}\}, t)\}] \qquad \dots[3.9]$$

We also define

$$\delta\Psi = \Psi - \langle\Psi\rangle. \qquad\qquad ...[3.10]$$

It is easy to show that the following correspondence exists between the different moments of the N_r and the one-, two-, ... particle functions:

$$\langle N_r(X, t)\rangle = F_r^{(1)}(X, t),$$

$$\langle N_r(X, t)N_{r'}(X', t)\rangle = F_{r,r'}^{(2)}(X, X', t) + \frac{V}{\mathcal{N}_r}\delta_{rr'}\delta(X-X')F_r^{(1)}(X, t),$$

$$\langle N_r(X, t)N_{r'}(X', t)N_{r''}(X''t)\rangle = F_{r,r',r''}^{(3)}(X, X', X'', t)$$
$$+ \sum_{\{\rho\}} \frac{V}{\mathcal{N}_r}\delta_{rr'}\delta(X-X')F_{r'r'}^{(2)}(X, X', t)$$
$$+ \frac{V^2}{\mathcal{N}_r^2}\delta_{rr'}\delta_{r'r''}\delta(X-X')\delta(X'-X'')F_r^{(1)}(X, t)$$

etc. $\qquad\qquad ...[3.11]$

where $\sum_{\{\rho\}}$ represents the sum over all distinct permutations of $r, r', \ldots,$ $F^{(1)}, F^{(2)}, \ldots$ are the one-particle, two-particle, ... distributions, and $F^{(s)} \equiv V^s \int d'\{X_i\} D_1$, where the prime indicates that the integration is carried out over all variables, except those of the s-particles under consideration. We have passed to the usual limit $\mathcal{N}_r, V \to \infty$, with $n_r = \mathcal{N}_r/V$ finite.

We will restrict the theory presented here to spatially homogeneous plasma. Further, we will assume $E_0, B_0 = 0$, but we shall indicate how to generalize the results for $B_0 \neq 0$. Finally, we consider only the electrostatic interaction between particles, although electromagnetic interaction can be easily treated within the Klimontovich formalism.

Using the invariance of the functions with respect to translations (because of the assumption of average spatial uniformity), we obtain the following relations between the Fourier transforms:

$$\langle\delta N_{r,k}(v, t)\delta N_{r',k'}(v', t)\rangle = V\left[P_{r,r',k}(v, v', t) + \frac{1}{n_r}\delta_{rr'}\delta(v-v')F_r(v)\right]\delta_{k+k'},$$

$$\langle\delta N_{r,k}(v, t)\delta N_{r',k'}(v', t)\delta N_{r'',k''}(v'', t)\rangle$$
$$= V\left[H_{r,r',r''}(v, v', v'', t) + \sum_{\{\rho\}}\frac{1}{n_{r'}}\delta_{r'r''}\delta(v'-v'')P_{rr'k}(v, v', t)\right.$$
$$\left.+ \frac{1}{n_r^2}\delta_{rr'}\delta_{r'r''}\delta(v-v')\delta(v'-v'')F_r(v, t)\right]\delta_{k+k'+k''},$$

etc. $\qquad\qquad ...[3.12]$

where we freely use the equivalence relation

$$\delta_{k+k'}V \leftrightarrow (2\pi)^3\delta(k+k') \qquad\qquad ...[3.12a]$$

in passing from box to continuum normalization. There, $P_{r,r'}(X, X', t) \equiv P_{r,r'}(x-x', v, v', t)$ and $H(X, X', X'', t) \equiv H(x-x'', x'-x'', v, v', v'', t)$ are respectively the two- and three-particle correlation functions. The Fourier transforms are defined by

$$h_k = \int d^3\zeta \, e^{-ik\cdot\zeta} h(\zeta), \qquad \qquad ...[3.13]$$

We now average equation [3.4] over phase-space, and then subtract the result from equation [3.4] to obtain, since $\langle E \rangle = 0$:

$$\frac{\partial}{\partial t} F_r(v, t) = -\left(\frac{q}{m}\right)_r \frac{\partial}{\partial v} \cdot \langle \delta E(x, t) \delta N_r(X, t) \rangle \qquad ...[3.14]$$

and

$$\left(\frac{\partial}{\partial t} + v \cdot \frac{\partial}{\partial x}\right) \delta N_r(X, t) + \left(\frac{q}{m}\right)_r \delta E(x, t) \cdot \frac{\partial F_r}{\partial v}(v, t)$$

$$= \left(\frac{q}{m}\right)_r \frac{\partial}{\partial v} \cdot [\langle \delta E(x, t) \delta N_r(X, t) \rangle - \delta E(x, t) \delta N_r(X, t)]. \quad ...[3.15]$$

We have simplified the notation by replacing $F_r^{(1)}$ by F_r, etc. The self-consistent electric field set up by the plasma fluctuations δN_r is determined by

$$\frac{\partial}{\partial x} \cdot \delta E(x, t) = \sum_{r'} 4\pi q_{r'} \, n_{r'} \int d^3v \delta N_{r'}(x, v, t). \qquad ...[3.16]$$

These equations can alternatively be represented by their projections in k-space:

$$\frac{\partial}{\partial t} F_r(v, t) = -\lim_{V \to \infty} V^{-1} \left(\frac{q}{m}\right)_r \int \frac{d^3k}{(2\pi)^3} \, ik \cdot \frac{\partial}{\partial v} \langle \delta\psi_k(t) \delta N_{r,k}(v, t) \rangle \quad ...[3.17]$$

$$\left(\frac{\partial}{\partial t} + ik \cdot v\right) \delta N_{r,k}(v, t) - i\left(\frac{q}{m}\right)_r \delta\psi_k(t) k \cdot \frac{\partial F_r}{\partial v}(v, t)$$

$$= i\left(\frac{q}{m}\right)_r \int \frac{d^3l}{(2\pi)^3} (k-l) \cdot \frac{\partial}{\partial v} \delta\psi_{k-l}(t) \delta N_{r,l}(v, t)$$

$$...[3.18a]$$

$$\delta\Psi_k(t) = \sum_{r'} 4\pi \frac{q_{r'} n_{r'}}{k^2} \int d^3v \delta N_{r',k}(v, t) \qquad ...[3.18b]$$

where $\delta E_k(t) = -ik\delta\Psi_k(t)$. The initial condition for equation [3.18a] is

$$\delta N_{r,k}(v, 0) = \frac{1}{n_r} \sum_{i \in r} e^{-ik\cdot x_{i0}} \delta(v - v_{i,0}). \qquad ...[3.19]$$

Equations [3.18a, b] are the non-linear Vlasov equations. Together with equation [3.17], they describe the microscopic evolution of the system in phase-space.

4

SOLUTION OF THE LINEAR EQUATIONS AND CONVENTIONAL KINETIC THEORY

We shall neglect for the moment the non-linear terms on the right-hand side of equation [3.18a] and the effect of the slow time variation of F. The implications of this procedure will be discussed later. The Laplace transform of a function $h(t)$ is defined by:

$$h(p) = \int_0^\infty e^{-pt}h(t)dt = \mathscr{L}h(t) \quad (\text{Re } p>0) \qquad \dots[4.1]$$

Equations [3.18a, b] are then equivalent to

$$(p+ik \cdot v)\delta N_{r,k}(v, p) - i\left(\frac{q}{m}\right)_r \delta\psi_k(p)k \cdot \frac{\partial F_r}{\partial v}(v) = \delta N_{r,k}(v, 0), \quad \dots[4.2a]$$

$$\delta\Psi_k(p) = \sum_s 4\pi \frac{q_s n_s}{k^2}\int d^3v_s \delta N_{s,k}(v, p). \qquad \dots[4.2b]$$

The solution of these equations has been given by Landau:

$$\delta\psi_k(p) = \sum_s 4\pi \frac{q_s n_s}{k^2}\int_{\mathscr{L}} d^3v \frac{\delta N_{s,k}(v, 0)}{p+ik \cdot v}\bigg/\varepsilon(k, p) \qquad \dots[4.3a]$$

$$\delta N_{r,k}(v, p) = \frac{\delta N_{r,k}(v, 0)}{p+ik \cdot v} + i\left(\frac{q}{m}\right)_r \frac{1}{p+ik \cdot v} k \cdot \frac{\partial F_r}{\partial v}(v)\delta\psi_k(p) \quad \dots[4.3b]$$

where the dielectric function $\varepsilon(k, p)$ is

$$\varepsilon(k, p) = 1 - 4\pi i \sum_r \frac{q_r n_r}{k^2}\int_{\mathscr{L}} d^3v \left(\frac{q}{m}\right)_r \frac{1}{p+ik \cdot v} k \cdot \frac{\partial F_r}{\partial v}(v) \quad \dots[4.4]$$

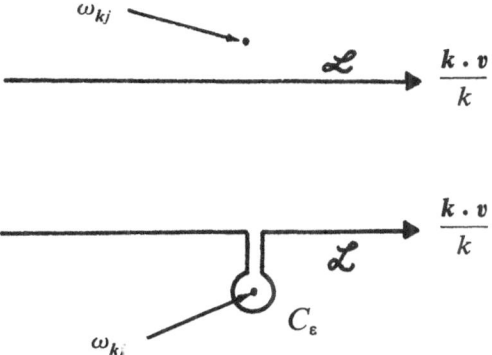

FIG. 4.1. Landau contour: (a) Im $\omega_{kj}>0$; (b) Im $\omega_{kj}<0$.

The Landau contour[20] (Fig. 4.1) provides the analytic continuation of the integrals in the half plane Re $p < 0$. The Laplace inversion of equation [4.3a] yields:

$$2\pi i \, \delta\Psi_k(t) = \int_{\sigma-i\infty}^{\sigma+i\infty} dp \, e^{pt} \, \frac{\mathcal{N}(p)}{\varepsilon(k, p)} \qquad \ldots[4.5]$$

in obvious notation. Note also that the contour is to the right of all the singularities of the integrand. For sufficiently large values of t ($t \gg \omega_p^{-1}$), the contribution from the semi-circle in the left half p plane (Re $p \to -\infty$) is vanishingly small so that asymptotically

$$\delta\Psi_k(t) =$$

$$4\pi \sum_r \frac{q_r n_r}{k^2} \int d^3v \, \delta N_{r,k}(v, 0) \left[\frac{e^{-ik \cdot vt}}{\varepsilon(k, -ik \cdot v)} + \sum_j \frac{e^{-i\omega_{k_j}t}}{(-i\omega_{k_j} + ik \cdot v)\partial\varepsilon_k/\partial(-i\omega_{k_j})} \right] \qquad \ldots[4.6]$$

where the ω_{kj} are the roots of the dispersion function:

$$\varepsilon(k, -i\omega_{kj}) = 0. \qquad \ldots[4.7]$$

Note that since they are functionals of F, they are in fact slowly varying functions of time. The system is stable if, for all j,

$$\gamma_{kj} = \text{Im } \omega_{kj} < 0. \qquad \ldots[4.8]$$

In conventional kinetic theory, the part of the solution [4.6] associated with the zeros of $\varepsilon(k, p)$ has been discarded as dying away rapidly. This procedure leads to a Markoffian equation for the one-particle distribution function as predicted by Bogoliubov's conjecture. Quasi-linear theory on the contrary emphasizes the role of the solutions $e^{-i\omega_{kj}t}$, as some modes are assumed (weakly) unstable and neglects the discrete particle contribution.

In the first case, the system is described by the B.G.L. equation

$$\frac{\partial F_r}{\partial t}(v, t) = \left(\frac{q_r}{m_r}\right)^2 \int d^3k \, k \cdot \frac{\partial}{\partial v} \frac{2}{\pi} \sum_s \frac{q_s^2 n_s}{k^4} \int d^3v' \frac{F_s(v', t)}{|\varepsilon(k, -ik \cdot v')|^2} \pi\delta(k \cdot v$$
$$- k \cdot v') k \cdot \frac{\partial F_r}{\partial v}(v, t) - i \left(\frac{q}{m}\right)_r \int d^3k \, k \cdot \frac{\partial}{\partial v} \frac{q_r}{2\pi^2 k^2} \frac{F_r(v, t)}{\varepsilon^*(k, ik \cdot v)} \qquad \ldots[4.9]$$

and in the latter case, one obtains the Drummond and Pines[5] equations:

$$\frac{\partial F_r}{\partial t}(v, t) = \left(\frac{q}{m}\right)_r^2 \int d^3k \, k \cdot \frac{\partial}{\partial v} \left[\frac{\gamma_k}{|k \cdot v - \omega_k|^2} k \cdot \frac{\partial}{\partial v} F_r(v, t) \right] I_k(t)$$

$$\frac{dI_k}{dt}(t) = 2\gamma_k(t) I_k(t), \qquad \ldots[4.10]$$

where γ_k corresponds to the unstable root (by assumption) of equation [4.7].

We note that the B.G.L. equation diverges should the Landau decrement (corresponding to the least stable mode) approach zero. This reflects failure of the Bogoliubov Ansatz in certain regimes of k-space: The fluctuation spectrum of the energy density does not in general relax "instantaneously" to a functional of the one-particle distribution function, and the system is not Markoffian in the F_r alone. The Drummond and Pines equations on the other hand are inappropriate for those regions of k-space where the Landau damping is large. Further, there is no apparent connection between equations [4.9] and [4.10].

We shall remedy the situation in the first part of this work, and derive a more general kinetic theory of plasma in the enlarged space $\{F_r^{(1)}, I_k\}$ by retaining the complete solution equation [4.6]. We shall consider the particle discreteness effects (first term of equation [4.6]) and collective plasma behaviour (second term of equation [4.6] on a par, and obtain a closed set of equations for the $\{F_r^{(1)}, I_k\}$ which automatically provide the proper transition between stable and unstable regimes.

The solution of equation [4.7], however, no longer yields the proper rate of decay or growth of the waves if $|\gamma_k|$ is small. In following chapters we therefore consider the effect both of the non-linear terms on the right-hand side of equation [3.18a], and of the slow time variation of $F_r^{(1)}$ in the left-hand side of that equation. They can indeed modify appreciably the rate of damping or growth of the waves in those regimes where the Landau decrement is small.

5

JUSTIFICATION OF AN EXPANSION OF THE NON-LINEAR EQUATIONS

Weakly turbulent and stable plasma are characterized by the existence of an expansion parameter $\lambda \ll 1$ which scales the size of the fluctuations. We therefore define λ by

$$\lambda^2 = \varepsilon_{max}/n\Theta \qquad \qquad ...[5.1]$$

for all t. There, ε is the energy density of the electric field associated with the fluctuations, and Θ is a measure of the mean kinetic energy. It has been shown in the literature (Drummond and Pines, 1962) that if equation [5.1] is satisfied initially, it will also be satisfied at later times if

$$\gamma_k \lesssim \lambda^2 |\operatorname{Re} \omega_k| \qquad \qquad ...[5.2]$$

in the unstable domains of k-space.

We are ultimately interested in equations for statistical averages (related to the moments of the δN_r), not the δN_r themselves. Since we can permute the operations of derivation and integration introduced by the right-hand sides of equations [3.14] and [3.15], and the averaging operation defined by equation [3.9] we can treat formally the fluctuations $\delta N_r(X, t)$ as small (we assume, as in the existing literature, that the statistical averages $F_r, P_{r,r'}, \ldots$

are sufficiently smooth functions). More precisely, equation [5.1] then implies that for wavelengths of the order of the Debye shielding length λ_D,‡ i.e. $k\lambda_D \sim 1$, we can set

$$\delta N_r(X, t) \lesssim \lambda F_r(v, t).$$

With help of equation [5.2], we can scale the different terms of equation [3.18]. It is easy to show that, with the fundamental time scale of the Vlasov equation in general of order ω_p^{-1}, the right-hand side of the equation is small (of order λ compared with the left-hand side), and thus can be treated by iteration.

On the other hand, the time scale of the equation for the evolution of the wave spectrum which is predicted by the linear theory is the Landau decrement γ_k. We emphasize that, although the nonlinearities must here be considered on the same level as the Landau decrement, we can construct this equation from the iterative solution of equation [3.18].

We note also that equation [3.14] shows that the time scale of evolution of F_r is of the order of $\omega_p^{-1}\lambda^{-2}$, i.e. very long compared with the Vlasov time scale, ω_p^{-1}. In a stable plasma $\lambda^2 = \varepsilon \equiv (n\lambda_D^3)^{-1}$, where ε is the "plasma expansion parameter".

6

WAVE KINETIC EQUATION TO LOWEST ORDER

We establish here to lowest order in the iteration scheme a set of coupled equations which describe the simultaneous evolution of waves and particles. In the proper limits, the equations reduce to the Balescu-Guernsey-Lenard kinetic equation, or the Pines-Schrieffer equations (see also Harris).

The Landau solution, equation [4.6] indicates that if the dispersion function $\varepsilon(k, -i\omega)$ has real (or nearly real) roots, there will be some set of particles which will resonate with the collective plasma oscillations and give rise to secularities. This already suggests failure in attempting to obtain a kinetic equation for the particles alone uniformly over the $\{k, v\}$ phase-space, and indeed, one must consider in general the time evolution of the fluctuation spectrum of the wave energy on the same time scale as the evolution of the F_r. As we shall show, this failure is just why the B.G.L. equation and the result of Rostoker (1961) for the coefficient of spatial diffusion across a magnetic field are divergent as one approaches marginal stability.

Before proceeding, we shall introduce and define the following notation. Let

$$\delta\psi_k(t) \equiv \delta\phi_k(t) + \delta\chi_k(t) \qquad \qquad ...[6.1]$$

‡ In our order of magnitude estimates, we shall often take $k \sim \lambda_D^{-1}$ and $\omega_k \sim \omega_p$. It is clear that the conclusions are still valid as long as λ is the smallest parameter in the problem (for example $\lambda \ll k\lambda_D \ll \lambda^{-1}$).

and

$$\delta\phi_k(t) \equiv \delta\phi_k^+(t) + \delta\phi_k^-(t), \quad \delta\chi_k(t) \equiv \delta\chi_k^+(t) + \delta\chi_k^-(t), \quad \dots[6.2]$$

where

$$\delta\phi_k^\sigma(t) = 4\pi \sum_r \frac{q_r n_r}{k^2} \int d^3v\, \delta N_{r,\,k}(v,0) \frac{e^{-i\omega_k^\sigma t}}{i k\cdot v - i\omega_k^\sigma} \left(\frac{\partial\varepsilon_k(-i\omega)}{\partial(-i\omega)}\right)^{-1}_{\omega=\omega_k^\sigma} \quad \dots[6.3]$$

and

$$\delta\chi_k^\sigma(t) = 4\pi \sum_r \frac{q_r n_r}{k^2} \int_{\mathscr{H}^\sigma} d^3v\, \delta N_{r,\,k}(v,0) \frac{e^{-i k\cdot vt}}{\varepsilon(k,-i k\cdot v)}. \quad \dots[6.4]$$

Here, ω_k^σ ($\sigma = \pm$) represent the two least stable roots of the dispersion equation; and, because their contribution is vanishingly small after a few plasma periods, we shall neglect the other roots of $\varepsilon(k,-i\omega)$. The waves $\delta\Psi_k^+$ (resp. $\delta\Psi_k^-$) correspond to disturbances propagating parallel to k (resp. antiparallel to k). Therefore H^+ (resp. H^-) corresponds to the half space $k\cdot v > 0$ ($k\cdot v < 0$). It can easily be shown that the time average $(\delta\Psi_k^+\delta\Psi_k^-)^t$ is zero, so that the effect of the correlation $\langle \delta\Psi_{-k}^+\delta\Psi_k^-\rangle$ is irrelevant on the slow time-scale evolution of the plasma.

We consider the time derivative of equation [6.1]

$$\frac{d}{dt}\delta\Psi_k^\sigma(t) = -i\omega_k^\sigma\delta\Psi_k^\sigma(t) + 4\pi\sum_r\frac{q_r n_r}{k^2}\int_{\mathscr{H}^\sigma} d^3v\,\delta N_{r,\,k}(0)\frac{i\omega_k - i k\cdot v}{\varepsilon(k,-i k\cdot v)}e^{-i k\cdot vt}$$

$$\dots[6.5]$$

where we have eliminated $\delta\phi_k^\sigma(t)$ using equations [6.1] and [6.4]. Clearly, this operation eliminates from our description the possible resonance (or secular behaviour) mentioned earlier. Note that we have neglected the slow time variation of F (and hence of $\varepsilon(k,-i k\cdot v)$ and ω_k^σ). Were it possible to account for these variations exactly, an operation similar to the elimination of $\delta\phi_k^\sigma$ would still have eliminated the resonance (or secular behaviour). This justifies our procedure in deriving equation [6.5]. We are using this equation to derive the rate of change of the fluctuation spectrum on the slow time scale; therefore, we synchronize the functions F to $F(t)$, etc.

We multiply equation [6.5] by $\delta\Psi_{-k} = \delta\Psi_k^*$, add the complex conjugate, and perform the ensemble average of the equation so obtained. There results:

$$\frac{1}{2}\frac{d}{dt}I_k^\sigma = \gamma_k^\sigma(t)I_k^\sigma(t) + \frac{2}{\pi}\sum_r\frac{q_r^2 n_r}{k^4}\int_{\mathscr{H}^\sigma} d^3v\, F_r(v,0)\frac{-\gamma_k^\sigma(t)}{|\varepsilon(k,-i k\cdot v)|^2}$$

$$+ \lim_{V\to\infty} V^{-1}\frac{1}{4\pi^2}\sum_r\frac{q_r n_r}{k^2}\int_{\mathscr{H}^\sigma} d^3v\,\frac{i\omega_k^\sigma - i k\cdot v}{\varepsilon(k,-i k\cdot v)}\langle\delta N_{rk}(v,0)\delta\phi_{-k}^\sigma(0)\rangle$$

$$e^{-i(k\cdot v - \omega_{-k}^\sigma)t} + \text{c.c.} \quad \dots[5.6]$$

where c.c. means the complex conjugate. $\sum_\sigma k^2 I_k(t)/8\pi$ is the electric field energy per mode and unit volume, and

$$I_k^\sigma(t) = \lim_{V \to \infty} (2\pi)^{-3} \frac{\langle |\delta\Psi_k^\sigma(t)|^2 \rangle}{V}. \qquad \ldots[6.7]$$

(We note that the terms involving the $P_{rs,k}$, occurring in the ensemble averages, present the same type of universal decay as discussed previously.)

By closing the contour of integration along the axis $k \cdot v/k$ in an obvious way, it is possible to show that the role of the last term of equation [6.6] is to provide the analytic continuation of the second term should $\gamma_k^\sigma(t)$ (as a function of k or t) pass through zero. Therefore, the wave kinetic equation is

$$\tfrac{1}{2}\frac{d}{dt} I_k^\sigma(t) = \gamma_k^\sigma(t) I_k^\sigma(t) + \frac{2}{\pi} \sum_r \frac{q_r^2 n_r}{k^4} \int_{S^\sigma} d^3v \, \frac{-\gamma_k^\sigma(t) F_r(v,t)}{|\,\varepsilon(k,-ik\cdot v)\,|^2} \qquad \ldots[6.8]$$

where S^σ is the contour of Fig. 6.1. The domain of integration is still the half space H^σ. As usual, we have synchronized the distribution function $F_r(v,0)$ (introduced by the ensemble average) to $F_r(v,t)$ so that equation [6.8] is valid on the slow time-scale. We shall actually show that, to higher order in the iteration, the plasma dynamics provides such a synchronization.

It is important to note that we made no restriction on the magnitude of γ_k^σ; therefore, equation [6.8] can describe any regime where nonlinear effects are negligible.

The first term of equation [6.8] describes the net result of induced emission and absorption of plasma waves by the particles; it can be either positive or negative according to which of these processes dominates. The second term, which is always positive, represents the spontaneous emission of longitudinal Cerenkov radiation.

a)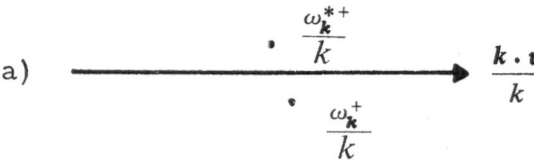

b)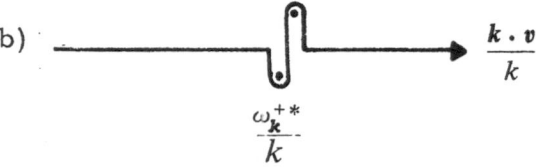

FIG. 6.1. The contour S^+ versus the zeros $k \cdot v = \omega_k^+$ of $\epsilon(k, -ik \cdot v)$
(a) if Im $\omega_k^+ < 0$;   (b) if Im $\omega_k^+ > 0$.

If we integrate equation [6.8] formally in time, we obtain:

$$I_k^\sigma(t) = I_k^\sigma(0)\,e^{2\int_0^t \gamma_k{}^\sigma(t')dt'} - \frac{4}{\pi}\sum_r \frac{q_r^2 n_r}{k^4}\int_{S^\sigma} d^3v \int_0^t e^{2\int_{t'}^t \gamma_k{}^\sigma dt''}\,\frac{\gamma_k^\sigma(t')F_r(v,t')dt'}{|\varepsilon(k,-ik\cdot v,t')|^2}$$

...[6.9]

If γ_k^σ is negative, and large compared with the inverse of the time scale of variation of $F_r,(|\gamma_k^\sigma| \gg \lambda^2|\omega_k|)$, the first term of equation [6.9], which is the transient solution, can be systematically neglected. Furthermore, we can disregard the time variation of F_r in the second term, since most of the contribution to the integral comes from the upper limit of integration. We obtain

$$I_k^\sigma(t) = \frac{2}{\pi}\sum_r \frac{q_r^2 n_r}{k^4}\int_{S^\sigma} d^3v\,\frac{F_r(v,t)}{|\varepsilon(k,-ik\cdot v)|^2}.$$

...[6.10]

The absorption "instantaneously" balances the emission, so that the plasma presents all the characteristics of a gas of uncorrelated dressed particles. This is the basis of Rostoker's superposition principle. This result is improperly used for all k in obtaining the B.G.L. equation.

The right-hand side of equation [6.10] can easily be computed for an electron plasma in thermal equilibrium; the result is:

$$\frac{1}{8\pi}\sum_\sigma k^2 I_k^\sigma = \frac{1}{(2\pi)^3}\frac{\theta_e}{2}\frac{1}{1+k^2\lambda_D^2},$$

...[6.11]

where θ_e is the electron temperature in energy units. Equation [6.11] is a well-known result.

In the limit $\gamma_k^\sigma \to 0$, on the other hand, the last term of equation [6.8] can be approximated by integrating across the resonance. If we explicitly represent the Landau decrement γ_k^σ in terms of F_r in the first term, we obtain:

$$\frac{1}{2}\frac{\partial \varepsilon_k(-i\omega)}{\partial \omega}\bigg|_{\omega_k\sigma}\frac{d}{dt}I_k^\sigma(t) = \sum_r \frac{\omega_{pr}^2}{k^2}\int d^3v\,\pi\delta(\omega_k^\sigma - k\cdot v)\,k\cdot\frac{\partial F_r}{\partial v}I_k^\sigma(t)$$
$$+ \frac{2}{\pi}\sum_r \frac{q_r^2 n_r}{k^4}\int d^3v\,\pi\delta(\omega_k^\sigma - k\cdot v)F_r(v,t)\frac{\partial \varepsilon_k(-i\omega)}{\partial \omega}\bigg|_{\omega_k\sigma}^{-1},$$

...[6.12]

where $\varepsilon_k(-i\omega) \equiv \varepsilon(k,-i\omega)$, and ω_{pr} is the plasma frequency of species r.

This result was first derived by Pines and Schrieffer (1962) and later by Harris (1967) via a quantum-mechanical treatment. (See also the derivation from the B.B.G.K.Y. hierarchy given by Frieman and Rutherford.)

In an electron plasma in thermal equilibrium $dI_k/dt = 0$, and

$$F_e(v) = C\exp\left\{-\tfrac{1}{2}\frac{mv^2}{\theta}\right\}.$$

..[6.13]

Introducing equation [6.13] in equation [6.12] yields:

$$\frac{1}{8\pi} \sum_\sigma \omega_k^\sigma \left.\frac{\partial\varepsilon(-i\omega)}{\partial\omega}\right|_{\omega_k^\sigma}^{-1} k^2 I_k = \frac{1}{(2\pi)^3}\,\theta. \qquad \ldots[6.14]$$

The left-hand side of equation [6.14] represents the total plasmon energy (the potential energy of the wave and the kinetic energy associated with the oscillation of the particles), and therefore agrees with the Rayleigh-Jeans distribution. Since, for plasmons $\partial\varepsilon(k, -i\omega_k^\sigma)/\partial\omega \simeq 2/\omega_k^\sigma$, and $k\lambda_D \ll 1$, equation [6.14] agrees also with the exact result, equation [6.11].

Equation [6.12] is, however, inappropriate in some important regimes of the plasma. Such is generally the case if the system is weakly unstable.

The wave amplitude can eventually reach a high level, well above the thermal value given by equation [6.14], and non-linear damping or growth rates can compete with the Landau decrement to determine the effective time scale of the wave equation. Such is also the case for waves of very large phase velocity, even in a stable plasma, since the Landau decrement usually decreases exponentially as $k \to 0$. Collisional damping of the waves is no longer negligible, and indeed for sufficiently small k (such that $|\gamma_k| \lesssim \varepsilon\omega_{pe}$), it fixes the time scale of the wave equation. Further, $\gamma_k^\sigma \sim \varepsilon\omega_{pe}$ implies $F(v_k) \sim \varepsilon V_{the}^{-3}$ (in a Maxwellian plasma at least) where v_k is the phase velocity of a wave. Therefore bremsstrahlung emission will compete with and overtake the longitudinal Cerenkov emission as the dominant process of wave generation. The former process indeed is higher order in the discreteness parameter ε, but involves two particles of average velocity ($F(v) \sim v^{-3}$). All of these effects will be discussed later.

7

PARTICLE KINETIC EQUATION

We introduce equations [6.1]–[6.3] in equation [4.3b] to obtain

$$\delta N_{r,k}(v, t)$$
$$= \delta N_{r,k}(v, 0)\,e^{-ik\cdot vt} + i\sum_\sigma \delta\phi_k^\sigma(0)\left[e^{-i\omega_k t} - e^{-ik\cdot vt}\right](g_{k,\omega_k r}\cdot k)F_r(v, t)$$
$$+ 4\pi i\sum_s \frac{q_s n_s}{k^2}\int d^3 v'\,\frac{\delta N_{sk}(0)}{\varepsilon(k, -ik\cdot v')}\left[e^{-ik\cdot v't} - e^{-ik\cdot vt}\right](g_{k, k\cdot v', r}\cdot k)F_r(v, t),$$
$$\ldots[7.1]$$

where

$$g_{k,\omega,r} \equiv \left(\frac{q}{m}\right)_r \frac{1}{ik\cdot v - i\omega}\frac{\partial}{\partial v}. \qquad \ldots[7.2]$$

We shall considerably shorten the algebra if we note that we have the freedom to deform the path of integration along the axis $v\cdot k/|v|$ below the poles of the operators $g_{k,\omega,r}$ in the right-hand side of equation [3.17] provided we assume that the analytic continuation of $\langle \delta\Psi_{-k}(t)\cdot\delta N_{r,k}(\mathbf{r},t)\rangle$ exists in a finite strip containing the $\mathbf{v}\cdot k/|\mathbf{v}|$ axis. (Note that equation [7.1] is well defined as $k\cdot\mathbf{v}\to\omega_k$.) In order to avoid confusion, we label these operators $g^+_{k,\omega,r}$; for example

$$g^+_{k,\,k\cdot v',\,r}\equiv(q/m)_r\,\frac{1}{\lambda+ik\cdot v-ik\cdot v'}\cdot\frac{\partial}{\partial v},\quad\lambda\to0^+\qquad\ldots[7.3]$$

(in the limit $\mathrm{Im}\,\omega\to0$, one can alternatively deform the contour or introduce the parameter λ to define the topology).

We eliminate $\delta\phi^\sigma_k(0)$ from equation [7.1], multiply by $\sum_\sigma\delta\Psi^\sigma_k(t)$, and introduce the result in equation [3.17] to obtain

$$\frac{\partial F_r}{\partial t}(v,t)$$

$$=\left(\frac{q}{m}\right)_r\int d^3k\,k\cdot\frac{\partial}{\partial v}\sum_\sigma I^\sigma_k(t)(g^+_{k,\,\omega_k\sigma,\,r}\cdot k)F_r(v,t)$$

$$+\left(\frac{q}{m}\right)_r\int d^3k\,k\cdot\frac{\partial}{\partial v}\frac{2}{\pi}\sum_s\frac{q^2_s n_s}{k^4}\sum_\sigma\int_{\mathscr{H}^\sigma}d^3v'\,\frac{F_s(v',t)}{|\,\varepsilon(k,-ik\cdot v')\,|^2}[(g^+_{k,\,k\cdot v',\,r}\cdot k)$$

$$-(g^+_{k,\,\omega_k\sigma,\,r}\cdot k)]F_r-i\left(\frac{q}{m}\right)_r\int d^3k\,k\cdot\frac{\partial}{\partial v}\frac{q_r}{2\pi^2k^2}\,\frac{F_r(v,t)}{\varepsilon^*(k,ik\cdot v)}$$

$$+\left(\frac{q}{m}\right)_r\int d^3k\,k\cdot\frac{\partial}{\partial v}\frac{2}{\pi}\sum_s\frac{q^2_s n_s}{k^4}\sum_\sigma\int_{\mathscr{H}^\sigma}d^3v'\,\frac{F_s(v',t)}{\varepsilon(k,-ik\cdot v')}\left(\frac{\partial\varepsilon_k(-i\omega)}{\partial(-i\omega)}\right)^{-1*}_{\omega_k\sigma}$$

$$\frac{e^{-ik\cdot v't+i\omega_k\sigma*t}}{i\omega^{\sigma*}_k-ik\cdot v'}[(g^+_{k,\,k\cdot v',\,r}\cdot k)-(g^+_{k,\,\omega_k\sigma,\,r}\cdot k)]F_r(v,t)$$

$$-i\left(\frac{q}{m}\right)_r\int d^3k\,k\cdot\frac{\partial}{\partial v}\frac{q_r}{2\pi^2k^2}\sum_\sigma\left(\frac{\partial\varepsilon_k(-i\omega)}{\partial(-i\omega)}\right)^{-1*}_{\omega_k\sigma}\frac{F_r(v,t)}{i\omega^{\sigma*}_k-ik\cdot v}e^{-ik\cdot vt+i\omega_k\sigma*t}.$$

$$\ldots[7.4]$$

The last two terms arise from products of the form

$$\langle\delta\phi_{-k}(t)\delta\chi_k(t)\rangle\cdot\langle\delta\phi_{-k}(t)\delta N_k(t)\rangle.$$

By using arguments similar to those developed elsewhere in this paper, we can show that these terms provide the analytic continuation of the two preceding ones when $\mathrm{Im}\,\omega^\sigma_k$ passes through zero from the stable side. With the deformation of the contours implied with the operators $g^+_{k,\,\omega,\,r}$, the second and fourth terms of equation [7.4] provide a transient contribution which is shown elsewhere to decay as $1/t^{\frac{1}{2}}$ or faster, according to the magnitude of $|\,\mathbf{v}\,|$ (see also Guernsey[21]).

As usual, the explicit contribution from the $P_{r,s,k}$ is strongly decaying. Note also that it is justified in this order to synchronize the distribution function $F_r(v,0)$ to $F_r(v,t)$.

The kinetic equation for the particles finally takes the form:

$$\frac{\partial}{\partial t} F_r(v, t) = \left(\frac{q}{m}\right)_r \int d^3k\, k \cdot \frac{\partial}{\partial v} \sum_\sigma (g^+_{k, \omega_k^\sigma, r} \cdot k) F_r(v, t) I_k^\sigma(t)$$

$$+ \left(\frac{q}{m}\right)_r \int d^3k\, k \cdot \frac{\partial}{\partial v} \frac{2}{\pi} \sum_s \frac{q_0^2 n_s}{k^4} \sum_\sigma \int_{S^\sigma} d^3v' \frac{F_s(v', t)}{|\varepsilon(k, -ik \cdot v')|^2}$$

$$[(g^+_{k, k \cdot v', r} \cdot k) - (g^+_{k\omega_k^\sigma, r} \cdot k)] F_r$$

$$- i \left(\frac{q}{m}\right)_r \int d^3k\, k \cdot \frac{\partial}{\partial v} \frac{q_r}{2\pi^2 k^2} \frac{F_r(v, t)}{\varepsilon^{-*}(k, ik \cdot v)}. \qquad \dots[7.5]$$

We repeat that the deformation $(+)$ is *below* the poles of the operators g (above the poles of g^*). The deformation $(-)$ is above the poles of ε^{-1} (below the poles of ε^{-1*}). It may also be worth noting that the reality of the integrand is a consequence of the relation $\omega_k^\sigma = -\omega_k^{\sigma*}$, and the equality $F_r(\omega_k^\sigma/k) = [F_r(\omega_k^{\sigma*}/k)]^*$ (since F_r is a real function).

The closed set of equations [6.8] and [7.5] provides complete statistical knowledge of the evolution of the plasma, if non-linear effects (to be treated later) are unimportant. The initial conditions are $F_r(v, 0)$ and $I_k(0)$. Later in this lecture, we show that, in terms of the usual functions of the B.B.G.K.Y. hierarchy

$$I_k^\sigma(0) = \frac{2}{\pi} \sum \frac{q_r^2 n_r}{k^4} \left[\int_{S^\sigma_-} d^3v\, F_r(v, 0) \frac{1}{|ik \cdot v - i\omega_k^\sigma|^2} \left| \frac{\partial \varepsilon_k(-i\omega)}{\partial(-i\omega)} \right|^{-2}_{\omega_k^\sigma} \right.$$

$$\left. + \int_{S^\sigma} d^3v\, F_r(v, 0) \frac{1}{|\varepsilon(k, -ik \cdot v)|^2} \right] + \frac{2}{\pi} \sum_{r,s} \frac{q_r q_s n_r n_s}{k^4} \int_{S^\sigma_-} d^3v$$

$$\int_{S^\sigma_-} d^3v'\, P_{r,s,k}(v, v', 0) \frac{1}{(ik \cdot v - i\omega_k^\sigma)} \cdot \frac{1}{(-ik \cdot v' + i\omega_k^{\sigma*})}$$

$$\left| \partial \varepsilon_k(-i\omega)/\partial(-i\omega) \right|^{-2}_{\omega_k^\sigma}, \qquad \dots[7.6a]$$

where the contour S^σ_- is defined by Fig. 7.1.

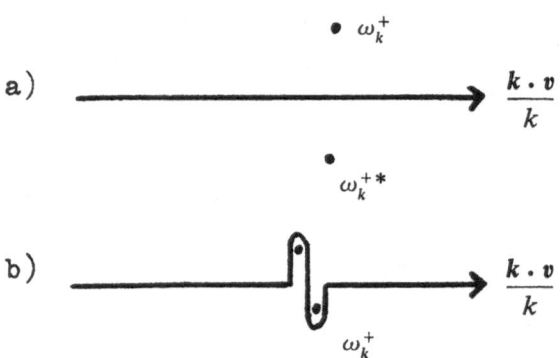

FIG. 7.1. The contour $S_-^{\sigma(+)}$, (a) Im $\omega_k^+ > 0$; (b) Im $\omega_k^+ < 0$.

For strongly damped modes, we can introduce the asymptotic solution equation [6.10] in equation [7.5], which thus reduces obviously to the Balescu-Guernsey-Lenard equation. On the other hand, as Im $\omega_k \to 0$, the term $(g^+_{k,\omega_k\sigma,r} \cdot k)F_r$ removes the characteristic divergence of the B.G.L. theory; this corresponds to a new phenomenon which will be interpreted below.

To pull out the physical content of equation [7.5], it is useful to split the k-space into two domains, Ω_1 and Ω_2, which are defined as follows:

$$\Omega_1 \equiv \{k: -\gamma_k \geqslant \alpha\lambda^2 \,|\,\omega_k\,|\}$$

$$\Omega_2 \equiv \{k: |\,\gamma_k\,| \leqslant \alpha\lambda^2 \,|\,\omega_k\,|\} \quad \lambda^{-1} \gg \alpha \gg 1. \quad ...[7.6b]$$

It can easily be shown that:

$$\frac{\partial F_r}{\partial t}(v, t) \cong \left(\frac{q}{m}\right)^2_r \sum_\sigma \int_{\Omega_2} d^3k\, k \cdot \frac{\partial}{\partial v}\left|\pi\delta(\omega^\sigma_k - k \cdot v)\, k \cdot \frac{\partial F_r}{\partial v}(v, t)\right| I^\sigma_k(t)$$

$$+ \left(\frac{q}{m}\right)^2_r \sum_\sigma \int_{\Omega_2} d^3k\, k \cdot \frac{\partial}{\partial v}\left[P \frac{k \cdot \dfrac{\partial F_r}{\partial v}(v, t)}{(k \cdot v - \omega^\sigma_k)^2}\right]\left[\gamma^\sigma_k I^\sigma_k + \frac{2}{\pi}\sum \frac{q^2_s n_s}{k^4}\right.$$

$$\left.\int d^3v'\pi\delta(\omega^\sigma_k - k \cdot v')\, F_s \left|\frac{\partial\varepsilon_k}{\partial\omega^\sigma_k}\right|^{-2}\right]$$

$$+ \left(\frac{q}{m}\right)_r \sum_\sigma \int_{\Omega_2} d^3k\, k \cdot \frac{\partial}{\partial v}\left[\frac{q_r}{2\pi^2 k^2}\,\pi\delta(\omega^\sigma_k - k \cdot v)\, F_r(v, t)\left(\frac{\partial\varepsilon_k}{\partial\omega^\sigma_k}\right)^{-1}\right]$$

$$+ \left(\frac{q}{m}\right)_r \sum_\sigma \int_{\Omega_2} d^3k\, k \cdot \frac{\partial}{\partial v}\frac{2}{\pi}\sum_s \frac{q^2_s n_s}{k^4}\int_{S^\sigma} d^3v'\left[P \frac{F_s(v', t)}{|\,\varepsilon(k, -ik \cdot v')\,|^2}\right]$$

$$[(g^+_{k, k \cdot v', r} \cdot k) - (g^+_{k, \omega_k\sigma, r} \cdot k)]F_r(v, t)$$

$$- \left(\frac{q}{m}\right)_r \int_{\Omega_2} d^3k\, k \cdot \frac{\partial}{\partial v}\frac{2}{\pi}\sum_s \frac{q_r q_s n_s}{k^4}\int d^3v'\left[P \frac{F_r(v, t)}{|\,\varepsilon(k, -ik \cdot v)\,|^2}\right]$$

$$(g^+_{k, k \cdot v, s} \cdot k)F_s(v', t)$$

$$+ \left(\frac{q}{m}\right)_r \int_{\Omega_1} d^3k\, k \cdot \frac{\partial}{\partial v}\frac{2}{\pi}\sum_s \frac{q^2_s n_s}{k^4}\int d^3v' \frac{\pi\delta(k \cdot v - k \cdot v')}{|\,\varepsilon(k, -ik \cdot v)\,|^2}$$

$$\left[\frac{q_r}{m_r}\, k \cdot \frac{\partial}{\partial v} - \frac{q_s}{m_s}\, k \cdot \frac{\partial}{\partial v'}\right]F_r(v, t)F_s. \quad ...[7.7]$$

Here, the symbol P means that the Cauchy principal value must be taken. The second term is the contribution from the singularity at $k \cdot v' = \omega_k$ in the second term of equation [7.5] and the principal value of the first term. Equation [7.7] is very similar to that given by Frieman and Rutherford (1964), if one neglects their non-linear terms in the wave amplitude (these will be treated later in this lecture). The third term of their equation (133), however, is a poor approximation of the fourth term of equation [7.7]. The above derivation has the advantage of being much simpler (compare equations [7.5] and [7.7]).

Usually, the terms involving $P \mid \varepsilon(k, -ik \cdot v') \mid^{-2}$ are negligible, because weak stability or instability occurs generally at small k and away from the resonance $\varepsilon(k, -ik \cdot v') \sim 1/(k\lambda_D)^2$. The interpretation of the other terms is as follows. The first term represents the change in F_r due to stimulated emission by, and absorption of, the resonating waves. The second and third terms describe the consequences of the time variation of I_k and of the spontaneous emission by the particle; inclusion of these processes removes the B.G.L. divergences (note that the drag term is also divergent in B.G.L. because of the absence of a contour deformation as $\gamma_k^\sigma \to 0$). The last term is best understood as incoherent dressed particle-particle scattering.

We show later on that the set of equations [6.12] and [7.7] conserves total energy of the system, field-particles:

$$\frac{\partial}{\partial t}\left[\sum_r \tfrac{1}{2}n_r m_r \int d^3v \, v^2 F_r(v, t) + \sum_\sigma \int_{\Omega_2} d^3k \, \frac{k^2 I_k}{8\pi}\right] = \frac{\partial}{\partial t}[\mathscr{E}_P + \mathscr{E}_F]. \qquad \ldots[7.8]$$

It is interesting to give an estimate of the order of magnitude of the different terms in equations [7.7] in the particular case of a system of interacting plasmons and electrons. If we neglect the fourth and fifth terms, we obtain the following estimates for the others, in the order they appear:

$$\varepsilon\langle k\lambda_D\rangle \frac{\theta_e}{\theta_T} \cdot \frac{\langle k^2 I_k\rangle}{\theta_T}; \quad \varepsilon\langle k\lambda_D\rangle^{-1} \frac{\theta_e}{\theta_T} \langle e^{-\frac{1}{4}k^2\lambda_D^2}\rangle; \quad \varepsilon\langle k\lambda_D\rangle \frac{\theta_e}{\theta_T}; \quad \varepsilon \ln \varepsilon \left(\frac{\theta_e}{mv^2}\right)^{\frac{3}{4}},$$
$$\ldots[7.9]$$

where $\langle k\lambda_D\rangle$ and v are respectively the radius of Ω_2 (in Debye units), and the velocity of the tail electrons; θ_e, θ_T and $\langle k^2 I_k\rangle$ are the effective temperature of the main body of the electron distribution, of the tail, and of the plasmons, respectively. The order of magnitude of $\langle k\lambda_D\rangle$ is obtained by equating the Landau decrement to $\varepsilon\omega_p$: thus, $\langle k\lambda_D\rangle \sim (\ln \varepsilon)^{-\frac{1}{2}}$. Introduction of this result in equation [7.9] shows that the last term is negligible for large velocities and/or large temperature ratios $k^2 I_k/\theta_T$, θ_e/θ_T. Equation [7.7] then reduces to

$$\frac{\partial F_e}{\partial t}(v, t)$$

$$\cong \left(\frac{q}{m}\right)^2_e \sum_\sigma \int_{\Omega_2} d^3k \, k \cdot \frac{\partial}{\partial v}\left[\pi\delta(\omega_k^\sigma - k \cdot v) \, k \cdot \frac{\partial F_e}{\partial v}(v, t)\right] I_k^\sigma(t)$$

$$+ \left(\frac{q}{m}\right)_e \sum_\sigma \int_{\Omega_2} d^3k \, k \, \frac{\partial}{\partial v}\left[\frac{q_e}{2\pi k^2} \, \pi\delta(\omega_k^\sigma - k \cdot v) F_e(v, t)\right]\left[\frac{\partial \varepsilon_k(-i\omega)}{\partial \omega}\right]^{-1}_{\omega_k^\sigma}.$$
$$\ldots[7.10]$$

Equations [6.12] and [7.10] show that high-energy electrons and plasmons on the one hand, and thermal electrons and ions on the other hand, will move almost independently towards separate quasi-equilibria. This conclusion agrees with experimental data (Eviatar[22]) as well as computer analysis (Dawson and Shanny[23]). In the long run, however, the last term of equation [7.7], which provides the link between these two regimes, will bring the tail of the distribution to the temperature of the main body of the plasma particles.

Equation [7.10] was also obtained by Pines and Schrieffer, and by Harris (*loc. cit.*). If we neglect interaction with thermal particles, it is easy to show that equations [6.13] and [6.14] with $\theta = \theta_T$, are exact equilibrium distributions for tail particles and plasmons. θ_T and C are to be determined by solving the specific initial value problem. However, the simplified equations [6.12] and [7.10] do not show an approach to local equilibrium by themselves. Indeed, any spherically symmetric, stable distribution $F_e(mv^2/2\theta)$, with I_k^σ given via

$$\omega_k^\sigma \frac{\partial \varepsilon_k(-i\omega)}{\partial \omega}\bigg|_{\omega_k\sigma} \frac{k^2}{8\pi} I_k^\sigma = \frac{1}{(2\pi)^3}\, \theta \tilde{F}_e\left(\frac{\omega_k^\sigma}{k}\right)\bigg/\tilde{F}_e'\left(\frac{\omega_k^\sigma}{k}\right) \qquad \text{...[7.11]}$$

where

$$\tilde{F}(u) = \int d^3v\, \delta\left(u - \frac{k\cdot v}{k}\right) F(v) \qquad \text{...[7.12]}$$

is a steady-state solution.

Equation [7.5] is inappropriate to describe the evolution of a weakly unstable plasma. Indeed, as discussed previously, the wave amplitude can reach a high level, and, when $k^2 I_k \gtrsim \theta\varepsilon^{-\frac{1}{2}}$, additional "collision" terms, which are quadratic in the wave amplitudes, must be added to equation [7.5]. These will be considered later in this paper.

8

NON-LINEAR MODIFICATIONS

With the iteration procedure described earlier, we obtain a wave kinetic equation which is appropriate to describe the evolution of the fluctuation spectrum $I_k^\sigma(t)$ over the entire k-space, if one excludes strong turbulence. The theory systematically includes the non-linear processes of radiation (by bremsstrahlung, wave-particle and wave-wave scattering) and of absorption of the waves (collisional damping, non-linear Landau damping).

The coefficient of collisional damping and the rate of bremsstrahlung emission are computed for a Maxwellian plasma in the dipole approximation. We recover in a simple fashion the conductivity formula of Oberman, Ron and Dawson.[17] Our derivation can be compared with Dupree's calculation.[13]

We demonstrate that the plasma dynamics provides the synchronization of the rate of Cerenkov radiation so that the evolution remains Markoffian in the enlarged space $\{I_k, F_s\}$ to this higher order in the expansion.

Finally, we generalize the equation of evolution of the one-particle distribution function to describe weakly turbulent plasma. The non-linear terms in the wave amplitude agree with the results of Frieman and Rutherford.[9]

8.1. Corrections due to the time dependence of the Vlasov operator

Non-linear effects enter the theory directly via the right-hand side of equation [3.18a], and indirectly because of the slow time variation of F_r in the left-hand side. We shall correct first for this time-dependence of the Vlasov operator. (The different corrections are additive to the order in which we are working.)

The solution [4.6] of the Vlasov equation is secular for those free streaming particles which are resonant with the eigenfrequencies of the (approximately) time-independent Vlasov operator. Were it possible to account exactly for the slow time variation of the operator, we could eliminate the resonant behaviour of the solution by a procedure similar to that given in equation [6.6]. The structure of the resonance will be slightly modified; this leads to a renormalization of the rate of Cerenkov emission. Such renormalizations are usually disregarded in kinetic theory.

To correct the first term of equation [6.6] for the slow time-dependence of F_r, we shall assume that the "homogeneous solutions" (the words are improper here) of the Vlasov equations are of the form

$$\delta\Psi_k(t) \sim e^{-i\int_0^t \omega(t')dt'} \qquad \ldots[8.1]$$

where the slow time-dependence is entirely contained in $\omega(t)$.

To calculate ω correctly to order λ^2, we follow the procedure of Bernstein and Engelman.[24] Integrating the left-hand side of equation [3.18a] forward in time, we obtain:

$$\delta N_{r,k}(v, t)$$
$$= \delta N_{r,k}(v, 0)e^{-ik\cdot vt} + i\left(\frac{q}{m}\right)_r \int_0^t dt' \delta\Psi_k(t') k \cdot \frac{\partial F_r}{\partial v}(v, t')e^{-ik\cdot v(t-t')}.$$
$$\ldots[8.2]$$

This result introduced into Poisson's equation yields

$$\mathscr{T}(t, 0)\delta\Psi_k(t) = 4\pi \sum \frac{q_r n_r}{k^2} \int d^3v\, \delta N_{r,k}(v, 0)e^{-ik\cdot vt} \qquad \ldots[8.3]$$

with

$$\mathscr{T}(t, 0)\delta\Psi_k(t)$$
$$\equiv \delta\Psi_k(t) - 4\pi i \sum \frac{q_r n_r}{k^2} \int_0^t dt' \int d^3v \left(\frac{q}{m}\right)_r k \cdot \frac{\partial F_r}{\partial v}(v,t')e^{-ik\cdot v(t-t')}\delta\Psi_k(t').$$
$$\ldots[8.4]$$

By direct substitution, we obtain that equation [8.1] satisfies equation [8.3] trivially for times $t \gg \omega_p^{-1}$, if ω is a root of the following dispersion relation accurate to order λ^2:

$$1 - 4\pi i \sum_r \frac{q_r n_r}{k^2} \int_{\mathscr{L}} d^3v \left(\frac{q}{m}\right)_r \frac{1}{-i\omega(t) + ik \cdot v} \left[k \cdot \frac{\partial F_r(v,t)}{\partial v} - \frac{\partial}{\partial t} \frac{k \cdot \dfrac{\partial F_r(v,t)}{\partial v}}{-i\omega(t) + ik \cdot v} \right]$$
$$= 0 \qquad\qquad [8.5]$$

Let $\delta\varepsilon(k, -i\omega(t))$ be the correction to the dielectric function, equation [4.4], and $\delta\omega_k(t)$ be the correction to the roots of the same equation. It is easily shown that

$$\delta\omega_k(t) = i \left(\frac{\partial\varepsilon_k(-i\omega_k)}{\partial(-i\omega_k)}\right)^{-1} \left\{\left(\frac{\partial}{\partial t}\right)_{\omega_k} + \tfrac{1}{2}\left(\frac{\partial}{\partial t}\right)_F\right\} \left(\frac{\partial\varepsilon_k(-i\omega_k)}{\partial(-i\omega_k)}\right)$$
$$= \operatorname{Re} \delta\omega_k(t) + i\Delta_k(t) \qquad\qquad \dots[8.6]$$

where $(\partial/\partial t)_x$ means that x is not to be differentiated. It is interesting to note that if $|\operatorname{Re} \omega_k| \gg |\operatorname{Im} \omega_k|$, $i\partial\varepsilon_k(-i\omega_k)/\partial(-i\omega_k)$ is almost real, and $\delta\omega_k$ is pure imaginary.

8.2. A diagrammatic systematization of the iteration

To nth order in the iteration, the Vlasov equations [3.18] read:

$$\delta N_{r,k}^{(n)} - i\delta\Psi_k^{(n)} \left(\frac{q}{m}\right)_r \frac{1}{p + ik \cdot v_r} k \cdot \frac{\partial F_r}{\partial v_r} = i\int \frac{d^3l}{(2\pi)^3} \left(\frac{q}{m}\right)_r \frac{1}{p + ik \cdot v_r} (k-l) \cdot \frac{\partial}{\partial v_r}$$
$$\mathscr{L}\left[\delta\Psi_{k-l}^{(n-1)} \delta N_{r,l}^{(1)} + \cdots + \delta\Psi_{k-l}^{(1)} \delta N_{r,l}^{(n-1)}\right]$$
$$\delta\Psi_k^{(n)} = \sum_s 4\pi \frac{q_s n_s}{k^2} \int d^3v_s \, \delta N_{s,k}^{(n)} \quad (n>1), \qquad \dots[8.7a]$$

where we have chosen the initial conditions:

$$\delta N_{r,k}^{(1)}(v_r, 0) = \delta N_{r,k}(v_r, 0); \quad \delta N_{r,k}^{(n)}(v_r, 0) = 0, \quad (n>1). \quad \dots[8.7b]$$

In the following, the order to which a solution is obtained is indicated either by a superscript "(n)" or by the number of subscripts. We shall also adopt either the time representation or the Laplace representation of any function. We shall further represent the original function and its transform by the same symbol, and drop the arguments p, v or t, v when no confusion is possible.

We recall that the solution of the linearized Vlasov equations is of the type

$$\delta\Psi_k^{\sigma,\,(1)}(t) = \delta\Psi_{k,\,1}^\sigma(t) + \delta\Psi_{k,\,2}^\sigma(t) \qquad \ldots[8.8]$$

where

$$\delta\Psi_{k,\,1}^\sigma(t) = \delta\phi_k^\sigma(t) = 4\pi \sum \frac{q_s n_s}{k^2} \int d^3v_s \, \frac{\delta N_{s,\,k}(v_s,\,0)}{ik\cdot v_s - i\omega_k^\sigma} \left(\frac{\partial\varepsilon_k(-i\omega_k^\sigma)}{\partial(-i\omega_k^\sigma)}\right)^{-1} e^{-i\omega_k^\sigma t}$$

$$\ldots[8.8a]$$

and

$$\delta\Psi_{k,\,2}^\sigma(t) = \delta\chi_k^\sigma(t) = 4\pi \sum \frac{q_s n_s}{k^2} \int d^3v_s \, \frac{\delta N_{s,\,k}(v_s,\,0)}{\varepsilon(k,\,-ik\cdot v_s)} e^{-ik\cdot v_s t}. \qquad \ldots[8.8b]$$

Also, under a velocity integral, the solution $\delta N_k^{(1)}$ is asymptotically equivalent to

$$\delta N_{r,\,k}^{(1)}(v_r,\,t) = \delta N_{r,\,k,\,1}(v_r,\,t) + \delta N_{r,\,k,\,2}(v_r,\,t) + \delta N_{r,\,k,\,3}(v_r,\,t) \qquad \ldots[8.9]$$

with

$$\delta N_{r,\,k,\,1}(v_r,\,t) = i \sum_\sigma \delta\Psi_{k,\,1}^\sigma(t)(g_{k,\,\omega_k^\sigma,\,r}^+ \cdot k) F_r(v_r,\,t) \qquad \ldots[8.9a]$$

$$\delta N_{r,\,k,\,2}(v_r,\,t) = i \sum_s 4\pi \frac{q_s n_s}{k^2} \int d^3v_s \, \frac{\delta N_{s,\,k}(v_s,\,0)}{\varepsilon(k,\,-ik\cdot v_s)} e^{-ik\cdot v_s t}$$
$$(g_{k,\,k\cdot v_s,\,r}^+ \cdot k) F_r(v_r,\,t) \qquad \ldots[8.9b]$$

$$\delta N_{r,\,k,\,3}(v_r,\,t) = \delta N_{r,\,k}(v_r,\,0) e^{-ik\cdot v_r t}. \qquad \ldots[8.9c]$$

Since the right-hand side of equation [8.7a], which we represent symbolically by $S_{r,\,k}^{(n)}$, is a functional of the lower order solutions $\delta\psi_k^{(1)} \cdots \delta N_k^{(n-1)}$, it is useful to systematize the calculation of $\delta N_{r,\,k}^{(n)}$ and $\delta\psi_k^{(n)}$ with the diagrams of Fig. 8.1. For example $\delta N_{r,\,k1,\,1}$ and $\delta\psi_{k1,\,1}$ are the functions set up by the non-linear term $S_{r,\,k}[\delta\psi_{k,\,1};\,\delta N_{r,\,k,\,1}]$. Similarly $\delta\psi_{k11,\,2}$ and $\delta\psi_{k2,\,11}$ (note the position of the comma) have their respective origin in $S_{r,\,k}[\delta\psi_{k11};\,\delta N_{r,\,k,\,2}]$ and $S_{r,\,k}[\delta\psi_{k,\,2};\,\delta N_{r11}]$.

Equations [8.7a and b] are linear in $\delta N_{r,\,k}^{(n)}$; the algebra involved in their solution is well-known, and will not be repeated here. We shall state the results later in this lecture, where we introduce the following compact notations (from B. B. Kadomtsev[8] 1965):

$$S^+(k,\,\omega_{i,\,j} \,|\, l,\,\omega_i) = (g_{k,\,\omega_{ij},\,r}^+ \cdot (k-l))(g_{l,\,\omega_i,\,r}^+ \cdot l)$$
$$s^+(k,\,\omega_{i,\,j} \,|\, l,\,\omega_i) \equiv S^+(k,\,\omega_{i,\,j} \,|\, l,\,\omega_i) + S^+(k,\,\omega_{i,\,j} \,|\, k-l,\,\omega_j)$$
$$= s^+(k,\,\omega_{i,\,j} \,|\, k-l,\,\omega_j) \qquad \ldots[8.10]$$

where

$$\omega_{i,j} = \omega_i + \omega_j. \qquad \ldots[8.11]$$

Also,

$$V(k, \omega_{i,j} \mid l, \omega_i) = 4\pi \sum_s \frac{q_s n_s}{k^2} \int d^3 v_s \, S^+(k, \omega_{i,j} \mid l, \omega_i) F_s \qquad \ldots[8.12]$$

and

$$v(k, \omega_{ij} \mid l, \omega_i) \equiv V(k, \omega_{i,j} \mid l, \omega_i) + V(k, \omega_{i,j} \mid k-l, \omega_j)$$
$$= v(k, \omega_{i,j} \mid k-l, \omega_i).$$

8.3. General form of wave kinetic equation

We follow the same procedure as was described before, to construct the wave kinetic equation. Consider

$$\delta \Psi_k^\sigma = \delta \Psi_k^{(1)\sigma} + \delta \Psi_k^{(2)\sigma} + \delta \Psi_k^{(3)\sigma} \qquad \ldots[8.13]$$

and

$$\frac{d}{dt} \delta \Psi_k^\sigma = -i\omega_k^\sigma \delta \Psi_k^\sigma + \left(\frac{d}{d_i} + i\omega_k^\sigma\right) \left[\delta \Psi_{k,2}^{(1)\sigma} + \delta \Psi_k^{(2)\sigma} + \delta \Psi_k^{(3)\sigma}\right] - i\omega_k^\sigma \delta \Psi_k^\sigma$$

$$\ldots[8.14]$$

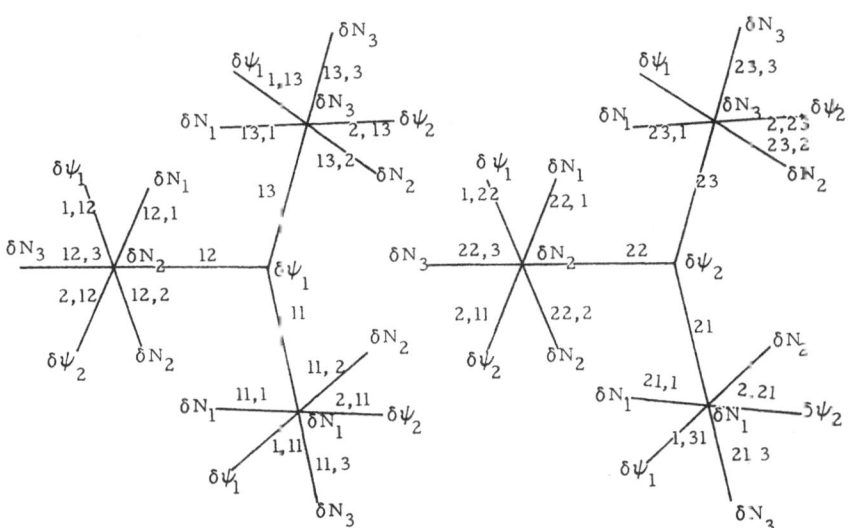

FIG. 8.1. Classification of non-linear interactions

where we have eliminated $\delta\Psi^\sigma_{k,1}$ ($\equiv \delta\phi^\sigma_k$) using equation [8.13]. According to the previously described discussion, the effect of the slow variation of F_r, ω_k and ε_k is adequately represented by the last term of equation [8.14], We multiply that equation by $\delta\Psi^\sigma_{-k} = \delta\Psi^{\sigma*}_k$, add the complex conjugate. and average the result over the statistical ensemble. We obtain:

$$\tfrac{1}{2}\frac{d}{dt}\langle\delta\Psi^{\sigma*}_k\delta\Psi^\sigma_k\rangle = \gamma^\sigma_k\langle\delta\Psi^{\sigma*}_k\delta\Psi^\sigma_k\rangle + \mathrm{Re}\,\langle\delta\Psi^{\sigma*}_k\left(\frac{d}{dt}+i\omega^\sigma_k\right)$$
$$[\delta\Psi^\sigma_{k,2}+\delta\Psi^{(2)\sigma}_k+\delta\Psi^{(3)\sigma}_k]\rangle + \Delta^\sigma_k\,\mathrm{Re}\,\langle\delta\Psi^{\sigma*}_k\delta\Psi^\sigma_k\rangle.$$
$$\dots[8.15]$$

We wish to retain nonlinear effects to the lowest non-vanishing order only. It will, therefore, be sufficient to consider

$$\tfrac{1}{2}\frac{d}{dt}\langle\delta\Psi^{\sigma*}_k\delta\Psi^\sigma_k\rangle = (\gamma^\sigma_k+\Delta^\sigma_k)\langle\delta\Psi^{\sigma*}_k\delta\Psi^\sigma_k\rangle + \left(\tfrac{1}{2}\frac{d}{dt}-\gamma^\sigma_k\right)\langle\delta\Psi^{\sigma*}_{k,2}\delta\Psi^\sigma_{k,2}\rangle$$
$$+\left(\tfrac{1}{2}\frac{d}{dt}-\gamma^\sigma_k\right)\langle\delta\Psi^{(2)\sigma*}_k\delta\Psi^{(2)\sigma}_k\rangle$$
$$+\,\mathrm{Re}\,\langle\delta\Psi^\sigma_{k,1}\left(\frac{d}{dt}+i\omega_k\right)\delta\Psi^{(3)\sigma}_k\rangle + \left(\tfrac{1}{2}\frac{d}{dt}-\gamma^\sigma_k\right)$$
$$2\,\mathrm{Re}\,\langle\delta\Psi^{\sigma*}_{k,2}\delta\Psi^{(3)\sigma}_k\rangle$$
$$+\left(\tfrac{1}{2}\frac{d}{dt}-\gamma^\sigma_k\right)2\,\mathrm{Re}\,\langle\delta\Psi^{\sigma*}_{k,2}\delta\Psi^{(2)\sigma}_k\rangle$$
$$+\,\mathrm{Re}\,\langle\delta\Psi^{\sigma*}_{k,1}\left(\frac{d}{dt}+i\omega^\sigma_k\right)[\delta\Psi^\sigma_{k,2}+\delta\Psi^{(2)\sigma}_k]\rangle. \qquad\dots[8.16]$$

The above decomposition of equation [8.15] is useful because it indicates how to obtain the different processes from Fig. 8.1.

The first two terms represent linear Landau damping (including the effect of the slow time variation of F_r) and Cerenkov emission of the waves. The third term (respectively the fourth) contain the different non-linear mechanism of emission (resp. of absorption). The fifth term, and also part of the fourth term, provide a correction to the radiation processes, but are always dominated by Cerenkov and/or bremsstrahlung emission. The remaining contributions give rise to dynamical processes which synchronize the emission rate of the Cerenkov radiation, therefore maintaining the Markoffian behaviour to this order of the expansion.

In summary, the wave kinetic equation will assume the following form:

$$\tfrac{1}{2}\frac{d}{dt}I^\sigma_k(t) = [\gamma^\sigma_k(t)+\Delta^\sigma_k(t)+\Gamma^\sigma_k(t)]I^\sigma_k(t)+C^\sigma_k(t)+B^\sigma_k(t) \qquad\dots[8.17]$$

where the new symbols $\Gamma_k^\sigma(t)$, $B_k^\sigma(t)$ represent respectively the nonlinear mechanisms of damping and radiation of the waves; $C_k^\sigma(t)$ is the Cerenkov emission rate.

It can be shown that $\delta\Psi_k^{(n)}$ can be split into two components $\delta\Psi_k^{(n)\sigma}$ ($\sigma = \pm$) by requiring that the disturbances under the integration signs propagate parallel (or anti-parallel) to k. One can then show that, to the required order, these waves are statistically uncorrelated:

$$\langle \delta\Psi_{-k}^{(n)+} \delta\Psi_k^{(m)-} \rangle = 0. \qquad \ldots[8.18]$$

In many instances, there will be no need to complicate the notation further to include $\pm$ possibilities, because statistical averaging naturally provides this decomposition. This will be shown in a specific example below, which also gives us an opportunity to make further comments on the procedure.

The typical example we choose can be written symbolically as follows:

$$\sum_\sigma (\delta\Psi_{k1,\,11}^\sigma + \delta\Psi_{k11,\,1}^\sigma) = S_k \Big[\sum_{\sigma_1\sigma_2\sigma_3} \delta\Psi_{k'-l,\,1}^{\sigma_1} \delta\Psi_{l,\,1}^{\sigma_2} \delta\Psi_{k-k',\,1}^{\sigma_3} \Big] \qquad \ldots[8.19]$$

so that

$$\langle \delta\Psi_{k,\,1}^{\pm\,*} \sum_\sigma (\delta\Psi_{k1,\,11} + \delta\Psi_{k11,\,1}) \rangle = S_k \Big[\sum_{\sigma_1\sigma_2\sigma_3} \langle \delta\Psi_{k,\,1}^{\pm\,*} \delta\Psi_{k'-l,\,1}^{\sigma_1} \delta\Psi_{l,\,1}^{\sigma_2} \delta\Psi_{k-k',\,1}^{\sigma_3} \rangle \Big]$$
$$\ldots[8.19a]$$

The ensemble average on the right-hand side can be written as a sum of irreducible functions:

$$\langle \delta\Psi_{k,\,1}^{\pm\,*} \delta\Psi_{k'-l,\,1}^{\sigma_1} \delta\Psi_{l,\,1}^{\sigma_2} \delta\Psi_{k-k',\,1}^{\sigma_3} \rangle$$
$$= \sum_{\{\rho\}} \big[\langle \delta\Psi_{k,\,1}^{\pm\,*} \rangle \langle \delta\Psi_{k'-l,\,1}^{\sigma_1} \delta\Psi_{l,\,1}^{\sigma_2} \delta\Psi_{k-k',\,1}^{\sigma_3} \rangle_{\text{irr.}}$$
$$+ \langle \delta\Psi_{k,\,1}^{\pm\,*} \delta\Psi_{k'-l,\,1}^{\sigma_1} \rangle_{\text{irr.}} \langle \delta\Psi_{l,\,1}^{\sigma_2} \delta\Psi_{k-k',\,1}^{\sigma_3} \rangle_{\text{irr.}}$$
$$+ \langle \delta\Psi_{k,\,1}^{\pm\,*} \delta\Psi_{k'-l,\,1}^{\sigma_1} \delta\Psi_{l,\,1}^{\sigma_2} \delta\Psi_{k-k',\,1}^{\sigma_3} \rangle_{\text{irr.}} \big], \qquad \ldots[8.20]$$

where the sum $\sum_{\{\rho\}}$ is over all distinct permutations. The first term of equation [8.20] is zero; the ratio of the third term to the second is of order λ^2. Thus, retaining only the second term yields, with the help of equations [3.12] and [3.12a] and equations [6.4] and [6.7]:

$$\lim_{V \to \infty} V^{-1}(2\pi)^{-3} \langle \Psi_{k,\,1}^{\pm\,*} \delta\Psi_{k'-l,\,1}^{\sigma_1} \delta\Psi_{l,\,1}^{\sigma_2} \delta\Psi_{k-k',\,1}^{\sigma_3} \rangle$$
$$= \sum_{\sigma_3} I_{k,\,1}^\pm I_{k-k',\,1}^{\sigma_3} (2\pi)^6 \big[\delta(l-k'+k)\delta_{\sigma_2\sigma_3}\delta_{\sigma_1\pm} + \delta(l-k)\delta_{\sigma_1\sigma_2}\delta_{\sigma_2\,\pm}.$$
$$\ldots[8.21]$$

We have rejected the possibility $I_{k,1} \pm I_{l,1}^{\sigma} \delta(k')$ on grounds of the spatial homogeneity of the ensemble of realizations, since $\delta\Psi_{k'-l1} \delta\Psi_{l,1}$ arises from the development of $\delta\Psi_{k',1}^{\sigma'}$. The definition of $I_{k,1}^{\sigma}$ is the following:

$$
\begin{aligned}
&I_{k,1}^{\sigma}(t) \\
&= \frac{2}{\pi} \sum_{r,s} \frac{q_r n_r q_s n_s}{k^4} \int d^3 v_r \int d^3 v_s\, P_{rs,k}(\boldsymbol{v}_r, \boldsymbol{v}_s, 0) \frac{e^{2\gamma_k^{\sigma} t}}{(i\boldsymbol{k}\cdot\boldsymbol{v}_r - i\omega_k^{\sigma})(-i\boldsymbol{k}\cdot\boldsymbol{v}_s + i\omega_k^{\sigma*})} \\
&\left| \frac{\partial\varepsilon_k(-i\omega_k)}{\partial(-i\omega_k)} \right|^{-2} + \frac{2}{\pi} \sum_r \frac{q_r^2 n_r}{k^4} \int d^3 v_r\, F_r(\boldsymbol{v}_r, 0) \frac{e^{2\gamma_k^{\sigma} t}}{|i\boldsymbol{k}\cdot\boldsymbol{v} - i\omega_k^{\sigma}|^2} \left| \frac{\partial\varepsilon_k(-i\omega)}{\partial-(i\omega)} \right|^{-2}.
\end{aligned}
$$
$$...[8.22]$$

Note the contour, which is different from S_-^{σ} in stable regimes. The theory will provide the proper deformations naturally, as was already the case with the simplified kinetic theory.

8.4. *Description of the non-linear mechanisms of absorption*

The fourth term of equation [8.16] describes the different non-linear processes of absorption which we shall consider here at length.

There are many contributions of different nature to the effective non-linear "decay" rate arising from the pieces of the solution of the non-linear Vlasov equations. It will be sufficient to show how the treatment goes for the following two pieces:

(i)

$$
\lim_{V \to \infty} V^{-1} (2\pi)^{-3} \left\langle \delta\Psi_{k,1}^{\sigma} \left(\frac{d}{dt} + i\omega_k^{\sigma} \right) [\delta\Psi_{k1,11} + \delta\Psi_{k11,1}] \right\rangle. \quad ...[8.23a]
$$

We have shown in equation [8.21] that the ensemble average selects two values of $l: l = k$ and $l = k' - k$. Thus we obtain in a straightforward manner:

$$
\begin{aligned}
&I_{k,1}^{\sigma} \int d^3 k' \sum_{\sigma'} \frac{2\gamma_{k'}^{\sigma'}}{\varepsilon(k, -i\omega_k^{\sigma} + 2\gamma_{k'}^{\sigma'})} I_{k',1}^{\sigma'} \frac{v(k, \omega_k^{\sigma} + 2i\gamma_{k'}^{\sigma'} \mid k'\ \omega_{k'}^{\sigma'})}{\varepsilon^-(k'', -i\omega_{k,-k'}^{\sigma;\sigma'})} \\
&\hspace{7cm} v(k'', \omega_{k,-k'}^{\sigma;\sigma'} \mid k, \omega_k^{\sigma}) \\
&- i I_{k,1}^{\sigma} \int d^3 k' \sum_{\sigma'} \frac{2\gamma_{k'}^{\sigma'}}{\varepsilon(k, -i\omega_k^{\sigma} + 2\gamma_{k'}^{\sigma'})} I_{k',1}^{\sigma'} \sum_r 4\pi \frac{q_r n_r}{k^2} \int d^3 v_r (g_{k,\omega_k^{\sigma}+2i\gamma_{k'}^{\sigma'},r}^+ \cdot k) \\
&\hspace{7cm} s^+ (k'', \omega_{k,-k'}^{\sigma\sigma'} \mid k, \omega_k^{\sigma}) F_r \quad ...[8.23b]
\end{aligned}
$$

where we have introduced the new compact notations

$$
k'' = k - k'; \qquad \omega_{k,-k'}^{\sigma;\sigma'} = \omega_k^{\sigma} + \omega_{-k'}^{\sigma'} = \omega_k^{\sigma} - \omega_{k'}^{\sigma'*}. \qquad ...[8.24]
$$

The deformation of the contour implicit in the definition of ε^- would be easily justified by retaining the complete p-dependence.

(ii)

$$\lim_{V \to \infty} V^{-1}(2\pi)^{-3} \langle \delta\Psi_{k,1}^{\sigma} \left(\frac{d}{dt}+i\omega_k^{\sigma}\right)[\delta\Psi_{k21,2}+\delta\Psi_{k2,21}+\delta\Psi_{k12,2}+\delta\Psi_{k2,12}]\rangle.$$

$$...[8.25a]$$

There, the ensemble average selects $l = k'-k$ only. Equation [3.12] further imposes $v_s = v_t$ in the average. (The contribution from the pair correlation function, which provides an analytic continuation, is considered later.) Thus we get

$$I_{k,1}\left(\frac{\partial\varepsilon_k(-i\omega_k^{\sigma})}{\partial(-i\omega_k^{\sigma})}\right)^{-1}\int d^3k' \frac{2}{\pi}\sum_s \frac{q_s^2 n_s}{k'^4}\sum_{\sigma'}\int_{\neq\sigma'}d^3v_s \frac{F_s}{|\varepsilon(k',-ik'\cdot v_s)|^2}$$

$$\frac{v(k,\omega_k^{\sigma}\,|\,k',k'\cdot v_s)}{\varepsilon^-(k'',-i\omega_k^{\sigma}+ik'\cdot v_s)}\,v(k'',\omega_k^{\sigma}-k'\cdot v_s\,|\,k,\omega_k^{\sigma})$$

$$-iI_{k,1}\left(\frac{\partial\varepsilon_k(-i\omega_k^{\sigma})}{\partial(-i\omega_k^{\sigma})}\right)^{-1}\int d^3k' \frac{2}{\pi}\sum_s \frac{q_s^2 n_s}{k'^4}\sum_{\sigma'}\int_{\neq\sigma'}d^3v_s \frac{F_s}{|\varepsilon(k',-ik'\cdot v_s)|^2}$$

$$\sum_r 4\pi\frac{q_r n_r}{k^2}\int d^3v_r(g_{k,\omega_k^{\sigma},r}^+\cdot k')\,s^+(k'',\omega_k^{\sigma}-k'\cdot v_s\,|\,k,\omega_k^{\sigma})F_r.$$

$$..[8.25b]$$

We shall now explain (the calculation is similar to that given elsewhere in the text) how to obtain proper contour deformations under the velocity integrals in equation [8.25b] ($H^{\sigma'} \to S^{\sigma'}$), and in equation [8.23b], where we replace $I_{k',1}^{\sigma'}$ by an expression similar to equation [8.22] for a moment. The contribution from

$$\left\langle \delta\Psi_{k1}^{\sigma*}\left(\frac{d}{dt}+i\omega_k^{\sigma}\right)[\delta\Psi_{k1,12}+\delta\Psi_{k12,1}+\delta\Psi_{k1,21}+\delta\Psi_{k21,1}]\right\rangle$$

(choose $l = k'-k$) and

$$\left\langle \delta\Psi_{k,1}^{*}\left(\frac{d}{dt}+i\omega_k^{\sigma}\right)[\delta\Psi_{k2,11}+\delta\Psi_{k11,2}]\right\rangle$$

(choose $l = k$ and $l = k'-k$) provides the deformation $H^{\sigma'} \to S^{\sigma'}$ in equation [8.25b] and the deformation $S_-^{\sigma'}$ in the second term of equation [8.22]. The contribution from the pair correlation functions in the same averages and in equation [8.25a] provides the deformation $S_-^{\sigma'}$ in the first term of equation [8.22].

(Most of the remaining contributions to the wave kinetic equation can be obtained in the same way. There are certain terms, however, (when $l = k$) which require special attention; but in general they do not contribute to the damping of the waves.)

We now eliminate $I_{k,1}^{\sigma}$ and $I_{k',1}^{\sigma'}$ in terms of I_k^{σ} and $I_{k'}^{\sigma'}$, in equation [8.23b] and similar expressions. The equivalent "damping" rate Γ_k^{σ} due to non-linear processes is, thus, given below:

Γ_k^σ

$$= \int d^3k' \sum_{\sigma'} I_{k'}^{\sigma'} \frac{2\gamma_{k'}^{\sigma'}}{\varepsilon(k, -i\omega_k^\sigma + 2\gamma_{k'}^{\sigma'})} \frac{v(k, \omega_k^\sigma + 2i\gamma_{k'}^{\sigma'} \mid k', \omega_{k'}^{\sigma'})}{\varepsilon^-(k'', -i\omega_{k, -k'}^{\sigma\sigma'})} v(k'', \omega_{k, -k'}^{\sigma, \sigma'} \mid k, \omega_k^\sigma)$$

$$- i \int d^3k' \sum_{\sigma'} I_{k'}^{\sigma'} \frac{2\gamma_{k'}^{\sigma'}}{\varepsilon(k, -i\omega_k^\sigma + 2\gamma_{k'}^{\sigma'})} \sum_r 4\pi \frac{q_r n_r}{k^2} \int d^3v_r (g_{k, \omega_{k^\sigma} + 2i\gamma_{k'}^{\sigma'}}^+ \cdot k')$$

$$s^+(k'', \omega_{k, -k'}^{\sigma\sigma'} \mid k, \omega_k^\sigma) F_r$$

$$+ \int d^3k' \sum_{\sigma'} \frac{2}{\pi} \sum_s \frac{q_s^2 n_s}{k'^4} \int_{S^{\sigma'}} d^3v_s \frac{F_s}{\mid \varepsilon(k', -ik' \cdot v_s) \mid^2}$$

$$\left[\left| \frac{\partial \varepsilon_k}{\partial(-i\omega_k^\sigma)} \right|^{-1} \frac{v(k, \omega_k^\sigma \mid k', k' \cdot v_s)}{\varepsilon^-(k'', -i\omega_k^\sigma + ik' \cdot v_s)} v(k'', \omega_k^\sigma - k' \cdot v_s \mid k, \omega_k^\sigma) \right.$$

$$\left. - \frac{2\gamma_{k'}^{\sigma'}}{\varepsilon(k', -i\omega_k^\sigma + 2\gamma_{k'}^{\sigma'})} \cdot \frac{v(k, \omega_k^\sigma + 2i\gamma_{k'}^{\sigma'} \mid k', \omega_{k'}^{\sigma'})}{\varepsilon^-(k'', -i\omega_{k, -k'}^{\sigma\sigma'})} v(k'', \omega_{k, -k'}^{\sigma\sigma'} \mid k\omega_k^\sigma) \right]$$

$$- i \int d^3k' \sum_{\sigma'} \frac{2}{\pi} \sum_s \frac{q_s^2 n_s}{k'^4} \int_{S^{\sigma'}} d^3v_s \frac{F_s}{\mid \varepsilon(k', -ik' \cdot v_s) \mid^2} \sum_r 4\pi \frac{q_r n_r}{k^2} \int d^3v_r$$

$$\left[\left| \frac{\partial \varepsilon_k}{\partial(-i\omega_k^\sigma)} \right|^{-1} (g_{k, \omega_{k^\sigma}}^+ \cdot k') s^+ \left. k'', \omega_k^\sigma - k' \cdot v_s \mid k, \omega_k^\sigma \right) \right.$$

$$\left. - \frac{2\gamma_{k'}^{\sigma'}}{\varepsilon(k, -i\omega_k^\sigma + 2\gamma_{k'}^{\sigma'})} (g_{k, \omega_{k^\sigma} + 2i\gamma_{k'}^{\sigma'}}^+ \cdot k') s^+(k'', \omega_{k, -k'}^{\sigma\sigma'} \mid k, \omega_k^\sigma) F_r \right]$$

$$- i \left| \frac{\partial \varepsilon_k}{\partial(-i\omega_k^\sigma)} \right|^{-1} \int d^3k' \sum_{\sigma'} \frac{2}{\pi} \sum_s \frac{q_s^2 n_s}{k^2 k'^2} \int_{S^{\sigma'}} d^3v_s$$

$$(g_{k, \omega_{k^\sigma}, s}^+ \quad k'') \left[\frac{v(k'', \omega_k^\sigma - k' \cdot v_s \mid k, \omega_k^\sigma)}{\varepsilon^-(k'', -i\omega_k^\sigma + ik' \cdot v_s)} \frac{F_s}{\varepsilon^*(k', ik' \cdot v_s)} \right.$$

$$\left. + \frac{k^2}{k''^2} \frac{v(k, \omega_k^\sigma \mid k', k' \cdot v_s)}{\varepsilon^-(k'', -i\omega_k^\sigma + ik' \cdot v_s)} \frac{F_s}{\varepsilon(k', -ik' \cdot v_s)} \right]$$

$$+ \left| \frac{\partial \varepsilon_k}{\partial(-i\omega_k^\sigma)} \right|^{-1} \int d^3k' \sum_{\sigma'} \frac{2}{\pi} \sum_s \frac{q_s^2 n_s}{k^2 k'^2} \left(\frac{q}{m} \right)_s \int_{S^{\sigma'}} d^3v_s$$

$$\left[k \cdot k' (3k^2 - k \cdot k') \frac{1}{(\lambda + ik \cdot v_s - i\omega_k^\sigma)^4} \frac{F_s}{\varepsilon(k', -ik' \cdot v_s)} \right.$$

$$\left. + (k \cdot k')^2 \frac{1}{(\lambda + ik \cdot v_s - i\omega_k^\sigma)^4} \frac{F_s}{\varepsilon(k', ((k-k') \cdot v_s + i\omega_{k'}^{\sigma'})} \right]. \qquad \ldots [8.26]$$

Higher-order contributions to the emission rate, which arise from elimination of $I_{k, 1}^\sigma$, are included in the results given elsewhere as is the specific origin of the different terms. (See Ref. 11, App. E.)

The interpretation of the previous equation is as follows:

The first two terms, which are the only ones considered in quasi-linear theory[8] and in the kinetic theory of weakly unstable plasma,[9] describe the scattering of the waves by particles (non-linear Landau damping). In contrast to the well-known cascade process of energy transfer toward the small eddies of conventional hydrodynamic turbulence,[25] in the case of the electron- and the ion-wave instabilities, the energy flux resulting from the interaction of the modes is directed toward small wave numbers. The total amount of energy in the waves decreases. Further, the interaction is usually *not* local (in *k*-space).

The remaining contributions to equation [8.26] describe the collisional damping of the waves. In the dipole approximation, this process is dominated by the scattering of electrons by ions. We shall see that the two last terms of equation [8.26] are negligible in that case.

For strongly stable modes *k'*, it is easy to show, with the help of equation [6.10], that the rate of nonlinear Landau damping is instantaneously balanced by a piece of the third and fourth terms of equation [8.26]. For weakly stable or unstable modes, on the other hand, the same piece removes divergences similar to those of the B.G.L. equation.

We shall also indicate that in the ion-wave problem only the first two terms of equation [8.26] play a significant role. Indeed, the Cerenkov emission rate (see equations [6.9] and [6.12] is very large, and the amplitude of the collective plasma oscillations grows to a high level, well above the thermal value. Therefore, the time scales of the equation are fixed by the Landau decrement and/or the scattering processes of the waves by particles. These time scales are much shorter than $\varepsilon^{-1}\omega_k^{-1}$, and thus the last four terms of equation [8.26] are negligible (always of order $\varepsilon^{-1}\omega_k^{-1}$) (discreteness effects).

The evolution of the spectrum of electron waves, on the other hand, generally requires consideration of the complete expression equation [8.26]. In this case, Cerenkov emission is usually weak, and the collective plasma oscillations generally do not reach so high a level. Further, the domain where they are eventually above their thermal value is usually small. Therefore, the processes of decay via particle-particle scattering usually compete to determine the time scale of the kinetic equation.

We now compute the rate of collisional damping of the waves in an equilibrium plasma. Since only plasmons are long-lived, we shall use the dipole approximation and neglect the effect of electron-electron scattering. We shall thus use the inequality

$$\omega_k^\sigma \simeq \omega_{pe} \gg kv_{the} \quad (k\lambda_D \ll 1), \qquad \qquad \ldots [8.27]$$

where ω_k^σ is the frequency of a typical plasmon, and v_{the} the electron thermal velocity.

We integrate the operators $g_k^+, \omega_k^\sigma, r$ by parts in the fourth term of equation [8.26] and obtain:

$$\lim_{k \to 0} \Gamma_k^\sigma \cong \left[\frac{\partial \varepsilon_k}{\partial(-i\omega_k^\sigma)}\right]^{-1} \int_{\Omega_1} d^3k' \sum_{\sigma'} \frac{2}{\pi} \frac{q_i^2 n_i}{k'^4} \int_{S^{\sigma'}} d^3v_i \frac{F_i}{|\varepsilon(k', ik' \cdot v_i)|^2}$$

$$v(k'', \omega_k^\sigma - k' \cdot v_i \,|\, k, \omega_k^\sigma) \left[\frac{v(k, \omega_k^\sigma \,|\, k', k' \cdot v_i)}{\varepsilon^-(k'', -i\omega_k^\sigma + ik' \cdot v_i)} - \frac{k''^2}{k^2}\left(\frac{q}{m}\right)_e \frac{k \cdot k'}{\omega_k^{\sigma 2}}\right].$$

$$...[8.28]$$

Note that, in the small mass ratio limit, only the electrons contribute to the matrix elements "v". The evaluation goes as follows:

$$v(k, \omega_k^\sigma \,|\, k', k' \cdot v) = \sum_r 4\pi \frac{q_r n_r}{k^2} \int d^3v_r [(g_{k, \omega_k^\sigma, r}^+ \cdot k'')(g_{k', k' \cdot v, r}^+ \cdot k')$$

$$+ (g_{k, \omega_k^\sigma, r}^+ \cdot k')(g_{k'', \omega_k^\sigma - k' \cdot v, r}^+ \cdot k'')] \cdot F_r$$

$$\simeq \left(\frac{q}{m}\right)_e 4\pi \frac{q_e n_e}{k^2} i \frac{k \cdot k''}{\omega_k^{\sigma 2}} [\int d^3v_e (g_{k', k' \cdot v, e}^+ \cdot k') F_e$$

$$- \int d^3v_e (g_{k'', \omega_k^\sigma - k' \cdot v, e}^+ \cdot k'') F_e]$$

$$= \left(\frac{q}{m}\right)_e \frac{k \cdot k'}{\omega_k^{\sigma 2}} \cdot \frac{k'^2}{k^2} \{D(-k', -i\omega_k^\sigma + ik' \cdot v)$$

$$- D(k', -ik' \cdot v)], \qquad ...[8.29]$$

where D is obtained from the dielectric function ε (equation [4.4]) merely by dropping the ion contribution. We have also:

$$v(k'', \omega_k^\sigma - k' \cdot v \,|\, k, \omega_k^\sigma) = \sum 4\pi \frac{q_r n_r}{k''^2} \int d^3v_r [(g_{k'', \omega_k^\sigma - k' \cdot v, r}^+ \cdot (-k'))(g_{k, \omega_k^\sigma, r}^+ \cdot k)$$

$$+ (g_{k'', \omega_k^\sigma - k' \cdot v}^+ \cdot k)(g_{-k', -k' \cdot v}^+ \cdot (-k'))] F_r.$$

The very first operator, applied to $g_{k, \omega_k^\sigma, r}^+$, yields higher order terms (in powers of k). Thus, in this term, we can invert the order of the scalar products $-k' \cdot \frac{\partial}{\partial v_r}$ and $k \cdot \frac{\partial}{\partial v_r}$. We add $(g_{k, \omega_k^\sigma, r}^+ \cdot (-k'))$ and $(g_{-k', -k' \cdot v, r}^+ \cdot (-k'))$ integrate the operator $g_{k'', \omega_k^\sigma - k' \cdot v, r}^+$ by parts and obtain, after splitting the resulting integrand into partial fractions,

$$v(k'', \omega_k^\sigma - k' \cdot v \,|\, k, \omega_k^\sigma)$$

$$\simeq \left(\frac{q}{m}\right)_e 4\pi \frac{q_e n_e}{k'^2} (-i) \frac{k \cdot k'}{\omega_k^{\sigma 2}} \int d^3v [(g_{-k', \omega_{k-k' \cdot v}, e}^+ \cdot (-k'))$$

$$- (g_{-k', -k' \cdot v, e}^+ \cdot (-k'))] F_e$$

$$= \left(\frac{q}{m}\right)_e \frac{k \cdot k'}{\omega_k^{\sigma 2}} [D(-k', -i\omega_k^\sigma + ik' \cdot v) - D(-k', ik' \cdot v)]. \quad ...[8.30]$$

Note that the only approximation made to arrive at equations [8.29] and [8.30] is $k \ll k'$.

These results are introduced in equation [8.28]. With the help of the equality

$$\int d^3v_i \frac{F_i(v_i)}{|\varepsilon(k, -ik \cdot v_i)|^2} = \frac{(k^2\lambda_D^2)^2}{(1+k^2\lambda_D^2)(2+k^2\lambda_D^2)} \qquad \ldots[8.31]$$

we obtain

$$\lim_{k \to 0} \Gamma_k^\sigma \cong \mathrm{Re}\left\{ \left|\frac{\partial \varepsilon_k}{\partial(-i\omega_k^\sigma)}\right|^{-1} \frac{2}{\pi} q_i^2 n_i \left(\frac{q}{m}\right)_e^2 \frac{1}{\omega_k^{\sigma 4}} \int d^3k' \frac{k \cdot k'}{k^2k'^2} \frac{1+k'^2\lambda_D^2}{2+k'^2\lambda_D^2} \right.$$
$$\left. [D^{-1}(-k', \omega_k^\sigma) - D^{-1}(k', 0)] \right\}, \qquad \ldots[8.32]$$

a formula which agrees with the result of Oberman, Ron, and Dawson.[17]

8.5. Description of the non-linear mechanisms of emission

The third term of equation [8.16] represents the different non-linear processes of emission of plasma waves, which we now consider. To render the calculation systematic, we shall merely refer the reader to the diagrams given elsewhere (Ref. 11, App. G). The different combinations, there numbered from I_1 to V_2, contribute to the kinetic equation. The others merely provide contour deformations; we shall not come back to that question. Four particular combinations (indicated by dotted lines) require special treatment, but we do not consider them here (see Ref. 11, App. G).

As in the previous section, we shall give only a few examples of calculation.

(i)

$$\lim_{V \to \infty} V^{-1}(2\pi)^{-3} \left(\frac{d}{dt} - \gamma_k^\sigma\right) \langle \delta\Psi_{k, 11}^{\sigma*} \, \delta\Psi_{k, 11}^\sigma \rangle.$$

With the help of equation [8.18] we obtain:

$$\lim_{V \to \infty} V^{-1}(2\pi)^{-3} \left(\frac{d}{dt} - \gamma_k^\sigma\right) \int \frac{d^3k'}{(2\pi)^3} \sum_{\sigma_{k'}\sigma_{k''}} \int \frac{d^3l}{(2\pi)^3} \sum_{\sigma_{i'}\sigma_{i''}} \varepsilon^{-1}(k, -i\omega_{k'k''}^{\sigma_{k'}\sigma_{k''}})$$
$$\varepsilon^{-1*}(k, i\omega_{i'', i''}^{\sigma_{i'}\sigma_{i''}}) V(k, \omega_{k'k''}^{\sigma_{k'}\sigma_{k''}} | k'\omega_{k'}^{\sigma_k}) V^*(k, \omega_{\Psi, \Psi}^{\sigma_{i'} \sigma_{i''*}} | l', \omega_{i'}^{\sigma_{i'}*})$$
$$\langle \delta\Psi_{k'', 1}^{\sigma_{k''}} \delta\Psi_{k', 1}^{\sigma_{k'}} \delta\Psi_{l'', 1}^{*\sigma_{i''}} \delta\Psi_{l', 1}^{*\sigma_{i'}} \rangle \rho_{\underset{i'i''}{\omega\sigma'\sigma''}}^\sigma \rho_{\underset{k'k''}{\omega\sigma'\sigma''}}^\sigma,$$

where

$$\rho_{\underset{k'k''}{\omega\sigma_{k'}\sigma_{k''}}}^\sigma = \begin{bmatrix} 1 & \text{if} & \text{sign } \omega_{k'k''}^{\sigma_{k'}\sigma_{k''}} = \text{sign } \omega_k^\sigma \\ & & \\ 0 & \text{if} & \text{sign } \omega_{k'k''}^{\sigma_{k'}\sigma_{k''}} = -\text{sign } \omega_k^\sigma \end{bmatrix} \quad k'' = k - k'.$$
$$\ldots[8.33]$$

The statistical average selects $l = k'$ and $l = k - k'$. Grouping both contributions yields:

$$\int d^3k' \sum_{\sigma'\sigma''} \frac{\gamma_{k'}^{\sigma'} + \gamma_{k''}^{\sigma''} - \gamma_k^\sigma}{|\varepsilon^-(k\ -i\omega_{k'k''}^{\sigma'\sigma''})|^2}\ v(k, \omega_{k'k''}^{\sigma'\sigma''} \,|\, k', \omega_{k'}^{\sigma'})\, V^*(k, \omega_{k',k''}^{\sigma'\sigma''*} \,|\, k', \omega_{k'}^{\sigma'*})$$

$$I_{k',1}^{\sigma'}\, I_{k'',1}^{\sigma''}\, \rho_{\omega_{k'k''}^{\sigma'\sigma''}}^\sigma,$$

(The deformation of the contour would be easily justified, again, by retaining the complete p-dependence as before.) It is possible to write the above expression in a symmetric form; indeed, replacing k' by k'' in the integrand, and taking half of the sum, we obtain:

$$\tfrac{1}{2} \int d^3k' \sum_{\sigma'\sigma''} \frac{\gamma_{k'}^{\sigma'} + \gamma_{k''}^{\sigma''} - \gamma_k^\sigma}{|\varepsilon^-(k, -i\omega_{k'k''}^{\sigma'\sigma''})|^2}\ |\, v(k, \omega_{k'k''}^{\sigma'\sigma''} \,|\, k', \omega_{k'}^{\sigma'})\,|^2\, I_{k',1}^{\sigma'}\, I_{k'',1}^{\sigma''}\, \rho_{\omega_{k'k''}^{\sigma'\sigma''}}^\sigma,$$

$$...[8.34]$$

(ii)

$$\lim_{V \to \infty} V^{-1} (2\pi)^{-3} \left(\frac{d}{dt} - \gamma_k^\sigma \right) \langle (\delta\Psi_{k,12} + \delta\Psi_{k,21}) \delta\Psi_{k,13} \rangle.$$

Here we use the technique of the Laplace transform to arrive at the result. The reason is that equation (E-4a) of Ref. 11 presents some new behaviour (double pole), and great care must be exercised. We have:

$$\lim_{V \to \infty} V^{-1}(2\pi)^{-3} \int \frac{d^3k'}{(2\pi)^3} \sum_{\sigma_{k''}} \int \frac{d^3l}{(2\pi)^3} \sum_{\sigma_{l''}} \sum_s 4\pi \frac{q_s n_s}{k'^2} \int d^3v_s\, \varepsilon^{-1*}(k', ik'\cdot v_s)$$

$$\frac{v''(k, \omega_{k''}^* + k'\cdot v_s \,|\, k', k'\cdot v_s)}{\varepsilon^*(k, i\omega_{k''}^{\sigma_{k''}*} + ik'\cdot v_s)}\, i \sum_t 4\pi \frac{q_t n_t}{k^2} \int d^3v_t \left(\frac{q}{m}\right)_t ik\cdot l''$$

$$\frac{1}{(p+ik\cdot v_t - ik'\cdot v_s - i\omega_{k''}^{\sigma''})^2} \frac{1}{(p+i\omega_{l''}^{\sigma i''} + il\cdot v_k - i\omega_{k''}^{*\sigma_{k''}} - ik'\cdot v_s)}$$

$$\frac{p-\gamma_k}{\varepsilon(k, p - i\omega_{k''}^{\sigma_{k''}*} - ik'\cdot v_s)}\, \langle \delta\Psi_{k'',1}^*(0)\, \delta N_{s,k'}^*(0)\, \delta\Psi_{l''}(0)\, \delta N_{t,l}(0) \rangle$$

$$\rho_{\omega_{k''}^\sigma + k'\cdot vs}^\sigma\, \rho_{\omega_{l''}^\sigma + l\cdot v_k}^\sigma$$

The statistical ensemble selects $l = k'$ and $v_s = v_t$. The non-oscillating pole is at $p = 2\gamma^\sigma \frac{k-k'}{k-k'}$. (The oscillating poles provide contour deformations.) Thus we obtain after integration of $(ik''\cdot v_t - i\omega_{k''}^{\sigma_{k''}})^{-2}$ by parts:

$$-i \int d^3k' \sum_{\sigma_{k'}} I_{k',1}^{\sigma_{k'}}(t)\, \frac{2}{\pi} \sum_s \frac{q_s^2 n_s}{k^2 k''^2} \int d^3v_s (g_{k',\omega_{k',\sigma'},s}^+ \cdot k)$$

$$\left[\frac{v^*(k, \omega_{k'}^{*\sigma_{k'}} + k''\cdot v_s \,|\, k'', k''\cdot v_s)}{\varepsilon^{-*}(k'',|\, k''\cdot v_s)} \frac{(\gamma_{k'}^{\sigma'} - \gamma_k^\sigma)\, F_s}{|\varepsilon^-(k, -i\omega_{k'}^{\sigma'} - ik'\cdot v_s)|^2}\, \rho_{\omega_{k',\sigma'} + k''\cdot vs}^\sigma \right].$$

$$...[8.35]$$

We eliminate $I_{k',1}^{\sigma_{k'}}$ and $I_{k'',1}^{\sigma_{k''}}$ in equations [8.34] and [8.35] and similar expressions. The rate of emission of plasma waves due to non-linear processes, is given below (here we drop the symbols ρ^σ for simplicity).

$$B_k^\sigma(t)$$

$$= \frac{1}{2} \int d^3k' \sum_{\sigma'\sigma''} I_{k'}^{\sigma'} I_{k''}^{\sigma''} \frac{(\gamma_{k'}^{\sigma'}+\gamma_{k''}^{\sigma''}-\gamma_k^\sigma)}{|\varepsilon^-(k,-i\omega_{k'k''}^{\sigma'\sigma''})|^2} |v(k,\omega_{k'k''}^{\sigma'\sigma''}|k',\omega_{k'}^{\sigma'})|^2$$

$$+ \int d^3k' \sum_{\sigma'} I_{k'}^{\sigma'} \frac{2}{\pi} \sum_\sigma \frac{q_s^2 n_s}{k''^4} \int d^3v_s \frac{F_s}{|\varepsilon^-(k'',-ik''\cdot v_s)|^2}$$

$$\left[\frac{(\gamma_{k'}^{\sigma'}-\gamma_k^\sigma)}{|\varepsilon^-(k,-i\omega_{k'}^{\sigma'}-ik''\cdot v_s)|^2} |v(k,\omega_{k'}^{\sigma'}+k''\cdot v_s|k',\omega_{k'}^{\sigma'})|^2 \right.$$

$$\left. - \frac{(\gamma_{k'}^{\sigma'}+\gamma_{k''}^{\sigma''}-\gamma_k^\sigma)}{|\varepsilon^-(k,-i\omega_{k'k''}^{\sigma'\sigma''})|^2} |v(k,\omega_{k'k''}^{\sigma'\sigma''}|k',\omega_{k'}^{\sigma'})|^2 \right]$$

$$+ \int d^3k' \frac{2}{\pi} \sum_s \frac{q_s^2 n_s}{k''^4} \int d^3v_s \frac{F_s}{|\varepsilon^-(k'',-ik''\cdot v_s)|^2} \frac{2}{\pi} \sum_t \frac{q_t^2 n_t}{k'^4}$$

$$\int d^3v_t \frac{F_t}{|\varepsilon^-(k',-ik'\cdot v_t)|^2} \frac{1}{2} \left[\frac{-\gamma_k^\sigma}{|\varepsilon^-(k,-ik'\cdot v_t-ik''\cdot v_s)|^2} \right.$$

$$|v(k,k'\cdot v_t+k''\cdot v_s|k',k'\cdot v_t)|^2 - \frac{\gamma_{k'}^{\sigma'}-\gamma_k^\sigma}{|\varepsilon^-(k,-i\omega_{k'}^{\sigma'}-ik''\cdot v_s)|^2}$$

$$|v(k,\omega_{k'}^{\sigma'}+k''\cdot v_s|k'\omega_{k'}^{\sigma'})|^2 - \frac{\gamma_{k''}^{\sigma''}}{|\varepsilon^-(k,-ik'\cdot v_t-i\omega_{k''}^{\sigma''})|^2}$$

$$\left. |v(k,k'\cdot v_t+\omega_{k''}^{\sigma''}|k',k'\cdot v_t)|^2 + \frac{\gamma_{k'}^{\sigma'}+\gamma_{k''}^{\sigma''}-\gamma_k^\sigma}{|\varepsilon^-(k,-i\omega_{k'k''}^{\sigma'\sigma''})|^2} |v(k,\omega_{k'k''}^{\sigma'\sigma''}|k',\omega_{k'}^{\sigma'})|^2 \right]$$

$$- i \int d^3k' \sum_{\sigma'} I_{k'}^{\sigma'} \frac{2}{\pi} \sum_s \frac{q_s^2 n_s}{k^2 k''^2} \int d^3v_s 2(g_{k',\omega_{k'\sigma'},s}^+ \cdot k)$$

$$\left[\frac{(\gamma_{k'}^{\sigma'}-\gamma_k^\sigma)F_s}{|\varepsilon^-(k,-i\omega_{k'}^{\sigma'}-ik''\cdot v_s)|^2} \cdot \frac{v^*(k,\omega_{k'}^{\sigma'*}+k''\cdot v_s|k'',k''\cdot v_s)}{\varepsilon^{-*}(k'',ik''\cdot v_s)} \right]$$

$$- i \int d^3k' \frac{2}{\pi} \sum_s \frac{q_s^2 n_s}{k^2 k''^2} \int d^3v_s \frac{2}{\pi} \sum_t \frac{q_t^2 n_t}{k'^4} \int d^3v_t \frac{F_t}{|\varepsilon^-(k',-ik'\cdot v_t)|^2}$$

$$2\left\{ (g_{k',k'\cdot v_t,s}^+ \cdot k) \left[\frac{-\gamma_k^\sigma F_s}{|\varepsilon^-(k,-ik'\cdot v_t-ik''\cdot v_s)|^2} \frac{v^*(k,k'\cdot v_t+k''\cdot v_s|k',k'\cdot v_t)}{\varepsilon^{-*}(k'',ik''\cdot v_s)} \right] \right.$$

$$\left. -(g_{k',\omega_{k'\sigma'},s}^+ \cdot k) \left[\frac{(\gamma_{k'}^{\sigma'}-\gamma_k^\sigma)F_s}{|\varepsilon^-(k,-i\omega_{k'}^{\sigma'}-ik''\cdot v_s)|^2} \frac{v^*(k,\omega_{k'}^{\sigma'*}+k'\cdot v_s|k'',k''\cdot v_s)}{\varepsilon^{-*}(k'',+ik''\cdot v_s)} \right] \right\}$$

$$+ \int d^3k' \sum_{\sigma'} I_{k'}^{\sigma'} \frac{2}{\pi} \sum_s \frac{q_s^2 n_s}{k^4} \left(\frac{q}{m}\right)_s \oint d^3v_s \frac{F_s}{|ik'\cdot v_s-i\omega_{k'}^{\sigma'}|^4} (k\cdot k')^2$$

$$\frac{\gamma_{k'}^{\sigma'} - \gamma_k^{\sigma}}{|\varepsilon^-(k, -i\omega_{k'}^{\sigma'} - ik'' \cdot v_s)|^2} + \int d^3k' \frac{2}{\pi} \sum_s \frac{q_t^2 n_t}{k'^4}$$

$$\int d^3v_t \frac{F_t}{|\varepsilon^-(k', -ik' \cdot v_t)|^2} \frac{2}{\pi} \sum_s \frac{q_s^2 n_s}{k^4} \left(\frac{q}{m}\right)_s \oint d^3v_s (k \cdot k')^2$$

$$\left[\frac{F_s}{|ik'' \cdot v_s - ik' \cdot v_t|^4} \frac{-\gamma_k^{\sigma}}{|\varepsilon^-(k, -ik' \cdot v_t - ik'' \cdot v_s)|^2} \right.$$

$$\left. - \frac{F_s}{|ik'' \cdot v_s - i\omega_{k'}^{\sigma'}|^4} \frac{\gamma_{k'}^{\sigma'} - \gamma_k^{\sigma}}{|\varepsilon^-(k, -i\omega_{k'}^{\sigma'} - ik'' \cdot v_s)|^2} \right]$$

$$+ \int d^3k' \frac{2}{\pi} \sum_t \frac{q_t^2 n_t}{k^2 k'^2} \left(\frac{q}{m}\right)_t \int d^3v_t \frac{2}{\pi} \sum_s \frac{q_s^2 n_s}{k^2 k''^2} \left(\frac{q}{m}\right)_s$$

$$\int^1 d^3v_s \, k \cdot k' \, k \cdot k'' \frac{F_t}{\varepsilon^-(k', -ik' \cdot v_t)} \frac{F_s}{\varepsilon^{-*}(k'', ^1 - ik'' \cdot v_s)}$$

$$\frac{1}{(\lambda + ik' \cdot (v_s - v_t))^2} \frac{1}{(\lambda + ik'' \cdot (v_s - v_t))^2} \frac{-\gamma_k^{\sigma}}{|\varepsilon^-(k, -ik' \cdot v_t - ik'' \cdot v_s)|^2}.$$

$$...[8.36]$$

For strongly damped modes, k' or/and k'', many simplifications occur, as usual. In particular, if both k' and $k'' \in \Omega_1$, (see equation [7.6b]), then among the three first contributions to equation [8.36], only the first piece of the third one survives. On the other hand, for weakly stable or unstable modes, the last pieces of the second and third terms remove divergences similar to those of the B.G.L. equation. This interconnection of the different contributions to equation [8.36] indicates that they are different aspects of the same process: emission by nonlinear interactions.

More precisely, there are, besides Cerenkov emission, three basic mechanisms of radiation of plasma waves: via wave-wave scattering, wave-particle scattering, and particle-particle scattering.

We could repeat the same arguments as were used previously to show that a complete analysis of the evolution of the spectrum of plasmons requires in general consideration of all radiation processes. In the limit of small damping rates these preserve total energy of the waves and particles participating in the collision. Indeed, the emission is restricted by the presence of δ-functions to those interacting waves and particles which satisfy the following conditions:

$$\delta(\omega_k^{\sigma} - \omega_{k'}^{\sigma'} - \omega_{k''}^{\sigma''}) \quad \delta(\omega_k^{\sigma} - \omega_{k'}^{\sigma'} - k'' \cdot v_s) \quad \delta(\omega_k^{\sigma} - k' \cdot v_t - k'' \cdot v_s),$$

$$...[8.37a]$$

which admit an explicit quantum-mechanical interpretation:

$$\hbar\omega_k^{\sigma} = \hbar\omega_{k'}^{\sigma'} + \hbar\omega_{k''}^{\sigma''} \quad \hbar\omega_k^{\sigma} = \hbar\omega_{k'}^{\sigma'} + \Delta p_s \cdot v_s \quad \hbar\omega_k^{\sigma} = \Delta p_t \cdot v_t + \Delta p_s \cdot v_s.$$

$$...[8.37b]$$

Here $\hbar\omega_k^\sigma$ is the energy transferred to the mode (k, ω_k^σ); $\Delta p_s = \hbar k''$ and $\Delta p_t = \hbar k'$ represent the momentum transferred by the colliding particles.

The first term of equation [8.36] is well defined in the limit $\gamma_k^\sigma, \gamma_{k'}^{\sigma'}, \gamma_{k''}^{\sigma''} \to 0$, provided

$$\frac{\partial}{\partial k'} \omega_{k'k''}^{\sigma'\sigma''} \neq 0, \quad \text{whenever} \quad \omega_{k'k''}^{\sigma'\sigma''} = \omega_k^\sigma. \qquad ...[8.38]$$

Other possibilities will be rejected here, although this difficulty occurs in the treatment of the ion-wave instability, when the modes $k \to 0$ are excited.

We now compute the rate emission of plasmons by longitudinal bremsstrahlung in the dipole approximation for a plasma in thermal equilibrium. All the considerations below equation [8.27] still apply (one can easily show that electron-electron scattering does not contribute to the emission in this limit), and equation [8.36] can be approximated as follows:

$$B_k^\sigma(t) = \int_{\Omega_1} d^3k' \frac{2}{\pi} \frac{q_e^2 n_e}{k''^4} \int d^3v_e \frac{F_e}{|\varepsilon^-(k'', -ik'' \cdot v_e)|^2} \frac{2}{\pi} \frac{q_i^2 n_i}{k'^4}$$

$$\int d^3v_i \frac{F_i}{|\varepsilon^-(k', -ik' \cdot v_i)|^2} |v(k, \omega_k^\sigma | k', k' \cdot v_i)|^2$$

$$\pi\delta(\omega_k^\sigma - k'' \cdot v_e - k' \cdot v_i) \left|\frac{\partial \varepsilon_k}{\partial \omega_k^\sigma}\right|^{-2}$$

$$- i \int_{\Omega_1} d^3k' \frac{2}{\pi} \frac{q_e^2 n_e}{k^2 k''^2} \int d^3v_e \frac{2}{\pi} \frac{q_i^2 n_i}{k'^4} \int d^3v_i \frac{2F_i (g_{k', k' \cdot v_i, e}^+ \cdot k)}{|\varepsilon^-(k', -ik' \cdot v_i)|^2}$$

$$\left[F_e \frac{v^*(k, \omega_k^\sigma | k', k' \cdot v_i)}{\varepsilon^{-*}(k'', ik'' \cdot v_e)} \pi\delta(\omega_k^\sigma - k'' \cdot v_e - k' \cdot v_i)\right] \left|\frac{\partial \varepsilon_k}{\partial \omega_k^\sigma}\right|^{-2}$$

$$+ \int_{\Omega_1} d^3k' \frac{2}{\pi} \frac{q_e^2 n_e}{k^4} \left(\frac{q}{m}\right)_e \oint d^3v_e \frac{2}{\pi} \frac{q_i^2 n_i}{k'^4} \int d^3v_i \frac{F_i}{|\varepsilon^-(k', -ik' \cdot v_i)|^2} (k \cdot k')^2$$

$$\left[F_e \frac{1}{|ik'' \cdot v_e - ik' \cdot v_i|^4} \pi\delta(\omega_k^\sigma - k'' \cdot v_e - k' \cdot v_i)\right] \left|\frac{\partial \varepsilon_k}{\partial \omega_k^\sigma}\right|^{-2}.$$

$$...[8.39]$$

The extra factor "2" in the first term arises because one can take either of the combinations ($s = e, t = i$ or $s = i, t = e$) in equation [8.36]. With the help of equation [8.29] we obtain, after passing to the limit $k \to 0$,

$$\frac{2}{\pi} \frac{q_e^2 n_e}{\omega_k^{\sigma 4}} \left(\frac{q}{m}\right)_e^2 \frac{2}{\pi} \frac{q_i^2 n_i}{k^2} \left|\frac{\partial \varepsilon_k}{\partial \omega_k^\sigma}\right|^{-2} \int d^3k' \left(\frac{k}{k} \frac{k'}{k'}\right)^2 \frac{1}{k'^2} \int d^3v_i \frac{F_i}{|\varepsilon^-(k', -ik' \cdot v_i)|^2}$$

$$\int d^3v_e \pi\delta(\omega_k^\sigma + k' \cdot v_e) F_e$$

$$\left[\left(1 - 2\,\text{Re}\,\frac{D(k', 0)}{D(k', -i\omega_k^\sigma)} + \left|\frac{D(k', 0)}{D(k', -i\omega_k^\sigma)}\right|^2\right) - 2\left(1 - \text{Re}\,\frac{D(k', 0)}{D(k', -i\omega_k^\sigma)}\right) + 1\right].$$

$$...[8.40]$$

Each term inside the bracket corresponds to one of the terms of equation [8.39]. Using the results equation [8.31] we finally have for a Maxwellian plasma:

$$\lim_{k \to 0} B_k^\sigma(t) = \left| \frac{\partial \varepsilon_k}{\partial \omega_k^\sigma} \right|^{-2} \frac{2}{\pi} q_i^2 n_i \left(\frac{q}{m} \right)_e^2 \frac{1}{2\pi^2} \frac{1}{\omega_k^{\sigma 5}} \frac{1}{k^2} \theta_e$$
$$\int_{\Omega_1} d^3 k' \left(\frac{\mathbf{k} \cdot \mathbf{k}'}{kk'} \right)^2 \frac{1+k'^2 \lambda_D^2}{2+k'^2 \lambda_D^2} \frac{\text{Im } D(k', -i\omega_k^\sigma)}{|D(k', -i\omega_k^\sigma)|^2}. \quad ...[8.41]$$

(This result is in agreement with that given by Birmingham, Dawson, and Oberman.[26]) This result can be compared with equation [8.32]. For small k, Cerenkov emission and Landau damping decrease exponentially fast. For wave numbers k such that

$$k\lambda_D \lesssim (\ln \varepsilon)^{-\frac{1}{2}} \quad\quad ...[8.42]$$

the equilibrium spectrum is determined by a balance between collisional damping of the waves and bremsstrahlung emission. We obtain

$$\lim_{k \to 0} \frac{1}{8\pi} \sum_\sigma \omega_k^\sigma \frac{\partial \varepsilon_k}{\partial \omega_k^\sigma} k^2 I_k^\sigma = \frac{1}{(2\pi)^3} \theta_e \quad\quad ...[8.43]$$

which is the Rayleigh-Jeans distribution.

8.6. *Synchronization of the rate of emission by Cerenkov radiation*

In the lowest-order theory, discussed before, it was justified to "synchronize" the Cerenkov emission rate as obtained via the Landau *initial value* problem. As we go to higher order in the expansion, however, there is *a priori* no reason to assume that the system cannot develop non-Markoffian behaviour. We shall not discuss here this important question in detail and *prove* that the plasma dynamics indeed synchronizes the emission rate, provided one excludes strong turbulence, but refer the reader to §12 of Ref. 11.

8.7. *Effects of mode-coupling on the evolution of the one-particle distributions*

The reasons which justify the inclusion in the wave kinetic equation of higher order processes in the discreteness parameter do not apply when the evolution of the one-particle distribution functions is considered. Collective plasma motions, however, can play a significant role in the evolution of a weakly unstable plasma, as the amplitude of the oscillations can reach a very high level, well above the thermal value (maximal ordering requires $\lambda^2 \sim \varepsilon^{\frac{1}{2}}$). Therefore, to iterate the Vlasov equation, we take simply $\delta\Psi_k \equiv \delta\Psi_{k,1}$ and $\delta N_k = \delta N_{k,1}$. We shall not carry through the algebra, which in the present case is much simplified. We shall merely, and for the sake of completeness, state the final results, and indicate briefly how they have been obtained. We have:

$$\frac{\partial F_r}{\partial t} = \left(\frac{q}{m}\right)_r \int d^3k \sum_\sigma k \cdot \frac{\partial}{\partial v_r} (g^+_{k,\,\omega_k,\,r} \cdot k) F_r I^\sigma_k(t)$$

$$- \left(\frac{q}{m}\right)_r \int d^3k \sum_\sigma k \cdot \frac{\partial}{\partial v_r} \frac{1}{(-i\omega^\sigma_k + ik \cdot v_r)} I^\sigma_k(t) \frac{\partial}{\partial t} (g^+_{k,\,\omega_{k^\sigma},\,r} \cdot k) F_r$$

$$+ \left(\frac{q}{m}\right)_r \int d^3k \sum_\sigma \left(\frac{\partial \varepsilon_k}{\partial \omega_k}\right)^{-1} k \cdot \frac{\partial}{\partial v_r} \left\{\frac{\partial}{\partial \omega_k} (g^+_{k,\,\omega_{k^\sigma},\,r} \cdot k) F_r\right\}$$

$$\int_{\Omega_2} d^3k' \sum_{\sigma'} \left[\frac{v(k'', \omega^{\sigma\sigma'}_{k,\,-k'} \mid k, \omega^\sigma_k)}{\varepsilon^-(k', -i\omega^{\sigma\sigma'}_{k,\,-k'})} v(k, \omega^\sigma_k \mid k', \omega^{\sigma'}_{k'}) - i \sum_s 4\pi \frac{q_s n_s}{k^2}\right.$$

$$\int d^3v_s (g^+_{k,\,\omega_{k^\sigma s}} \cdot k') s^+ (k'', \omega^{\sigma\sigma'}_{k,\,-k'} \mid k, \omega^\sigma_k) F_s \bigg] I^\sigma_k I^{\sigma'}_{k'} - i\left(\frac{q}{m}\right)_r$$

$$\int d^3k \sum_\sigma k \cdot \frac{\partial}{\partial v_r} \int_{\Omega_2} d^3k' \sum_{\sigma'} \left[\frac{v(k'', \omega^{\sigma\sigma'}_{k,\,-k'} \mid k, \omega^\sigma_k)}{\varepsilon^-(k'', -i\omega^{\sigma\sigma'}_{k,\,-k'})} s^+(k, \omega^\sigma_k \mid k', \omega^{\sigma'}_{k'}) F_r\right.$$

$$- i(g^+_{k,\,\omega_{k^\sigma},\,r} \cdot k') s^+ (k'' \omega^{\sigma\sigma'}_{k,\,-k'} \mid k, \omega^\sigma_k) F_r\bigg] I^\sigma_k I^{\sigma'}_{k'}$$

$$- \tfrac{1}{2}\left(\frac{q}{m}\right)_r \int d^3k\, k \cdot \frac{\partial}{\partial v_r} \int d^3k' \sum_{\sigma'\sigma''} \left[\frac{v(k, \omega^{\sigma'\sigma''}_{k'k''} \mid k'\omega^{\sigma'}_k)}{\varepsilon^-(k, -i\omega^{\sigma'\sigma''}_{k'k''})} \{(g^+_{k,\,\omega_{k^\sigma},\,r} \cdot k)\right.$$

$$- (g^+_k, \omega^{\sigma'\sigma''}_{k'k''})\} F_r + is(k, \omega^{\sigma'\sigma''}_{k'k''} \mid k', \omega^{\sigma'}_{k'}) F_r \frac{v^*(k, \omega^{\sigma'\sigma''}_{k'k''} \mid k'\omega^{\sigma'}_{k'})}{\varepsilon^{-*}(k, i\omega^{\sigma'\sigma''}_{k'k''})}\bigg] I^{\sigma'}_{k'} \Gamma^{\sigma''}_{k''}$$

$$+ \left(\frac{q}{m}\right)_r \int d^3k\, k \cdot \frac{\partial}{\partial v_r} \sum_s \frac{2}{\pi} \frac{q^2_s n_s}{k^4} \int d^3v_s \frac{F_s}{\mid \varepsilon^-(k, -ik \cdot v_s)\mid^2}$$

$$[(g^+_{k,\,k\cdot v_s,\,r} \cdot k) - (g^+_{k,\,\omega_{k^\sigma},\,r} \cdot k)] F_r$$

$$+ i\left(\frac{q}{m}\right)_r \int d^3k\, k \cdot \frac{\partial}{\partial v_r} \frac{q^2_r}{2\pi^2 k^2} \frac{F_r}{\varepsilon^-(k, -ik \cdot v_r)}. \qquad \ldots[8.44]$$

To obtain this equation, we have proceeded as follows. Consider the ensemble average

$$\langle \delta N_k \delta\psi_{-k}\rangle = \langle \delta N^{(1)}_k \delta\psi_{-k}\rangle + \langle \delta N^{(2)}_k \delta\psi_{-k}\rangle + \langle \delta N^{(3)}_k \delta\psi_{-k}\rangle,$$

which appears in the right-hand side of equation [3.17]. To obtain the nonlinear terms of equation [8.44] we identify $\delta N^{(1)}_k$ with $\delta N_{k,1} = i(g^+_{k,\omega_k} \cdot k) F_r \delta\psi_{k,1}$. Eliminating $\delta\psi_{k,1}$ in terms of $\delta\psi_k$ yields

$$\langle \delta N_k \delta\psi_{-k}\rangle = i(g^-_{k\omega_k} \cdot k) F_r \{\langle \delta\psi_k \delta\psi_{-k}\rangle - \langle \delta\psi^{(2)}_k \delta\psi_{-k}\rangle - \langle \delta\psi^{(3)}_k \delta\psi_{-k}\rangle\}$$
$$+ \langle \delta N^{(2)}_k \delta\psi_{-k}\rangle + \langle \delta N^{(3)}_k \delta\psi_{-k}\rangle, \qquad \ldots[8.45]$$

where, except for the first term, we are allowed to expand $\delta\psi_{-k}$ in a power series. The procedure outlined here above is similar to the elimination of $\delta\psi_{k,1}$ in equation [8.14].

The first and last two terms of equation [8.44] appear already in the kinetic equation [7.5]. The others, which are negligible in the domain Ω_1 of stability, have been obtained by Frieman and Rutherford in their quasi-linear theory (equations (64) and (68)) via a multiple time scale analysis. As in the previous chapters, we could show that the elimination of $\delta\psi_{k,1}$ in equation [8.45], which amounts to a renormalization, removes the fast time scale secularities which appear via the iterative solution of the Vlasov equations. We can proceed with equation [8.44] as with equation [7.5]. We note that the integrand of the third term is proportional to the non-linear Landau damping of the waves (the two first terms of equation [8.24] and finally obtain:

$$
\begin{aligned}
\frac{\partial F_r}{\partial t} &= \left(\frac{q}{m}\right)_r^2 \int_{\Omega_2} d^3k \sum_\sigma \mathbf{k}\cdot\frac{\partial}{\partial v_r}\left[\pi\delta(\omega_k^\sigma - \mathbf{k}\cdot\mathbf{v}_r)\,\mathbf{k}\cdot\frac{\partial F_r}{\partial v_r}\right] I_k^\sigma(t) \\
&+ \left(\frac{q}{m}\right)_r^2 \int_{\Omega_2} d^3k \sum_\sigma \mathbf{k}\cdot\frac{\partial}{\partial v_r}\left[P\,\frac{\mathbf{k}\cdot\partial F_r/\partial v_r}{(\mathbf{k}\cdot\mathbf{v}-\omega_k^\sigma)^2}\right]\tfrac{1}{2}\frac{dI_k^\sigma}{dt} \\
&+ \left(\frac{q}{m}\right)_r \int_{\Omega_2} d^3k \sum_\sigma \mathbf{k}\cdot\frac{\partial}{\partial v_r}\left[\frac{q_r^2}{2\pi^2 k^2}\,\pi\delta(\omega_k^\sigma - \mathbf{k}\cdot\mathbf{v}_r)F_r(v,t)\right]\left(\frac{\partial\varepsilon_k}{\partial\omega_k}\right)^{-1} \\
&+ \left(\frac{q}{m}\right)_r \int_{\Omega_2} d^3k \sum_\sigma \mathbf{k}\cdot\frac{\partial}{\partial v_r}\frac{2}{\pi}\sum_s \frac{q_s^2 n_s}{k^4}\int_{S^\sigma} d^3 v_s\left[P\,\frac{F_s}{\mid\varepsilon(k,-ik\cdot v_s)\mid^2}\right] \\
&\qquad\qquad [(g_{k,\,k\cdot v_s,\,r}^+ \cdot \mathbf{k}) - (g_{k,\,\omega_k^\sigma,\,r}^+\cdot\mathbf{k})]F_r \\
&- \left(\frac{q}{m}\right)_r \int_{\Omega_2} d^3k\,\mathbf{k}\cdot\frac{\partial}{\partial v_r}\frac{2}{\pi}\sum_s \frac{q_r q_s n_s}{k^4}\int d^3 v_s\left[P\,\frac{F_r}{\mid\varepsilon(k,-ik\cdot v_r)\mid^2}\right] \\
&\qquad\qquad (g_{k,\,k\cdot v_r,\,s}^+\cdot\mathbf{k})F_s \\
&+ \left(\frac{q}{m}\right)_r \int_{\Omega_1} d^3k\,\mathbf{k}\cdot\frac{\partial}{\partial v_r}\frac{2}{\pi}\sum_s \frac{q_s^2 n_s}{k^4}\int d^3 v_s\,\frac{\pi\delta(\mathbf{k}\cdot\mathbf{v}_r - \mathbf{k}\cdot\mathbf{v}_s)}{\mid\varepsilon(k,-ik\cdot v_r)\mid^2} \\
&\qquad\qquad \left[\frac{q_r}{m_r}\mathbf{k}\cdot\frac{\partial}{\partial v_r} - \frac{q_s}{m_s}\mathbf{k}\cdot\frac{\partial}{\partial v_s}\right]F_r F_s \\
&- i\left(\frac{q}{m}\right)_r \int_{\Omega_2} d^3k \sum_\sigma \mathbf{k}\cdot\frac{\partial}{\partial v_r}\int_{\Omega_2} d^3k' \sum_{\sigma'}\left[\frac{v(k'',\omega_{k,-k'}^{\sigma\sigma'}\mid k\omega_k^\sigma)}{\varepsilon^-(k'',-i\omega_{k,-k'}^{\sigma\sigma'})}\right] \\
&\qquad s^+(k,\omega_k^\sigma\mid k',\omega_{k'}^{\sigma'})F_r - i(g_{k,\,\omega_k^\sigma,\,r}^+\cdot\mathbf{k})s^+(k'',\omega_{k,-k'}^{\sigma\sigma'}\mid k,\omega_k^\sigma)F_r]\,I_k^\sigma I_{k'}^{\sigma'} \\
&- \tfrac{1}{2}\left(\frac{q}{m}\right)_r \int_{\Omega_2} d^3k\,\mathbf{k}\cdot\frac{\partial}{\partial v_r}\int_{\Omega_2} d^3k' \sum_{\sigma'\sigma''}\left\{\left[P\,\frac{\mid v(k,\omega\mid k',\omega_{k'}^{\sigma'})\mid^2}{\mid\varepsilon^-(k,-i\omega_{k'k''}^{\sigma'\sigma''})\mid^2}\right]\right. \\
&\qquad\qquad [(g_{k\omega_k}^+\cdot\mathbf{k}) - (g_{k,\,\omega^{\sigma'\sigma''}_{k'k''}}^+\cdot\mathbf{k})]F_r \\
&\qquad\left. + is^+(k,\omega_{k'k''}^{\sigma'\sigma''}\mid k',\omega_{k'}^{\sigma'})F_r\,\frac{v^*(k,\omega_{k'k''}^{\sigma'\sigma''}\mid k',\omega_{k'}^{\sigma'})}{\varepsilon^-(k,i\omega_{k'k''}^{\sigma'\sigma''})}\right\} I_{k'}^{\sigma'} I_{k''}^{\sigma''}.
\end{aligned}
$$

$$\dots[8.46]$$

Equations [8.44] or [8.46] and equation [8.17] are the final equations of the theory. They constitute a closed, statistically complete description of the evolution of a homogeneous plasma, provided a kinetic regime has been reached, which excludes strong turbulence.

9

FORMULATION OF THE PROBLEM OF A TEST PARTICLE
IN A MAGNETIC FIELD

Within the framework of the Klimontovich formalism, we give a simple and original derivation of the kinetic theory of a test-particle in a uniform magnetic field, and homogeneous background plasma. The theory is generalized, as in the previous chapters, to include the effects of both waves and (field) particles. We shall indicate how the Bogoliubov ansatz yields to divergences similar to those of the B.G.L. equation, while our theory automatically provides the crossing from stable to unstable regimes.

As a very important application, we shall compute the coefficient of spatial diffusion across a uniform magnetic field, and in various limits, we recover the previous result of Rostoker,[3] and Drummond and Rosenbluth.[14]

Besides the $s-1$ species of field particles, $r = 2, \ldots, s$, for which the initial information is statistical in nature (smooth initial condition), there is one species, labelled "1", consisting of a single test-particle, whose initial position X_{10} is regarded as given in $[x, v]$ space. At later times, however, and because of incoherent scattering by field particles and waves, one can only assign a probability of presence to the test-particle in $[x, v]$.

Let

$$N_1(X, t) = V\delta(X - X_1(t)) \qquad \ldots[9.1]$$

be the phase-space function of the test-particle. $X \equiv [x, v]$ is a field point in configuration space, $X_1(t)$ describes the trajectory of the test-particle. Note that

$$X_1(t) = X_1(\{X_{i0}\}, X_{10}, t) \qquad ..[9.2]$$

depends explicitly on the initial position and time t. Since X_{10} is the *same* for every realization of the ensemble, we introduce the following function in phase-space:

$$D_2(X_1, \{X_i\}, t; X_{10}, \{X_{i0}\}\, 0) \qquad \ldots[9.3]$$

which is the joint probability density that the system is at $(X_{10}, \{X_{i0}\})$ initially, and at $(X_1, \{X_i\})$ at time t. D_1 satisfies the Liouville equation $[3.7]$ in the variables $X_1, \{X_i\}$ and t, with the singular initial conditions:

$$D_2(X_1, \{X_i\}, 0; X_{10}, \{X_{i0}\}, 0) = D_1(X_1, \{X_i\}, 0)\delta(\{X_i\} - \{X_{i0}\})\delta(X_1 - X_{10}).$$
$$..[9.4]$$

Let $\psi(X; X_1, \{X_i\}; X_{10}, \{X_{i0}\}, 0)$ be any phase-space function. We define the average over the statistical ensemble via:

$$F_1(X_{10})\langle\psi(X, t \mid X_{10})\rangle_{(2)} \equiv V\int d^3X_1 d^3\{X_i\}d\{X_{i0}\}\psi(X; X_1, \{X_i\};$$
$$X_{10}, \{X_{i0}\}, 0)\, D_2(X_1, \{X_i\}, t; X_{10}, \{X_{i0}\}, 0). \ldots[9.5]$$

Let

$$\langle N_1(X, t)\rangle_{(2)} = W_{1\,|\,1}(Xt\,|\,X_{1\,0})\ddagger$$

$$\langle N_r(X, t)\rangle_{(2)} = W_{1\,|\,r}(Xt\,|\,X_{1\,0})$$

$$\langle \delta N_r(X, t)\delta N_{r'}(X', t)\rangle_{(2)} = G_{1\,|\,rr'}(X, X', t\,|\,X_{1\,0}).$$

$$\left.\right\}\quad ...[9.6]$$

It is easy to show, with the help of equations [9.4] and [9.5], that

$$W_{1\,|\,1}(X, 0\,|\,X_{1\,0}) = V\delta(X-X_{1\,0})$$

$$W_{1\,|\,r}(X, 0\,|\,X_{1\,0}) = F_r(X, 0)+P_{1\,|\,r}(X, 0\,|\,X_{1\,0})$$

$$G_{1\,|\,rr'}(X, X', 0\,|\,X_{1\,0}) = P_{rr'}(X, X', 0)+H_{1\,|\,rr'}(X, X', 0\,|\,X_{1\,0})$$
$$-P_{1\,|\,r}(X0\,|\,X_{1\,0})P_{1\,|\,r'}(X'0\,|\,X_{1\,0})$$

$$...[9.7]$$

with $P_{1\,|\,r}F_1(X_{1\,0}) = P_{1,r}$, etc. $W_{1\,|\,1}(Xt\,|\,X_{1\,0})$ is the probability of having particle "1" at X and t, if it is certainly at $X_{1\,0}$ at $t = 0$. The interpretation of the other functions is clear.

The phase function $N_1(X, t)$ obeys the Liouville equation

$$\frac{d}{dt}N_1 \equiv \left(\frac{\partial}{\partial t}+\dot{x}\frac{\partial}{\partial x}+\dot{v}\frac{\partial}{\partial v}\right)N_1 = 0, \qquad ...[9.8]$$

from which we obtain (as previously described):

$$\left[\frac{\partial}{\partial t}+v\cdot\frac{\partial}{\partial x}+\left(\frac{q}{m}\right)_1\frac{v}{c}\times B_0\cdot\frac{\partial}{\partial v}\right]W_{1\,|\,1}(Xt\,|\,X_{1\,0})$$

$$= -\frac{q_1}{m_1}\frac{\partial}{\partial v}\cdot[\langle\delta E\cdot\delta N_1\rangle_{(2)}+(E_p+e)W_{1\,|\,1}] \qquad ...[9.9]$$

$$\left[\frac{\partial}{\partial t}+v\cdot\frac{\partial}{\partial x}+\left(\frac{q}{m}\right)_1\frac{v}{c}\times B_0\cdot\frac{\partial}{\partial v}\right]\delta N_1(X, t)$$

$$= \frac{q_1}{m_1}\frac{\delta}{\partial v}\cdot(\langle\delta E\delta N_1\rangle_{(2)}-(E_p+e)\delta N_1-\delta E W_{1\,|\,1}-\delta E\delta N_1) \quad ...[9.10]$$

where B_0 is the externally imposed homogeneous magnetic field. The fluctuations δE and the inhomogeneous average field (the same for all the realizations of the ensemble) are given by

$$\nabla_x\cdot\delta E = \sum_r 4\pi q_r n_r \int \delta N_r(X, t)d^3v+4\pi q_1 V^{-1}\int \delta N_1(X, t)d^3v ...[9.11]$$

$$\nabla_x\cdot(E_p+e) = \sum_r 4\pi q_r n_r \int \omega_{1\,|\,r}(X, t\,|\,X_{1\,0})d^3v+4\pi q_1 V^{-1}\int W_{1\,|\,1}(Xt\,|\,X_{1\,0})$$

$$...[9.12]$$

‡ The functions $W_{1\,|\,1}$, $W_{1\,|\,r}$ are related to the functions $W_{1,1}$, $W_{1,r}$ introduced by Rostoker,[3] via $W_{1,1}(Xt, X_{1\,0}) = F(X_{1\,0})W_{1\,|\,1}(Xt\,|\,X_{1\,0})$ etc.

and we have defined

$$\omega_{1\,|\,r}(Xt\,|\,X_{1\,0}) \equiv W_{1\,|\,r}(X, t\,|\,X_{1\,0}) - F_r(X, t). \qquad ...[9.13]$$

The fields e and E_p are respectively the bare Coulomb field of the test-particle and the polarization field due to the associated mean disturbance of the field particles. In equation [9.9] Coulomb self-interaction must be carefully excluded.

The system of equations [9.9]–[9.12] is not closed, and a further equation is needed to describe the evolution of $\omega_{1\,|\,r}$. This is easily obtained by averaging equation [3.3] via equation [9.5] and subtracting off equation [3.14]:

$$\left[\frac{\partial}{\partial t} + v \cdot \frac{\partial}{\partial x} + \left(\frac{q}{m}\right)_r \frac{v}{c} \times B_0 \cdot \frac{\partial}{\partial v}\right] \omega_{1\,|\,r}(Xt\,|\,X_{1\,0}) + \left(\frac{q}{m}\right)_r (E_p + e) \cdot \frac{\partial}{\partial v} F_r$$

$$= -\left(\frac{q}{m}\right)_r (E_p + e) \cdot \frac{\partial}{\partial v} \omega_{1\,|\,r} - \left(\frac{q}{m}\right)_r \frac{\partial}{\partial v} \cdot \{\langle \delta E \delta N_r \rangle_{(2)} - \langle \delta E \delta N_r \rangle\}.$$

$$...[9.14]$$

We scale the different terms of equation [9.14], and obtain the following order of magnitude ratios (equations [9.11] and [9.12] relate the size of δE, E_p to δN_r, $\omega_{1\,|\,r}$):

$$1 : \frac{v\tau}{\lambda_D}\alpha : \Omega\tau : \frac{v\tau}{\lambda_D}\alpha^{-1} : \lambda^2 \frac{v\tau}{\lambda_D}\alpha^{-1} : \lambda^2 \frac{v\tau}{\lambda_D}\alpha^{-1}. \qquad ...[9.15]$$

Here $\alpha \equiv \lambda_D/L$, and L is the length scale of the inhomogeneity. We have used the initial conditions, equation [9.7] in our estimates. Of course, τ is the time scale under consideration. We take $L \sim \lambda_D$. Thus the last two terms are negligible for times $t \ll \lambda^{-2}\omega_p^{-1}$ (take $v \sim v_{th}$). Since $\delta N_1(X, 0) = 0$, the third term on the right-hand side of equation [9.10] is dominant for some interval of time. We regard $W_{1\,|\,1}$ and δN_1 as Green functions; thus $\partial/\partial v \sim 1/v_{th}$, and formally $\delta N_1 \sim \frac{q}{m} v_{th}^{-1} \delta E W_{1\,|\,1} t$. We introduce this result in the other terms of the right-hand side of equation [9.10] and form the ensemble average occurring in equation [9.9]. We conclude that the third term is dominant for times $t \ll \lambda^{-2}\omega_p^{-1}$. The validity of our theory is not restricted to such intervals, however. Indeed, one could show that δN_1 and $\omega_{1\,|\,r}$ are oscillating functions (and thus bounded), so that the effect of the nonlinear terms stays negligible.

We thus consider the following simplified set of equations:

$$\left[\frac{\partial}{\partial t} + v \cdot \frac{\partial}{\partial x} + \Omega_1 \frac{\partial}{\partial \phi}\right] W_{1\,|\,r}(Xt\,|\,X_{1\,0})$$

$$= -\left(\frac{q}{m}\right)_1 \frac{\partial}{\partial v} \cdot [\langle \delta E \delta N_1 \rangle + (E_p + e) W_{1\,|\,1}] \qquad ...[9.16]$$

$$\left[\frac{\partial}{\partial t}+\boldsymbol{v}\cdot\frac{\partial}{\partial\boldsymbol{x}}+\Omega_1\frac{\partial}{\partial\phi}\right]\delta N_1(Xt)=-\left(\frac{q}{m}\right)_1\frac{\partial}{\partial\boldsymbol{v}}\delta\cdot E W_{1\,|\,1} \qquad \ldots[9.17]$$

$$\nabla_{\boldsymbol{x}}\cdot E_p=\sum_r 4\pi q_r n_r\int\omega_{1\,|\,r}(Xt\,|\,X_{1\,0})d^3v$$

$$\nabla_{\boldsymbol{x}}\cdot\boldsymbol{e}=4\pi q_1 V^{-1}\int W_{1\,|\,1}(Xt\,|\,X_{1\,0})d^3v \qquad \ldots[9.18]$$

$$\left[\frac{\partial}{\partial t}+\boldsymbol{v}\cdot\frac{\partial}{\partial\boldsymbol{x}}+\Omega_r\frac{\partial}{\partial\phi}\right]\omega_{1\,|\,r}(Xt\,|\,X_{1\,0})+\left(\frac{q}{m}\right)_r E_r\cdot\frac{\partial F_r}{\partial\boldsymbol{v}}=-\left(\frac{q}{m}\right)_r\boldsymbol{e}\cdot\frac{\partial F_r}{\partial\boldsymbol{v}}$$

$$\ldots[9.19]$$

where $\Omega_r=(q_r B_0)/(m_r c)$ is the cyclotron frequency of species r, and the angular co-ordinate ϕ is defined in Fig. 9.1.

We shall show that equation [9.16] can be written in the following form:

$$\left[\frac{\partial}{\partial t}+\boldsymbol{v}\cdot\frac{\partial}{\partial\boldsymbol{x}}+\Omega_1\frac{\partial}{\partial\phi}\right]W_{1\,|\,1}(Xt\,|\,X_{1\,0})=\frac{1}{m_1}\frac{\partial}{\partial\boldsymbol{v}}\cdot\left(\Pi^{(1)}\cdot\frac{\partial}{\partial\boldsymbol{v}}+F^{(1)}\right)W_{1\,|\,1},$$

$$\ldots[9.20]$$

where $\Pi^{(1)}$ and $F^{(1)}$ are respectively the diffusion tensor and the drag force introduced in previous chapters.

9.1. *The explicit form of the diffusion tensor*

The equations of the unperturbed orbit of the test-particle as a parametric function of the configuration variable X define the propagator P:

$$P(Xt\,|\,t')X=X^*(Xt\,|\,t') \qquad \ldots[9.21]$$

where:

$$\boldsymbol{v}^*(Xt\,|\,t')=v_z\hat{e}_z-v_\perp[\hat{e}_x\sin(\phi-\Omega_1(t-t'))-\hat{e}_y\cos(\phi-\Omega_1(t-t'))]$$

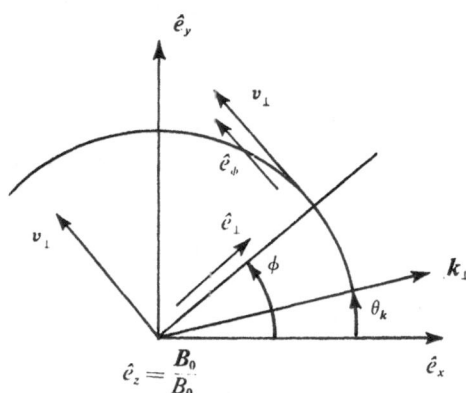

FIG. 9.1. Different frames of reference.

$$x^*(Xt \mid t') = x - (t-t')v_z\hat{e}_z + \frac{v_\perp}{\Omega_1}\left[\hat{e}_x\cos(\phi - \Omega_1(t-t')) + \hat{e}_y\sin(\phi - \Omega_1(t-t'))\right]$$

$$- \frac{v_\perp}{\Omega_1}\left[\hat{e}_x\cos\phi + \hat{e}_y\sin\phi\right]. \qquad ...[9.22]$$

At time t, the test-particle distribution function has spread over some volume of configuration space because of the effect of the right-hand side of the diffusion equation, equation [9.20]. We shall verify a posteriori that this right-hand side is small, of order $\lambda^2\omega_p$ in dimensionless units, and thus has negligible effects for times $t \ll \lambda^{-2}\omega_p^{-1}$. The formal solution of equation [9.17] is:

$$\delta N_1(X, t) = -\int_0^t dt' \left(\frac{q}{m}\right)_1 P(Xt \mid t') \frac{\partial}{\partial v} \cdot \delta E(X, t') W_{1\mid 1}(Xt' \mid X_{10})$$

$$...[9.23]$$

We shall assume and justify later that only a small region $[t-\Delta t, t]$, (with $\Delta t \ll \lambda^{-1}\omega_p^{-1}$) of the domain of integration contributes to the statistical average $\langle \delta E \delta N_1 \rangle$ of equation [9.16]. Therefore we use the approximation relation

$$P(Xt \mid t') W_{1\mid 1}(Xt' \mid X_{10}) \equiv W_{1\mid 1}(Xt \mid X_{10}) \qquad ...[9.24]$$

since for small intervals of time $W_{1\mid 1}$ is an integral of the motion.

We introduce equation [9.23] in equation [9.16] and use equations [9.20] and [9.24] to identify the diffusion tensor:

$$\Pi^{(1)}(X, t) \equiv \frac{q_1^2}{m_1} \left\langle \delta E(X, t) \int_0^t dt' P(Xt \mid t')\delta E(X, t') \right\rangle. \qquad ...[9.25]$$

Since the operations of translation and rotation commute, P can be written as the product of a translator and a rotator: $P = TR$. Since plane waves are eigenfunctions of T, we shall introduce the Fourier transform of $\delta E(x, t')$. The appropriate representation of the rotation in the cylindrical coordinate system of Fig. 9.1 is

$$R(t, t') = \begin{bmatrix} \cos\Omega_1(t-t') & -\sin\Omega_1(t-t') & 0 \\ \sin\Omega_1(t-t') & \cos\Omega_1(t-t') & 0 \\ 0 & 0 & 1 \end{bmatrix} \qquad ...[9.26]$$

The diffusion tensor thus takes the form:

$$\Pi(X, t) = \left(\frac{q^2}{m}\right)_1 V^{-1} \int \frac{d^3k}{(2\pi)^3} \int_0^t dt' \exp\left[-ik \cdot x + ik \cdot x^*(Xt \mid t')\right]$$

$$k \cdot R(t, t') \cdot k\langle \delta\Psi_{-k}(t)\delta\Psi_k(t')\rangle. \qquad ...[9.27]$$

It is well known that the generalization of the function $\delta\chi_k$ (introduced in a previous chapter) in the presence of the magnetic field is:

$$\delta\chi_k(t) = 4\pi \sum_{r \neq 1} \frac{q_r^2 n_r}{k^2} \int d^3v' \sum J_m^2\left(\frac{k_\perp v_\perp'}{\Omega_r}\right) \frac{\delta N_{r,k}^m(0)}{\varepsilon(k, -i[k \cdot v']_m)} e^{-i[k \cdot v']_m t}$$

...[9.28]

where the J_m are the Bessel functions of the first kind. The Fourier-Bessel transform $\delta N_{r,k}^m$ is defined by

$$J_m\left(\frac{k_\perp v_\perp}{\Omega_1}\right) \delta N_{r,k}^m(v_\perp v_z t) = \frac{1}{2\pi} \int_0^{2\pi} d\varphi \, \exp\left[i \frac{k_\perp v_\perp}{\Omega_r} \cos(\varphi - \theta_k) - im(\varphi - \theta_k)\right] i^{-m} \delta N_{r,k},$$

...[9.29]

with

$$[k \cdot v]_m \equiv k_z v_z + m\Omega_r.$$

We introduce the equations of the unperturbed orbit in equation [9.27] and recall the identity

$$\exp\left[i \frac{k_\perp v_\perp}{\Omega_1} \cos(\varphi - \theta_k - \Omega_1(t - t'))\right] \equiv \sum_{-\infty}^{\infty} e^{in[\varphi - \theta_k - \Omega_1(t-t')]} i^n J_n\left(\frac{k_\perp v_\perp}{\Omega_1}\right).$$

...[9.30]

A straightforward integration leads to:

$$\Pi = \frac{q_1^2}{m_1} V^{-1} \int \frac{d^3k}{(2\pi)^3} \exp\left[-i \frac{k_\perp v_\perp}{\Omega_1} \cos(\varphi - \theta_k)\right] \sum i^n J_n\left(\frac{k_\perp v_\perp}{\Omega_1}\right) e^{in(\varphi - \theta_k)} k$$

$$\sum_\sigma \left| \left[\Psi_k^n(\omega_k^\sigma, v) \left(\langle \delta\phi_{-k}^\sigma(t) \, \delta\phi_k^\sigma(t)\rangle + 4\pi \sum_s \frac{q_s n_s}{k^2} \int_{H^\sigma} d^3v' \sum_{-\infty}^{\infty} J_m^2\left(\frac{k_\perp v_\perp'}{\Omega_1}\right) \right. \right. \right.$$

$$\frac{e^{-i(\omega_k^\sigma - [k \cdot v']_m)t}}{\varepsilon^*(k, i[k \cdot v']_m)} \langle \delta N_{s,-k}(0) \, \delta\phi_k^\sigma(0)\rangle \right) + 4\pi \sum_r \frac{q_r n_r}{k^2} \int_{H^\sigma} d^3v'' \sum_{-\infty}^{\infty} J_p^2\left(\frac{k_\perp v_\perp''}{\Omega_r}\right)$$

$$\frac{1}{\varepsilon(k, -i[k \cdot v'']_p)} \Psi^n([k \cdot v'']_p, v) \left(e^{+i(\omega_k^{\sigma*} - [k \cdot v'']_p)t} \langle \delta\phi_{-k}^\sigma(0) \, \delta N_k^p(0)\rangle \right.$$

$$+ 4\pi \sum_s \frac{q_s n_s}{k^2} \int_{H^\sigma} d^3v' \sum_{-\infty}^{\infty} J_m^2\left(\frac{k_\perp v_\perp'}{\Omega_s}\right) \frac{e^{+i([k \cdot v']_n - [k \cdot v'']_m)t}}{\varepsilon^*(k, i[k \cdot v']_m)}$$

$$\left. \left. \left. \langle \delta N_{s,-k}^m(0) \, \delta N_{r,k}^p(0)\rangle \right) \right] \right|$$

...[9.31]

where the vectors $\Psi_k^n(\omega, v)$ are defined as follows:

$$\Psi_k^n(\omega, v) = \begin{bmatrix} \dfrac{k_\perp}{2} e^{i(\varphi-\theta_k)} \dfrac{1-e^{i\omega t - i[k\cdot v]_{n+1}t}}{i[k\cdot v]_{n+1} - i\omega} + \dfrac{k_\perp}{2} e^{-i(\varphi-\theta_k)} \dfrac{1-e^{i\omega t - i[k\cdot v]_{n-1}t}}{i[k\cdot v]_{n-1} - i\omega} \\[3mm] \dfrac{k_\perp}{2i} e^{i(\varphi-\theta_k)} \dfrac{1-e^{i\omega t - i[k\cdot v]_{n+1}t}}{i[k\cdot v]_{n+1} - i\omega} - \dfrac{k_\perp}{2i} e^{-i(\varphi-\theta_k)} \dfrac{1-e^{i\omega t - i[k\cdot v]_{n-1}t}}{i[k\cdot v]_{n-1} - i\omega} \\[3mm] k_z \dfrac{1-e^{i\omega t - i[k\cdot v]_n t}}{i[k\cdot v]_n - i\omega} \end{bmatrix}.$$

$$...[9.32]$$

By arguments that by now are familiar, we could show that the terms $(i[k\cdot v]_r - i\omega)^{-1} e^{-i([k\cdot v]_s t + i\omega t)}$ appearing in $\Psi_k^n(\omega, v)$ provide the analytic continuation to $1/(i[k\cdot v]_r - i\omega)$ if Im ω passes through zero. This provides a deformation of the path of integration along k_z analogous to that introduced below equation [7.3] to provide the topology of the S contour (Fig. 6.1).

We use relation

$$\langle \delta N_{r,k}^m \delta N_{s,-k}^p \rangle J_m\left(\frac{k_\perp v_\perp'}{\Omega_r}\right) J_p\left(\frac{k_\perp v_\perp''}{\Omega_s}\right) = V(-1)^p \left[\frac{1}{2\pi} \frac{1}{n_r} \delta_{m,-p} \delta_{r,s} F_r + P_{k,r,t}^{m,p}\right]$$

$$...[9.33]$$

which generalizes equation [3.12]. We find

$$\Pi(X, t) = \frac{q_1^2}{m_1} \int d^3k \exp\left[-i\frac{k_\perp v_\perp}{\Omega_1} \cos(\varphi - \theta_k)\right] \sum_n i J_n\left(\frac{k_\perp v_\perp}{\Omega_1}\right) e^{in(\varphi-\theta_k)}$$

$$\sum_\sigma k \left[\Psi_k^{n,+}(\omega_k^\sigma, v) I_k^\sigma(t) + \frac{2}{\pi} \sum_r \frac{q_r^2 n_r}{k^4} \int_{H^\sigma} d^3v' \sum_{-\infty}^\infty J_m^2\left(\frac{k_\perp v_\perp'}{\Omega_r}\right)\right.$$

$$\left.\frac{F_r(v', t)}{|\varepsilon(k, -i[k\cdot v']_m)|^2} (\Psi_k^{n,+}([k\cdot v']_m, v) - \Psi_k^{n,+}(\omega_k^\sigma, v))\right]$$

$$...[9.34]$$

where the $\Psi_k^{n,+}$ are obtained from the Ψ_k^n by merely dropping the rapidly oscillating terms. The symbol "$+$" retains its previous definition. The contribution from the $P_{k,r,s}^{m,p}$ to the diffusion tensor disappears asymptotically by phase mixing, and has been dropped from equation [9.34].

The very fact that the diffusion tensor relaxes to an asymptotic quasi-stationary form in a time of the order of τ_{PM}, the phase-mixing time scale $[\tau_{PM} \sim (k_z v_z)^{-1}$, or ω_k^{-1}, according to which is larger] is the *a posteriori* justification of equation [9.24], (where we neglect the perturbation of the orbit); in physical terms, the Markoffian behaviour of $W_{1|1}(X t | X_1^0)$ is concomitant with phase-mixing processes. They also justify the synchronization of $F_r(v', t)$. Further, one can verify that the order of magnitude of the right-hand side of equation [9.16] is as assumed above in equation [9.25].

Finally we perform the angular integration in k-space, and obtain the components of the diffusion tensor in the following form:

$$\Pi_{\rho\rho} = \frac{q_1^2}{m_1} \int d^3k \sum_n k_\perp^2 J_n'^2 \left(\frac{k_\perp v_\perp}{\Omega_1}\right) \Xi_k^{n,\,+} (v_\perp, v_z, t)$$

$$\Pi_{\rho\varphi} = \Pi_{\varphi\rho} = i \frac{q_1^2}{m_1} \int d^3k \sum_n k_\perp^2 \frac{n\Omega_1}{k_\perp v_\perp} J_n \left(\frac{k_\perp v_\perp}{\Omega_1}\right) J_n' \left(\frac{k_\perp v_\perp}{\Omega_1}\right) \Xi_k^{n,\,+} = 0$$

$$\Pi_{\rho z} = -\Pi_{z\rho} = -i \frac{q_1^2}{m_1} \int d^3k \sum_n k_\perp k_z J_n \left(\frac{k_\perp v_\perp}{\Omega_1}\right) J_n' \left(\frac{k_\perp v_\perp}{\Omega_1}\right) \Xi_k^{n,\,+}$$

$$\Pi_{\varphi\varphi} = \frac{q_1^2}{m_1} \int d^3k \sum_n k_\perp^2 \left(\frac{n\Omega_1}{k_\perp v_\perp}\right)^2 J_n^2 \left(\frac{k_\perp v_\perp}{\Omega_1}\right) \Xi_k^{n,\,+}$$

$$\Pi_{\varphi z} = +\Pi_{z\varphi} = \frac{q_1^2}{m_1} \int d^3k \sum_n k_\perp k_z \left(\frac{n\Omega_1}{k_\perp v_\perp}\right) J_n^2 \left(\frac{k_\perp v_\perp}{\Omega_1}\right) \Xi_k^{n,\,+}$$

$$\Pi_{zz} = \frac{q_1^2}{m_1} \int d^3k \sum_n k_z^2 J_n^2 \left(\frac{k_\perp v_\perp}{\Omega_1}\right) \Xi_k^{n,\,+}$$

$$...[9.35]$$

where the scalar $\Xi_k^{n,\,+}$ is defined by

$$\Xi_k^{n,\,+} (v_\perp, v_z, t) = \sum_\sigma \left[\frac{I_k^\sigma(t)}{(i[k \cdot v]_n - i\omega_k^\sigma)^+} + \frac{2}{\pi} \sum_{m,r} \frac{q_r^2 n_r}{k^4} \int_{S^\sigma} d^3v' J_m^2 \left(\frac{k_\perp v_\perp'}{\Omega_1}\right) \right.$$

$$\left. \frac{F_r(v, \,'t)}{|\varepsilon(k, -i[k \cdot v']_m)|^2} \left\{ \frac{1}{(i[k \cdot v]_n - i[k \cdot v']_m)^+} - \frac{1}{(i[k \cdot v]_n - i\omega_k^\sigma)^+} \right\} \right].$$

$$...[9.36]$$

As previously described we can approximate equation [9.36] by

$$\Xi_k^{n,\,+} (v_\perp, v_z, t) = \sum_\sigma \left[\pi\delta(\omega_k^\sigma - [k \cdot v]_n) I_k^\sigma(t) + P \frac{1}{([k \cdot v]_n - \omega_k^\sigma)^2} \frac{2}{\pi} \sum_r \frac{q_r^2 n_r}{k^4} \right.$$

$$\int_{H^\sigma} d^3v' \sum_m J_m^2 \left(\frac{k_\perp v_\perp'}{\Omega_r}\right) F_r \pi\delta(\omega_k^\sigma - [k \cdot v']_n) \left|\frac{\partial \varepsilon_k}{\partial \omega_k}\right|^{-2}$$

$$+ \frac{2}{\pi} \sum_r \frac{q_r^2 n_r}{k^4} \int_{H^\sigma} d^3v' \sum_m J_m^2 \left(\frac{k_\perp v_\perp'}{\Omega_r}\right) P \frac{F_r}{|\varepsilon(k, -i[k \cdot v']_m)|^2}$$

$$\left. \{\pi\delta([k \cdot v]_n - [k \cdot v']_m) - \pi\delta(\omega_k^\sigma - [k \cdot v]_n)\} \right]$$

$$...[9.37]$$

where the symbol P means that the Cauchy principal value must be taken under an integral.

The third term in equation [9.37] has been given by Rostoker for a stable plasma. It is not well-defined for weakly stable and unstable modes. This is another manifestation of the breakdown of the conventional B.G.L. kinetic theory in both stable and unstable regimes. The reason is, of course, the failure of the Bogoliubov ansatz. The first term of equation [9.3] has been given by Drummond and Rosenbluth[14] for an unstable plasma. The second and fourth terms provide a smooth transition between the different regimes.

Equations [9.35] and [9.37] describe the different processes by which the test-particle is scattered off its helicoidal orbit: by resonating waves, by the wake of other particles, and via "collisions" with these particles.

9.2. *The drag force*

The polarization field E_p, which describes the plasma response to the test-particle disturbance, yields the drag force on the test particle (after extracting self-field effects).

From equation [9.18] we obtain

$$e(x, t+t' \mid X_{10}) = q_1 V^{-1} \int d^3 X' \frac{x-x'}{|x-x'|^3} W_{1|1}(X', t+t' \mid X_{10}).$$

$$...[9.38]$$

We shall assume and justify later that the polarization field relaxes to a functional of e in an interval of time $t' \ll \lambda^{-1}\omega_p^{-1}$. We use equation [9.24] to put equation [9.38] in the form

$$e(x, t+t' \mid X_{10}) = q_1 V^{-1} \int d^3 X' \frac{x-x^*(X't \mid t+t')}{|x-x^*(X't \mid t+t')|^3} W_{1|1}(X't \mid X_{10})$$

$$...[9.38a]$$

so that

$$e_k(t+t' \mid X_{10})$$
$$= -4\pi q_1 V^{-1} i \int d^3 X' R(\varphi', t \mid t+t') \cdot \frac{k}{k^2} e^{-ik \cdot x^*(X', t \mid t+t')} W_{1|1}$$

$$...[9.39]$$

where the operator R takes care of the rotation of the vector $x-x^*$ as previously. Note that

$$R(\varphi', t \mid t+t') \cdot k = (k_\perp \cos(\varphi'+\Omega_1 t-\theta_k), -k_\perp \sin(\varphi'+\Omega_1 t-\theta_k), k_z).$$

It is easily shown, using the equations of the unperturbed orbit equation [9.22] and the identity equation [9.30], that the Laplace transform (variable t') of equation [9.39] is:

$$e_k(p, t \mid X_{1\,0})$$

$$= -4\pi i q_1 V^{-1} \int d^3 X' W_{1\mid 1}(X't \mid X_{1\,0}) \exp\left[-ik \cdot x' + i\frac{k_\perp v'_\perp}{\Omega_1}\cos(\varphi' - \theta_k)\right]$$

$$\sum_{-\infty}^{\infty} e^{-in(\varphi' - \theta k)} i^{-n} J_n\left(\frac{k_\perp v'_\perp}{\Omega_1}\right)\left[\frac{1}{2}\left(\frac{e^{i(\varphi' - \theta_k)}}{p + i[k \cdot v']_{n-1}} + \frac{e^{-i(\varphi' - \theta_k)}}{p + i[k \cdot v']_{n+1}}\right)\frac{k_\perp}{k^2},\right.$$

$$\left. -\frac{1}{2i}\left(\frac{e^{i(\varphi' - \theta_k)}}{p + i[k \cdot v']_{n-1}} - \frac{e^{-i(\varphi' - \theta_k)}}{p + i[k \cdot v']_{n+1}}\right)\frac{k_\perp}{k^2}, \frac{1}{p + i[k \cdot v']_n}\frac{k_z}{k^2}\right].$$

$$...[9.39a]$$

We solve the Vlasov equations [9.18] and [9.19] in the usual way, and invert the Laplace transform to obtain for the z component, for example:

$$E_{p,k,z}(t + t' \mid X_{1\,0}) = -e_{k,z}(t + t' \mid X_{1\,0}) - 4\pi i q_1 V^{-1} \int d^3 X' W_{1\mid 1}(X't \mid X_{1\,0})$$

$$\exp\left[-ik \cdot x' + i\frac{k_\perp v'_\perp}{\Omega_1}\cos(\varphi' - \theta_k)\right]\frac{k_z}{k^2}\sum_{-\infty}^{\infty} e^{-in(\varphi' - \theta k)} i^{-n} J_n\left(\frac{k_\perp v'_\perp}{\Omega_1}\right)$$

$$\left[\frac{e^{-i[k \cdot v']_n t'}}{\varepsilon(k, -ik \cdot v')} - \sum_\sigma \frac{e^{-i\omega_k^\sigma t'}}{[\partial\varepsilon_k/\partial(-i\omega_k^\sigma)](-i\omega_k^\sigma + i[k \cdot v']_n)}\right], \quad ...[9.40]$$

where we have discarded the effect of the initial condition $\omega_{1\mid r,k}(t)$ which does not contribute to $E_{p,z}$ for $\omega_p t' \gg 1$ (phase mixing). The term $e^{-i\omega_k t'}$ will merely provide a contour deformation for the first term, so that

$$E_p(x, t + t' \mid X_{1\,0})$$

$$= -e(x, t + t' \mid X_{1\,0}) - 4\pi i q_1 V^{-1} \int \frac{d^3 k}{(2\pi)^3}\int d^3 X' W_{1\mid 1}(X't \mid X_{1\,0})$$

$$\sum_{-\infty}^{\infty} e^{-in(\varphi' - \theta k)} i^{-n} J_n\left(\frac{k_\perp v'_\perp}{\Omega_1}\right)\exp\left[-ik \cdot (x - x') + i\frac{k_\perp v'_\perp}{\Omega_1}\cos(\varphi' - \theta_k)\right]$$

$$\left[\frac{k_\perp}{2k^2}\left(e^{i(\varphi' - \theta_k)}\frac{e^{-i[k \cdot v']_{n-1} t'}}{\varepsilon^-(k, -i[k \cdot v']_{n-1})} + e^{-i(\varphi' - \theta_k)}\frac{e^{-i[k \cdot v']_{n+1} t'}}{\varepsilon^-(k, -i[k \cdot v'])}\right),\right.$$

$$\frac{k_\perp}{2ik^2}\left(e^{i(\varphi' - \theta_k)}\frac{e^{-i[k \cdot v']_{n-1} t'}}{\varepsilon^-(k, -i[k \cdot v']_{n-1})} - e^{-i(\varphi' - \theta_k)}\frac{e^{-i[k \cdot v']_{n+1} t'}}{\varepsilon^-(k, -i[k \cdot v']_{n+1})}\right),$$

$$\left.\frac{k_z}{k^2}\frac{e^{-i[k \cdot v']_n t'}}{\varepsilon^-(k, -i[k \cdot v']_n)}\right]. \quad ...[9.41]$$

It is not possible to reduce the result further without explicit knowledge of $W_{1\mid 1}$. If the spread of the test-particle distribution is small ($t \simeq 0$), so that a dipole approximation is sufficient, we can easily calculate the field at the test-particle. Introducing $x = x^*(X, 0 \mid t')$ and $W_{1\mid 1}(X'0 \mid X_{1\,0}) = V\delta(X' - X_{1\,0})$, we obtain after averaging over θ_k:

$$\mathscr{F}_\rho^{(1)}[x^*(X_{10}0\,|\,t')]$$
$$= 4\pi q_1^2 \int \frac{d^3k}{(2\pi)^3} \sum_{-\infty}^{\infty} \frac{k_\perp}{k^2} \frac{\mathrm{Re}\,\varepsilon(k,\,-i[k\cdot v_{10}]_n)}{|\varepsilon^-(k,\,-i[k\cdot v_{10}]_n)|^2} J_n\left(\frac{k_\perp v_{110}}{\Omega_1}\right) J_n'\left(\frac{k_\perp v_{11c}}{\Omega_1}\right)$$

$$\mathscr{F}_\varphi^{(1)}[x^*(X_{10}0\,|\,t')]$$
$$= 4\pi q_1^2 \int \frac{d^3k}{(2\pi)^3} \sum_{-\infty}^{\infty} \frac{1}{k^2} \frac{n\Omega_1}{v_{110}} \frac{\mathrm{Im}\,\varepsilon(k,\,-i[k\cdot v_{10}]_n)}{|\varepsilon^-(k,\,-i[k\cdot v_{10}]_n)|^2} J_n^2\left(\frac{k_\perp v_{110}}{\Omega_1}\right)$$

$$\mathscr{F}_z^{(1)}[x^*(X_{10}0\,|\,t')]$$
$$= 4\pi q_1^2 \int \frac{d^3k}{(2\pi)^3} \sum_{-\infty}^{\infty} \frac{k_z}{k^2} \frac{\mathrm{Im}\,\varepsilon(k,\,-i[k\cdot v_{10}]_n)}{|\varepsilon^-(k,\,-i[k\cdot v_{10}]_n)|^2} J_n^2\left(\frac{k_\perp v_{110}}{\Omega_1}\right).$$

$$...[9.42]$$

Therefore the drag force is independent of t'. The relaxation is again concomitant with phase-mixing processes. This justifies the assumption below equation [9.38]. The result of equation [9.42] has been given by Rostoker (*loc. cit.*). It is valid as long as the diffusion of the test particle is negligible. The reason is clearly that, in the general case, a dipole approximation is inappropriate.

9.3. *The coefficient of spatial diffusion across a uniform magnetic field*

The spatial diffusion of particles of species "1" across the magnetic field is characterized by

$$\langle r_\perp^2 \rangle = V^{-1} \int d^3X W_{1|1}(Xt\,|\,X_{10}) r_\perp^2(t), \qquad ...[9.43]$$

where

$$r_\perp^2(t) = [x - x^*(X_{10}0\,|\,t)]^2 + [y - y^*(X_{10}0\,|\,t)]^2. \qquad ...[9.44]$$

It will be sufficient for our purpose to consider that the spread of $W_{1|1}$ in $[x, v]$ space is negligible.

We integrate equation [9.20] along the unperturbed orbit and obtain, to first order

$$W_{1|1}(Xt\,|\,X_{10})$$
$$= W_{1|1}^0(Xt\,|\,X_{10}) + \int_0^t dt' P(Xt\,|\,t') \left[\frac{1}{m_1}\frac{\partial}{\partial v}\cdot\left(\Pi\cdot\frac{\partial}{\partial v}+\mathscr{F}\right) W_{1|1}^0(Xt'\,|\,X_{10})\right]$$

$$...[9.45]$$

where $W_{1|1}^0$ is the zero order solution. The first term obviously does not contribute to the diffusion in zero order. Therefore:

$$\langle r_\perp^2 \rangle = \int d^3X \int_0^t dt' \left\{ \left[x - x_{10} - \frac{v_{\perp10}}{\Omega_1}(\cos(\varphi_{10}+\Omega_1 t) - \cos\varphi_{10}) \right]^2 \right. $$

$$\left. + \left[y - y_{10} - \frac{v_{\perp10}}{\Omega_1}(\sin(\varphi_{10}+\Omega_1 t) - \sin\varphi_{10}) \right]^2 \right\}$$

$$\frac{1}{m_1}\frac{\partial}{\partial v} \cdot \left\{ \mathbf{\Pi} \cdot \frac{\partial}{\partial v} + \mathscr{F}_d \right\} \delta(X^*(Xt \mid t') - X^*(X_{10}0 \mid t')). \qquad ...[9.46]$$

We have used

$$W^0_{1|1}(Xt' \mid X_{10}) = V\delta(X - X^*(X_{10}0 \mid t')) \qquad ...[9.47]$$

and equation [9.21]. The quantities $X^*(Xt \mid t')$ and $X^*(X_{10}0 \mid t')$ are computed from equation [9.22] where we set the appropriate values for $v_\perp$, v_z, and φ. Integrating over the spatial co-ordinates we obtain:

$$\langle r_\perp^2 \rangle = \int d^3v \int_0^t dt'$$

$$\left[\left\{ \frac{v_\perp}{\Omega_1}[\cos(\varphi - \Omega_1(t-t')) - \cos\varphi] - \frac{v_{\perp10}}{\Omega_1}[\cos(\varphi_{10}+\Omega_1 t') - \cos(\varphi_{10}+\Omega_1 t)] \right\}^2 \right.$$

$$\left. + \left\{ \frac{v_\perp}{\Omega_1}[\sin(\varphi - \Omega_1(t-t')) - \sin\varphi] - \frac{v_{\perp10}}{\Omega_1}[\sin(\varphi_{10}+\Omega_1 t') - \sin(\varphi_{10}+\Omega_1 t)] \right\}^2 \right]$$

$$\frac{1}{m_1} \cdot \frac{\partial}{\partial v} \cdot \left(\mathbf{\Pi} \cdot \frac{\partial}{\partial v} + \mathscr{F}_d \right) \delta(v_z - v_{z10}) \frac{\delta(v_\perp - v_{\perp10})}{v_\perp} \delta(\varphi - \varphi_{10} - \Omega_1 t).$$

$$...[9.48]$$

Since the bracket is zero if $v_\perp = v_{\perp,10}$, it is clear that the only contributions arise when it is differentiated twice. Note that $\mathbf{\Pi}_{\rho\varphi} = 0$, so that one obtains finally:

$$\langle r_\perp^2 \rangle = \int_0^t dt' \frac{4}{m\Omega_1^2}(\mathbf{\Pi}_{\rho\rho} + \mathbf{\Pi}_{\varphi\varphi})[1 - \cos\Omega_1(t'-t)] \qquad ...[9.49a]$$

or

$$\langle r_\perp^2 \rangle \cong \frac{4}{m_1\Omega_1^2}(\mathbf{\Pi}_{\rho\rho} + \mathbf{\Pi}_{\phi\phi})t = D_\perp t, \qquad ...[9.49]$$

which defines the coefficient of spatial diffusion across a uniform magnetic field.

This calculation (described previously) has been given by Rostoker (*loc. cit.*), who also showed that the stable part of the spectrum yields the "classical" diffusion coefficient. The contribution from the weakly stable or unstable modes is now given via equations [9.36] and [9.37], where the I_k are obtained by solving the initial value problem (equations [8.17] and [8.48]). It is clear from these equations that even in weakly stable plasma, the fluctuation energy can be much larger than its thermal value, especially if the plasma has fallen from instability to stability, or if spontaneous emission of plasmons is high due to a superabundance of energetic particles.

10

CONCLUSIONS

In the present work we have shown that in plasma, where collective motions are made possible by the long range of the Coulomb interaction, there is properly no general kinetic equation for the one-particle distribution function alone. Instead, one must describe the time evolution of the electric field fluctuations simultaneous with the time evolution of the one-particle distribution.

This enlarged kinetic description reduces to the usual one (B.G.L.) in the strongly stable regimes of k-space, i.e. when particle emission "instantaneously" balances the absorption of the waves. It also embraces not only the weakly unstable situation where we recover the results of quasi-linear theories, but it is also the first systematic description of those regimes which are dominated by collisional absorption of the waves and/or emission by particle-particle and wave-particle scattering. The theory also provides the continuous transition from instability to stability. Therefore, an initially weakly unstable distribution can be followed all the way to thermal equilibrium.

We have shown that the tail of the electron distribution and the fluctuation spectrum of plasmons are in strong thermal contact and we have indicated under which circumstances there is only weak coupling with the main body of the distribution. The calculation of the rate of decay of the waves also yields the general expression of the electrical conductivity of plasma, including the effect of the scattering of electrons by waves (non-linear Landau damping) and at ions (collisional damping). This expression reduces to known results in the dipole approximation and for a plasma in thermal equilibrium.

We have also resolved, in the framework of our kinetic theory, the outstanding disparity for the coefficient of spatial diffusion across a homogeneous magnetic field in the weakly stable region[3] and the unstable regime.[14] We have shown that the diffusion may be abnormally large in stable plasma if spontaneous emission of plasmons is high due to a superabundance of energetic particles or if the plasma has fallen from instability to stability.

REFERENCES

1. A. LENARD. 1960. *Ann. Phys.*, **10**, 390.
2. N. N. BOGOLIUBOV. 1946. *Problems of a Dynamical Theory in Statistical Physics* (translation by E. K. Gora in *Studies of Statistical Mechanics*, Vol. I (North-Holland, 1962)).

3. N. ROSTOKER. 1961. *Nucl. Fusion* 1, 101.
4. N. ROSTOKER. 1964. *Physics Fluids*, 4, 479.
5. W. E. DRUMMOND and D. PINES. 1962. *Nucl. Fusion Suppl.*, 3, 1049.
6. A. A. VEDENOV and E. P. VELIKHOV. 1963. *Soviet Phys. JETP*, 16, 682.
7. A. A. VEDENOV, E. P. VELIKHOV and R. Z. SAGDEEV. 1961. *Soviet Phys. Usp.*, 4, 332.
8. B. B. KADOMTSEV. 1965. *Plasma Turbulence.* Academic Press.
9. E. FRIEMAN and P. RUTHERFORD. 1964. *Ann. Phys.* 28, 134.
10. S. IORDANSKII and A. KULIKOVSKII. 1964. *Soviet Phys. JETP*, 19, 499.
11. A. ROGISTER and C. OBERMAN. 1967. Princeton University, PPL MATT-583.
12. IU. L. KLIMONTOVICH. 1958. *Soviet Phys. JETP*, 34, 114.
13. T. DUPREE. 1963. *Physics Fluids*, 6, 1714.
14. W. E. DRUMMOND and M. N. ROSENBLUTH. 1962. *Physics Fluids*, 5, 1507.
15. D. PINES and J. R. SCHRIEFFER. 1962. *Phys. Rev.*, 125, 804.
16. E. HARRIS. 1967. *Physics Fluids*, 10, 238.
17. C. OBERMAN, A. RON and J. DAWSON. 1962. *Physics Fluids*, 5, 1514.
18. R. BALESCU. 1960. *Physics Fluids*, 3, 52.
19. R. GUERNSEY. 1960. Ph.D. dissertation, University of Michigan, Ann Arbor.
20. L. LANDAU. 1946. *J. Phys. U.S.S.R.*, 10, 25.
21. R. GUERNSEY. 1967. *Can. J. Phys.*, 45, 179.
22. A. EVIATAR. 1966. *J. geophys. Res.*, 71, 2715.
23. J. DAWSON and R. SHANNY. 1967. Princeton University, PPL MATT-568.
24. I. BERNSTEIN and F. ENGELMAN. 1966. *Physics Fluids*, 9, 937.
25. C. C. LIN and W. H. REID. 1963. *Handbuch der Physik.* Springer-Verlag, Berlin.
26. T. BIRMINGHAM, J. DAWSON and C. OBERMAN. 1965. *Physics Fluids*, 8, 297.
27. MAYER and MAYER. 1962. *Studies in Statistical Mechanics.* Wiley and Sons.

3

PLASMA WAVES AND OSCILLATIONS

R. J. Tayler

Astronomy Centre, University of Sussex

1

Introduction

Earlier in this Summer School you have learnt something of the different levels of description of a plasma; the independent particle treatment, the use of kinetic equations, the two fluid equations and magnetohydrodynamics. If quantum and relativistic effects are neglected, a complete description of a plasma is in principle given by Newton's laws of motion for all of the particles combined with the statement that the electromagnetic fields are produced by the charges of the particles, with the possible addition of vacuum fields produced by external coils and batteries. Thus, in the case that only electromagnetic forces act on the particles, each particle has an equation of motion of the form

$$m_i \frac{dv_i}{dt} = e_i \left[E(x_i) + \frac{v_i \times B(x_i)}{c} \right], \qquad \dots[1.1]$$

where the suffix i denotes the particle. Alternatively we can use the N-particle Liouville equation

$$\frac{\partial F}{\partial t} + \sum_i v_i \cdot \frac{\partial F}{\partial x_i} + \sum_i \frac{e_i}{m_i} \left[E(x_i) + \frac{v_i \times B(x_i)}{c} \right] \cdot \frac{\partial F}{\partial v_i} = 0, \qquad \dots[1.2]$$

which is exactly equivalent to [1.1]. In this equation F is a product of δ-functions in ordinary space and velocity space, and E and B are the microscopic electric and magnetic fields.

In this present course we are concerned with collective properties of a plasma; many-particle rather than single-particle phenomena. These oscillations and waves do not depend on the precise positions and velocities of all the particles. We thus do not need to try the (usually impossible) task of solving equation [1.2] but must instead try to obtain from this equation simpler equations which describe the average properties of the plasma. The first procedure is to introduce the one-particle distribution function

$$f_1(x_1, v_1, t) = \int dx_2 dx_3 \dots dv_2 dv_3 \dots F(x_i, v_i, t), \qquad \dots[1.3]$$

103

and to try to obtain from [1.2] an equation determining the behaviour of f_1. It is not in general possible to obtain an equation for f_1 alone, as simple integration of [1.2] gives an equation in which f_1 is coupled to the two-particle distribution function

$$f_2(x_1, v_1, x_2, v_2, t) = \int dx_3 \dots dv_3 \dots F(x_i, v_i, t). \qquad \dots[1.4]$$

However, in some approximations which have been described in earlier lectures a closed kinetic equation for f_1 can be obtained. These closed equations include the Boltzmann equation with collision term

$$\frac{\partial f}{\partial t} + v \cdot \frac{\partial f}{\partial x} + \frac{e}{m}\left(E + \frac{v \times B}{c}\right) \cdot \frac{\partial f}{\partial v} = \left(\frac{\partial f}{\partial t}\right)_c, \qquad \dots[1.5]$$

where here E and B are macroscopic fields and $(\partial f/\partial t)_c$ is the rate of change of f due to close two-particle collisions, and f is taken to be a continuous function, the Vlasov equation

$$\frac{\partial f}{\partial t} + v \cdot \frac{\partial f}{\partial x} + \frac{e}{m}\left(E + \frac{v \times B}{c}\right) \cdot \frac{\partial f}{\partial v} = 0, \qquad \dots[1.6]$$

in which two-particle correlations are assumed negligible and the Fokker-Planck equation. A kinetic equation of type [1.5] or [1.6] is required for each species of particle in the plasma.

Although equations [1.5] and [1.6] are much simpler than [1.2], they still contain much detailed information about the velocities of the individual particles in the plasma. In many cases we are interested in phenomena which depend only on the behaviour of a few of the moments of f such as the density ρ, mean velocity $\bar{v}$ and the pressure p:

$$\rho = \int mf d^3v, \qquad \dots[1.7]$$

$$\bar{v} = \int fv d^3v \,/\, \int f d^3v, \qquad \dots[1.8]$$

$$p = \tfrac{1}{3} \int mf(v - \bar{v})^2 d^3v. \qquad \dots[1.9]$$

Equations [1.5] and [1.6] can be integrated over velocity space in an attempt to obtain hydrodynamic equations for these moments. In this case also, the equations contain higher moments and can only be closed by making approximations. Hydrodynamic equations can be obtained for each species of particle (if only electrons and ions are present, these are the two fluid hydrodynamic equations), or they can be added to form a single set of magnetohydrodynamic equations. In some circumstances, these reduce in a reasonable approximation to the equations of ideal magnetohydrodynamics.

In this particular course we are concerned with plasma waves governed either by the hydrodynamic equations or by the Vlasov kinetic equation. In studying these, we shall not exhaust the possible collective behaviour of a plasma. Clearly, the more we smear out the original distribution function F, the more the predicted oscillations are characteristic of the behaviour of the whole plasma. If more detail is allowed in our description of the distribution function, we may discover oscillations in which a particular subgroup of particles play a key role, this is what happens when we consider waves predicted by the Vlasov equation. This means that it is not possible to state how many different classes of wave may be propagated in a plasma. There is no completely clear division between individual particle behaviour and collective behaviour. All that can be said is that each approximate description of a plasma predicts a definite set of waves and that, as mentioned above, only waves characteristic of the plasma as a whole are predicted by the hydrodynamic equations.

In this course we shall mainly be concerned with the propagation of small amplitude plane waves in a spatially homogeneous medium, although some space will be devoted to large amplitude waves and spatial inhomogeneities. Thus we mainly consider the propagation of disturbances of the form $\exp(i k \cdot x - i\omega t)$, where we regard the (real) wave frequency as given, and solve for possible values of the wave vector k. In contrast, the simplest studies of plasma instabilities usually regard the wave vector k as given, and try to obtain values for the growth rate $-i\omega$. Clearly there is really no sharp distinction between the two problems; and which procedure is most appropriate in a particular physical problem depends on the initial conditions and the boundary conditions. In fact it is useful to consider the behaviour of the dispersion equation $\phi(k, \omega) = 0$ for all complex values of k and ω. In this way it is possible to distinguish the intrinsic behaviour of the system from that which appears in the particular coordinate system we have chosen to study. In this course I shall have no time to discuss the distinction between amplifying and evanescent waves and convective and nonconvective instabilities, but they can be found in Buneman[1] and Sturrock[2], for example.

In principle, we are concerned with oscillations about states of equilibrium; but in many cases in plasma physics we are concerned with phenomena in which the time-scale for approach to complete equilibrium is very long compared with the time-scale of the problem being studied. Thus in many cases we are concerned with a "collisionless" plasma governed by the Vlasov equation, in which collisions are negligibly important compared with the effects of large-scale fields. In this case there is no reason why the initial particle distribution functions should be Maxwellian; in the absence of a magnetic field, there is no restriction on the shape of the distribution function but a large-scale magnetic field does impose some order on a collection of collisionless particles. At this stage it is perhaps worth mentioning one property of a magnetic field which blurs the distinction between individual particle behaviour and collective behaviour. All particles gyrate around the lines of force with frequency eB/mc and, as this frequency depends only on the position and not on the velocity of the particle, this individual-particle property plays an important role in wave propagation.

There will be insufficient time in this course to describe the multitude of waves which can be propagated in plasmas under different conditions; for these reference must be made to standard texts such as Stix[3] and Ginzburg.[4] In what follows an attempt has been made to discuss principles rather than fine details.

2

MAGNETOHYDRODYNAMIC WAVES

2.1. *Ideal magnetohydrodynamics*

In this section we consider the propagation of waves in a perfectly conducting fluid which contains a magnetic field B. All other dissipative terms in the equations (due to viscosity and thermal conductivity) are also neglected, and the waves are considered to be of low enough frequency for the displacement current and the electric force terms in the equation of motion to be ignored. With these approximations, and using Gaussian units for the electromagnetic quantities, the equations of magnetohydrodynamics may be written:

$$\rho \frac{dv}{dt} = -\operatorname{grad} p + \frac{\operatorname{curl} B \times B}{4\pi} + \rho F, \qquad \ldots[2.1]$$

$$\frac{\partial \rho}{\partial t} + \operatorname{div} \rho v = 0, \qquad \ldots[2.2]$$

$$\frac{1}{p} \frac{dp}{dt} = \frac{\gamma}{\rho} \frac{d\rho}{dt}, \qquad \ldots[2.3]$$

and

$$\frac{\partial B}{\partial t} = \operatorname{curl}(v \times B). \qquad \ldots[2.4]$$

In these equations ρ, p and γ are the fluid density, pressure and ratio of specific heats, v the fluid velocity, F any body force of non-electromagnetic origin (e.g. gravity) and d/dt denotes the total derivative $\partial/\partial t + v \cdot \operatorname{grad}$.

We first consider the propagation of small amplitude waves in a medium of uniform pressure p_0, density ρ_0 and containing a uniform magnetic field $B_0 b$, where b is a unit vector. Suppose that the quantities depart from their equilibrium values by amounts of the form

$$q_1 \exp(ik \cdot x - i\omega t). \qquad \ldots[2.5]$$

Then the linearized equations are:

$$\omega \rho_0 v_1 = p_1 k - [(k \times B_1) \times B_0 b]/4\pi, \qquad \ldots[2.6]$$

$$\omega \rho_1 = \rho_0 k \cdot v_1, \qquad \ldots[2.7]$$

$$p_1 = c_s^2 \rho_1, \qquad \ldots[2.8]$$

and $$\omega B_1 = -B_0 k \times (v_1 \times b),\qquad\qquad\qquad ...[2.9]$$

where $c_S = (\gamma p_0/\rho_0)^{\frac{1}{2}}$ is the velocity of sound in the undisturbed medium.

These equations can be combined to give

$$\omega^2 v_1 = c_S^2 (k \cdot v_1) k + c_H^2 \{k \times [k \times (v_1 \times b)]\} \times b,\qquad ...[2.10]$$

where $c_H = (B_0^2/4\pi\rho_0)^{\frac{1}{2}}$ is the hydromagnetic (Alfvén) velocity in the undisturbed medium. This equation can be rewritten

$$\omega^2 v_1 = (c_H^2 + c_S^2)(k \cdot v_1) k + c_H^2 [(k \cdot b)^2 v_1 - (k \cdot b)(v_1 \cdot b) k - (k \cdot v_1)(k \cdot b) b].$$
$$...[2.11]$$

If we now form scalar products of [2.11] with k, b and $k \times b$, we obtain

$$[\omega^2 - k^2(c_S^2 + c_H^2)] k \cdot v_1 + k^2 c_H^2 (k \cdot b) b \cdot v_1 = 0,\qquad ...[2.12]$$

$$\omega^2 b \cdot v_1 - c_S^2 (k \cdot b) k \cdot v_1 = 0,\qquad\qquad ...[2.13]$$

and $$[\omega^2 - c_H^2 (k \cdot b)^2] k \times b \cdot v_1 = 0.\qquad\qquad ...[2.14]$$

From equations [2.12], [2.13] and [2.14] we can now deduce the essential character of the small amplitude magnetohydrodynamic waves. Equation [2.14] gives the Alfvén waves. These possess motions which are transverse both to the direction of propagation k and to the uniform magnetic field b and the phase velocity is

$$(\omega/k)_A = c_H \cos\theta,\qquad\qquad\qquad ...[2.15]$$

where θ is the angle between the direction of propagation and the magnetic field. Equations [2.12] and [2.13] are coupled and lead to a quadratic equation for ω^2 given k or for k^2 given ω and the direction of propagation. This is

$$\omega^4 - k^2 \omega^2 (c_S^2 + c_H^2) + k^2 c_S^2 c_H^2 (k \cdot b)^2 = 0,$$

giving the phase velocities

$$(\omega/k)_\pm = 2^{-\frac{1}{2}} \{(c_S^2 + c_H^2) \pm [(c_S^2 + c_H^2)^2 - 4c_S^2 c_H^2 \cos^2\theta]^{\frac{1}{2}}\}^{\frac{1}{2}}.\quad ...[2.16]$$

The waves corresponding to these values of the phase velocity are called the fast and slow magnetosonic waves. In the case of propagation along the field, they have phase velocities c_S and c_H, thus reducing to ordinary sound waves and Alfvén waves; for propagation directly across the field, the phase velocity is $(c_S^2 + c_H^2)^{\frac{1}{2}}$, the hydromagnetic sound speed.

These results exhibit the first important property of a conducting medium containing a magnetic field, it is anisotropic. Waves propagating in different directions have different properties. To bring out one important consequence of this anisotropy consider the group velocity of Alfvén waves; in a medium of the type considered here the group velocity is also the velocity of energy propagation. The group velocity is defined by

$$v_{gr} = d\omega/dk;\qquad\qquad\qquad ...[2.17]$$

that is it is the velocity whose components are $\partial\omega/\partial k_x$, $\partial\omega/\partial k_y$, $\partial\omega/\partial k_z$. Consider Alfvén waves propagating in one direction with

$$\omega = c_H \mathbf{k} \cdot \mathbf{b},$$

then

$$v_{gr} = c_H \mathbf{b}. \qquad \qquad ...[2.18]$$

From equation [2.18] we see that the group velocity of Alfvén waves is in the direction of the zero order magnetic field and this is also the direction in which energy is propagated.

Lighthill[5] has stressed the importance of this property of Alfvén waves. In the approximation which we are considering in this section, the energy content in an Alfvén wave disturbance is propagated directly along the lines of force and it does not suffer the spherical attenuation of, for example, ordinary sound in a fluid without a magnetic field. The larger the sound velocity becomes compared with the hydromagnetic velocity, the more pronounced becomes this property and in the limit of an incompressible fluid only Alfvén waves can be propagated. This lack of purely geometrical attenuation makes Alfvén waves very important in many astrophysical problems as a disturbance at one point can be propagated to another point along the lines of magnetic force and not be dissipated in the surrounding medium. Later we shall consider what happens to this property of Alfvén waves when various dissipative processes are allowed for.

It is more difficult to consider the group velocities of the fast and slow magnetosonic waves in the general case, but it is reasonably simple when either the hydromagnetic velocity or the sound velocity is very much greater than the other. In this case, if we denote the velocities by $c_>$ and $c_<$, we have

$$\omega^2 \approx k^2 c_>^2 + [k^2 - (\mathbf{k} \cdot \mathbf{b})^2] c_<^2 \qquad ...[2.19]$$

and

$$\omega^2 \approx (\mathbf{k} \cdot \mathbf{b})^2 c_<^2 \left[1 - \frac{c_<^2}{c_>^2} \left(1 - \frac{(\mathbf{k} \quad \mathbf{b})^2}{k^2} \right) \right], \qquad ...[2.20]$$

showing that to a first approximation the fast wave has a spherically symmetrical propagation of energy while the slow wave mainly propagates energy along the field. For further details see, for example, Thompson.[6]

It is of course possible to include the displacement current in this discussion. This leads to alterations of order c_H^2/c^2 in the phase velocity of the waves. The displacement current then formally becomes important when the density is very low but it is likely that some other factors omitted in this section are also important then.

2.2. *Wave propagation in a medium with varying properties*

In this section we are concerned with wave propagation in a medium in which the physical properties vary with position. Before considering the general theory, we study a simple example in which the equations can be solved exactly.

Thus we study the propagation of magnetosonic waves in a horizontally stratified medium where the stratification is caused by a gravitational field. In the equilibrium state the physical quantities have the values

$$\rho = \rho_0 \exp(-2az), \qquad \qquad ...[2.21]$$

$$p = p_0 \exp(-2az) \qquad \qquad ...[2.22]$$

and

$$\mathbf{B} = [B_0 \exp(-az), 0, 0]. \qquad \qquad ...[2.23]$$

The gravitational field is

$$\mathbf{g} = [0, 0, -g] \qquad \qquad ...[2.24]$$

and the system is in equilibrium provided that

$$a = \rho_0 g / (B_0^2/4\pi + 2p_0). \qquad \qquad ...[2.25]$$

The equilibrium is not necessarily stable, but relative values of B_0^2 and p_0 can be chosen so that it is stable. Note that it is not on the face of it a very good example of a spatially inhomogeneous medium, as the local velocity of sound and hydromagnetic velocity do not vary from point to point. Nevertheless it does possess one property of interest as we shall see.

Small departures from this equilibrium are considered, in which each physical quantity has a variation of the form

$$q_1(z) \exp(-i\omega t), \qquad \qquad ...[2.26]$$

so that we study propagation entirely in the z direction but cannot assume a simple harmonic variation in z. The perturbed equations have the form:

$$-i\omega \rho_0 \exp(-2az) v_1$$
$$= -D p_1 - a B_0 \exp(-az) B_1/4\pi - B_0 \exp(-az) D B_1/4\pi - \rho_1 g. \quad ...[2.27]$$

$$-i\omega \rho_1 - 2a\rho_0 \exp(-2az) v_1 + \rho_0 \exp(-2az) D v_1 = 0, \qquad ...[2.28]$$

$$-i\omega p_1 - 2a p_0 \exp(-2az) v_1 = c_S^2(-i\omega \rho_1 - 2a\rho_0 \exp(-2az) v_1) \qquad ...[2.29]$$

and

$$-i\omega B_1 = a B_0 \exp(-az) v_1 - B_0 \exp(-az) D v_1, \qquad \qquad ...[2.30]$$

where v_1 is the z component of the perturbed velocity, B_1 is the x component of the perturbed magnetic field and all other components of both vectors are zero, and D denotes d/dz.

All quantities other than v_1 can be eliminated from these equations to give the single equation

$$\omega^2 v_1 = -(c_S^2 + c_H^2) D^2 v_1 + 2a(c_S^2 + c_H^2) D v_1, \qquad \qquad ...[2.31]$$

showing that the spatial variations in the equilibrium quantities just lead to one additional term in this equation. It is now clear that we can write

$v_1 \propto \exp(ikz)$, where k may be complex, to obtain the dispersion equation

$$\omega^2 = k^2[c_S^2 + c_H^2] + 2aik[c_S^2 + c_H^2], \qquad \dots[2.32]$$

showing the modification produced by the spatial inhomogeneity in the relation between ω and k. We see that k cannot be real for real ω. Thus either attenuation or amplification occurs.

If we assume that a is small compared with $|k|$, so that the medium is slowly varying in the wavelength of a magnetosonic wave, we can write

$$k \approx -ia \pm \omega/(c_S^2 + c_H^2)^{\frac{1}{2}}. \qquad \dots[2.33]$$

This shows that superimposed on the ordinary magnetosonic oscillation of wave number $\omega/(c_S^2 + c_H^2)^{\frac{1}{2}}$ there is a growth in the amplitude of v_1 with $v_1 \propto \exp(az)$. From equations [2.28], [2.29] and [2.30] it can be seen that none of ρ_1, p_1 and B_1 increases in amplitude with z, but they all in fact grow in proportion to the corresponding equilibrium quantities.

This property, that in the absence of dissipation the amplitude of an oscillation must grow as it moves into a region of decreasing density, is really quite obvious and follows from the conservation of energy, but it is perhaps of interest to study one problem in which the equations can be solved exactly. It is also clear that the condition on the density variations will be modified if we have spherical waves propagating outwards in a medium of spherical symmetry. In the particular problem studied above, an exercise for the student is the corresponding problem for propagation at an arbitrary angle to the vertical.

In the more general case when the equations cannot be solved exactly the problem is much more difficult. If the scale of variation of the equilibrium quantities is large compared with the wavelength of the oscillation, the method of geometrical optics can be used. This is discussed for plasma waves by Ginzburg[4] and for magnetohydrodynamic waves by Lighthill.[7] In the general case of the propagation of waves in a stratified medium, we are concerned not merely with deciding how the amplitude of a particular wave changes as it propagates. We are concerned also with the possibility that it may be reflected and prevented from entering a particular medium or that it might change its character. Thus Lighthill is particularly concerned with the propagation of waves in the solar atmosphere and with the possibility that gravity waves in the lower atmosphere might transform into magnetohydrodynamic waves higher up. In other cases magnetohydrodynamic waves moving into a medium of decreasing density might transform into electromagnetic waves, although that is not permitted by the equations of the present section.

In the approximation of geometrical optics it is supposed that any perturbed physical quantity has the form

$$q_1(z) \exp i(k_x x + k_y y + k_z(z) - \omega t), \qquad \dots[2.34]$$

where again we are supposing that the medium is horizontally stratified. If the equation to be solved has, for example, the form

$$\nabla^2 f + (\omega^2/c^2)\alpha(z)f = 0,$$

or
$$\frac{d^2 f}{dz^2} + (\omega^2/c^2)\varepsilon(z)f = 0, \qquad \qquad ...[2.35]$$

where

$$\varepsilon(z) = \alpha(z) - (k_x^2 + k_y^2)c^2/\omega^2,$$

we attempt a solution in the form

$$f = [f_0 + (c/\omega)f_1 + (c/\omega)^2 f_2 + \ldots]\exp(-i\omega\psi(z)/c). \qquad ...[2.36]$$

Substitution of [2.36] into [2.35] then leads to an equation of the form

$$A\omega^2/c^2 + B\omega/c + C + Dc/\omega \ldots = 0, \qquad \qquad ...[2.37]$$

where A, B etc. are expressions involving the f_i, ψ and their derivatives. We then demand that equation [2.37] holds for all values of ω/c and equate A, B ... to zero. Then we obtain successive equations of the form

$$\{\varepsilon - (d\psi/dz)^2\}f_0 = 0, \qquad \qquad ...[2.38]$$

$$df_0/dz + (f_0 d^2\psi/dz^2)/2d\psi/dz = 0, \qquad \qquad ...[2.39]$$

$$df_1/dz + (f_1 d^2\psi/dz^2)/2d\psi/dz = (d^2 f_0/dz^2)/2id\psi/dz, \qquad ...[2.40]$$

and so on. The solutions of equations [2.38] and [2.39] then give

$$f_0(z) = C/(d\psi/dz)^{\frac{1}{2}} = C/\{\varepsilon(z)\}^{\frac{1}{4}} \qquad \qquad ...[2.41]$$

where C is a constant. The third equation then gives

$$f_1(z) = \frac{1}{\{\varepsilon(z)\}^{\frac{1}{4}}} \int_{z_0}^{z} \frac{d^2 f_0/dz^2}{2i\{\varepsilon(z)\}^{\frac{1}{4}}}\, dz. \qquad \qquad ...[2.42]$$

This procedure is of course the well known J.W.K.B. approximation. The first approximation of geometrical optics (neglect of f_1 and higher terms) then requires that

$$c\,|f_1|/\omega \ll f_0. \qquad \qquad ...[2.43]$$

Let us now use this method to study the propagation of Alfvén waves in a medium with variable pressure and density. If we suppose that the fluid is stratified in the z direction and that there is a uniform magnetic field in the z direction and we consider propagation along the field, we do not need to specify what additional force is causing the stratification and the perturbed equations reduce to

$$d^2 v_1/dz^2 + [\omega^2/c_H^2(z)]v_1 = 0, \qquad \qquad ...[2.44]$$

where this equation holds for either the x or y component of the perturbed velocity. Equation [2.44] has precisely the form [2.35], and it can be solved by the method of geometrical optics, provided the density varies slowly over a wavelength, to give

$$v_1(z) \approx C(c_H/\omega)^{\frac{1}{2}} \exp\left(-\int i\omega dz/c_H\right), \qquad \ldots[2.45]$$

showing the variation of both amplitude and phase with density.

2.3. *Effect of transport processes on magnetohydrodynamic waves*

In this section we consider problems in which the collision frequencies which are important in transport processes are higher than the gyration frequencies of the particles. In this case the transport processes are isotropic and we can write the basic equations in the form

$$\rho \frac{d\boldsymbol{v}}{dt} = -\operatorname{grad} p + \frac{\operatorname{curl} \boldsymbol{B} \times \boldsymbol{B}}{4\pi} + \tfrac{1}{3}\mu \operatorname{grad} \operatorname{div} \boldsymbol{v} + \mu \nabla^2 \boldsymbol{v}, \qquad \ldots[2.46]$$

$$\frac{\partial \rho}{\partial t} + \operatorname{div} \rho \boldsymbol{v} = 0, \qquad \ldots[2.2]$$

$$\tfrac{3}{2}\frac{dp}{dt} = -\tfrac{3}{2}p \operatorname{div} \boldsymbol{v} - p_{\alpha\beta}\frac{\partial v_\beta}{\partial x_\alpha} + \lambda \nabla^2 T + \frac{c^2}{16\pi^2\sigma}(\operatorname{curl} \boldsymbol{B})^2, \qquad \ldots[2.47]$$

$$p = R\rho T/M, \qquad \ldots[2.48]$$

and

$$\frac{\partial \boldsymbol{B}}{\partial t} + \frac{c^2}{4\pi\sigma} \operatorname{curl} \operatorname{curl} \boldsymbol{B} = \operatorname{curl}(\boldsymbol{v} \times \boldsymbol{B}), \qquad \ldots[2.49]$$

where μ, λ and σ are the coefficients of viscosity, thermal conductivity and electrical conductivity, the ratio of specific heats has been taken to be 5/3 (for a fully ionized gas), R is the gas constant, M is the mean mass of the particles in the plasma in terms of the mass of the hydrogen atom and

$$p_{\alpha\beta} = (p + \tfrac{2}{3}\mu \operatorname{div} \boldsymbol{v})\delta_{\alpha\beta} - \mu\left(\frac{\partial v_\alpha}{\partial x_\beta} + \frac{\partial v_\beta}{\partial x_\alpha}\right). \qquad \ldots[2.50]$$

We now consider again perturbations of an equilibrium in which there is a uniform magnetic field and the fluid has a uniform density, pressure and temperature T_0. The linearized equations then are:

$$\omega\rho_0\boldsymbol{v}_1 = p_1\boldsymbol{k} - [(\boldsymbol{k} \times \boldsymbol{B}_1) \times \boldsymbol{B}_0\boldsymbol{b}]/4\pi - (i\mu/3)(\boldsymbol{k} \cdot \boldsymbol{v}_1)\boldsymbol{k} - i\mu k^2\boldsymbol{v}_1,\ldots[2.51]$$

$$\omega\rho_1 = \rho_0\boldsymbol{k} \cdot \boldsymbol{v}_1 \qquad \ldots[2.7]$$

$$\omega p_1 = (5p_0/3)\boldsymbol{k} \cdot \boldsymbol{v}_1 - (2i\lambda/3)k^2 T_1, \qquad \ldots[2.52]$$

$$p_1 = (RT_0/M)\rho_1 + (R\rho_0/M)T_1 \qquad \ldots[2.53]$$

and

$$\omega\boldsymbol{B}_1 - (ic^2/4\pi\sigma)\boldsymbol{k} \times (\boldsymbol{k} \times \boldsymbol{B}_1) = -B_0\boldsymbol{k} \times (\boldsymbol{v}_1 \times \boldsymbol{b}). \qquad \ldots[2.54]$$

It is again possible to combine these equations to obtain one equation for $\boldsymbol{v}_1$ and this has the form

$$\omega^2 v_1 = c_S^2 \left[\frac{(2i\lambda/5)k^2 T_0 + \omega p_0}{(2i\lambda/3)k^2 T_0 + \omega p_0}\right](k \quad v_1)k + \frac{c_H^2}{[1+ik^2c^2/4\pi\sigma\omega]}$$

$$\{k\times[k\times(v_1\times b)]\}\times b - \frac{i\mu\omega}{3\rho_0}(k\cdot v_1)k - \frac{i\mu\omega}{\rho_0}k^2 v_1.\ddagger \ldots[2.55]$$

It is clear that equation [2.55] is only a slight generalisation of equation [2.10] and it can again be separated to give Alfvén and magnetosonic waves, showing how they are separately affected by the finite transport processes. Thus, forming the scalar product of [2.55] with $k\times b$, we obtain

$$[\omega^2 - c_H^2(k\cdot b)^2/\{1+ik^2c^2/4\pi\sigma\omega\} + i\mu\omega k^2/\rho_0]k\times b\cdot v_1 = 0. \quad \ldots[2.56]$$

This gives the dispersion equation for Alfvén waves

$$[\omega^2 + i\mu\omega k^2/\rho_0][1+ik^2c^2/4\pi\sigma\omega] = k^2 c_H^2 \cos^2\theta. \quad \ldots[2.57]$$

Regarded as an equation for k^2 given ω, equation [2.57] has two roots. One of these gives the modification to Alfvén wave propagation by viscosity and resistivity while the other mode, which does not exist in the absence of viscosity and resistivity, is not a propagating mode.

Provided the viscous and resistive terms are not too large, the modified dispersion equation for the Alfvén wave is

$$k^2(c_H^2\cos^2\theta - i\mu\omega/\rho_0 - ic^2\omega/4\pi\sigma) \approx \omega^2, \quad \ldots[2.58]$$

which shows the characteristic damping associated with finite viscosity and finite resistivity. Thus, if we write

$$k = k_0 + ik_1,$$

we have

$$k_1 = (k_0^3/2)[\mu/\rho_0\omega + c^2/4\pi\sigma\omega], \quad \ldots[2.59]$$

where the characteristic length-scale associated with viscous dissipation is $(\mu/\rho_0\omega)^{\frac{1}{2}}$ and that with resistive dissipation is $(c^2/4\pi\sigma\omega)^{\frac{1}{2}}$.

By taking scalar products of [2.55] with k and b we can also obtain the equations governing the modified magnetosonic waves. Thus

$$\left\{\omega^2 - k^2 c_S^2\left[\frac{(2i\lambda/5)k^2 T_0 - \omega p_0}{(2i\lambda/3)k^2 T_0 - \omega p_0}\right] - \frac{k^2 c_H^2}{(1+ik^2c^2/4\pi\sigma\omega)} + \frac{4i\mu\omega k^2}{3\rho_0}\right\}k\cdot v_1$$

$$+ \frac{k^2 c_H^2}{(1+ik^2c^2/4\pi\sigma\omega)}(k\cdot b)b\cdot v_1 = 0, \quad \ldots[2.60]$$

$$(\omega^2 + i\mu\omega k^2/\rho_0)b\cdot v_1 - \left\{c_S^2\left[\frac{(2i\lambda/5)k^2 T_0 + \omega p_0}{(2i\lambda/3)k^2 T_0 + \omega p_0}\right] - \frac{i\mu\omega}{3\rho_0}\right\}(k\cdot b)k\cdot v_1 = 0.$$

$$\ldots[2.61]$$

‡ Note that care must be exercised in dealing with complex vectors, the most obvious reason being that $a\cdot a$ need not be real and positive and can vanish.

It is clear that in the general case this leads to a very complicated dispersion equation, but in the particular cases of propagation along and across the field it is easy to see how the magnetosonic and the sound waves are modified by dissipation.

In many cases in plasma physics we are concerned with situations in which the particles' gyration radii are small compared with mean free paths between collisions. If both of these are small enough for hydrodynamic equations to give a valid description of the plasma, the transport coefficients become anisotropic; it is then more difficult for particles to move across field lines than along them. In the case where the mean free path and gyration radius are comparable, the transport coefficients are extremely complicated (see e.g. Chapman and Cowling).[8] The case in which normal dissipative processes are unimportant will be discussed in Section 3.

2.4. *Magnetohydrodynamic waves in a bounded medium*

At present we have discussed the propagation of waves in an infinite medium. In this section we consider the propagation of waves down a cylinder, the magnetohydrodynamic waveguide. The main difference between wave propagation in bounded and unbounded media is that, for all except the simplest rectilinear boundaries, the various modes are coupled by the boundary conditions. In addition, only certain frequencies of propagation may be possible.

Suppose a uniform plasma with a field parallel to the z axis is contained within a cylinder of radius R. In this case it is convenient to use cylindrical polar coordinates (r, θ, z) and to take the typical perturbed quantity in the form

$$q_1(r) \exp i(m\theta + kz - \omega t). \qquad \qquad ...[2.62]$$

If the plasma is contained within a perfectly conducting tube, we must demand that the radial components of the velocity and the magnetic field vanish at the wall.

The perturbed equations can be written

$$-i\omega\rho_0 v_{1r} = -Dp_1 + (B_0/4\pi)[ikB_{1r} - DB_{1z}], \qquad ...[2.63]$$

$$-i\omega\rho_0 v_{1\theta} = -(im/r)p_1 + (B_0/4\pi)[ikB_{1\theta} - (im/r)B_{1z}], \qquad ...[2.64]$$

$$-i\omega\rho_0 v_{1z} = -ikp_1, \qquad ...[2.65]$$

$$-i\omega\rho_1 + \rho_0[(1/r)D(rv_{1r}) + (im/r)v_{1\theta} + ikv_{1z}] = 0, \qquad ...[2.66]$$

$$p_1 = c_S^2\rho_1, \qquad ...[2.67]$$

$$-i\omega B_{1r} = ikB_0 v_{1r}, \qquad ...[2.68]$$

$$-i\omega B_{1\theta} = ikB_0 v_{1\theta} \qquad ...[2.69]$$

and $\qquad -i\omega B_{1z} = (B_0/r)D(rv_{1r}) - (imB_0/r)v_{1\theta}, \qquad ...[2.70]$

where D denotes d/dr.

Equations [2.63] and [2.68], and [2.64] and [2.69], can be combined in pairs to give

$$(k^2 c_H^2 - \omega^2) v_{1r} = (i\omega/\rho_0) D[p_1 + B_0 B_{1z}/4\pi] \qquad \ldots [2.71]$$

and $\qquad (k^2 c_H^2 - \omega^2) v_{1\theta} = (\omega m/\rho_0 r)[p_1 + B_0 B_{1z}/4\pi]. \qquad \ldots [2.72]$

From equations [2.71] and [2.72] we see that either

$$\omega^2 = k^2 c_H^2 \quad \text{and} \quad p_1 + B_0 B_{1z}/4\pi = 0, \qquad \ldots [2.73]$$

or we can use equation [2.71] and [2.72] to eliminate v_{1r} and $v_{1\theta}$ from the other equations. The mode given by [2.73] represents the propagation of Alfvén waves down the tube. For this mode, further study of the equations shows that

$$p_1 = \rho_1 = v_{1z} = B_{1z} = 0, \qquad \ldots [2.74]$$

while v_{1r} and $v_{1\theta}$ must further satisfy

$$D(rv_{1r}) + imv_{1\theta} = 0. \qquad \ldots [2.75]$$

Equation [2.75] is the only further constraint on the velocity components, and hence on the magnetic field components, other than the requirement that v_{1r} vanishes on $r = R$.

The solution of the equations when $\omega^2 \neq k^2 c_H^2$ is discussed for example by Tayler[9] (note however that there ω is replaced by $i\omega$). The solution for v_{1r} is shown to be

$$v_{1r} \propto J_m'(\kappa r), \qquad \ldots [2.76]$$

where J_m is the Bessel function and the dash denotes differentiation with respect to the argument and

$$\kappa^2 = \frac{(\omega^2 - k^2 c_H^2)(\omega^2 - k^2 c_S^2)}{[\omega^2 (c_S^2 + c_H^2) - k^2 c_S^2 c_H^2]}. \qquad \ldots [2.77]$$

If the roots of $J_m'(x) = 0$ are j_{mn}, the dispersion equation for the mode is

$$\frac{j_{mn}^2}{R^2} = \frac{(\omega^2 - k^2 c_H^2)(\omega^2 - k^2 c_S^2)}{[\omega^2 (c_S^2 + c_H^2) - k^2 c_S^2 c_H^2]}. \qquad \ldots [2.78]$$

There are an infinite number of real roots j_{mn}.

For given positive values of k^2, j_{mn}^2 and R^2, there are two values of ω^2 for which waves will propagate. One of these lies in the range

$$k^2 c_S^2 c_H^2/(c_S^2 + c_H^2) < \omega^2 < k^2 c_<^2 \qquad \ldots [2.79]$$

and the other in the range

$$k^2 c_>^2 < \omega^2 < \infty, \qquad \ldots [2.80]$$

where as before $c_<$ and $c_>$ are the lesser and greater of c_H and c_S.

2.5. *Large amplitude magnetohydrodynamic waves*

Consider first a rather special case. Alfvén waves in an incompressible fluid are not restricted to be of small amplitude. If we consider non-linear terms we cannot use complex exponentials for the perturbed quantities, because the real and imaginary parts no longer satisfy the equations individually. The perturbed equations can be written (again we take $v_0 = 0$ and $B_0 = $ constant etc.)

$$\frac{\partial v_1}{\partial t} + v_1 \cdot \operatorname{grad} v_1 = -(1/\rho_0) \operatorname{grad}(p_1 + B_1^2/8\pi)$$

$$+ (B_0/4\pi\rho_0)(b \cdot \operatorname{grad} B_1) + (1/4\pi\rho_0) B_1 \cdot \operatorname{grad} B_1, \quad \dots [2.81]$$

$$\operatorname{div} v_1 = \operatorname{div} B_1 = 0 \qquad \qquad \dots [2.82]$$

and $$\frac{\partial B_1}{\partial t} + v_1 \cdot \operatorname{grad} B_1 = B_0 b \cdot \operatorname{grad} v_1 + B_1 \cdot \operatorname{grad} v_1. \qquad \dots [2.83]$$

We can now see that if $p_1 + B_1^2/8\pi$ is constant, and if $v_1 = \pm B_1/(4\pi\rho_0)^{\frac{1}{2}}$, both equations [2.81] and [2.83] reduce to

$$\frac{\partial B_1}{\partial t} = \pm \frac{B_0}{(4\pi\rho_0)^{\frac{1}{2}}} \, b \cdot \operatorname{grad} B_1. \qquad \dots [2.84]$$

If we take b in the z direction, then [2.84] becomes

$$\frac{\partial B_1}{\partial t} = \pm c_H \frac{\partial B_1}{\partial z}, \qquad \qquad \dots [2.85]$$

which has solutions

$$B_1 = B_0 f(x, y, z \pm c_H t), \qquad \qquad \dots [2.86]$$

where f is an arbitrary vector function apart from the requirement that

$$\operatorname{div} f = 0. \qquad \qquad \dots [2.87]$$

We thus have waves of arbitrary amplitude moving in either the positive or negative z direction, but we cannot of course superpose solutions.

This is a rather special case, in which waves can be propagated with an arbitrary amplitude, and even in this case the effects of compressibility, however small, will cause this approximation to break down. More often we are concerned with the gradual steepening of a small amplitude wave into a shock wave. This could be caused either by the non-linear terms in a uniform medium or, as in the example studied in Section 2.2, by a small amplitude wave running into a region of decreasing density. Thompson[6] gives a general discussion, which shows that compression waves in magnetohydrodynamics do steepen into shock discontinuities.

In this section we shall consider only the simplest problem, a steady shock wave with conditions uniform both in front of and behind the shock. Further, we consider only the case of a one dimensional shock propagating normal to the magnetic field. In the first approximation we regard the shock transition as a discontinuity. The values of the physical variables in the uniform regions on the two sides of the shock are then related by the magnetohydrodynamic Rankine-Hugoniot conditions first discussed by de Hoffman and Teller.[20]

In the simplest case these relations can be found as follows. If the shock motion is in the x direction and the field is in the y direction and we use the frame in which the shock front is at rest, we have the equations

$$\frac{d}{dx}(\rho v_x) = 0, \qquad \ldots[2.88]$$

$$B_y \frac{dv_x}{dx} + v_x \frac{dB_y}{dx} = \frac{c^2}{4\pi\sigma}\frac{d^2 B_y}{dx^2} \qquad \ldots[2.89]$$

and
$$\rho v_x \frac{dv_x}{dx} = -\frac{dp}{dx} - \frac{B_y}{4\pi}\frac{dB_y}{dx} + \frac{4\mu}{3}\frac{d^2 v_x}{dx^2}, \qquad \ldots[2.90]$$

where we include the transport coefficients σ and μ because they must become important in the shock transition region. We can now integrate equations [2.88]–[2.90] through the transition region to obtain relations between the values of the physical quantities on the two sides of the shock. If we do this, we obtain

$$[\rho v_x] = 0, \qquad ..[2.91]$$

$$[v_x B_y] = (c^2/4\pi\sigma)[dB_y/dx] = 0 \qquad ..[2.92]$$

and
$$[\rho v_x^2 + p + B_y^2/8\pi] = (4\mu/3)[dv_x/dx] = 0, \qquad ..[2.93]$$

where the square brackets denote the change in the quantity within them across the shock, and where the right-hand sides of equations [2.92] and [2.93] are zero, because conditions are uniform sufficiently far away from the shock on either side. These are three conditions between four quantities on each side of the two sides of the shock, and one more relation is required if the conditions at infinity on one side of the shock are to be expressed in terms of those on the other side.

This further relation is obtained by multiplying equation [2.90] by v_x before integrating. This gives

$$\rho v_x [\tfrac{1}{2}v_x^2 + p/\rho + B^2/4\pi\rho]_1^2 + \frac{4\mu}{3}\int_1^2 \left(\frac{dv_x}{dx}\right)^2 dx$$
$$+ \frac{c^2}{16\pi^2\sigma}\int_1^2 \left(\frac{dB_y}{dx}\right)^2 dx - \int_1^2 p\frac{d}{dx}\left(\frac{1}{\rho}\right)\rho v_x dx = 0. \qquad \ldots[2.94]$$

In [2.94] the last three terms represent the rate at which heat is deposited in the fluid by viscous and resistive dissipation and the rate at which the gas is doing work. They are therefore equally the rate at which the gas is increasing its internal energy

$$\rho v_x [p/(\gamma-1)\rho].$$

Thus equation [2.94] can be rewritten

$$[\tfrac{1}{2}v_x^2 + \gamma p/(\gamma-1)\rho + B^2/4\pi\rho] = 0. \qquad \ldots[2.95]$$

We can now express ρ_2, p_2, v_{x2}, B_{y2} behind the shock in terms of ρ_1, p_1, v_{x1} and B_{y1} in front of the shock. By manipulation of these equations one can obtain the speed of the shock (in the frame in which the fluid ahead of the shock is at rest) in terms of the hydromagnetic and sound speeds in the gas ahead of the shock and the density jump across the shock. This gives

$$v_S^2 = \left(\frac{2}{\gamma-1}\right) \frac{c_S^2 + c_H^2[1+(1-\gamma/2)(\rho_2/\rho_1-1)]}{[\rho_1/\rho_2-(\gamma-1)/(\gamma+1)]}. \qquad \ldots[2.96]$$

We can express v_S as a Mach number, as is usual in hydrodynamics, but in this case we have several possible speeds c_S, c_H, $(c_S^2+c_H^2)^{\tfrac{1}{2}}$, in terms of which we can express v_S. For propagation across the field the relevant speed is perhaps $(c_S^2+c_H^2)^{\tfrac{1}{2}}$, and we can write

$$M^2 = \frac{v_S^2}{c_S^2+c_H^2} = \left(\frac{2}{\gamma-1}\right)\left[\frac{1+\{c_H^2/(c_S^2+c_H^2)\}(1-\gamma/2)(\rho_2/\rho_1-1)}{\rho_1/\rho_2-(\gamma-1)/(\gamma+1)}\right].$$

$$\ldots[2.97]$$

The Mach speed approaches infinity when $\rho_2/\rho_1 = (\gamma+1)/(\gamma-1)$, and this gives the maximum density compression that can be obtained in the shock; this result is the hydrodynamic result and is independent of the strength of the magnetic field.

The problem of oblique propagation is considerably more complicated and it is discussed by Thompson.[6] The actual structure of the shock transition is determined by the values of the transport coefficients and the width of the transition region is of the order of the appropriate mean free path. Shock structure is discussed, for example by Marshall[11] and Germain.[12]

3

Oscillations of a Two-Fluid Plasma

3.1. *Basic equations*

In this section we will suppose that we have a fully ionized plasma in which the electrons and ions separately obey hydrodynamic equations and that they are coupled both by long-range electromagnetic forces and by close collisions, although in the applications we make we will neglect the collisions. The discussion given in this section can be extended to a multicomponent plasma and to a partially ionized plasma; here, of course, the collisions must be included if the neutral gas is to be coupled to the ionized gas.

The equations of motion for the two fluids are taken to be

$$n_e m_e \left[\frac{\partial v_e}{\partial t} + v_e \cdot \text{grad } v_e \right] = -\text{grad } p_e - F - n_e e[E + (v_e \times B)/c] \qquad \text{...[3.1]}$$

and

$$n_i m_i \left[\frac{\partial v_i}{\partial t} + v_i \cdot \text{grad } v_i \right] = -\text{grad } p_i + F + n_i Ze[E + (v_i \times B)/c], \qquad \text{...[3.2]}$$

where n_e, m_e, p_e and v_e are the number density, mass, pressure, and velocity of the electrons and the corresponding quantities with suffix i of the ions. The electron charge is $-e$ and the ion charge is Ze. F is the frictional force between the electrons and ions and it can conveniently be approximated by

$$F = \nu m_e n_e [v_e - v_i], \qquad \text{...[3.3]}$$

where ν is an approximate collision frequency for electrons with ions.

We can now introduce instead of the velocities v_i and v_e, the mass velocity v and the current j defined by

$$v = (m_e n_e v_e + m_i n_i v_i)/(m_e n_e + m_i n_i) \qquad \text{...[3.4]}$$

and

$$j = -n_e e v_e + n_i Ze v_i. \qquad \text{...[3.5]}$$

We also define

$$\rho = n_e m_e + n_i m_i \qquad \text{...[3.6]}$$

and

$$p = p_e + p_i. \qquad \text{...[3.7]}$$

The continuity equations for the electrons and ions, which we have not written down, immediately combine to give

$$\frac{\partial \rho}{\partial t} + \text{div } \rho v = 0, \qquad \text{...[3.8]}$$

the usual equation of mass conservation. The combination of the equations of motion is not so simple because of the quadratic terms in the velocities; but, if we are considering only wave propagation in media in which velocities are zero in the undisturbed state, we may neglect the quadratic terms and obtain

$$\rho \frac{\partial v}{\partial t} = -\text{grad } p + qE + j \times B/c, \qquad \text{...[3.9]}$$

where

$$q = e[Zn_i - n_e] \qquad \qquad ...[3.10]$$

and

$$\frac{\partial j}{\partial t} = \frac{e}{m_e} \text{ grad } p_e - \frac{Ze}{m_i} \text{ grad } p_i + \left(\frac{n_e e^2}{m_e} + \frac{n_i Z^2 e^2}{m_i} \right) E$$
$$+ \left(\frac{n_e e^2}{m_e} v_e + \frac{n_i Z^2 e^2}{m_i} v_i \right) \times B/c + \left(\frac{e}{m_e} + \frac{Ze}{m_i} \right) F. \qquad ...[3.11]$$

Equation [3.11], which is the generalized Ohm's law, is still very complicated but it can be simplified if two further approximations are made. These are that m_e is very small compared to m_i and that the plasma is quasi-neutral so that q is small and can be neglected whenever it multiplies anything which vanishes in the equilibrium state. With these assumptions, in equation [3.11] we can write

$$v_i \approx v, \quad v_e \approx v - j/n_e e, \quad eF/m_e \approx -vj, \qquad ...[3.12]$$

and with neglect of terms with m_i in the denominator equation [3.11] becomes

$$\frac{\partial j}{\partial t} = \frac{e}{m_e} \text{ grad } p_e + \frac{n_e e^2}{m_e} [E + v \times B/c] - \frac{e}{m_e c} j \times B - vj,$$

or

$$\frac{m_e}{n_e e^2} \frac{\partial j}{\partial t} + \frac{j}{\sigma} = E + \frac{v \times B}{c} + \frac{1}{n_e e} \left[\text{grad } p_e - \frac{j \times B}{c} \right], \qquad ...[3.13]$$

where $\sigma = n_e e^2/m_e v$ is the electrical conductivity of the plasma.

This generalized Ohm's law has three terms which we have not included before. These are the electron inertia term $(m_e/n_e e^2) \partial j/\partial t$ and the last two terms on the right-hand side (electron pressure gradient and Hall effect) which essentially arise from the Larmor motions of electrons and ions. The Hall effect alone gives rise to an anisotropic electrical conductivity. If electron inertia and pressure gradient are negligible equation [3.13] can be solved to obtain

$$j = \sigma \left[E + \frac{v \times B}{c} \right], \qquad ...[3.14]$$

where σ is a tensor; and in this form it can be seen that, in a strong magnetic field, the conductivity across the field is reduced compared with that along the field. Although we call σ the conductivity tensor, some of its components do not lead to dissipation of energy, and it is this fact that will be important when we consider the propagation of waves. In what follows we shall neglect the effect of the Ohmic term which has already been considered in Section 2.3 and consider the effect of the additional terms in equation [3.13].

The terms due to the finite Larmor motions which appear in the generalized Ohm's law have their analogues in the equation of motion if the interactions between electrons and ions are treated more carefully than has been done above, where in equation [3.3] we have neglected the frictional force between particles of the same species and have used a very simple approximation for the interaction between the two species. If the problem is considered more carefully, the collisions introduce viscous terms into the equations of motion and, in the presence of a magnetic field, the viscosity is anisotropic. The viscosity is largely caused by the ions and in a strong magnetic field it depends on both the gyration frequency and gyration radius of the ions. If the ions are cold, the terms are unimportant. They have been considered, for example, by Roberts and Taylor.[13]

If neutral particles are also present in the plasma, they are only coupled to the electrons and ions through collision terms. It is then necessary to keep three velocities (v, j and r_n) and to have three coupled equations of motion. Discussions of the two and three fluid equations may be found, for example, in Van Kampen and Felderhof[14] and Schlüter.[15]

3.2. *Wave propagation in a uniform medium*

We now consider an equilibrium in which there is a uniform magnetic field $B_0 b$, density ρ_0 pressure p_0 and electron and ion number densities n_{e0} and n_{i0}. The propagation of waves is considered in which all of perturbed quantities are proportional to $\exp(ik \cdot x - i\omega t)$ and, as before, the displacement current is neglected. The perturbed equations are

$$\omega \rho_0 v_1 = p_1 k - (B_0/4\pi)(k \times B_1) \times b, \qquad \qquad ...[3.15]$$

$$-\omega[B_1 + (c^2/\omega_{pe}^2) k \times (k \times B_1)] = B_0 k \times (v_1 \times b) - (iB_0 c/4\pi n_{e0} e) k \\ \times [(k \times B_1) \times b], \qquad ...[3.16]$$

$$p_1 = c_S^2 \rho_1 \qquad \qquad ...[3.17]$$

and

$$\rho_1 = (\rho_0/\omega) k \cdot v_1, \qquad \qquad ...[3.18]$$

where

$$\omega_{pe}^2 = 4\pi n_{e0} e^2/m_e. \qquad \qquad ...[3.19]$$

These equations can be combined to give

$$\omega^2 v_1 = c_S^2 (k \cdot v_1) k + (B_0 \omega/4\pi \rho_0) b \times (k \times B_1) \qquad ...[3.20]$$

and

$$-(1 - k^2 c^2/\omega_{pe}^2) B_1 = (B_0/\omega) k \times (v_1 \times b) - (iB_0 c/4\pi n_{e0} e\omega)(b \cdot k) k \times B_1. \\ ...[3.21]$$

It can be seen from equation [3.21] that if the Hall effect is unimportant, the electron inertia term by itself replaces c_H^2 by $c_H^2/(1 - k^2 c^2/\omega_{pe}^2)$. This term therefore becomes very important for short wavelengths; however, it is very likely that other effects neglected in the present equations may then become important. The Hall effect is unimportant for propagation perpendicular to the field ($k \cdot b = 0$), but is particularly important for propagation along the field.

In the latter case equations [3.20] and [3.21] become

$$\omega^2 v_1 = k^2 c_S^2 (b \cdot v_1) b - (k B_0 \omega / 4\pi\rho_0) B_1 \qquad \ldots[3.22]$$

and

$$-(1 - k^2 c^2 / \omega_{pe}^2) B_1 = (k B_0 / \omega)[v_1 - (b \cdot v_1) b] - (i B_0 c k^2 / 4\pi n_{e0} e\omega) b \times B_1. \qquad \ldots[3.23]$$

Substitution of [3.22] into [3.23], then leads to one equation for v_1 which is

$$-(1 - k^2 c^2 / \omega_{pe}^2)[k^2 c_S^2 (b \cdot v_1) b - \omega^2 v_1] \\ = k^2 c_H^2 [v_1 - (b \cdot v_1) b] - (i k^2 c^2 \omega \Omega_e / \omega_{pe}^2) b \times v_1, \qquad \ldots[3.24]$$

where

$$\Omega_e = -e B_0 / m_e c. \qquad \ldots[3.25]$$

In equation [3.25] we define the electron gyration frequency with the sign of the charge included, because in what follows in Section 4 we wish to define the gyration frequency in terms of the charge of a particle when the sign of the charge is not known.

Scalar product of equation [3.24] with b now gives

$$(1 - k^2 c^2 / \omega_{pe}^2)(k^2 c_S^2 - \omega^2) b \cdot v_1 = 0, \qquad \ldots[3.26]$$

giving, as before, the relation

$$k^2 c_S^2 = \omega^2, \qquad \ldots[3.27]$$

for sound waves. Taking the vector product of [3.24] with b gives

$$[\omega^2 (1 - k^2 c^2 / \omega_{pe}^2) - k^2 c_H^2] b \times v_1 = -(i k^2 c^2 \omega \Omega_e / \omega_{pe}^2) b \times (b \times v_1). \qquad \ldots[3.28]$$

Here we must resist the temptation to say that $b \times v_1$ and $b \times (b \times v_1)$ are perpendicular, and that equation [3.28] can only be satisfied by $b \times v_1 = 0$. As mentioned earlier, if v_1 is complex, it is possible for $b \times v_1$ to be perpendicular to itself without vanishing. The dispersion equation can be obtained by taking a further vector product of [3.28] with b to give

$$[\omega^2 (1 - k^2 c^2 / \omega_{pe}^2) - k^2 c_H^2] b \times (b \times v_1) = (i k^2 c^2 \omega \Omega_e / \omega_{pe}^2) b \times v_1. \qquad \ldots[3.29]$$

The dispersion equation obtained from equations [3.28] and [3.29] is then

$$\omega^2 (1 - k^2 c^2 / \omega_{pe}^2) = k^2 c_H^2 \pm k^2 c^2 \omega \Omega_e / \omega_{pe}^2. \qquad \ldots[3.30]$$

This equation shows that the two Alfvén waves which propagate along the field in the absence of the Hall effect, the genuine Alfvén wave and the degenerate magnetosonic wave, are coupled by the Hall effect to give two waves of different polarization with different propagation properties. This means that, if the Hall effect is important, there are no waves which suffer no geometrical attenuation. This particular problem has been discussed by Lighthill.[5]

Equation [3.30] at present makes no reference to the ions: but it can be simplified if rewritten in terms of the ion gyration frequency. Thus, in the approximation of this section

$$c_H^2/\Omega_i = cB_0/4\pi Zn_{i0}e = cB_0/4\pi n_{e0}e = -c^2\Omega_e/\omega_{pe}^2. \qquad ...[3.31]$$

Thus equation [3.30] can be rewritten

$$\omega^2(1 - k^2c^2/\omega_{pe}^2) = k^2c_H^2(1 \mp \omega/\Omega_i). \qquad ...[3.32]$$

If we neglect the electron inertia term, we see that as $\omega \to \Omega_i$

$$k^2 \to \infty \quad \text{or} \quad \omega^2/2c_H^2, \qquad ...[3.33]$$

and that, for $\omega > \Omega_i$, one of the waves cannot propagate. Here we see that, as mentioned in the introduction, this individual particle property Ω_i gives rise to collective phenomena because all particles have the same value of Ω_i. Further important effects of the gyration and plasma frequencies will be considered in the following sections. Note that, as mentioned earlier, the Hall effect acts quite differently from Ohmic dissipation.

4

OSCILLATIONS OF A COLD PLASMA

4.1. Introduction

In this section we suppose that the displacement current and electric field are important, so that for the first time we can study the plasma effects in the propagation of electromagnetic waves; but we assume that collisions are unimportant, and that the temperature of the plasma is low enough for us to assume that all particles of any one species have the same velocity. In Section 5 we shall consider problems in which the particle distribution function in velocity space is vitally important.

As collisions are neglected, any neutral particles cannot contribute to the waves considered here. We can, however, consider an arbitrary number of species of particle in a uniform external magnetic field, and for generality we will suppose that each species has a mass velocity parallel to the magnetic field in the steady state. Later we will attempt to obtain the propagation properties of a hot plasma, by introducing a large number of streams of a given type of particle with different zero-order velocities. We shall, however, find that we do not obtain all of the properties of a hot plasma in this way.

Suppose that the rth type of particle has mass m_r, charge e_r, number density n_r and velocity v_r. Then the equations of the problem are

$$m_r \frac{dv_r}{dt} = e_r E + e_r v_r \times B/c, \qquad \qquad ...[4.1]$$

$$\frac{\partial n_r}{\partial t} + \text{div}\,(n_r v_r) = 0, \qquad \qquad ...[4.2]$$

$$\text{curl}\,E = -\frac{1}{c}\frac{\partial B}{\partial t} \qquad \qquad ...[4.3]$$

and
$$\text{curl}\,B = \frac{4\pi}{c}\sum_r n_r e_r v_r + \frac{1}{c}\frac{\partial E}{\partial t}, \qquad \qquad ...[4.4]$$

where the other Maxwell equations are implied by the above equations and equations [4.1] and [4.2] apply for each species of particle separately. In the steady state we assume that $n_r = N_r$, $v_r = V_r b$ and $B = B_0 b$, where

$$\sum N_r e_r = \sum N_r e_r V_r = 0, \qquad \qquad ...[4.5]$$

and N_r, V_r and B_0 are constants and b is a constant vector.

The perturbed quantities v_{r1}, n_{r1}, B_1, E_1 are all taken proportional to exp $i(k \cdot x - \omega t)$, and the unit suffixes are omitted from the perturbed quantities in what follows as this leads to no confusion. The perturbed equations are

$$v_r = \frac{ie_r}{m_r \omega_r}E + \frac{i\Omega_r}{\omega_r}v_r \times b + \frac{ie_r V_r}{m_r c \omega_r}b \times B, \qquad \qquad ...[4.6]$$

$$n_r = N_r k \cdot v_r/\omega_r, \qquad \qquad ...[4.7]$$

$$B = ck \times E/\omega \qquad \qquad ...[4.8]$$

and
$$ik \times B = \frac{4\pi}{c}\sum n_r e_r V_r b + \frac{4\pi}{c}\sum N_r e_r v_r - \frac{i\omega}{c}E, \qquad \qquad ...[4.9]$$

where
$$\omega_r = \omega - k \cdot bV_r, \quad \text{and} \quad \Omega_r = e_r B_0/m_r c. \qquad \qquad ...[4.10]$$

Scalar and vector products of [4.6] with b then give (using [4.8])

$$v_r \cdot b = (ie_r/m_r \omega_r)E \cdot b \qquad \qquad ...[4.11]$$

and

$$v_r \times b = \frac{ie_r}{m_r \omega_r}E \times b + \frac{i\Omega_r}{\omega_r}[v_r \cdot bb - v_r] + \frac{ie_r V_r}{m\omega\omega_r}[(b \cdot E)(k \times b) - (b \cdot k)(E \times b)]$$

$$= \frac{ie_r}{m_r \omega}E \times b - \frac{e_r \Omega_r}{m_r \omega_r^2}(E \cdot b)b - \frac{i\Omega_r}{\omega_r}v_r + \frac{ie_r V_r}{m_r \omega\omega_r}(b \cdot E)(k \times b). \quad ...[4.12]$$

Substitution of [4.12] into [4.6] then gives

$$v_r = \frac{e_r}{m_r(\omega_r^2 - \Omega_r^2)}\left[\frac{i\omega_r^2}{\omega}E - \frac{\Omega_r \omega_r}{\omega}E \times b - \frac{i\Omega_r^2}{\omega_r}(E \cdot b)b \right.$$

$$\left. - \frac{V_r \Omega_r}{\omega}(E \cdot b)k \times b + i\frac{V_r \omega_r}{\omega}(E \cdot b)k\right] \qquad \qquad ...[4.13]$$

where we note the possible occurrence of a resonance if for any species of particle $\omega_r^2 = \Omega_r^2$.

Taking the scalar product of equation [4.13] with k then gives

$$k \cdot v_r = \frac{e_r}{m_r(\omega_r^2 - \Omega_r^2)} \left[\frac{i\omega_r^2}{\omega} (k \cdot E) - \frac{\Omega_r \omega_r}{\omega} k \cdot E \times b \right.$$
$$\left. - \frac{i\Omega_r^2}{\omega_r} (E \cdot b)(k \cdot b) + i \frac{V_r \omega_r k^2}{\omega} (E \cdot b) \right]. \qquad ...[4.14]$$

Combination of equations [4.7], [4.8], [4.13] and [4.14] with [4.9] then gives

$$c^2(k \cdot E)k + (\omega^2 - k^2 c^2)E = \sum \frac{\omega_{pr}^2}{\omega_r^2 - \Omega_r^2} \left[\omega_r^2 E + i\Omega_r \omega_r (E \times b) \right.$$
$$+ iV_r \Omega_r (b \cdot E)k \times b + V_r \omega_r (b \cdot E)k + V_r \omega_r (k \cdot E)b$$
$$\left. + iV_r \Omega_r (k \cdot E \times b)b - \frac{\Omega_r^2 \omega^2}{\omega_r^2} (b \cdot E)b + k^2 V_r^2 (b \cdot E)b \right]$$
$$..[4.15]$$

where $\qquad\qquad \omega_{pr}^2 = 4\pi N_r e_r^2 / m_r. \qquad\qquad ...[4.16]$

4.2. Wave propagation—special cases

(a) $B_0 = V_r = 0$. No zero order magnetic field and no stream velocities. In this case equation [4.15] becomes

$$c^2(k \cdot E)k + (\omega^2 - k^2 c^2)E = \sum \omega_{pr}^2 E. \qquad ...[4.17].$$

Scalar and vector products of [4.16] with k then give

$$(\omega^2 - \sum \omega_{pr}^2)k \cdot E = 0 \qquad ...[4.18]$$

and $\qquad (\omega^2 - k^2 c^2 - \sum \omega_{pr}^2)k \times E = 0, \qquad ..[4.19]$

yielding the two well known dispersion equations for longitudinal plasma oscillations

$$\omega^2 = \sum \omega_{pr}^2 \qquad ...[4.20]$$

and transverse electromagnetic waves

$$\omega^2 = k^2 c^2 + \sum \omega_{pr}^2. \qquad ...[4.21]$$

In equations [4.20] and [4.21]

$$\omega_p^2 \equiv \sum \omega_{pr}^2 \qquad ...[4.22$$

is the square of the plasma frequency. To a high degree of approximation $\omega_p = \omega_{pe}$. From equation [4.21] it can be seen that electromagnetic waves cannot propagate in a plasma if their frequency is less than the plasma frequency. Waves approaching a region where the plasma frequency increases to exceed the wave frequency are reflected; this is the basis of the technique of ionospheric sounding. In some astronomical objects in which thermal equilibrium is achieved, the density of electrons may be high enough for the Planck distribution of photons to be seriously distorted by lack of propagation below the plasma frequency (if $\hbar\omega_p > kT$).

(b) $B_0 = 0$, but $V_r \neq 0$. In this case equation [4.14] becomes

$$c^2(k \cdot E)k + (\omega^2 - k^2c^2)E$$
$$= \sum \omega_{pr}^2 \left[E + \frac{V_r}{\omega_r} E \cdot bk + \frac{V_r}{\omega_r} E \cdot kb + \frac{k^2V_r^2}{\omega_r^2} E \cdot bb \right]. \qquad ...[4.23]$$

In this case it is probably simplest to introduce a particular coordinate system. Thus, take

$$b = (0, 0, 1) \qquad ...[4.24]$$
and $$k = k(0, \sin \theta, \cos \theta) \qquad ...[4.25]$$

where $0 < \theta < \pi/2$. Then three equations are obtained:

$$[\omega^2 - k^2c^2 - \sum \omega_{pr}^2] E_x = 0, \qquad ...[4.26]$$

$$[\omega^2 - k^2c^2 \cos^2 \theta - \sum \omega_{pr}^2] E_y + [k^2c^2 \cos \theta \sin \theta - \sum \omega_{pr}^2 (kV_r/\omega_r) \sin \theta] E_z$$
$$= 0, \qquad ...[4.27]$$

$$[k^2c^2 \sin \theta \cos \theta - \sum \omega_{pr}^2 (kV_r/\omega_r) \sin \theta] E_y + [\omega^2 - k^2c^2 \sin^2 \theta - \sum \omega_{pr}^2 [1 +$$
$$(2kV_r/\omega_r) \cos \theta + k^2V_r^2/\omega_r^2]] E_z = 0. \qquad ...[4.28]$$

Equation [4.26] shows that there is always one mode which obeys the dispersion equation [4.21] for plasma-modified electromagnetic waves. The other equations can be simplified if either $\theta = 0$ or $\theta = \pi/2$. In the first case, equation [4.27] also leads to the dispersion equation [4.21], while equation [4.28] becomes

$$\omega^2 \left[1 - \sum \frac{\omega_{pr}^2}{\omega_r^2} \right] E_z = 0, \qquad ...[4.29]$$

where in this case $\omega_r = \omega - kV_r$. This gives the dispersion equation

$$1 = \sum (\omega_{pr}^2/\omega_r^2), \qquad ...[4.30]$$

which is a simple generalization of equation [4.20].

When the direction of propagation is perpendicular to the beams, equations [4.27] and [4.28] remain coupled and give the dispersion equation

$$(\omega^2 - \sum \omega_{pr}^2)(\omega^2 - k^2c^2 - \sum \omega_{pr}^2[1 + k^2V_r^2/\omega^2]) = (\sum \omega_{pr}^2 kV_r/\omega)^2, \qquad ...[4.31]$$

which shows how the two modes [4.20] and [4.21] are coupled by the stream velocities when the propagation is across the beams.

The main interest in this case concentrates on equation [4.30], in which there are potential singularities when any $\omega_r = 0$. In the case of equation

[4.31] for propagation across the beams it is possible to see that corrections to relations [4.20] and [4.21] are of order V/c. Thus, if we neglect all terms in the sums other than that due to the electrons, we can obtain

$$\omega^2 \approx \omega_{pe}^{2}(1 - V_e^2/c^2) \qquad \text{...[4.32]}$$

and

$$\omega^2 \approx \omega_{pe}^2(1 + V_e^2/c^2) + k^2 c^2, \qquad \text{...[4.33]}$$

where of course these relations are true in the frame in which the zero-order current vanishes and can be modified if another frame of reference is taken.

Equation [4.30] does contain some qualitatively new results. In particular, it predicts two-stream instability and its generalizations. We discuss the simplest case, in which there are two oppositely streaming and equal electron beams with enough ions to neutralize the system at rest. If the electron beams have number densities $N_e/2$ and velocities $\pm V$, the dispersion equation is

$$1 = \frac{\omega_p^2}{2(\omega - kV)^2} + \frac{\omega_p^2}{2(\omega + kV)^2} + \frac{Zm_e}{m_i}\frac{\omega_p^2}{\omega^2}, \qquad \text{...[4.34]}$$

where

$$\omega_p^2 = 4\pi N_e e^2/m_e. \qquad \text{...[4.35]}$$

The last term in equation [4.34] is important only when ω^2 is very small. In the first instance we neglect it, and equation [4.34] can be written in the two alternative forms

$$k^4 V^4 - k^2 V^2(\omega_p^2 + 2\omega^2) + \omega^2(\omega^2 - \omega_p^2) = 0, \qquad \text{...[4.36]}$$

or

$$\omega^4 - \omega^2(\omega_p^2 + 2k^2 V^2) + k^2 V^2(k^2 V^2 - \omega_p^2) = 0. \qquad \text{...[4.37]}$$

If we consider waves of given frequency ω and study their propagation properties, equation [4.36] shows that, if $\omega_p^2 > \omega^2$, one of the roots for k^2 is negative. Thus, if $\omega^2 < \omega_p^2$, one of the waves does not propagate, just as in the case of relation [4.21]. If, however, we consider k given, and solve [4.37] for ω^2, it can be seen that one of the roots for ω^2 is negative, if $k^2 V^2 < \omega_p^2$. This is the two-stream instability. In this simplest version sufficiently long wave perturbations are unstable, whatever the velocity of the streams. This simple result is modified if the beams have thermal spread, and instability occurs only if the relative stream velocity exceeds something of the order of the thermal velocity within the streams.

We can now ask what is the influence of the term we have so far omitted in equation [4.34]. If the full equation is written

$$(1 - Zm_e\omega_p^2/m_i\omega^2)(\omega^2 - k^2 V^2)^2 = \omega_p^2(\omega^2 + k^2 V^2), \qquad \text{...[4.38]}$$

it can be seen that, when $0 < \omega^2 < Zm_e\omega_p^2/m_i$, it is impossible for k^2 to be positive. More careful study of the equation confirms that for small enough ω^2 neither of the two modes propagate, and for higher ω^2 one mode propagates; while, if $\omega^2 > \omega_p^2(1 + Zm_e/m_i)$, both modes propagate. Regarded as an equation for ω^2 given k^2, [4.38] is a cubic and for small k^2 two of the waves are growing waves.

(c) $B_0 \neq 0$, but $V_r = 0$. In this case equation [4.15] becomes:

$$c^2(\mathbf{k} \cdot \mathbf{E})\mathbf{k} + (\omega^2 - k^2 c^2)\mathbf{E}$$
$$= \sum \frac{\omega_{pr}^2}{\omega^2 - \Omega_r^2} [\omega^2 \mathbf{E} + i\omega\Omega_r(\mathbf{E} \times \mathbf{b}) - \Omega_r^2(\mathbf{E} \cdot \mathbf{b})\mathbf{b}]. \qquad \dots[4.39]$$

Again we introduce the coordinate system of equations [4.24] and [4.25] to obtain the three equations

$$\left[\omega^2 - k^2 c^2 - \sum \frac{\omega_{pr}^2 \omega^2}{\omega^2 - \Omega_r^2}\right] E_x - \sum \frac{i\Omega_r \omega \omega_{pr}^2}{\omega^2 - \Omega_r^2} E_y = 0, \qquad \dots[4.40]$$

$$\sum \frac{i\Omega_r \omega \omega_{pr}^2}{\omega^2 - \Omega_r^2} E_x + \left[\omega^2 - k^2 c^2 \cos^2\theta - \sum \frac{\omega_{pr}^2 \omega^2}{\omega^2 - \Omega_r^2}\right] E_y + k^2 c^2 \sin\theta \cos\theta\, E_z = 0$$
$$\dots[4.41]$$

and

$$k^2 c^2 \sin\theta \cos\theta\, E_y + [\omega^2 - k^2 c^2 \sin^2\theta - \sum \omega_{pr}^2] E_z = 0. \qquad \dots[4.42]$$

In general, all three components of the vector $\mathbf{E}$ are coupled, but the equations decouple if either $\theta = 0$ or $\theta = \pi/2$. We will discuss these two cases. If $\theta = 0$, equation [4.42] gives the dispersion equation [4.20] for longitudinal plasma oscillations, which are unaffected by the field. Equations [4.40] and [4.41] give two oppositely polarized waves with equations

$$\omega^2 - k^2 c^2 - \sum \omega_{pr}^2 \omega/(\omega \pm \Omega_r) = 0. \qquad \dots[4.43]$$

In these dispersion equations we have resonances at $\mp\Omega_r$, where it must be remembered that Ω_e is negative. Note that replacement of ω by $-\omega$ interchanges the polarizations, so that all waves can be considered by taking $-\infty < \omega < \infty$ and one sign in equation [4.43]. If, for example, we consider a mixture of electrons and one species of ion and take the positive sign in [4.43], there are two bands of values for ω for which the wave does not propagate; the one band is for ω immediately below $-\Omega_i$ and the other for ω immediately above $-\Omega_e$. Note that in this case $\omega_{pe}^2/\Omega_e = -\omega_{pi}^2/\Omega_i$, as can easily be checked. If, instead of regarding ω as specified, we specify positive k^2, it is easy to verify that there are four real roots for ω so that there are no unstable waves.

In the case of frequencies very much less than the gyration frequencies, $|\omega| \ll \Omega_i, -\Omega_e$, there is a transition to Alfvén waves. Thus, if the inequality is satisfied,

$$k^2 c^2 \approx \omega^2 \pm (\omega_{pe}^2 \omega/\Omega_e)(1 \pm \omega/\Omega_e) \pm (\omega_{pi}^2 \omega/\Omega_i)(1 \pm \omega/\Omega_i)$$
$$= \omega^2(1 + \omega_{pe}^2/\Omega_e^2 + \omega_{pi}^2/\Omega_i^2)$$
$$= \omega^2(1 + c^2/c_H^2). \qquad \dots[4.44]$$

If the hydromagnetic velocity is very much less than the velocity of light, this gives the standard dispersion equation for Alfvén waves.

Consider next the case of propagation across the field. In this case equation [4.42] reduces to equation [4.21] for transverse electromagnetic waves unaffected by the magnetic field. The other equations give

$$k^2 c^2 = \omega^2 (1 - \sum \{\omega_{pr}^2/(\omega^2 - \Omega_r^2)\})$$
$$- [\sum \{\omega_{pr}^2 \Omega_r/(\omega^2 - \Omega_r^2)\}]^2 / (1 - \sum \{\omega_{pr}^2/(\omega^2 - \Omega_r^2)\}). \quad ...[4.45]$$

Consider the behaviour of this equation for large and small ω^2. For $\omega^2 \to \infty$, $k^2 c^2 \approx \omega^2$, the dispersion equation for electromagnetic waves. For $\omega^2 \to 0$, $k^2 c^2 \approx \omega^2 [1 + c^2/c_H^2]$, which is the limiting case of the magnetosonic wave when the velocity of sound tends to zero. At first sight it appears that there are resonances at all the gyration frequencies, but in fact, near the gyration frequency of any particle, the potentially singular terms cancel out. There are, however, resonances when

$$1 = \sum \omega_{pr}^2/(\omega^2 - \Omega_r^2), \quad ...[4.46]$$

and there are therefore as many resonances as there are species of particle. In the case when only electrons and ions are present, the condition for resonance is

$$\omega^4 - \omega^2 (\Omega_e^2 + \Omega_i^2 + \omega_{pe}^2 + \omega_{pi}^2) + \Omega_i^2 \Omega + \omega_{pi}^2 \Omega_e^2 + \omega_{pe}^2 \Omega_i^2 = 0. \quad ...[4.47]$$

Because of the mass ratio of ion and electron, this is approximately

$$\omega^4 - \omega^2 (\Omega_e^2 + \omega_{pe}^2) + \Omega_e^2 (\Omega_i^2 + \omega_{pi}^2) = 0. \quad ..[4.48]$$

In the special case when $\Omega_e \ll \omega_{pe}$ (and hence $\Omega_i \ll \omega_{pi}$), the two roots of [4.48] are

$$\omega^2 \approx \Omega_e^2 + \omega_{pe}^2 \quad ...[4.49]$$

and
$$\omega^2 \approx \omega_{pi}^2 \Omega_e^2/\omega_{pe}^2 = \Omega_e \Omega_i. \quad ...[4.50]$$

The frequencies $\omega = (\Omega_e^2 + \omega_{pe}^2)^{\frac{1}{2}}$ and $\omega = (\Omega_e \Omega_i)^{\frac{1}{2}}$ are called the upper and lower hybrid resonances.

No attempt will be made here to discuss the general dispersion equation when both magnetic field and beam velocities are present. For more detail on these problems reference should be made to the standard textbooks such as those by Ginzburg[4] and Stix.[3]

4.3. Transition to a hot plasma

Consider the case of plasma oscillations when beams of particles are present, equation [4.30]. Suppose we have a large number of beams of electrons with different number densities and different beam velocities. If the number of beams is large enough and the velocity spacing of the beams is small enough, instead of regarding them as distinct beams we can consider them as providing an approximation to a velocity distribution function of a hot plasma. Thus, if we neglect the ions for the moment, we should be able to replace the summation in equation [4.30] by an integral and write

$$1 = \omega_p^2 \int_{-\infty}^{\infty} f_0(v)\,dv/(\omega - kv)^2, \qquad \ldots [4.51]$$

where $f_0(v)$ is a one dimensional velocity distribution function satisfying

$$\int_{-\infty}^{\infty} f_0(v)\,dv = 1 \qquad \ldots [4.52]$$

and ω_p is the plasma frequency derived from the total density of electrons. Equation [4.51] can be formally integrated by parts to give

$$1 + (\omega_p^2/k) \int_{-\infty}^{\infty} (\partial f_0/\partial v)\,dv/(\omega - kv) = 0. \qquad \ldots [4.53]$$

It is clear that this procedure can be generalized to take account of velocity distributions in three dimensions.

This dispersion equation was originally obtained by Vlasov from a study of the kinetic theory of a plasma, and it is clear that, if it were the correct equation, it could be obtained from the cold plasma equations alone. However, equations [4.51] and [4.53] do not give entirely correct results for the oscillation properties of a plasma. The trouble lies in the fact that, if ω and k are both real, there is a singularity in the integrand in both [4.51] and [4.53]. When there are only a finite number of beams, these resonances, as we have seen, do not cause any serious difficulty; but when the finite sum is replaced by an integral, the integral is not well defined.

When Vlasov obtained equation [4.53] he proposed that the Cauchy principal value of the integral should be taken in equation [4.53]. The principal value is defined by

$$P \int_{-a}^{b} (F(x)/x)\,dx = \lim_{\varepsilon \to 0} \left[\int_{-a}^{-\varepsilon} (F(x)/x)\,dx + \int_{\varepsilon}^{b} (F(x)/x)\,dx \right], \qquad \ldots [4.54]$$

provided that the limit in equation [4.54] exists.

Landau subsequently showed that one property of plasma waves was not correctly described by the equation

$$1 + (\omega_p^2/k) P \int_{-\infty}^{\infty} (\partial f_0/\partial v)\,dv/(\omega - kv) = 0. \qquad \ldots [4.55]$$

Plasma waves suffer a damping, known as Landau damping, which is not predicted by the equations we have so far used, or by equation [4.55]. In Section 5 we shall consider wave propagation from the standpoint of the Vlasov equation, and shall see how this collisionless damping discovered by Landau arises. We shall also discuss some other problems, in which the detailed structure of the particle distribution function is important.

5

OSCILLATIONS OF A HOT PLASMA

5.1. Introduction

In this section we suppose that the behaviour of a hot plasma is governed by the Vlasov equation for the one-particle distribution function.

$$\frac{\partial f_r}{\partial t} + v \cdot \frac{\partial f_r}{\partial x} + \frac{e_r}{m_r} \left[E + \frac{v \times B}{c} \right] \cdot \frac{\partial f_r}{\partial v} = 0, \qquad \ldots[5.1]$$

where there is one such equation for each species of particle, labelled by suffix r. In considering oscillations in such a plasma, we will suppose that we have an equilibrium state which is uniform in space and time, which may or may not contain a uniform magnetic field. If there is no magnetic field in the equilibrium state, the Vlasov equations are satisfied identically by any distribution functions $f_{rC}(v)$, and the only constraints that these functions have to satisfy are that the charge and current densities vanish. Thus

$$\sum e_r \int f_{r0}(v) d^3v = 0 \qquad \ldots[5.2]$$

and

$$\sum e_r \int f_{r0}(v) v d^3v = 0. \qquad \ldots[5.3]$$

In the case that a zero order field $B_0 b$ is present in the steady state, the distribution functions are constrained to have the form

$$f_{r0} = f_{r0}(v_\parallel, v_\perp^2), \qquad \ldots[5.4]$$

where

$$v = v_\parallel b + v_\perp \qquad \ldots[5.5]$$

and equations [5.2] and [5.3] must of course still be satisfied.

In using equation [5.1] we are neglecting all effects due to close collisions in the plasma. It may be that we wish to consider high frequency oscillations in a plasma in which collisions have played an important role in determining the equilibrium conditions, so that the zero-order distribution function might be essentially Maxwellian. If, alternatively, insufficient time has elapsed since the system was set up for an approach to a Maxwellian distribution, the more general distribution functions can characterize the equilibrium state.

In the previous section we have written down a dispersion equation for longitudinal oscillations in the absence of a magnetic field. In the presence of a three-dimensional distribution of velocities, this equation can be generalized to

$$1 + \sum \frac{\omega_{pr}^2}{k^2 N_r} P \int \frac{k \cdot \partial f_{0r}/\partial v}{\omega - k \cdot v} d^3v = 0, \qquad \ldots[5.6]$$

where N_r is the number density of the rth species of particle in the equilibrium state. We have stated that this equation is not quite correct and that it fails to predict some important phenomena. Landau[16] pointed out that the correct interpretation of the integrals in equation [5.6] cannot be obtained without considering the way in which the oscillation is set up. In other words, it is important to consider the initial conditions and to solve an initial value problem.

If the initial perturbation is well-behaved (we shall discuss this more carefully subsequently) it is still possible to obtain well defined dispersion equations. However, in this case there are separate equations for growing waves, undamped waves, and damped waves. These equations are

Growing waves:

$$1+\sum \frac{\omega_{pr}^2}{k^2 N_r} \int \frac{(\partial g_{0r}/\partial v)\,dv}{(\omega/k)-v} = 0, \qquad \ldots[5.7]$$

Undamped waves:

$$1+\sum \frac{\omega_{pr}^2}{k^2 N_r} P \int \frac{(\partial g_{0r}/\partial v)\,dv}{(\omega/k)-v} = \sum \frac{\pi i \omega_{pr}^2}{k^2 N_r} \left(\frac{\partial g_{0r}}{\partial v}\right)_{\omega/k}, \qquad \ldots[5.8]$$

and

Damped waves:

$$1+\sum \frac{\omega_{pr}^2}{k^2 N_r} \int \frac{(\partial g_{0r}/\partial v)\,dv}{(\omega/k)-v} = \sum \frac{2\pi i \omega_{pr}^2}{k^2 N_r} \left(\frac{\partial g_{0r}}{\partial v}\right)_{\omega/k}, \qquad \ldots[5.9]$$

where g_{0r} is obtained from f_{0r} by integrating over the two velocity components perpendicular to the direction of propagation. It can be seen that, if the waves are unstable, the dispersion relation is exactly as obtained by Vlasov but there is a correction to the relation for waves which, according to the Vlasov theory, would have been undamped but may now be damped.

If it is not considered necessary for the initial perturbation to be well-behaved, an arbitrary function of ω/k is introduced on the right-hand sides of equations [5.8] and [5.9]. This leads Van Kampen and Felderhof,[14] for example, to say that there is no dispersion equation for longitudinal plasma waves, as we no longer obtain a unique relation between ω and k, and an almost arbitrary relationship can be found between ω and k.

The essential reason for the difference between the Vlasov and Landau results can be seen as follows. For electrostatic oscillations, the linearized Vlasov equation for a single species of particle has the form

$$\frac{\partial f_1}{\partial t}+v \cdot \frac{\partial f_1}{\partial x}+\frac{e}{m}\, E_1 \cdot \frac{\partial f_0}{\partial v} = 0. \qquad \ldots[5.10]$$

Vlasov assumed that to solve this equation he should put both f_1 and E_1 proportional to $\exp i(k \cdot x - \omega t)$. This procedure gives a particular integral of equation [5.10], but there is also a complementary function for which f_1 is proportional to $\exp i(k \cdot x - k \cdot vt)$, where here we are assuming that k is given. This first-order distribution function represents freely moving particles and it is those which are moving with the phase velocity of the Vlasov wave which lead to the modification of the Vlasov dispersion equation.

We have mentioned above that some of the waves, which according to Vlasov are undamped, suffer damping—Landau damping. This is a rather particular type of damping, about which there has been considerable discussion in the literature. We have included no collisions or other dissipative processes in the Vlasov equation, so that we are here considering a collisionless damping rather than the more common sort due to dissipative processes. If the asymptotic behaviour of the perturbation in the distribution function f_1 and the electric field E_1 is considered, it is found that E_1 asymptotically tends to zero while the complementary function of [5.10] does not decay. Thus, if initially we have an electric field produced by density fluctuations in the plasma, this electric field will decay (in general) because the particles contributing to the initial density fluctuation have different velocities and the fluctuation gradually disperses. However, there is no real dissipation and the velocity distribution function does not tend to the Maxwellian form.

Having made these introductory remarks about Landau damping, we now consider the problem in somewhat more detail in the succeeding sections.

5.2. Oscillations in the absence of an external magnetic field

Consider first the case of longitudinal oscillations, in which the perturbed electric field is parallel to the vector k. In this case the perturbed magnetic field vanishes. Equation [5.10] then holds for any particle species. In solving equation [5.10] we will suppose that E_1 is proportional to exp $i(k \cdot x - \omega t)$ but will merely assume that f_1 varies like exp $ik \cdot x$. Only the shape of the velocity distribution parallel to the direction of propagation is important. Choosing the z axis to be the direction of propagation and introducing

$$g_0 = \int f_0(v) dv_x dv_y, \quad g_1 = \int f_1 dv_x dv_y, \qquad \ldots[5.11]$$

equation [5.10] becomes

$$\frac{\partial g_1}{\partial t} + ikv_z g_1 + \frac{e}{m} E_1 \frac{\partial g_0}{\partial v_z} = 0. \qquad \ldots[5.12]$$

Equation [5.12] can be solved for g_1 in terms of E_1. Thus, if

$$E_1 = E_{10} \exp(ikz - i\omega t), \qquad \ldots[5.13]$$

$$g_1 = \frac{eE_{10}(\partial g_0/\partial v)}{im(\omega - kv)} \exp(ikz - i\omega t) + C(v) \exp ik(z - vt), \qquad \ldots[5.14]$$

where, in [5.14] and in what follows, we have written v for v_z. We will explicitly consider only the cases of growing and damped waves; but it is only a simple generalization to obtain the equation for undamped waves.

Poisson's equation can now be written

$$ikE_1 = 4\pi \sum_r e_r \int_{-\infty}^{\infty} g_{1r} dv. \qquad \qquad ...[5.15]$$

Substitution of the values given by [5.14] for each species of particle into [5.15] then gives

$$ikE_{10} \exp(-i\omega t) = \sum \frac{\omega_{pr}^2 E_{10}}{iN_r} \exp(-i\omega t) \int_{-\infty}^{\infty} \frac{\partial g_{0r}/\partial v}{\omega - kv} dv$$
$$+ 4\pi \sum_r e_r \int C_r(v) \exp(-ikvt) dv. \qquad ...[5.16]$$

If $C_r(v)$ is divided into a part which has a singularity at $v = \omega/k$ and a non-singular part

$$C_r(v) = C_{r0}(v) E_{10} + C_{r1}(v) E_{10}/(\omega - kv), \qquad ...[5.17]$$

the second integral in [5.16] can be evaluated by contour integration to give

$$\int C_r(v) \exp(-ikvt) dv = 0, \qquad\qquad (\mathscr{I}(\omega) > 0)$$
$$\int C_r(v) \exp(-ikvt) dv = (2\pi i/k) E_{10} C_{r1}(\omega/k) \exp(-i\omega t), \quad (\mathscr{I}(\omega) < 0).$$
$$...[5.18]\ddagger$$

Then equation [5.16] becomes

$$1 + \sum \frac{\omega_{pr}^2}{kN_r} \int \frac{\partial g_{0r}/\partial v}{\omega - kv} dv = 0, \qquad\qquad (\mathscr{I}(\omega) > 0)$$
$$= \frac{8\pi^2}{k^2} \sum e_r C_{r1}(\omega/k), \qquad\qquad (\mathscr{I}(\omega) < 0)$$
$$...[5.19]$$

which are essentially Van Kampen's equations for growing and damped waves. The first of these two equations is of course [5.7], but the second equation is more general than Landau's equation [5.9].

Landau's equation can be recovered as follows. The perturbed distribution function at time $t = 0$ is

$$g_{1r} = E_{10} \exp(ikz) \left[\frac{e_r \partial g_{0r}/\partial v}{im_r(\omega - kv)} + C_{r0}(v) + \frac{C_{r1}(v)}{\omega - kv} \right]. \qquad ...[5.20]$$

If we now demand that the initial perturbation has no singularity at the phase velocity of the particular wave we are considering, we shall choose

$$C_{r1}(v) = ie_r(\partial g_{0r}/\partial v)/m_r \qquad\qquad ...[5.21]$$

and if this is substituted into [5.19] the equation reduces to Landau's equation [5.9]. It is in the sense of equation [5.21] that we say that the initial perturbation must be well-behaved if Landau's result is to be obtained.

‡ These equations are not correct for completely arbitrary C_{r1}. If $C_{r1}(v) \propto \exp(-v^2/v_0^2)$, there is in addition a transient term proportional to $\exp(-k^2 v_0^2 t^2/4)$.

We shall now consider two consequences of these dispersion equations. We shall discuss the conditions under which growing waves can occur and the rate at which stable waves are damped. However, first we make a few remarks about undamped waves. If ω and k are both real, as they are for undamped waves, the left-hand side of equation [5.8] is real, while the right-hand side is purely imaginary. This means that the two sides of the equation must vanish simultaneously. That is, Vlasov's equation

$$1+\sum \frac{\omega_{pr}^2}{k^2 N_r} P \int \frac{(\partial g_{0r}/\partial v)\,dv}{(\omega/k)-v} = 0 \qquad \text{...[5.22]}$$

must be satisfied, together with the subsidiary condition

$$\sum \frac{e_r^2}{m_r}\left(\frac{\partial g_{0r}}{\partial v}\right)_{\omega/k} = 0. \qquad \text{...[5.23]}$$

Thus undamped waves cannot possibly be obtained unless the combined distribution function $\sum (e_r^2 g_{0r}/m_r)$ has a turning value at $v = \omega/k$.

Consider next the possible existence of growing waves. First we make the definitions

$$g \equiv \sum \frac{e_r^2}{m_r} g_{0r} \qquad \text{...[5.24]}$$

and
$$\partial g/\partial v = G. \qquad \text{...[5.25]}$$

We now show that there can be no growing waves if G is zero for only one value of v; that is, if the combined distribution function g is single humped. Suppose that G is positive for $v < V$, zero at $v = V$ and negative for $v > V$, and define

$$H \equiv 1+\frac{4\pi}{k^2}\int_{-\infty}^{\infty} \frac{G\,dv}{(\omega/k)-v}. \qquad \text{...[5.26]}$$

We now show that H cannot vanish for imaginary ω. Let $\omega/k = x+iy$. Then

$$\mathscr{R}(H) = 1+\frac{4\pi}{k^2}\int_{-\infty}^{\infty} \frac{(x-v)\,G\,dv}{(x-v)^2+y^2} \qquad \text{...[5.27]}$$

and
$$\mathscr{I}(H) = -\frac{4\pi iy}{k^2}\int_{-\infty}^{\infty} \frac{G\,dv}{(x-v)^2+y^2}. \qquad \text{...[5.28]}$$

For given y and G of the character postulated, it is always possible to choose an x so that $\mathscr{I}(H)$ vanishes, but it is then possible to show that $\mathscr{R}(H)$ cannot vanish simultaneously. Thus, if $\mathscr{I}(H) = 0$, it is possible to add any multiple of it to $\mathscr{R}(H)$. In particular

$$\mathscr{R}(H) = \mathscr{R}(H)+[i(V-x)/y]\mathscr{I}(H) = 1+\frac{4\pi}{k^2}\int_{-\infty}^{\infty} \frac{(V-v)\,G(v)\,dv}{(x-v)^2+y^2},$$
$$\text{...[5.29]}$$

which is obviously positive and shows that single humped distributions cannot be unstable.

This argument can immediately be extended to show that all distributions isotropic in velocity space have stable longitudinal oscillations. For suppose we write

$$v = v_{\parallel} + v_{\perp}, \qquad \qquad ...[5.30]$$

where $\parallel$ denotes the direction of propagation and

$$f_0 = f_0(v_{\parallel}^2 + v_{\perp}^2). \qquad \qquad ...[5.31]$$

Using cylindrical coordinates in velocity space

$$g(v_{\parallel}) = 2\pi \int_0^{\infty} f_0(v_{\parallel}^2 + v_{\perp}^2) v_{\perp} dv_{\perp}$$

and

$$\partial g/\partial v_{\parallel} = 2\pi \int_0^{\infty} (\partial f_0/\partial v_{\parallel}) v_{\perp} dv_{\perp}$$

$$= 2\pi \int_0^{\infty} (\partial f_0/\partial v_{\perp}) v_{\parallel} dv_{\perp}$$

$$= -2\pi v_{\parallel} f_0(v_{\parallel}^2). \qquad \qquad ...[5.32]$$

The right-hand side of [5.32] vanishes only when $v_{\parallel} = 0$ (except in the rather pathological case when f_0 vanishes on a sphere in velocity space), and thus g is a single-humped distribution and cannot possess unstable oscillations.

Consider next the case of damped oscillations. If the damping is small, we write equation [5.9] in the form

$$H \equiv 1 + \frac{4\pi}{k^2} \int \frac{G(v) dv}{\omega/k - v} - \frac{8\pi^2 i}{k^2} G(\omega/k) = 0, \qquad ...[5.33]$$

put $\omega/k = x + iy$ where y is small and negative and perform a Taylor expansion of H. Thus

$$H \approx 1 + \frac{4\pi}{k^2} P \int \frac{G(v) dv}{x - v} + \frac{4\pi^2 i}{k^2} G(x) + \frac{4\pi i y}{k^2} P \int \frac{G'(v) dv}{x - v}$$

$$- \frac{4\pi^2 y}{k^2} G'(x) - \frac{8\pi^2 i}{k^2} G(x) + \frac{8\pi^2 y}{k^2} G'(x), \qquad ...[5.34]$$

where the prime denotes differentiation with respect to the argument of G. Equating the real and imaginary parts of H to zero gives the approximate equations

$$1 + \frac{4\pi}{k^2} P \int \frac{G(v) dv}{x - v} = 0 \qquad \qquad ...[5.35]$$

and

$$y = \pi G(x)/P \int \frac{G'(v) dv}{x - v}. \qquad \qquad ...[5.36]$$

Equation [5.35] is precisely Vlasov's equation for undamped waves, and this can be solved to give a value for x. Then the damping can be evaluated from equation [5.36].

In order to make any further progress, the form of the zero order distribution function must be specified. For simplicity we suppose that only the motion of the electrons need be considered (which is true to a high degree of approximation for high frequency waves) and that the distribution function is Maxwellian with temperature T_e. In addition, we suppose x is much greater than $(\kappa T_e/m_e)^{\frac{1}{2}}$, where κ is Boltzmann's constant. If this is so, we write [5.35] in the form

$$
1 = \frac{4\pi}{k^2} P \int \frac{g(v)\,dv}{(x-v)^2}
$$
$$
\approx \frac{4\pi}{k^2} P \int \frac{g(v)}{x^2}\,dv \left[1 + 2\frac{v}{x} + 3\frac{v^2}{x^2}\right]. \qquad ...[5.37]
$$

When the Maxwellian expression for $g(v)$,

$$
(N_e e^2/m_e)(m_e/2\pi\kappa T_e)^{\frac{1}{2}} \exp(-m_e v^2/2\kappa T_e),
$$

is introduced in [5.37], there results

$$
k^2/\omega_{pe}^2 = k^2/\omega_R^2 + (3k^4/\omega_R^4)(\kappa T_e/m_e)
$$

or
$$
\omega_R^2 = \omega_{pe}^2 + 3(\kappa T_e/m_e)k^2, \qquad ...[5.38]
$$

where $\omega_R = kx$.

The expression for y can be rewritten

$$
y = \pi G(x)/2P \int \frac{g(v)\,dv}{(x-v)^3}, \qquad ...[5.39]
$$

and the integral in the denominator can be evaluated by expansion in powers of v/x. If this is done, only the first term in the expansion is retained and a value for x is introduced from [5.38], the result is

$$
\omega_I = ky \approx -\sqrt{\frac{\pi}{8}} \frac{\omega_{pe}^4}{k^3 v_{th}^3} \exp\left(-\frac{3}{2} - \frac{\omega_{pe}^2}{2k^2 v_{th}^2}\right), \qquad ...[5.40]
$$

where $v_{th} = (\kappa T_e/m_e)^{\frac{1}{2}}$. This can be rewritten

$$
\omega_I \approx -\left(\frac{\pi}{8}\right)^{\frac{1}{2}} \frac{\omega_{pe}}{(k\lambda_D)^3} \exp\left(-\frac{3}{2} - \frac{1}{2k^2\lambda_D^2}\right), \qquad ...[5.41]
$$

where $\lambda_D = (\kappa T_e/4\pi N_e m_e e^2)^{\frac{1}{2}}$ is the Debye length. If the damping given by [5.41] is to be small, the wavelength of the oscillation must be large compared with the Debye length. Oscillations of wavelength comparable to the Debye length are damped very rapidly, unless the initial conditions are such that there are a large number of particles with velocities close to the phase velocity of the wave and we obtain a Van Kampen mode.

Because of shortage of space we shall not now consider the corresponding problem for transverse waves, but will turn to the problem of oscillations in an external magnetic field.

5.3. *Oscillations in an external magnetic field*

In this section we shall study only the oscillations of a Maxwellian plasma in an external field $B_0 b$. The linearized Vlasov equation for any species of particle has the form

$$\frac{\partial f_1}{\partial t} + v \cdot \frac{\partial f_1}{\partial x} + \frac{e}{m} E_1 \cdot \frac{\partial f_0}{\partial v} - \frac{eB_0}{mc} b \cdot v \times \frac{\partial f_1}{\partial v} = 0, \qquad \ldots[5.42]$$

where for a Maxwellian (or general isotropic) distribution $v \times B_1 \cdot \partial f_0/\partial v \equiv 0$. To solve equation [5.42] we perform a Fourier transform in space and a Laplace transform in time, and write

$$f_1(x, v, t) = \int d^3k \exp ik \cdot x F^*(k, v, t)$$

$$F(k, v, s) = \int dt \exp(-st) F^*(k, v, t) \qquad \ldots[5.43]$$

where $\mathscr{R}(s) > 0$. Then the linearized Vlasov equation can be written

$$(s + ik \cdot v)F - \Omega b \cdot v \times \partial F/\partial v = -(e/m)\mathscr{E} \cdot \partial f_0/\partial v + F^*(k, v, 0), \qquad \ldots[5.44]$$

where $\mathscr{E}$ is the transform of E_1 and the last term shows the influence of the initial conditions.‡

We now choose in velocity space a rectangular coordinate system such that

$$\left. \begin{aligned} b &= (0, 0, 1) \\ k &= (k \sin \theta, 0, k \cos \theta) \\ v &= (w \cos \phi, w \sin \phi, u). \end{aligned} \right\} \qquad \ldots[5.45]$$

Then equation [5.44] becomes

$$\frac{\partial F}{\partial \phi} - \frac{[s + i(kw \sin \theta \cos \phi + ku \cos \theta)]}{\Omega} F = \frac{e}{m\Omega} \mathscr{E} \cdot \frac{\partial f_0}{\partial v} - \frac{F^*}{\Omega}. \qquad \ldots[5.46]$$

This equation must then be solved subject to the requirement that F be periodic in ϕ with period 2π. The solution is obtained by introducing

$$v' = (w \cos \phi', w \sin \phi', u) \qquad \ldots[5.47]$$

and $$G = \exp\left[\frac{s + iku \cos \theta}{\Omega}(\phi - \phi') + \frac{ikw \sin \theta}{\Omega}(\sin \phi - \sin \phi')\right].$$

$$\ldots[5.48]$$

‡ This discussion is based on Bernstein (17).

Then

$$F(k, v, s) = \int_{\pm\infty}^{\phi} d\phi' \left[\frac{e}{m\Omega} \mathscr{E} \cdot \frac{\partial f_0(v')}{\partial v'} - \frac{F^*}{\Omega} \right] G, \qquad \ldots[5.49]$$

where the lower limit of the integral is taken as $+\infty$ or $-\infty$, according to the sign of Ω, to make the integral converge.

The Fourier-Laplace transforms of the Maxwell equations are

$$ik \cdot \mathscr{B}(k, s) = 0, \qquad\qquad\qquad \ldots[5.50]$$

$$ick \times \mathscr{E}(k, s) = -s\mathscr{B}(k, s) + \mathscr{B}^*(k, 0), \qquad \ldots[5.51]$$

$$ick \times \mathscr{B}(k, s) = \sum 4\pi e_r \int d^3v v F_r(k, v, s) + s\mathscr{E}(k, s) - \mathscr{E}^*(k, 0) \quad \ldots[5.52]$$

and $\qquad ik \cdot \mathscr{E}(k, s) = \sum 4\pi e_r \int d^3v F_r(k, v, s). \qquad \ldots[5.53]$

If the cross product of [5.51] with ick is taken and equation [5.52] is used, there results

$$(s^2 + c^2 k^2)\mathscr{E} - c^2 k(k \cdot \mathscr{E}) + \mathscr{E} \cdot Q = a, \qquad \ldots[5.54]$$

where Q is the tensor

$$Q = \sum \frac{4\pi e_r^2 s}{m_r \Omega_r} \int d^3v \int_{\pm\infty}^{\phi} d\phi' G_r \frac{\partial f_{r0}(v')}{\partial v'} v \qquad \ldots[5.55]$$

and

$$a = s\mathscr{E}^* + ick \times \mathscr{B}^* + \sum \frac{4\pi e_r s}{\Omega_r} \int d^3v v \int d\phi' G_r F_r^*. \qquad \ldots[5.56]$$

The problem is now reduced to solving equation [5.54] for $\mathscr{E}$. We introduce a matrix R such that

$$R_{ij} = (s^2 + k^2 c^2)\delta_{ij} - c^2 k_i k_j + Q_{ji}. \qquad \ldots[5.57]$$

Then

$$R \cdot \mathscr{E} = a,$$

or $\qquad\qquad \mathscr{E} = R^A \cdot a / |R|, \qquad\qquad \ldots[5.58]$

where the elements of R^A are the cofactors of their counterparts in R.

In order to go from equation [5.58] to the possible frequencies of the electric field in the plasma, we employ the inversion theorem for Laplace transforms

$$\mathscr{E}^*(k, t) = \frac{1}{2\pi i} \int_{s_0-i\infty}^{s_0+i\infty} ds \mathscr{E}(k, s) e^{st}. \qquad \ldots[5.59]$$

The contour of integration is a straight line parallel to the imaginary s-axis and to the right of all the singularities of $\mathscr{E}(k, s)$. The integral can be evaluated in terms of the poles of the integrand, assuming that there are no branch points etc. It can be shown that the elements of R and R^A have no singularities. Thus singularities arise, if

(a) a has singularities,

(b) $|R(s)| = 0$.

In case (a) the poles arise from the initial disturbance and these correspond to the Van Kampen modes in the absence of a magnetic field. The analogue of the Landau modes is obtained by assuming that a has no singularities and that the proper plasma frequencies are solutions of the equation

$$|R(s)| = 0. \qquad \ldots [5.60]$$

Equation [5.54] does not in general separate into equations for longitudinal and transverse waves. Scalar and vector products of the equation with k give

$$s^2 k \cdot \mathscr{E} + \mathscr{E} \cdot Q \cdot k = k \cdot a \qquad \ldots [5.61]$$

and
$$(s^2 + k^2 c^2) k \times \mathscr{E} + k \times (\mathscr{E} \cdot Q) = k \times a. \qquad \ldots [5.62]$$

In a plasma of moderate density and containing a moderate magnetic field there are two classes of wave; almost transverse waves, with frequency comparable with ck, and almost longitudinal waves with frequency com parable with $(\omega_{pe}^2 + \Omega^2)^{\frac{1}{2}}$ ($\ll ck$). In that case, equations [5.61] and [5.62] almost decouple to give approximate equations of the form

$$(s^2 + k^2 c^2) k \times \mathscr{E} = k \times a \qquad \ldots [5.63]$$

and
$$s^2 k \cdot \mathscr{E} + [(k \cdot Q \cdot k)/k^2] k \cdot \mathscr{E} = k \cdot a. \qquad \ldots [5.64]$$

The dispersion equation for longitudinal waves then has the form

$$k^2 s^2 + k \cdot Q \cdot k = 0. \qquad \ldots [5.65]$$

We consider this in the case in which the ions may be considered at rest because the frequency of the oscillation is high. In this case the equation has the form

$$k^2 s^2 + \frac{s \omega_p^2}{N\Omega} \int d^3 v \int_{-\infty}^{\phi} d\phi' k \cdot vk \cdot \frac{\partial f_0}{\partial v'} G = 0, \qquad \ldots [5.66]$$

where we omit the suffix e in quantities such as Ω_e in what follows and it must be noted that, with our definition, Ω_e is negative. If the equilibrium distribution is Maxwellian, three of the integrals in [5.66] can be evaluated. The detailed discussion may be found in Bernstein.[17] The result is

$$1 + k^2 a^2 = -(s/\Omega) \int_0^\infty dy \exp [sy/\Omega - \lambda(1 - \cos y) - \mu y^2/2], \qquad \ldots [5.67]$$

where $a^2 = \kappa T/4\pi Ne^2$, $\lambda = k^2 \rho^2 \sin^2 \theta$, $\mu = k^2 \rho^2 \cos^2 \theta$, θ is the angle between k and b and $\rho^2 = \kappa T/m\Omega^2$. If we now use the standard result

$$\exp (\lambda \cos y) = \sum_{-\infty}^{\infty} I_n(\lambda) \exp (iny), \qquad \ldots [5.68]$$

where $I_n(\lambda)$ is a modified Bessel function, equation [5.67] becomes

$$1+k^2a^2 = s \exp(-\lambda) \sum_{-\infty}^{\infty} I_n(\lambda) \int_0^{\infty} dt \exp\left[-\tfrac{1}{2}\Omega^2\mu t^2 - (s+in\Omega)t\right].$$
$$...[5.69]$$

Bernstein shows this dispersion equation reduces to that of Landau in the absence of a magnetic field, that there is Landau damping of waves, and that no growing waves exist. We will not here discuss the general theory, but will consider propagation orthogonal to the field. If we put $\theta = \pi/2$, equation [5·69] becomes

$$1+k^2a^2 = s \exp(-\lambda) \sum I_n(\lambda)/(s+in\Omega)$$
$$= \exp(-\lambda)\left[I_0(\lambda)+2 \sum_1^{\infty} (s/\Omega)^2 I_n(\lambda)/\{(s/\Omega)^2+n^2\}\right]. \qquad ...[5.70]$$

This equation is even in s; which means that, if there are no growing modes, there cannot be any damped modes either. We prove directly in this case that there can be no growing modes. If $s = i\beta+\gamma$, equation [5.70] can be written

$$\left.\begin{aligned}
1+k^2a^2 &= e^{-\lambda} \sum I_n(\lambda) \frac{\gamma^2+\beta^2+n\Omega\beta}{\gamma^2+(\beta+n\Omega)^2}, \\[2mm]
0 &= e^{-\lambda} \sum I_n(\lambda) \frac{n\Omega\gamma}{\gamma^2+(\beta+n\Omega)^2}.
\end{aligned}\right\} \qquad ...[5.71]$$

If γ is non-zero, β/γ times the second line of [5.71] can be added to the first line to give

$$1+k^2a^2 = 1-e^{-\lambda} \sum I_n(\lambda) \frac{n^2\Omega^2}{\gamma^2+(\beta+n\Omega)^2}, \qquad ...[5.72]$$

where we have used [5.68] with $y = 0$. It is clear that [5.72] is impossible, which shows that s cannot have a real part. Thus, although Landau damping occurs for propagation at an arbitrary angle to the field, it disappears for propagation perpendicular to the field.

If we write $s/\Omega = iq$, equation [5.70] can be written

$$1+k^2a^2 = e^{-\lambda}\left\{I_0(\lambda)+q \sum_1^{\infty} I_n(\lambda)\left[\frac{1}{q-n}+\frac{1}{q+n}\right]\right\} \qquad ...[5.73]$$

and this equation clearly has roots q between successive harmonics of the electron gyration frequency. A feature of this result is that there is a non-uniformity of behaviour for waves propagating perpendicular to the magnetic field. Thus the dispersion relation obtained by first putting $\theta = \pi/2$ and then letting $\Omega \to 0$ gives frequencies roughly at multiples of the gyration frequency, instead of the Landau frequencies which are obtained if first we put $\Omega = 0$. In fact, as $\Omega \to 0$, the waves different in character from those of Landau only occur for a range $\delta\theta$ about $\pi/2$ where $\delta\theta \to 0$ with Ω.

5.4. *Anisotropic hydromagnetic waves*

In Section 5.2 we considered the propagation of low frequency waves in a plasma in which the hydrodynamic equations were supposed to provide an adequate description. One of the effects of a strong magnetic field in a plasma in which collisions are infrequent is to cause the pressure along and across the field to be different. Chew, Goldberger and Low[18] showed that, to some approximation, hydromagnetic equations could be derived in which the scalar pressure is replaced by a diagonal tensor, and in which the components of the pressure satisfy the double adiabatic equations

$$\frac{d}{dt}(p_{\parallel}B^2/\rho^3) = 0 \qquad \qquad ...[5.74]$$

and
$$\frac{d}{dt}(p_{\perp}/B\rho) = 0, \qquad \qquad ...[5.75]$$

where $p_{\parallel}$ and $p_{\perp}$ are the pressure components along and across the field and B is the magnitude of the field. These equations are not completely accurate, and in particular they neglect the effect of heat flow along the field lines. However, they give qualitatively the correct results for the propagation of hydromagnetic waves, and they demonstrate the existence of two new types of unstable waves.

We study an equilibrium in which the magnetic field is $B_0 b$, the pressure components are $p_{\parallel 0}$ and $p_{\perp 0}$, and the density is ρ_0, and we take

$$\begin{aligned} b &= (0, 0, 1), \\ k &= k(0, \sin\theta, \cos\theta). \end{aligned} \qquad \right\} \qquad ...[5.76]$$

The components of the pressure tensor are

$$\begin{aligned} P_{xx} &= p_{\parallel}(B_x^2/B^2) + p_{\perp}(B_y^2 + B_z^2)/B^2, \\ P_{xy} &= P_{yx} = (p_{\parallel} - p_{\perp})B_x B_y/B^2, \end{aligned} \qquad ...[5.77]$$

with similar expressions for the other components. The perturbed components of the pressure tensor are then

$$\begin{aligned} P_{xx1} &= P_{yy1} = p_{\perp 1}, P_{zz1} = p_{\parallel 1}, \\ P_{zx1} &= P_{xz1} = (p_{\parallel 0} - p_{\perp 0})B_{1x}/B_0, \\ P_{zy1} &= P_{yz1} = (p_{\parallel 0} - p_{\perp 0})B_{1y}/B_0, P_{xy1} = P_{yx1} = 0. \end{aligned} \qquad ...[5.78]$$

The perturbed equations then have the form

$$-i\omega\rho_0 v_{1x} = -ik\cos\theta\,(p_{\parallel 0} - p_{\perp 0})B_{1x}/B_0 + ik\cos\theta\,B_{1x}B_0/4\pi, \quad ...[5.79]$$

$$\begin{aligned} -i\omega\rho_0 v_{1y} = &-ik\sin\theta\,p_{\perp 1} - ik\cos\theta\,(p_{\parallel 0} - p_{\perp 0})B_{1y}/B_0 \\ &-ik\sin\theta\,B_{1z}B_0/4\pi + ik\cos\theta\,B_{1y}B_0/4\pi, \end{aligned} \qquad ...[5.80]$$

$$-i\omega\rho_0 v_{1z} = -ik\sin\theta\,(p_{\parallel 0} - p_{\perp 0})B_{1y}/B_0 - ik\cos\theta\,p_{\parallel 1}, \qquad ...[5.81]$$

$$-i\omega\rho_1 + ik\sin\theta\,\rho_0 v_{1y} + ik\cos\theta\,\rho_0 v_{1z} = 0, \qquad ...[5.82]$$

$$-i\omega B_{1x} = ik\cos\theta\,B_0 v_{1x}, \qquad ...[5.83]$$

$$-i\omega B_{1y} = ik \cos \theta \, B_0 v_{1y}, \qquad \qquad \dots [5.84]$$

$$-i\omega B_{1z} = -ik \sin \vartheta \, B_0 v_{1y}, \qquad \qquad \dots [5.85]$$

$$p_{\|1}/p_{\|0} + 2B_{1z}/B_0 - 3\rho_1/\rho_0 = 0 \qquad \qquad \dots [5.86]$$

and $p_{\perp 1}/p_{\perp 0} - B_{1z}/B_0 - \rho_1/\rho_0 = 0.$ $\qquad \qquad \dots [5.87]$

Equations [5.79] and [5.83] immediately combine to give a dispersion equation for Alfvén waves modified by the anisotropy of the pressure,

$$\omega^2 = k^2 c_H^2 \cos^2 \theta \, [1 + 4\pi(p_{\perp 0} - p_{\|0})/B_0^2], \qquad \dots [5.88]$$

showing that these waves become unstable if

$$p_{\|0} > (p_{\perp 0} + B_0^2/4\pi). \qquad \qquad \dots [5.89]$$

This is known as the 'firehose instability'. We can also obtain two coupled equations for v_{1y} and v_{1z} in the form

$$[\omega^2 - k^2 c_H^2 - 2k^2(p_{\perp 0}/\rho_0)\sin^2 \theta - k^2 \cos^2 \theta \, (p_{\perp 0} - p_{\|0})/\rho_0] v_{1y}$$
$$- k^2 \sin \theta \cos \theta \, (p_{\perp 0}/\rho_0) v_{1z} = 0 \qquad \dots [5.90]$$

and $\quad [\omega^2 - 3k^2 \cos^2 \theta \, p_{\|0}/\rho_0] v_{1z} - k^2 \sin \theta \cos \theta \, (p_{\perp 0}/\rho_0) v_{1y} = 0. \dots [5.91]$

For propagation parallel to the field, these modes decouple to give one mode with dispersion equation

$$\omega^2 = k^2 c_H^2 [1 + 4\pi(p_{\perp 0} - p_{\|0})/B_0^2] \qquad \dots [5.92]$$

and one with

$$\omega^2 = 3k^2 p_{\|0}/\rho_0, \qquad \qquad \dots [5.93]$$

the sound waves. For propagation perpendicular to the field, we obtain the magnetosonic wave with dispersion equation

$$\omega^2 = k^2 c_H^2 + 2k^2 p_{\perp 0}/\rho_0. \qquad \qquad \dots [5.94]$$

In the case of arbitrary direction of propagation, the two modes are coupled and the dispersion equation is

$$\omega^4 - \omega^2 [k^2 c_H^2 + (k^2 p_{\perp 0}/\rho_0)(2 - \cos^2 \theta) + (2k^2 p_{\|0}/\rho_0)\cos^2 \theta]$$
$$+ k^4 [3c_H^2 p_{\|0}\rho_0 \cos^2 \theta + 3p_{\perp 0} p_{\|0} \cos^2 \theta \, (2 - \cos^2 \theta)$$
$$- 3p_{\|0}^2 \cos^4 \theta - \cos^2 \theta \, (1 - \cos^2 \theta) p_{\perp 0}^2]/\rho_0^2 = 0. \qquad \dots [5.95]$$

Unstable waves will occur if the roots of equation [5.95] for ω^2 are not both real and positive. It can be shown that for both roots to be real and positive for all values of θ it is required that

$$p_{\perp 0}^2/6(p_{\perp 0} + B_0^2/8\pi) < p_{\|0} < p_{\perp 0} + B_0^2/4\pi. \qquad \dots [5.96]$$

Thus, in addition to the firehose instability, we derive the possibility of instability if $p_{\perp 0}$ is too large. This is discussed in detail by Lüst.[19]

The precise form of the left-hand inequality in [5.96] is altered if heat flow along the field lines is permitted. If both electrons and ions have similar distribution functions which are Gaussion both along and across the field, the modification of the inequality is the removal of the 6 from the denominator. This is discussed by Chandrasekhar, Kaufman and Watson.[20]

ACKNOWLEDGMENT

I am grateful to Mr P. R. Owen for help in checking much of the algebra in these lectures, but the responsibility for any errors remaining is wholly mine.

REFERENCES

1. O. BUNEMAN. 1961. How to distinguish between attenuating and amplifying waves. In *Plasma Physics* (ed. J. E. Drummond), 143-166. McGraw-Hill.
2. P. A. STURROCK. 1961. Amplifying and evanescent waves, convective and nonconvective instabilities. In *Plasma Physics* (ed. J. E. Drummond), 124-142. McGraw-Hill.
3. T. H. STIX. 1962. *The Theory of Plasma Waves*. McGraw-Hill.
4. V. L. GINZBURG. 1964. *The Propagation of Electromagnetic Waves in Plasmas*. Pergamon, Oxford.
5. M. J. LIGHTHILL. 1960. Studies on magnetohydrodynamic waves and other anisotropic wave motions. *Phil. Trans. R. Soc.*, A, **252**, 397-430.
6. W. B. THOMPSON. 1964 *An Introduction to Plasma Physics*, 2nd edition. Pergamon, Oxford.
7. M. J. LIGHTHILL. 1967. Predictions on the velocity field coming from acoustic noise and a generalized turbulence in a layer overlaying a convectively unstable atmosphere region. In *I.A.U. Symposium No. 28*, ed. R. N. Thomas, 429-453. Academic Press, London.
8. S. CHAPMAN and T. G. COWLING. 1952. *The Mathematical Theory of Non-uniform Gases*, 2nd edition. Cambridge University Press.
9. R. J. TAYLER. 1957. The influence of an axial magnetic field on the stability of a constricted gas discharge. *Proc. phys. Soc.*, B, **70**, 1049-1063.
10. F. DE HOFFMAN and E. TELLER. 1950. Magnetohydrodynamic shocks. *Phys. Rev.*, **80**, 692-703.
11. W. MARSHALL. 1955. The structure of magnetohydrodynamic shock waves. *Proc. R. Soc.*, A, **233**, 367-376.
12. P. GERMAIN. 1960. Shock waves and shock-wave structure in magnetofluid dynamics. *Rev. mod. Phys.*, **32**, 951-958.
13. K. V. ROBERTS and J. B. TAYLOR. 1962. Magnetohydrodynamic equations for finite Larmor radius. *Phys. Rev. Lett.*, **8**, 197-198.
14. N. G. VAN KAMPEN and B. Y. FELDERHOF. 1967. *Theoretical Methods in Plasma Physics*. North Holland, Amsterdam.
15. A. SCHLÜTER. 1950. Dynamik des Plasmas I. *Naturf.*, 5A, 72-78.
—— 1951. Dynamik des Plasmas II. *Ibid.*, 6A, 73-78.
16. L. LANDAU. 1946. On the vibrations of the electronic plasma. *Fiz. Zh.*, (*J. Phys. U.S.S.R.*), **10**, 25-34.
17. I. B. BERNSTEIN. 1958. Waves in a plasma in a magnetic field. *Phys. Rev.*, **109**, 10-21.
18. G. F. CHEW, M. L. GOLDBERGER and F. E. LOW. 1956. The Boltzmann equation and the one-fluid hydromagnetic equations in the absence of particle collisions. *Proc. R. Soc.*, A, **236**, 112-118.
19. R. LÜST. 1959. Über die ausbreitung von Wellen in einem plasma. *Fortschr. Phys.*, **7**, 503-558.
20. S. CHANDRASEKHAR, A. N. KAUFMAN and K. M. WATSON. 1958. The stability of the pinch. *Proc. R. Soc.*, A, **245**, 435-455.

4

PLASMA INSTABILITIES

EDWARD G. HARRIS

University of Tennessee, Knoxville, Tennessee, U.S.A.

1

INTRODUCTION

THE subject of plasma instabilities has undergone a development in the last decade which at first sight seems chaotic. This apparent chaos is due to the rapid growth in the number of known instabilities. In a recent review by Lehnert[1] thirty-one plasma instabilities are listed. Of these, all but eleven have been discovered since 1958. (When we say "discovered" we mean discovered theoretically. Many of them have not been unambiguously identified experimentally.) However, along with this growth in the number of known instabilities, there has been a growth in understanding of the relationships existing within families of instabilities. For instance, there are a number of instabilities with frequencies near the ion cyclotron frequency and its harmonics. These were discovered by different people at different times and have a variety of names, such as "anisotropy", "loss cone", "drift cyclotron" etc. All of these may be described in terms of emission and absorption of plasma waves by energetic particles. When the emission exceeds the absorption, the wave grows (is unstable). Conditions under which emission may exceed absorption may be achieved in various ways, such as making the distribution functions anisotropic, putting a beam through the plasma, having spatial gradients of density or temperature, etc. In a sense, all of these instabilities depend upon an "inverted population" such as one has in masers and lasers. However, the distribution functions, $f_s(x, v)$, in plasmas depend on as many as six variables, in contrast to masers and lasers where the distribution functions depend on the energy only. It follows that there are many more ways of "inverting" the population in plasmas.

We shall try to present the subject of plasma instabilities in such a way as to emphasize such unity as exists. Perhaps within a few years it will be possible to write on a "unified theory of plasma instabilities".

It will be convenient to divide instabilities into two broad classes: (*a*) macroscopic and (*b*) microscopic. The macroscopic, or hydromagnetic,

145

instabilities imply the displacement of macroscopic portions of plasma. They may be analysed theoretically through the use of hydrodynamic equations. In other words, it is assumed, as an approximation, that all the particles in a given macroscopic volume execute the same average motion. This class of instabilities is the oldest and also the easiest to visualize. Our discussion of this class of instabilities will be qualitative and intuitive to a large extent. The microscopic or kinetic instabilities can be defined as those for which the differences in the motion of different particles in the same volume is important. The Vlasov equations are necessary for the theoretical analysis of this class of instabilities.

Almost all instability theories assume that the departure of the system from equilibrium is infinitesimal and then ask whether this infinitesimal departure grows or decays. It is clear that such theories are incapable of answering many of the questions which are of major importance. An unstable disturbance will not remain infinitesimal. One would like to know what happens when it becomes large. We have considered it appropriate to include a discussion of non-linear effects in these lectures. This is a subject which is just entering its adolescence. Another aspect of reality which is often ignored by plasma theorist is the finite size of the plasma. These lectures will conclude with a discussion of some recent theoretical work on micro-instabilities in finite plasmas.

A serious omission from these lectures is an account of experimental studies of instabilities. I hope that other lecturers will fill this gap. The recent review paper by Lehnert is a valuable guide to the literature of both theoretical and experimental studies of plasma instabilities.[1] Other good general treatments of instabilities are those of Rostoker[2] and Vedenov Velikhov and Sagdeev.[3]

2

MACROSCOPIC INSTABILITIES[4,5]

2.1. *The hydromagnetic equations*

The hydromagnetic (or magnetohydrodynamic, MHD) equations are usually‡ adequate for the theoretical investigation of this class of instabilities. These equations are

$$\frac{\partial \rho}{\partial t} + \nabla \cdot \rho v = 0 \qquad \qquad ...[2.1]$$

$$\rho \left[\frac{\partial v}{\partial t} + v \cdot \nabla v \right] = -\nabla \cdot P + \rho e E + \frac{1}{c} (J \times B). \qquad ...[2.2]$$

‡ Sometimes it is necessary to use the two-fluid equations.

These equations must be supplemented by Maxwell's equations

$$\nabla \cdot E = 4\pi \rho_e \qquad \ldots [2.3]$$

$$\nabla \cdot B = 0 \qquad \ldots [2.4]$$

$$\nabla \times E = -\frac{1}{c}\frac{\partial B}{\partial t} \qquad \ldots [2.5]$$

$$\nabla \times B = \frac{4\pi}{c}J + \frac{1}{c}\frac{\partial E}{\partial t}. \qquad \ldots [2.6]$$

Equations [2.1] to [2.6] are not yet complete. The assumption of perfect conductivity is often made; that is, it is assumed that the electric field must vanish in a frame of reference moving with the fluid. Thus

$$E + \frac{1}{c}v \times B = 0. \qquad \ldots [2.7]$$

This is equivalent to saying that electrons and ions move with the drift velocity

$$v = c\frac{E \times B}{B^2} \qquad \ldots [2.8]$$

perpendicular to B. This assumption is not valid if frequencies of the motion are comparable to collision frequencies, or cyclotron frequencies of the particles. To complete the sets of equations an equation relating P and ρ is needed. The magnetohydrodynamic (MHD) approximation assumes

$$P = 1P \qquad \ldots [2.9]$$

$$\frac{d}{dt}\left(\frac{P}{\rho^\gamma}\right) = 0. \qquad \ldots [2.10]$$

The gas constant γ is usually taken to be 5/3, the value appropriate for a gas of particles with only translational degrees of freedom. The Chew-Goldberger-Low (CGL) approximation assumes

$$P = 1P_\perp + (P_\parallel - P_\perp)\frac{BB}{B^2} \qquad \ldots [2.11]$$

$$\frac{d}{dt}\left(\frac{P_\parallel B^2}{\rho^3}\right) = 0 \qquad \ldots [2.12]$$

$$\frac{d}{dt}\left(\frac{P_\perp}{\rho B}\right) = 0. \qquad \ldots [2.13]$$

Now, equations [2.1] to [2.7], together with either equations [2.9] and [2.10] or equations [2.11], [2.12] and [2.13], give a complete set.

Let us consider the $(J \times B)$ term in equation [2.2]. We shall use equation [2.6] to eliminate J and neglect the displacement current. Then

$$\frac{1}{c}(\boldsymbol{J} \times \boldsymbol{B}) = \frac{1}{4\pi}(\nabla \times \boldsymbol{B}) \times \boldsymbol{B} = \frac{1}{4\pi}\boldsymbol{B} \cdot \nabla\boldsymbol{B} - \nabla\left(\frac{B^2}{8\pi}\right). \qquad \ldots[2.14]$$

The second term is like a scalar pressure of $B^2/8\pi$. The first term is a force which tends to straighten the field lines out. In a vacuum the magnetic field is in equilibrium under these two opposing forces.

2.2. Kink and sausage instabilities of the pinch

We can qualitatively discuss some instabilities on the basis of equation [2.14]. First, consider a plasma cylinder carrying a current as shown in Fig. 1. For simplicity, assume that the current is carried on the surface of the cylinder. The external magnetic field exerts a pressure of $B^2/8\pi$ on the surface of the cylinder and this is balanced by an internal pressure P. This equilibrium is not stable, however. Consider a perturbation which causes the radius of the cylinder to become smaller in some region as shown in Fig. 1(b). Then, since B varies inversely as the radius, the magnetic pressure will be greatest when the radius is smallest and the perturbation will grow. This is the "sausage" instability.

Next, consider the perturbation of the plasma cylinder shown in Fig. 1(c). The lines of B crowd together on the lower side of the kink. This increases the magnetic pressure there causing the perturbation to grow. This is the "kink" instability.

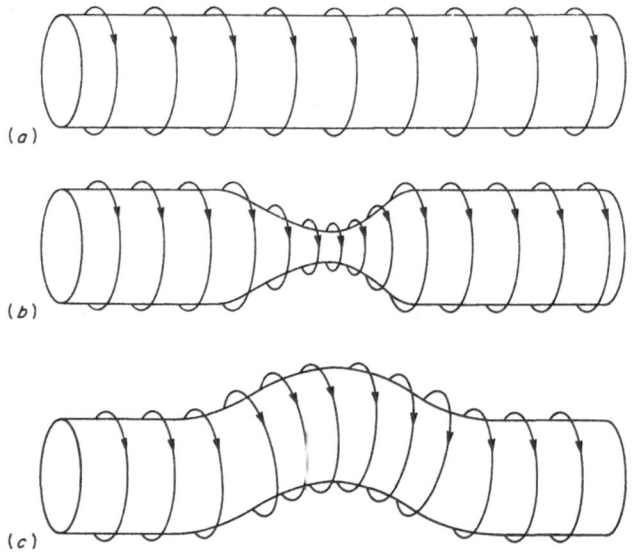

FIG. 1. Instabilities of the pinch. (a) The equilibrium configuration. (b) The sausage instability. (c) The kink instability.

There are some fairly obvious ways of making the pinch more stable. If a magnetic field parallel to the axis is trapped within the cylinder, then the sausage and kink instabilities would bend and compress the field lines, bringing into play forces which would oppose the growth of the instability. Unfortunately, it has not proved possible to stabilize the pinch completely. The reason is that the idealization of surface currents is never realized. It turns out that in a finite thickness current layer there are always instabilities which cannot be avoided.

2.3. *The Kelvin-Helmholtz instability*

There are a number of plasma instabilities which are closely related to instabilities of classical hydrodynamics. One of these is the Kelvin-Helmholtz instability, which occurs when there is relative motion between fluids separated by an interface. It is this instability which causes waves to grow when a wind blows over the surface of a body of water. A similar instability occurs when plasma flows perpendicular to a magnetic field.[6] The magnetic field behaves like a fluid with mass density $B^2/4\pi c^2$. The interface between plasma and field is unstable.

2.4. *The Rayleigh-Taylor instability*

Another classical instability is the Rayleigh-Taylor instability, which occurs when a heavy fluid is supported against gravity by a lighter fluid. A similar instability occurs when a plasma is supported against a gravitational field by a magnetic field as shown in Fig. 2. This instability is understandable on the basis of the guiding centre motion of the ions and electrons. We shall assume that the plasma-vacuum interface is sharp. Under the influence of the perpendicular gravitational and magnetic fields the ions and electrons have the drift velocities

$$v_{i,\,e} = c\,\frac{m_{i,\,e}\,g \times B}{e_{i,\,e}\,B^2}. \qquad\qquad ...[2.15]$$

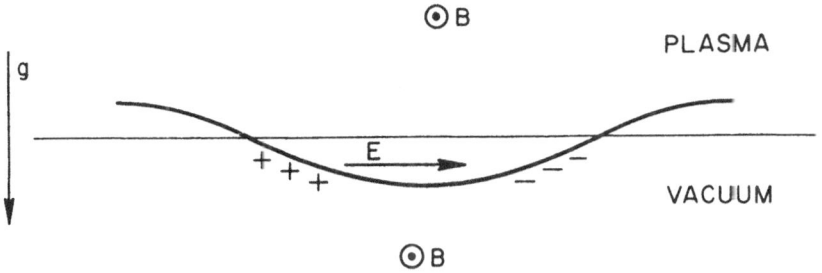

FIG. 2. Instability of a plasma supported by a magnetic field.

Because they have opposite charges, they drift in opposite directions. If the interface is perturbed as shown in Fig. 2, then the ions and electrons accumulate on the interface and produce an electric field as shown. Equation [2.8] shows that this electric field will cause the ions and electrons to drift together in such a direction as to cause the perturbation to grow. It can be shown[7, 8] that the perturbation grows as exp γt with

$$\gamma = \sqrt{(gk)}, \qquad \qquad ...[2.16]$$

where $k = 2\pi/\lambda$ is the wave number of the perturbation.

Equation [2.16] is only valid when the wavelength λ is much greater than the thickness of the interface. The case when there is a gentle gradient in density rather than a discontinuity is also of interest. This may be treated by the two fluid equations for a cold plasma. We shall only quote the results.[9] The dispersion relation is found to be

$$1 + \sum_s \frac{\omega_{ps}^2}{\omega_{cs}^2} = \frac{n'}{kn} \sum_s \frac{\omega_{ps}^2}{\omega_{cs}} \frac{1}{(\omega + kv_s)}. \qquad ...[2.17]$$

Here, s denotes the species (electrons and ions); ω_{ps} and ω_{cs} are plasma and cyclotron frequencies. The particles drift with velocities $v_s = g/\omega_{cs}$ in a direction perpendicular to g and to B. The particle density is n, and n' is its derivative in the direction of g. The frequency ω and wavenumber k of a wave which propagates perpendicularly to g and B are connected by equation [2.17]. In the case where only one type of ion is present, equation [2.17] is a quadratic equation for ω. It may be solved with the result

$$\omega = -\tfrac{1}{2}k(v_i + v_e) \pm \left[\tfrac{1}{4}k^2(v_i + v_e)^2 - \frac{n'\omega_{pi}^2}{Kn\omega_{ci}}(v_i + v_e) \right]^{\frac{1}{2}} \qquad ...[2.18]$$

where

$$K = 1 + \sum_s \frac{\omega_{ps}^2}{\omega_{cs}^2} = 1 + \frac{c^2}{V_a^2} \qquad \qquad ...[2.19]$$

and

$$V_a = (B^2/4\pi nM)^{\frac{1}{2}}$$

is the Alfvén velocity.

Equation [2.18] will have real roots if

$$\frac{n'}{n} > -\tfrac{1}{4} \frac{k^2 g}{\omega_{pi}^2} \left(1 + \frac{m}{M} \right) \left[1 + \frac{c^2}{V_a^2} \right]. \qquad ...[2.20]$$

This is the stability criterion. We have discussed this problem without derivation, because the same problem is treated in the chapter on microinstabilities.

Real gravitational fields are seldom of much concern in laboratory plasmas. However, if a plasma is accelerated, then it is in an effective gravitatational field and an instability may develop. This has been observed in z-pinch[10] and θ-pinch[11] experiments in which the plasma is very rapidly compressed, causing a large acceleration of a plasma vacuum interface.

2.5. Flute instabilities

Consider a plasma confined by a curved magnetic field as shown in Fig.3.

A particle spiraling along the magnetic field feels a centrifugal force of $mv_\parallel^2/R$. This force causes ions and electrons to drift in opposite directions. The resulting space charge produces an electric field which causes the particles to drift in the direction of this centrifugal force. This is quite analogous to the Rayleigh-Taylor instability with $v_\parallel^2/R$ playing the role of the gravitational field g. If g is directed out of the plasma, then the plasma is expected to be unstable. If g is directed into the plasma, then the plasma should be stable.

This is an instability which is expected to occur in mirror machines, and in fact it is sometimes observed.[12] There are circumstances under which the flute instability does not occur in mirror machines, probably because of finite Larmor radius effects[13] and the tying of the magnetic field lines in conducting plates at the ends of the machine.

One way of stabilizing against flute instabilities is to construct a magnetic field in which the lines are all curved away from the plasma. Then the centrifugal force is always directed into the plasma. This can be done in a simple cusp configuration[14] such as is shown in Fig. 4. However, one pays for stability with a much higher leak rate. A much more favourable configuration is the combination of the mirror field of Fig. 5a with the field produced by a number of conductors parallel to the axis as shown in Fig. 5b. It may be shown that, if there is a minimum in $B = |B|$, then the field lines curve away from this minimum. Plasma trapped in the region near this minimum is then stably confined. Ioffe[12] has experimentally verified the theoretical predictions in the field configuration of Fig. 5.

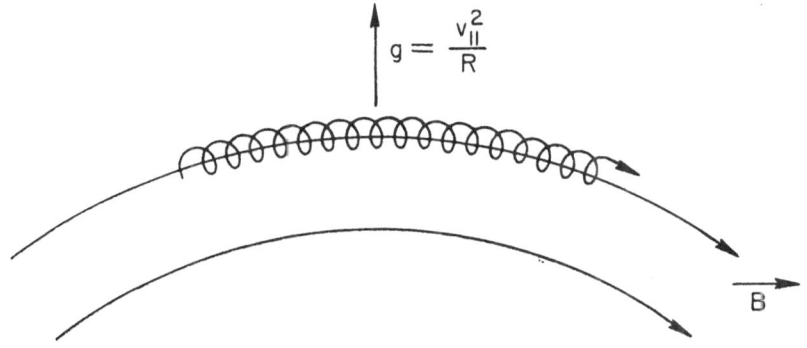

FIG. 3. The flute instability.

2.6. *The interchange stability criterion in fields with closed lines of* **B**

We shall now derive a very useful stability criterion for systems in which the lines close on themselves.

Consider two neighbouring flux tubes as shown in Fig. 6. Let the cross-

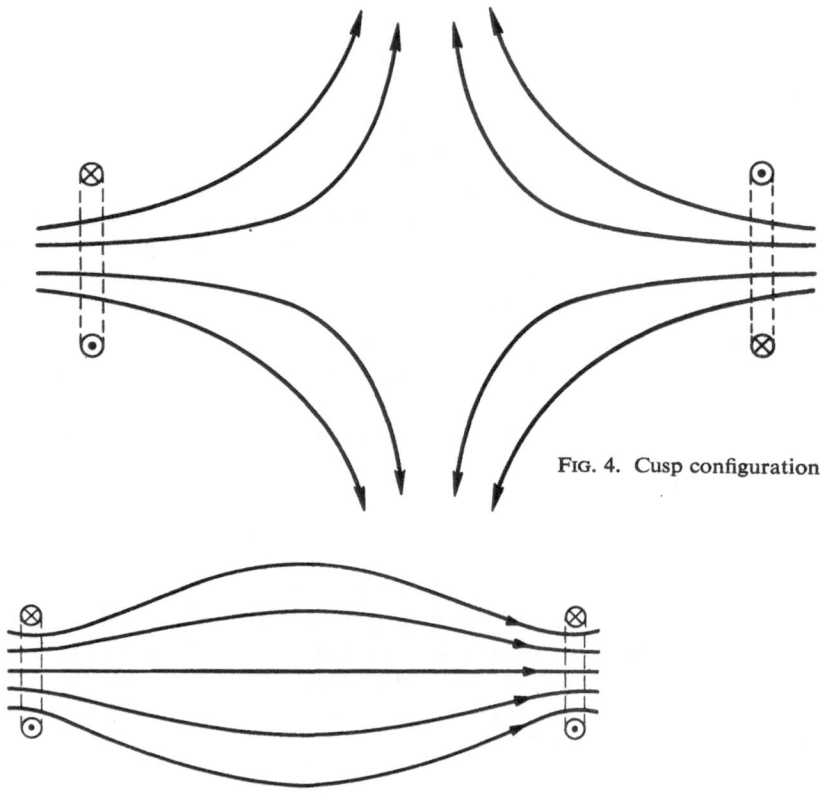

FIG. 4. Cusp configuration

(*a*) MAGNETIC MIRROR FIELD

FIG. 5

Magnetic mirror
machine with
Ioffe bars.

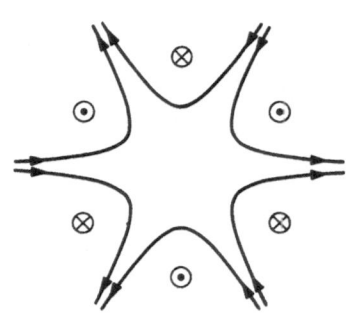

(*b*) STABILIZING FIELD

sectional areas be S_1 and S_2, the pressures P_1 and P_2, magnetic field B_1 and B_2 and fluxes ϕ_1 and ϕ_2. We have the relations

$$\phi_1 = B_1 S_1 \qquad \qquad \qquad ...[2.21]$$

$$V_1 = \phi_1 U_1 \qquad \qquad \qquad ...[2.22]$$

where

$$U_1 = \oint_1 \frac{dl}{B_1} \qquad \qquad \qquad ...[2.23]$$

with similar relations connecting ϕ_2, v_2 and U_2. Now, we ask for the change in the energy

$$W = \int d^3x \left[\frac{P}{\gamma-1} + \frac{1}{8\pi} B^2 \right] \qquad \qquad ...[2.24]$$

when the flux tubes are interchanged. This is $\delta W = \delta W_p + \delta W_B$, where

$$\delta W_p = \frac{1}{\gamma-1} \{ (P_1' V_1 + P_2' V_2) - (P_1 V_1 + P_2 V_2) \} \qquad ...[2.25]$$

$$\delta W_B = \frac{1}{8\pi} \{ \oint_1 B_1'^2 S_1 dl + \oint_2 B_2'^2 S_2 dl - \oint_1 B_1^2 S_1 dl - \oint_2 B_2^2 S_2 dl \}.$$
$$...[2.26]$$

Here P_1' and B_1' are the pressure and field in the first flux tube after the interchange, and similarly for P_2' and B_2'. Assuming an adiabatic expansion we have

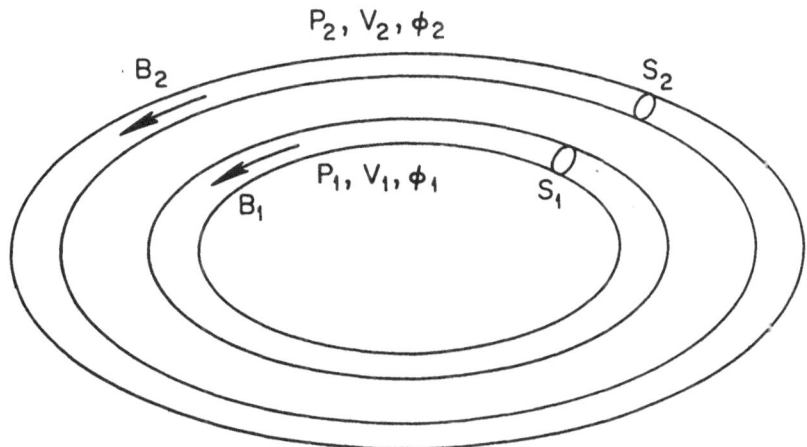

FIG. 6. Flux tubes.

$$P'_1 = P_2 \left(\frac{V_2}{V_1}\right)^\gamma \qquad \qquad ...[2.27]$$

$$P'_2 = P_1 \left(\frac{V_1}{V_2}\right)^\gamma . \qquad \qquad ...[2.28]$$

Using this in equation [2.25] and neglecting terms which are of higher order than the second in $\delta P = P_2 - P_1$ and $\delta V = V_2 - V_1$, we obtain

$$\delta W_p = \delta P \delta V + \frac{\gamma P_1}{V_1} (\delta V)^2 . \qquad \qquad ...[2.29]$$

Now

$$B'_1 S_1 = B_2 S_2 = \phi_2, \qquad \qquad ...[2.30]$$

$$B'_2 S_2 = B_1 S_1 = \phi_1 . \qquad \qquad ...[2.31]$$

So

$$\delta W_B = \frac{1}{2c} (\phi_2^2 - \phi_1^2) \left[\frac{I_1}{\phi_1} - \frac{I_2}{\phi}\right] = \frac{-1}{2c} \delta(\phi^2) \delta\left(\frac{I}{\phi}\right), \qquad ...[2.32]$$

where

$$I = \frac{c}{4\pi} \oint B dl \qquad \qquad ...[2.33]$$

is the current enclosed by the path.

An instability criterion is often derived in the following way.[8] Consider the interchange of tubes which contain equal flux; that is, choose $\phi_2 - \phi_1 = \delta\phi = 0$. Then $\delta W_B = 0$. If δW_p is negative, then the energy is lowered by the interchange and the system is unstable. Using equations [2.22] and [2.29] the instability criterion can be written as

$$\delta P \delta U + \frac{\gamma P}{U} (\delta U)^2 < 0 \qquad \qquad ...[2.34]$$

or

$$\frac{\delta \log P}{\delta \log U} + \gamma < 0. \qquad \qquad ...[2.35]$$

This is clearly a sufficient condition for instability. However, there may be other choices of $\delta\phi$ which lower δW even more than the choice $\delta\phi = 0$. What one should do is to choose $\delta\phi$ so as to minimize δW. If this minimum δW is negative, then the system is unstable. We can write

$$\delta W = \phi \left[\delta P \, \delta U + \frac{\gamma P}{U} (\delta U)^2\right] + \delta\phi \left[U \delta P + 2\gamma P \, \delta U - \frac{\delta I}{c}\right] + (\delta\phi)^2 \left[\frac{\gamma P U}{\phi} + \frac{I}{c\phi}\right].$$
$$...[2.36]$$

This can be simplified by using $\delta I = cU\delta P$ which follows from

$$\delta I = \int\int_{\Delta A} J\,dA, \qquad \ldots[2.37]$$

where ΔA is the area between the field lines B_1 and B_2. The current density is obtained from the equilibrium condition

$$0 = -\nabla P + \frac{1}{c}\,J \times B \qquad \ldots[2.38]$$

$$J = c\,\frac{B \times \nabla P}{B^2}. \qquad \ldots[2.39]$$

δW is a minimum when

$$\delta\phi = -\frac{\gamma P\,\delta U}{\left[\dfrac{\gamma PU}{\phi} + \dfrac{I}{c\phi}\right]}, \qquad \ldots[2.40]$$

and this minimum value is

$$\delta W = \frac{(\gamma PU + I/c)\delta P\,\delta U + I/c\,\dfrac{\gamma P}{U}\,(\delta U)^2}{\left[\dfrac{\gamma PU}{\phi} + I/c\,\phi\right]}. \qquad \ldots[2.41]$$

A sufficient condition for stability is

$$\left(1 + \frac{c\gamma PU}{I}\right)\delta P\,\delta U + \frac{\gamma P}{U}\,(\delta U)^2 > 0 \qquad \ldots[2.42]$$

or

$$\frac{\delta\log P}{\delta\log U} + \frac{\gamma}{1 + \dfrac{c\gamma PU}{I}} > 0. \qquad \ldots[2.43]$$

We shall now consider some applications of this stability criterion. First, we shall assume that $\beta = 8\pi P/B^2 \ll 1$. The term $c\gamma PU/I$ is negligible, and the stability criterion is

$$-\frac{d\log P}{d\log U} < \gamma. \qquad \ldots[2.44]$$

This may also be written as

$$-\nabla P \cdot \nabla U < \frac{\gamma P}{U}\,|\nabla U|^2. \qquad \ldots[2.45]$$

(a) *Plasma in the field of a straight current.*[4] If the plasma has a sharp boundary so that ∇P is very large, then equation [2.45] can be satisfied only if $\nabla P \cdot \nabla U$ is positive; that is, U must increase in the same direction in which P increases. For a sharp boundary

$$U \text{ (inside)} > U \text{ (outside)} \qquad \qquad ...[2.46]$$

for stability.

Next, we consider the case of a plasma confined in the field of a straight current I_0 along the axis. Then

$$B = \frac{2}{cr} I_0 \qquad \qquad ...[2.47]$$

$$U = \oint \frac{dl}{B} = \frac{2\pi r}{B} = \frac{\pi c}{I_0} r^2 \qquad \qquad ...[2.48]$$

$$d \log U = \frac{2}{r} dr = 2d \log r. \qquad \qquad ...[2.49]$$

The stability criterion is

$$-\frac{d \log P}{d \log r} < 2\gamma. \qquad \qquad ...[2.50]$$

By changing this inequality to an equality one finds

$$P \sim \frac{1}{r^{2\gamma}}. \qquad \qquad ...[2.51]$$

If the pressure decreases with r more rapidly than this, then the plasma is unstable.

(b) *Plasma in the field of a dipole.*[4] In this case the field lines are not actually closed; but, if the surface of the dipole is insulated so that no currents flow over it, then our criterion still applies. Since the length of a line is proportional to r, and B falls off as r^{-3}, then

$$U \sim r^4 \qquad \qquad ...[2.52]$$

$$d \log U = 4d \log r \qquad \qquad ...[2.53]$$

and the stability criterion is

$$-\frac{d \log P}{d \log r} < 4\gamma. \qquad \qquad ...[2.54]$$

The limiting stable pressure distribution is

$$P \sim \frac{1}{r^{4\gamma}}. \qquad \qquad ...[2.55]$$

Presumably, in the Van Allen belts (the belts of energetic plasma trapped in the earth's magnetic dipole field), equation [2.54] is satisfied; otherwise convection currents would occur and persist until a stable distribution was reached.

(c) *Plasma in a toroidal multipole.*[15] In a toroidal multipole there are N internal rings carrying current in the same direction. ($N = 2$, quadrupole; $N = 4$, octopole etc.) The cross section of an octopole is sketched in Fig. 7. It is useful to define a flux function ψ such that, if $\hat{z}$ denote the direction of the ring currents,

$$B = \nabla \times \psi \hat{z}. \qquad \qquad ...[2.56]$$

Then each field line is characterized by a value of ψ. The separatrix (i.e. the lines that pass through the point where $B = 0$) is assigned the value zero. Equation [2.34] can be written as

$$\frac{dP}{d\psi}\frac{dU}{d\psi} + \frac{\gamma P}{U}\left(\frac{dU}{d\psi}\right)^2 < 0 \qquad \qquad ...[2.57]$$

for instability. (Since the pressure is assumed to be a scalar, it must be constant along a field line; hence $P = P(\psi)$.) The last term in equation [2.57] is usually negligible, and the plasma pressure decreases with ψ; so the plasma is unstable where $dU/d\psi$ is positive. (*Note*: In the paper by Ohkawa and Rostoker[15] U is defined with the opposite sign to the U of these lectures.)

Calculations show that there is a critical ψ_c at which U has a minimum. Inside the corresponding field line the plasma is stable; outside it is unstable. This has been experimentally verified.[16]

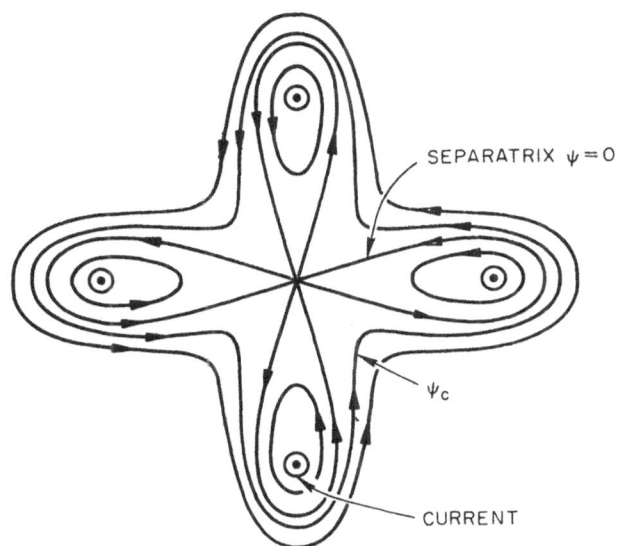

FIG. 7. Field lines in a toroidal octopole.

3

MICROINSTABILITIES

3.1. *Dispersion relation for a homogeneous plasma in a uniform magnetic field*[17, 18]

The hydromagnetic equations are not adequate for the study of this class of instabilities. Instead we must use the Vlasov equations

$$\frac{\partial f_s}{\partial t} + v \cdot \frac{\partial f_s}{\partial x} + \frac{e_s}{m_s}\left[E + \frac{1}{e} v \times B\right] \cdot \frac{\partial f_s}{\partial v} = 0 \qquad \ldots[3.1]$$

$$\nabla \cdot E = \sum_s 4\pi e_s \int d^3 v f_s \qquad \ldots[3.2]$$

$$\nabla \times B = \sum_s \frac{4\pi e_s}{c} \int d^3 v v f_s + \frac{1}{c}\frac{\partial E}{\partial t} \qquad \ldots[3.3]$$

$$\nabla \cdot B = 0 \qquad \ldots[3.4]$$

$$\nabla \times E = -\frac{1}{c}\frac{\partial B}{\partial t}. \qquad \ldots[3.5]$$

The subscript s denotes the species (electrons or ions). Since we shall be concerned with small departures from equilibrium, we linearize the equations by writing

$$f_s = f_{s0} + f_{s1} \qquad \ldots[3.6]$$

$$B = B_0 + B_1 \qquad \ldots[3.7]$$

$$E = 0 + E_1, \qquad \ldots[3.8]$$

where f_{s1}, B_1 and E_1 are small quantities, whose squares and products may be neglected. For the moment we shall consider only spatially homogeneous plasmas; so we may Fourier analyse all of the perturbed quantities. Thus

$$f_{s1}, B_1 E_1 \sim e^{i(k \cdot x - \omega t)}. \qquad \ldots[3.9]$$

It will be assumed that ω has a small positive imaginary part so that the perturbations vanish at $t = -\infty$. This will give us the prescription for handling some integrals which would otherwise be improper. It is equivalent to Landau's[19] prescription.

When equations [3.6], [3.7] and [3.8] are used in equation [3.1], we find

$$v \cdot \frac{\partial f_{s0}}{\partial x} + \frac{e_s}{m_s c} v \times B_0 \cdot \frac{\partial f_{s0}}{\partial v} = 0 \qquad \ldots[3.10]$$

$$\frac{\partial f_{s1}}{\partial t} + v \cdot \frac{\partial f_{s1}}{\partial x} + \frac{e_s}{m_s c}(v \times B_0) \cdot \frac{\partial f_{s1}}{\partial v} = -\frac{e_s}{m_s}\left[E_1 + \frac{1}{c} v \times B_1\right] \cdot \frac{\partial f_{s0}}{\partial v}. \qquad \ldots[3.11]$$

Equation [3.10] is satisfied if $f_{s0}(x, v)$ is a function of the constants of the motion of a particle in the unperturbed fields. Suitable constants are v_z, $v_\perp$, $x+v_y/\omega_{cs}$, $y-v_x/\omega_{cs}$. (We are assuming that $B_0 = Be_z$; then $v_1 = (v_x^2+v_y^2)^{\frac{1}{2}}$ and $x+v_y/\omega_{cs}$ and $y-v_x/\omega_{cs}$ are the coordinates of the guiding centre of the particle.) Since we want the plasma to be spatially uniform, we shall take

$$f_{s0} = f_{s0}(v_\perp, v_z). \qquad \text{...} [3.12]$$

Next, we shall give a prescription for solving equation [3.11]. There are other methods of solution than this, but this method has the advantage of greater generality. For our present purposes we shall write equation [3.11] as

$$\frac{\partial f_1}{\partial t} + v \cdot \frac{\partial f_1}{\partial x} + \frac{1}{m} F \cdot \frac{\partial f_1}{\partial v} = S(x, v, t). \qquad \text{...} [3.13]$$

Now, consider the equations of motion of a particle

$$\frac{dx}{dt} = v \qquad \text{..} [3.14]$$

$$\frac{dv}{dt} = -\frac{1}{m} F. \qquad \text{...} [3.15]$$

The solution of these equations will have the form

$$x = x(x_0, v_0, t) \qquad \text{...} [3.16]$$

$$v = v(x_0, v_0, t) \qquad \text{...} [3.17]$$

where x_0 and v_0 are the values of x and v at the time $t = 0$. Equations [3.16] and [3.17] can be inverted to obtain

$$x_0 = x_0(x, v, t) \qquad \text{..} [3.18]$$

$$v_0 = v_0(x, v, t). \qquad \text{..} [3.19]$$

Now, we claim that the solution of equation [3.12] is

$$f_1(x, v, t) = \int_{-\infty}^{t} dt' S(x_0(x, v, t-t'), v_0(x, v, t-t'), t'). \qquad \text{...} [3.20]$$

This is easily seen to be the solution if equation [3.20] is substituted into equation [3.12] and use is made of the relations

$$x_0(x, v, 0) = x \qquad \text{...} [3.21]$$

$$v_0(x, v, 0) = v \qquad \text{...} [3.22]$$

$$\frac{\partial x_0}{\partial t} + v \cdot \frac{\partial}{\partial x} x_0 + \frac{1}{m} F \cdot \frac{\partial}{\partial v} x_0 = \frac{d}{dt} x_0 = 0 \qquad \text{...} [3.23]$$

$$\frac{\partial}{\partial t} v_0 + v \cdot \frac{\partial}{\partial x} v_0 + \frac{1}{m} F \cdot \frac{\partial}{\partial v} v_0 = \frac{d}{dt} v_0 = 0. \qquad \text{..} [3.24]$$

Equations [3.21] and [3.22] follow from the definitions of x_0 and v_0 as initial values; Equations [3.23] and [3.24] are true, since x_0 and v_0 are constants of integration. By a change of variable equation [3.19] may be written as

$$f_1(x, v, t) = \int_0^\infty dt' S(x_0(x, v, t'), v_0(x, v, t'), t-t'). \qquad ...[3.25]$$

The particle equations of motion are easily solved for a particle in a uniform magnetic field. Equations [3.17] and [3.18] become

$$x_0 = x + v_y/\omega_{cs} - \frac{1}{\omega_{cs}} (v_x \sin \omega_{cs}t + v_y \cos \omega_{cs}t) \qquad ...[3.26]$$

$$y_0 = y - v_x/\omega_{cs} + \frac{1}{\omega_{cs}} (v_x \cos \omega_{cs}t - v_y \sin \omega_{cs}t) \qquad ...[3.27]$$

$$z_0 = z - v_z t \qquad ...[3.28]$$

$$v_{x0} = v_x \cos \omega_{cs}t - v_y \sin \omega_{cs}t \qquad ...[3.29]$$

$$v_{y0} = v_x \sin \omega_{cs}t + v_y \cos \omega_{cs}t \qquad ...[3.30]$$

$$v_{z0} = v_z. \qquad ...[3.31]$$

Using equation [3.5] to write

$$B_1 = \frac{c}{\omega} (k \times E_1), \qquad ...[3.32]$$

$S(x, v, t)$ may be written as

$$S_s(x, v, t) = -\frac{e_s}{m_s} \frac{1}{\omega} [(\omega - k \cdot v)E_1 + (v \cdot E_1)k] \cdot \frac{\partial f_{s0}}{\partial v}. \qquad ...[3.33]$$

Writing E_1 as

$$E_1(x, t) = E_1 e^{i(k \cdot x - \omega t)} \qquad ...[3.34]$$

we have

$$S_s(x, v, t) = -\frac{e_s}{m_s} \frac{e^{i(k \cdot x - \omega t)}}{\omega} [(\omega - k \cdot v)E_1 + (v \cdot E_1)k] \cdot \frac{\partial f_{s0}}{\partial v}. \qquad ...[3.35]$$

This is to be substituted into equation [3.25] and the integration over t carried out. First, consider the factor

$$e^{ik \cdot x_0(x, v, t') - \omega(t-t')} = e^{i(k \cdot x - \omega t)} e^{i \frac{k_\perp v_\perp}{\omega_{cs}} \sin \alpha} \times e^{i(\omega - k_z v_z)t' - i \frac{k_\perp v_\perp}{\omega_{cs}} \sin(\omega_{cs}t + \alpha)},$$

$$...[3.36]$$

where we have used equations [3.26], [3.27] and [3.28] and have let α be the angle between $k_\perp$ and $v_\perp$. We shall find it convenient to take $k_\perp$ along the x-axis. We now use the identity

$$e^{ia \sin x} = \sum_{n=-\infty}^{+\infty} J_n(a) e^{inx} \qquad ...[3.37]$$

Where $J_n(a)$ is the Bessel function. Using this identity twice in equation [3.34], we find

$$e^{ik \cdot x_0 - i\omega(t-t')} = e^{i(k \cdot x - \omega t)} \sum_{m=-\infty}^{+\infty} \sum_{n=-\infty}^{+\infty} J_m\left(\frac{k_\perp v_\perp}{\omega_{cs}}\right) J_n\left(\frac{k_\perp v_\perp}{\omega_{cs}}\right)$$

$$e^{i(m-n)\alpha} e^{i(\omega - k_z v_z - n\omega_{cs})t'}. \qquad ...[3.38]$$

We can write

$$[(\omega - k \cdot v_0) E_1 + (v_0 \cdot E_1)k] \cdot \frac{\partial f_{s0}}{\partial v}$$

$$= \frac{1}{v_\perp} \frac{\partial f_{s0}}{\partial v_\perp} [(\omega - k_z v_z) E_\perp v_\perp \cos(\omega_{cs} t' + \beta) + E_z v_z k_\perp v_\perp \cos(\omega_{cs} t' + \alpha)]$$

$$+ \frac{\partial f_{s0}}{\partial v_z} [(\omega - k_\perp v_\perp \cos(\omega_{cs} t' + \alpha)) E_z + k_z v_\perp E_\perp \cos(\omega_{cs} t' + \beta)]. \qquad ...[3.39]$$

Here β is the angle between $v_\perp$ and $E_\perp$. Equations [3.38] and [3.39] are now used in equation [3.35] and the integration in equation [3.25] can be performed with the result

$$f_{1s}(x, v, t) = -\frac{ie_s}{m_s \omega} \sum_{m=-\infty}^{+\infty} \sum_{n=-\infty}^{+\infty} \frac{J_m e^{i(m-n)\alpha}}{(\omega - k_z v_z - n\omega_{cs})}$$

$$\left\{ E_{1x} \frac{n}{\lambda_s} J_n U - i E_{1y} J_n' U + E_{1z} J_n W \right\}, \qquad ...[3.40]$$

where

$$U = (\omega - k_z v_z) \frac{\partial f_{s0}}{\partial v_\perp} + k_z v_\perp \frac{\partial f_{s0}}{\partial v_z} \qquad ...[3.41]$$

$$W = \frac{n\omega_c v_z}{v_\perp} \frac{\partial f_{s0}}{\partial v_\perp} + (\omega - n\omega_{cs}) \frac{\partial f_{s0}}{\partial v_z} \qquad ...[3.42]$$

$$\lambda_s = \frac{k_\perp v_\perp}{\omega_{cs}} \qquad ...[3.43]$$

The argument of J_n is understood to be λ_s. The prime on J_n' denotes the derivative of J_n with respect to its argument. In deriving equation [3.40] we have used the identities

$$J_{n+1}(\lambda)+J_{n-1}(\lambda) = \frac{2n}{\lambda} J_n(\lambda) \qquad \ldots[3.44]$$

and

$$J_{n+1}(\lambda)-J_{n-1}(\lambda) = 2\frac{d}{d\lambda} J_n(\lambda). \qquad \ldots[3.45]$$

Equation [3.40] can now be used to calculate the current:

$$J = \sum_s e_s \int d^3 v f_{s1}(v) v. \qquad \ldots[3.46]$$

By inspection we see that this can be written

$$J = \sigma \cdot E_1 \qquad \ldots[3.47]$$

where $\sigma(k, \omega)$ is the conductivity. Multiplying f_{1s} by $v = (v_\perp \cos\alpha, v_\perp \sin\alpha, v_z)$ and carrying out the velocity space integration gives

$$\sigma(k, \omega) = -i \sum_s \frac{e_s^2}{m_s \omega} \sum_{n=-\infty}^{+\infty} \int d^3 v \frac{S_s}{(\omega - k_z v_z - n\omega_{cs})} \qquad \ldots[3.48]$$

where

$$S_s = \begin{bmatrix} v_\perp U \left(\dfrac{nJ_n}{\lambda_s}\right)^2 & -iv_\perp U \dfrac{n}{\lambda_s} J_n J_n' & v_\perp W \dfrac{n}{\lambda_s} J_n^2 \\[2ex] iv_\perp U \dfrac{n}{\lambda_s} J_n J_n' & v_\perp U (J_n')^2 & iv_\perp W J_n J_n' \\[2ex] v_z U \dfrac{n}{\lambda_s} J_n^2 & -iv_z U J_n J_n' & v_z W J_n^2 \end{bmatrix} \qquad \ldots[3.49]$$

We may now use the current given by equation [3.47] in Maxwell's equations. Using equation [3.9] in equations [3.3] and [3.5] gives

$$(k \times B_1) = \frac{4\pi}{ic} \sigma \cdot E_1 - \frac{i\omega}{c} E_1 \qquad \ldots[3.50]$$

$$k \times E_1 = \frac{\omega}{c} B_1. \qquad \ldots[3.51]$$

Eliminating B_1 gives

$$k \times (k \times E_1) = -\frac{\omega^2}{c^2} \varepsilon \cdot E_1 \qquad \ldots[3.52]$$

where the dielectric tensor $\varepsilon(k, \omega)$ is defined by

$$\varepsilon(k, \omega) = 1 - \frac{4\pi}{i\omega} \sigma(k, \omega). \qquad \ldots[3.53]$$

Equation [3.52] may be written as

$$\left[k^2 1 - kk - \frac{\omega^2}{c^2} \, \varepsilon(k, \omega) \right] \cdot E_1 = 0. \qquad \ldots [3.54]$$

These are three homogeneous equations for the components of E_1. The condition that a non-trivial solution exists is that the determinant cf the co-efficients vanishes. This gives a relation between ω and k, the dispersion relation.

Because of the complexity of $\varepsilon(k, \omega)$ it is very difficult to do anything without simplifying the dispersion relation. We shall consider two simplifications. First, if k is parallel to B_0, the dispersion relation becomes quite simple. We shall show that there are then three frequencies. Two of these correspond to circularly polarized transverse waves; the third is a longitudinal (electrostatic) wave. If k is not parallel to B_0, then the waves cannot be separated into longitudinal and transverse waves exactly. However, this separation can be made approximately. This is discussed in Section 3.3.

3.2. Instabilities with k parallel to B_0

We shall assume that $k = ke_z$. Then $k_\perp = 0$ and by equation [3.43]. $\lambda_s = 0$. Since $J_n(0) = \delta_{n0}$ this leads to a great simplification in σ and ε, Equations [3.44] and [3.45] may be used in interpreting $nJ_n(\lambda)/\lambda$ and $J_n(\lambda)$ in the $\lambda \to 0$ limit. In this limit we find

$$\varepsilon(k, \omega) = \varepsilon_{ij}(k, \omega) \qquad \ldots [3.55]$$

where

$$\varepsilon_{11}(k, \omega) = 1 + \sum_s \frac{4\pi e_s^2}{m_s \omega^2} \int d^3v \, \frac{v_\perp}{4} \left[(\omega - kv_z) \frac{\partial f_{s0}}{\partial v_\perp} + kv_\perp \frac{\partial f_{s0}}{\partial v_z} \right]$$
$$\left[\frac{1}{(\omega - kv_z + \omega_{cs})} + \frac{1}{(\omega - kv_z - \omega_{cs})} \right] = \varepsilon_{22}(k, \omega) \quad \ldots [3.56]$$

$$\varepsilon_{33} = 1 + \sum_s \frac{4\pi e_s^2}{m_s \omega} \int d^3v \, \frac{v_z \, \partial f_{s0}/\partial v_z}{(\omega - kv_z)} \qquad \ldots [3.57]$$

$$= 1 + \sum_s \frac{4\pi e_s^2}{m_s k^2} \int d^3v k \, \frac{\partial f_{s0}/\partial v_z}{(\omega - kv_z)} \qquad \ldots [3.58]$$

$$\varepsilon_{12}(k, \omega) = -\varepsilon_{21}(k, \omega) = -i \sum_s \frac{4\pi e_s^2}{m_s \omega^2} \int d^3v \, \frac{v_\perp}{4} \left[(\omega - kv_z) \frac{\partial f_{s0}}{\partial v_\perp} + kv_\perp \frac{\partial f_{s0}}{\partial v_z} \right]$$
$$\times \left[\frac{1}{(\omega - kv_z + \omega_{cs})} - \frac{1}{(\omega - kv_z - \omega_{cs})} \right]. \qquad \ldots [3.59]$$

The remaining ε_{ij} are zero.

Writing out equation [3.54] one finds

$$\left(k^2 - \frac{\omega^2}{c^2}\varepsilon_{11}\right)E_x - \frac{\omega^2}{c^2}\varepsilon_{12}E_y = 0, \qquad \ldots [3.60]$$

$$-\frac{\omega^2}{c^2}\varepsilon_{21}E_x + \left(k^2 - \frac{\omega^2}{c^2}\varepsilon_{11}\right)E_y = 0, \qquad \ldots [3.61]$$

$$-\frac{\omega^2}{c^2}\varepsilon_{33}E_z = 0. \qquad \ldots [3.62]$$

One solution is $E_x = E_y = 0$, $E_z \neq 0$, $\varepsilon_{33}(k, \omega) = 0$. This is the longitudinal mode with k and E parallel to B_0. For if $k \times E = 0$, so it is purely electrostatic. This mode is discussed in Section 2a.

The other two solutions have $E_z = 0$, $E_y = \pm iE_x$. These are circularly polarized transverse waves. They are discussed in Section 2b.

2a. *Negative Landau damping; the two stream instability; ion sound wave instability.* We shall now explore some of the consequences of $\varepsilon_{33} = 0$ with ε_{33} given by equation [3.5]. Before doing so, we shall remark that, if we had assumed $B_0 = 0$ at the beginning and had then derived the dispersion relation for longitudinal waves, we would have found

$$\varepsilon(k, \omega) = 0 = 1 + \sum_s \frac{4\pi e_s^2}{m_s k^2} \int d^3v \, \frac{k \cdot \dfrac{\partial f_{s0}}{\partial v}}{(\omega - k \cdot v)} \qquad \ldots [3.63]$$

which is the same as equation [3.58] when $k = ke_z$. The magnetic field plays no role in the phenomena discussed in this section. The function

$$\varepsilon(k, \omega) = \frac{1}{k^2} \, k \cdot \mathbf{\varepsilon}(k, \omega) \cdot k \qquad \ldots [3.64]$$

is called the dielectric constant of the plasma. It is equal to ε_{33} when $k = ke_z$ so we shall drop the subscripts on ε_{33}. It is convenient to define

$$\int dv_x dv_y f_{s0}(v) = n_s F_s(v_z) \qquad \ldots [3.65]$$

then

$$\varepsilon(k, \omega) = 1 + \sum_s \frac{\omega_{ps}^2}{k} \int dv_z \, \frac{F_s'(v_z)}{(\omega - kv_z)}, \qquad \ldots [3.66]$$

where again the prime denotes differentiation.

Let us remember that ω is assumed to have a small positive imaginary part. This makes the integral in equation [3.66] proper. What we shall do is to write $\omega + i\eta$ for ω, do the integrals, and then take the limit as η goes to zero through positive values. This gives $\varepsilon(k, \omega)$ for real ω. Then values of $\varepsilon(k, \omega)$ for complex ω can be found by analytic continuation. We shall use the Plemelj formula

$$\frac{1}{\omega - kv_z + i\eta} \xrightarrow[\eta = 0+]{} P \frac{1}{\omega - kv_z} - i\pi\delta(\omega - kv_z). \qquad ...[3.67]$$

To write

$$\varepsilon(k, \omega) = \varepsilon_1(k, \omega) + i\varepsilon_2(k, \omega), \qquad ...[3.68]$$

where

$$\varepsilon_1(k, \omega) = 1 + \sum_s \frac{\omega_{ps}^2}{k} P \int dv_z \frac{F_s'(v_z)}{\omega - kv_z} \qquad ...[3.69]$$

$$\varepsilon_2(k, \omega) = -\pi \sum_s \frac{\omega_{ps}^2}{k} \int dv_z F_s'(v_z) \delta(\omega - kv_z)$$

$$= -\pi \sum_s \frac{\omega_{ps}^2}{k^2} F_s'\left(\frac{\omega}{k}\right). \qquad ...[3.70]$$

In equations [3.67] and [3.69] the P denotes that the principal value is to be taken.

Equation [3.70] shows that the contribution of each species to ε_2 is proportional to the slope of the distribution function at ω/k, the phase velocity of the wave. For waves with phase velocities much greater than the thermal velocities of the particles, this will be very small. This suggests that we write

$$\omega = \Omega_k + i\gamma_k \qquad ...[3.71]$$

and treat γ_k and ε_2 as small quantities. Then

$$\varepsilon(k, \omega) = 0 = \varepsilon_1(k, \Omega_k + i\gamma_k) + i\varepsilon_2(k, \Omega_k + i\gamma_k)$$

$$\simeq \varepsilon_1(k, \Omega_k) + i\gamma_k \frac{\partial \varepsilon_1}{\partial \Omega_k} + i\varepsilon_2(k, \Omega_k). \qquad ...[3.72]$$

From this we find that the real part of the frequency is a solution of

$$\varepsilon_1(k, \Omega_k) = 0,$$

and the imaginary part is given by

$$\gamma_k = -\frac{\varepsilon_2(k, \Omega_k)}{\dfrac{\partial \varepsilon_1}{\partial \Omega_k}(k, \Omega_k)}, \qquad ..[3.73]$$

which is a very useful relation.

Now, we shall find approximate expressions for ε_1, Ω_k and γ_k. First we shall consider high frequency oscillations with $\omega \simeq \omega_{pe}$. For these modes we can neglect the ions. Also, we expand the denominator in equation [3.69] and find

$$\varepsilon_1(k, \omega) = 1 + \frac{\omega_{pc}^2}{k\omega} \int dv_z F'(v_z) \left[1 + \frac{kv_z}{\omega} + \left(\frac{kv_z}{\omega}\right)^2 + \left(\frac{kv_z}{\omega}\right)^3 + \dots \right]$$

$$= 1 - \frac{\omega_{pc}^2}{\omega^2} \left[1 + \frac{2k\langle v_z \rangle}{\omega} + \frac{3k^2 \langle v_z^2 \rangle}{\omega^2} + \dots \right]. \qquad \dots[3.74]$$

Setting $\varepsilon_1(k, \Omega_k) = 0$ and solving approximately gives

$$\Omega_k \simeq \omega_{pe} \left(1 + \frac{k\langle v_z \rangle}{\omega_{pe}} + \tfrac{3}{2} \frac{k^2 \langle v_z^2 \rangle}{\omega_{pe}^2} \right). \qquad \dots[3.75]$$

Also

$$\frac{\partial \varepsilon_1}{\partial \Omega_k}(k, \Omega_k) \simeq \frac{2}{\Omega_k}. \qquad \dots[3.76]$$

And

$$\gamma_k = \frac{\pi \Omega_k}{2} \frac{\omega_{pe}^2}{k^2} F_e'\left(\frac{\Omega_k}{k}\right). \qquad \dots[3.77]$$

If F_e is a monotonically decreasing function, then F_e' is negative; and hence γ_k is negative, indicating damping. This is the familiar Landau damping.[19] However, if $F_e(v_z)$ has the form shown in Fig. 8, then F_e' will be positive near the "bump on the tail", and a wave with a phase velocity where $F_e'(v_z)$ is positive will grow exponentially. We will call this negative Landau damping.

The physical explanation of Landau damping is as follows. Particles with velocities slightly smaller than the phase velocity will be accelerated by the

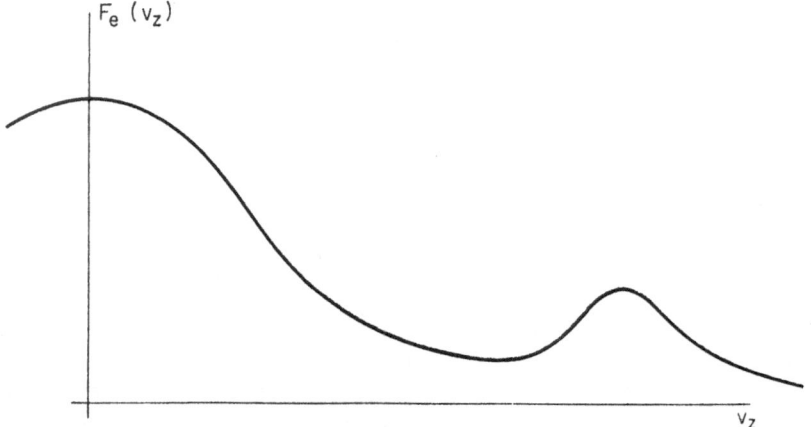

FIG. 8. A bump on the tail distribution.

wave and absorb energy from it. Particles with velocities slightly greater than the wave will be decelerated and lose energy to the wave. If there are more particles which can gain than can lose energy (i.e. $F_e' < 0$), then the wave is damped. If there are more particles which can lose than can gain energy (i.e. $F_e' > 0$), then the wave grows. This explanation will be developed further later.

An extreme case of this instability is the two-stream instability. Suppose that $F_e(v_z)$ consist of two rather narrow peaks. For simplicity, say

$$F_e(v_z) = (1-\beta)\delta(v_z) + \beta\delta(v_z - V). \qquad \ldots [3.78]$$

Now, the analysis which led to equation [3.77] is not valid. However, the integral in equation [3.66] is easily carried out, with the result

$$\varepsilon(k, \omega) = 0 = 1 - \frac{\omega_{pe}^2(1-\beta)}{\omega^2} - \frac{\omega_{pe}^2\beta}{(\omega - kV)^2}. \qquad \ldots [3.79]$$

This is a fourth degree equation for ω. It is easily shown that when

$$kV < \beta\omega_{pe}\left[1+\left(\frac{1-\beta}{\beta}\right)^{\frac{1}{3}}\right]^{\frac{3}{2}}, \qquad \ldots [3.80]$$

two of the roots are complex. Since the coefficients in equation [3.79] are real, the roots are complex conjugates. The root with Im $\omega > 0$ represents an exponentially growing wave.

Next, we shall consider some low frequency oscillations, the ion sound waves and instabilities involving them. We shall assume that

$$\langle v_z^2 \rangle_{ions} \ll \left(\frac{\Omega_k}{k}\right) \ll \langle v_z^2 \rangle_{electrons}. \qquad \ldots [3.81]$$

The reason for this will be made clear later. In the ion term in equation [3.63] we will integrate by parts, then neglect kv_z in comparison with ω, thereby obtaining ω_{pi}^2/ω^2 for the ion contribution. In the electron term we neglect ω in comparison with kv_z. If the electrons are assumed to have a Maxwellian distribution with temperature T_e, then the electrons contribute a term $\omega_{pc}^2 m/k^2 T_e = 1/k^2 L_D^2$ to ε_1. The result is

$$\varepsilon_1(k, \omega) = 1 + \frac{1}{k^2 L_D^2} - \frac{\omega_{pi}^2}{\omega^2}, \qquad \ldots [3.82]$$

from which

$$\Omega_k = \frac{\omega_{pi} k L_D}{\sqrt{(1+k^2 L_D^2)}} = \frac{\sqrt{(T_e/Mk)}}{\sqrt{(1+k^2 L_D^2)}}. \qquad \ldots [3.83]$$

This is the ion sound wave frequency. We also find

$$\frac{\partial\varepsilon_1}{\partial\Omega_k} = \frac{2}{\Omega_k}\left(1+\frac{1}{k^2 L_D^2}\right) \qquad \ldots [3.84]$$

$$\gamma_k = \frac{\pi \Omega_k}{2} \sum_s \frac{\omega_{ps}^2}{\left(k^2 + \frac{1}{L_D^2}\right)} F_s'\left(\frac{\Omega_k}{k}\right). \qquad \dots [3.85]$$

In Fig. 9 we have sketched $F_e(v_z)$ and $F_i(v_z)$ for Maxwellian distributions centred on the origin. If $T_e \gg T_i$ there will be a range of phase velocities $\Omega_k/k \simeq (T_e/M)^{\frac{1}{2}}$ for which $F_e'(\Omega_k/k)$ is small because of the flat region near $v_z = 0$, and $E_i'(\Omega_k/k)$ is small because of the tail of $F_i(v_z)$. If this is true, then

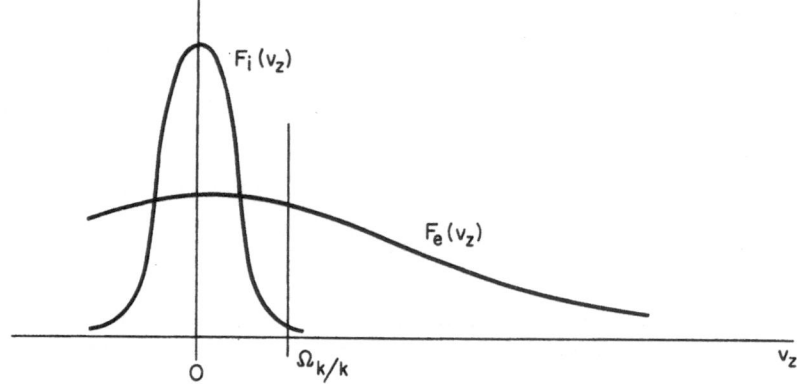

FIG. 9. Ion and electron distribution functions. No drift velocity.

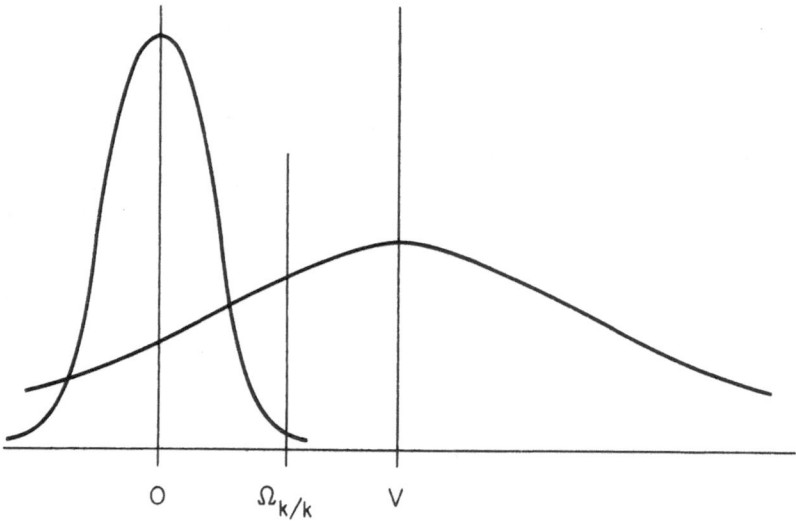

FIG. 10. Ion and electron distribution functions when the electrons have a drift velocity V relative to the ions.

equation [3.75] is satisfied. It is seen from equation [3.85] that the damping is small in this case. Generally $T_e > 5T_i$ is necessary for the damping to be negligibly small.

Ion sound waves can become unstable if there is a drift of the electrons relative to the ions. Then a sketch of F_i and F_e will be as shown in Fig. 10. If V is greater than the ion thermal velocity then $F_i'(\Omega_k/k)$ in equation [3.85] is negligible, but the term in $F_e'(\Omega_k/k)$ gives a positive contribution to γ_k and the amplitude of ion sound waves grows exponentially.[20]

2b. *Alfvén*[21,22,23] *and Whistler instabilities.*[24] We now return to equations [3.60], [3.61] and [3.62] and examine the solution with $E_z = 0$. We easily find that a solution is

$$E_y = \pm iE_x \qquad \qquad \ldots [3.86]$$

when ω is a solution of

$$\frac{k^2c^2}{\omega^2} = \varepsilon_{11} \pm i\varepsilon_{12} = 1 + \sum_s \frac{4\pi e_s^2}{m_s\omega^2} \int d^3v \frac{v_\perp}{2}$$

$$\times \left[(\omega - kv_z)\frac{\partial f_{s0}}{\partial v_\perp} + kv_\perp \frac{\partial f_{s0}}{\partial v_z} \right] \frac{1}{\omega - kv_z \pm \omega_{cs}}. \qquad \ldots [3.87]$$

It will be convenient to write this dispersion relation in the form

$$K(k, \omega) = K_1(k, \omega) + iK_2(k, \omega) = 0, \qquad \ldots [3.88]$$

where

$$K(k, \omega) = 1 - \frac{k^2c^2}{\omega^2} - \sum_s \frac{4\pi e_s^2}{m_s\omega^2} \int d^3v$$

$$\left[(\omega - kv_z)f_{s0} - \frac{kv_\perp^2}{2}\frac{\partial f_{s0}}{\partial v_z} \right] \frac{1}{\omega - kv_z \pm \omega_{cs}} \qquad \ldots [3.89]$$

where an integration by parts has been made. The Plemelj formula, equation [3.67] is used to separate $K(k, \omega)$ into its real and imaginary parts.

As a preliminary step we shall examine equation [3.89] in the cold plasma limit. Letting

$$f_{s0} = n_s\delta(v) \qquad \qquad \ldots [3.90]$$

we find

$$K(k, \omega) = 0 = 1 - \frac{k^2c^2}{\omega^2} - \sum_s \frac{\omega_{ps}^2}{\omega(\omega \pm \omega_{cs})}. \qquad \ldots [3.91]$$

Near $\omega \simeq 0$ we may expand and obtain

$$K = 1 - \frac{k^2c^2}{\omega^2} \mp \frac{1}{\omega}\left(\frac{\omega_{pe}^2}{\omega_{ce}} + \frac{\omega_{pi}^2}{w_{ci}} \right) + \left(\frac{\omega_{pe}^2}{\omega_{ce}^2} + \frac{\omega_{pi}^2}{\omega_{ci}^2} \right)$$

$$= 1 - \frac{k^2c^2}{\omega^2} + \frac{c^2}{V_a^2} = 0, \qquad \ldots [3.92]$$

since

$$\frac{\omega_{pe}^2}{\omega_{ce}} + \frac{\omega_{pi}^2}{\omega_{ci}} = -\frac{4\pi ne^2}{m}\frac{mc}{eB} + \frac{4\pi ne^2}{M}\frac{Mc}{eB} = 0 \qquad \ldots[3.93]$$

and

$$\frac{\omega_{pe}^2}{\omega_{ce}^2} + \frac{\omega_{pi}^2}{\omega_{ci}^2} = \frac{c^2}{B^2}\, 4\pi n(m+M) = \frac{4\pi\rho c^2}{B^2} = \frac{c^2}{V_a^2}, \qquad \ldots[3.94]$$

where V_a is the Alfvén velocity. Equation [3.92] gives the frequency

$$\Omega = \frac{kV_a}{\sqrt{(1+V_a^2/c^2)}} \simeq kV_a. \qquad \ldots[3.95]$$

This mode is an Alfvén wave.

Next, we consider equation [3.91] with the positive sign taken:

$$K(k,\omega) = 0 = 1 - \frac{k^2 c^2}{\omega^2} - \frac{\omega_{pe}^2}{\omega(\omega - |\omega_{ce}|)} - \frac{\omega_{pi}^2}{\omega(\omega + \omega_{ci})}. \qquad \ldots[3.96]$$

If we look in the frequency range

$$\omega_{ci} \ll \omega \ll |\omega_{ce}|, \qquad \ldots[3.97]$$

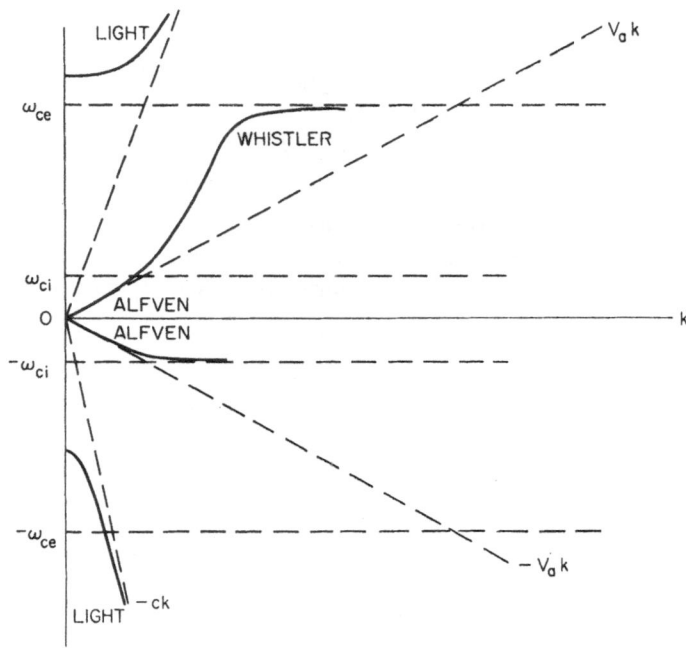

FIG. 11. Roots of the cold plasma dispersion relation for $k \parallel B_0$.

we have

$$K(k, \omega) \simeq 1 - \frac{k^2 c^2}{\omega^2} + \frac{\omega_{pe}^2}{\omega \mid \omega_{ce} \mid}. \qquad \ldots [3.98]$$

Neglecting unity in comparison with the other two terms gives

$$\Omega = \frac{c^2 \mid \omega_{ce} \mid}{\omega_{pe}^2} k^2. \qquad \ldots [3.99]$$

This is the frequency of a "Whistler", or as it is called in solid state physics, a "Helicon".

After some simple calculations one can sketch the behaviour of the roots of equation [3.96]. This is shown in Fig. 11.

Next, we shall find the thermal corrections to the Alfvén frequency by treating ω and $k v_z$ as small in comparison with ω_{cs} and expanding the denominator in equation [3.89]. In place of equation [3.92] we find

$$K(k, \omega) = 0 = 1 + \frac{c^2}{V_a^2} - \frac{k^2 c^2}{\omega^2}\left[1 - \left(\frac{P_\parallel - P_\perp}{\rho V_a^2}\right)\right], \qquad \ldots [3.100]$$

where

$$P_\parallel = \sum_s m_s \int d^3 v f_{s0} v_z^2 \qquad \ldots [3.101]$$

and

$$P_\perp = \sum_s m_s \int d^3 v f_{s0} \frac{v_\perp^2}{2} \qquad \ldots [3.102]$$

are the parallel and perpendicular components of pressure. We have assumed that $\langle v_z \rangle = 0$. From equation [3.100] we find

$$\omega^2 = \frac{k^2 V_a^2}{1 + V_a^2/c^2}\left[1 - \left(\frac{P_\parallel - P_\perp}{\rho V_a^2}\right)\right]. \qquad \ldots [3.103]$$

Clearly ω will be purely imaginary, if

$$P_\parallel - P_\perp > \rho V_a^2 = \frac{B^2}{4\pi}. \qquad \ldots [3.104]$$

This instability is known as the "fire-hose" instability. It is really a hydromagnetic instability, and was first derived from the fluid equations. The following physical interpretation can be given. An Alfvén wave causes a curvature of the lines of B. Particles spiralling along the lines exert a centrifugal force $m v_\parallel^2/R$ in a direction which would cause the curvature of the lines to increase. Summing over the particles gives $P_\parallel/R$ for this force. Opposing this are the restoring force $\nabla \mu \cdot Bn \sim P_\perp/R$, due to the perpendicular pressure, and $B \cdot \nabla B/4\pi \sim B^2/4\pi R$ due to the tension of the magnetic field lines. In an isotropic plasma the pressure forces cancel, leaving the magnetic tension to provide the restoring force, which causes the oscillations of the field lines. In an anisotropic plasma the centrifugal force may exceed the restoring forces, so that the perturbation grows.

Next, we shall examine instabilities which arise from $K_2(k, \omega)$. We shall assume that

$$K_1(k, \omega) = 0 \qquad \text{...[3.105]}$$

has the real root $\omega = \Omega_k$ and treating γ_k and K_2 as small quantities, we find

$$\gamma = -\frac{K_2(k, \Omega_k)}{\dfrac{\partial K_1}{\partial \Omega_k}(k, \Omega_k)} \qquad \text{...[3.106]}$$

by steps analogous to those which led to equation [3.73]. Using the Plemelj formula in equation [3.89] gives

$$K_2(k, \Omega) = +\pi \sum_s \frac{\omega_{ps}^2}{k\omega^2}\left[(\mp \omega_{cs})F_s\left(\frac{\Omega \pm \omega_{cs}}{k}\right) - \frac{k\langle v_\perp^2\rangle}{2}F_s'\left(\frac{\Omega \pm \omega_{cs}}{k}\right)\right]$$
$$\text{...[3.107]}$$

where

$$\int d^2v_\perp f_{s0}(v_\perp, v_z) = n_s F_s(v_z) \qquad \text{...[3.108]}$$

$$\int d^2v_\perp v_\perp^2 f_{s0}(v_\perp, v_z) = n_s \langle v_\perp^2\rangle F_s(v_z) \qquad \text{...[3.109]}$$

and again the prime denotes the derivative with respect to the argument.

First, we shall examine $K_2(k, \omega)$ when f_{s0} is the Maxwell-Boltzmann (M-B) distribution

$$f_{s0} = \frac{n_s}{\pi^{\frac{3}{2}}\alpha^3}e^{-v^2/\alpha^2}. \qquad \text{...[3.110]}$$

We find

$$K_2(k, \Omega_k) = \frac{\pi}{\omega}\sum_s \frac{\omega_{ps}^2}{k}F_s\left(\frac{\Omega_k + \omega_{cs}}{k}\right) \qquad \text{...[3.111]}$$

since

$$\frac{\partial K_1}{\partial \Omega_k} \simeq \frac{1}{\Omega_k^2}\frac{\omega_{pe}^2}{|\omega_{ce}|} \qquad \text{...[3.112]}$$

for Alfvén waves, and

$$\frac{\partial K_1}{\partial \Omega_k} \simeq \frac{1}{\Omega_k^2}\frac{\omega_{pe}^2}{|\omega_{ce}|} \qquad \text{...[3.113]}$$

for whistlers, we see that equation [3.106] gives damping for both modes. There are various ways in which the distribution function can be distorted so as to make $K_2(k, \Omega_k)$ change sign. For instance, suppose that f_{s0} is the two-temperature M-B distribution:

$$f_{s0} = \frac{n_s}{\pi^{\frac{3}{2}}\alpha_{\perp s}^2\alpha_{zs}} e^{-\frac{v_\perp^2}{\alpha_{\perp s}^2}-\frac{v_z^2}{\alpha_{zs}^2}}. \qquad ...[3.114]$$

Then, equation [3.111] is replaced by

$$K_2(k, \Omega_k) = \frac{\pi\omega_{ps}^2}{\omega_s k} \frac{1}{(\sqrt{\pi})\alpha_{sz}} e^{-\frac{1}{k^2\alpha_{sz}^2}(\Omega_k\pm\omega_{cs})^2} \left[\Omega_k\pm\omega_{cs}\left(\frac{\alpha_\perp^2}{\alpha_z^2}-1\right)\right] \qquad ..[3.115]$$

and this may be negative indicating an exponential growth.

Another source of instability is a beam of particles through the plasma. Let the distribution function of the particles in the beam be

$$f_{b0} = \frac{n_b}{\pi^{\frac{3}{2}}\alpha_b^3} e^{-(v-V)^2/\alpha_b^2} \qquad ...[3.116]$$

with $V = Ve_z$. Then the contribution of the beam particles of K_2 is

$$\frac{\pi}{\omega} \frac{\omega_{pb}^2}{k} \frac{1}{(\sqrt{\pi})\alpha_b} e^{-\frac{1}{k^2\alpha^2}(\omega\pm\omega_{cb}-kV)^2} [\omega-kV]. \qquad ...[3.117]$$

For $\omega < kV$, this gives a negative contribution which, with the proper choice of parameters, can outweigh the damping due to the rest of the plasma.

3.3. Quasi-electrostatic instabilities

Although no rigorous decoupling of longitudinal and transverse waves can be made, an approximate decoupling can be made when the plasma pressure is much less than the magnetic field pressure. Physically, this is because oscillations can occur without appreciably perturbing the magnetic field in such a plasma. Then $k \times E_1 = \omega/cB_1 \simeq 0$ and the waves are longitudinal. In this case we can obtain the dielectric constant from equation [3.64] and determine the frequency from $\varepsilon(k, \omega) = 0$. Alternatively, we can go back to equation [3.2] and assume that

$$E = -\nabla\phi = -ik\phi. \qquad ...[3.118]$$

Then

$$k^2\phi = \sum_s 4\pi e_s \int d^3v f_{s1}. \qquad ...[3.119]$$

Using equation [3.111] in [3.40] gives

$$f_{1s}(x, v, t) = -\frac{e_s}{m_s} \phi(x, t) \sum_{m=-\infty}^{+\infty} \sum_{n=-\infty}^{+\infty} \frac{J_n J_m e^{i(m-n)\alpha}}{(\omega-k_z v_z - n\omega_{cs})}$$
$$\left[\frac{n\omega_{cs}}{v_\perp}\frac{\partial f_{s0}}{\partial v_\perp}+k_z\frac{\partial f_{s0}}{\partial v_z}\right]. \qquad ...[3.120]$$

The integration over α causes terms with $m \neq n$ to vanish and equation [3.119] becomes

$$\varepsilon(k, \omega)\phi = 0, \qquad \ldots[3.121]$$

where

$$\varepsilon(k, \omega) = 1 + \sum_s \frac{4\pi e_s^2}{m_s k^2} \sum_{n=-\infty}^{+\infty} \int d^3v \frac{J_n^2\left(\frac{k_\perp v_\perp}{\omega_{cs}}\right)}{(\omega - k_z v_z - n\omega_{cs})} \left[\frac{n\omega_{cs}}{v}\frac{\partial f_{s0}}{\partial v} + k_z \frac{\partial f_{s0}}{\partial v_z}\right]. \qquad \ldots[3.122]$$

Since its original derivation,[17] this dispersion relation has been the subject of many investigations to determine the circumstances under which instabilities occur. We shall make some simplifying assumptions which permit us to gain some insight onto the nature of the instabilities.

3a. *Instabilities near harmonics of the ion cyclotron frequency.* We shall divide ε into its real and imaginary parts by using equation [3.67]. Then

$$\varepsilon_1(k, \omega) = 1 + \sum_s \frac{4\pi e_s^2}{m_s k^2} \sum_{n=-\infty}^{+\infty} P \int d^3v \frac{J_n^2\left(\frac{k_\perp v_\perp}{\omega_{cs}}\right)}{(\omega - k_z v_z - n\omega_{cs})}$$
$$\left[\frac{n\omega_{cs}}{v_\perp}\frac{\partial f_{s0}}{\partial v_\perp} + k_z \frac{\partial f_{s0}}{\partial v_z}\right] \qquad \ldots[3.123]$$

$$\varepsilon_2(k, \omega) = -\pi \sum_s \frac{4\pi e_s^2}{m_s k^2} \sum_{n=-\infty}^{+\infty} \int d^3v J_n^2\left(\frac{k_\perp v_\perp}{\omega_{cs}}\right) \delta(\omega - k_z v_z - n\omega_{cs})$$
$$\left[\frac{n\omega_{cs}}{v_\perp}\frac{\partial f_{s0}}{\partial v_\perp} + k_z \frac{\partial f_{s0}}{\partial v_z}\right]. \qquad \ldots[3.124]$$

Now, suppose that the electrons are "cold" but the ions are "hot". By this we mean that

$$\frac{k_\perp v_\perp}{\omega_{cs}} \simeq 0 \qquad \ldots[3.125]$$

for electrons but not for the ions. The electrons contribute much more to ε_1 than the ions because m_s^{-1} is a factor in equation [3.123]. For this reason, we neglect the ion contribution to ε_1. An integration by parts gives

$$\varepsilon_1(k, \omega) = 1 - \frac{\omega_{pe}^2}{\omega^2}\frac{k_z^2}{k^2}. \qquad \ldots[3.126]$$

We determine the real part of the frequency from

$$\varepsilon_1(k, \Omega_k) = 0 \qquad \ldots[3.127]$$

and the imaginary part from equation [3.73]. Equation [3.126] gives

$$\Omega_k = \omega_{pe}\frac{k_z}{k}. \qquad \ldots[3.128]$$

Equation [3.73] gives

$$\gamma_k = +\frac{\pi\Omega_k}{2}\sum_s \frac{4\pi e_s^2}{m_s k^2}\sum_{n=-\infty}^{+\infty}\int d^3v\, J_n^2\left(\frac{k_\perp v_\perp}{\omega_{cs}}\right)\delta(\Omega_k - k_z v_z - n\omega_{cs})$$
$$\left[\frac{n\omega_{cs}}{v_\perp}\frac{\partial f_{s0}}{\partial v_\perp} + k_z\frac{\partial f_{s0}}{\partial v_z}\right]. \qquad ..[3.129]$$

We are interested in the distribution functions, $f_{s0}(v_\perp, v_z)$, which make γ_k positive (indicating an instability) for some k. As the first example[25] we shall consider the two-temperature M-B distribution of equation [3.114]. Using the formula

$$\int_0^\infty \rho\,d\rho\, J_n^2(s\rho)\,e^{-\rho^2} = \tfrac{1}{2}e^{-s^2/2}I_n(s^2/2). \qquad ..[3.130]$$

Where I_n is the modified Bessel Function we find

$$\gamma_k = \frac{\pi}{2}\,\Omega_k \sum_s \frac{\omega_{ps}^2}{k^2\alpha_{zs}^2}\sum_{n=-\infty}^{+\infty}e^{-\frac{k_\perp^2\rho_s^2}{2}}I_n\left(\frac{k_\perp^2\rho_s}{2}\right)$$
$$\left\{Z_i'\left(\frac{\Omega_k - n\omega_{cs}}{k_z\alpha_{zs}}\right) - \frac{2n\omega_{cs}}{k_z\alpha_{zs}}\left(\frac{\alpha_{zs}^2}{\alpha_{\perp s}^2}\right)Z_i\left(\frac{\Omega_k - n\omega_{cs}}{k_z\alpha_z}\right)\right\}.$$
$$...[3.131]$$

In this equation

$$\rho_s = \frac{\alpha_{\perp s}}{\omega_{cs}} \qquad ..[3.132]$$

is the radius of gyrations of a particle of species s with the perpendicular velocity $\alpha_{\perp s}$. We have let

$$Z_i(z) = (\sqrt{\pi})\,e^{-z^2} \qquad ...[3.133]$$

$$Z_i'(z) = \frac{dZ}{dz} = -2(\sqrt{\pi})z e^{-z^2}. \qquad ...[3.134]$$

(The reason for this notation is that $Z_i(z)$ is the imaginary part of the Fried-Conte function defined by

$$Z(z) = \frac{1}{\sqrt{\pi}}\int_{-\infty}^{+\infty}dx\,\frac{e^{-x^2}}{x-z} \qquad ...[3.135]$$

for Im $z > 0$ and as its analytic continuation for Im $z < 0$. The functions Z and Z' have been tabulated.[26])

We can sketch the contributions to γ_k as a function of Ω_k. This is done in Fig. 12. The negative bump near $\Omega_k = 0$ in Fig. 12 is the electron contribution. We have drawn it tall, but narrow, because of our assumption that the electrons are cold (α_{ze} is small). Near multiples of ω_{ci} there are two bumps. The ions contribute a bump shown as a dotted curve. This comes from the Z_i term in equation [3.131] and is always negative. The ions also contribute through the Z'_i term in equation [3.131]. This contribution (shown as a solid line) is positive on the low-frequency side of $n\omega_{ci}$ and negative on the high-frequency side. If, for a given k and corresponding Ω_k the sum of these contributions is positive, then the wave is unstable; if negative, it is damped.

Now, by equation [3.128], Ω_k varies from zero to ω_{pe}. Unless $\omega_{pe} > n\omega_{ci}$, there will be no wave which can go unstable because of the positive contribution to γ_k near $n\omega_{ci}$. Thus

$$\omega_{pe} > n\omega_{ci} \qquad \qquad ...[3.136]$$

is an approximate necessary condition for instability with frequency $\Omega_k \simeq n\omega_{ci}$.

The stabilizing terms have as a factor $n(\alpha_{zi}/\alpha_{\perp i})^2$, and will clearly predominate for sufficiently large n. In fact, a simple calculation shows that waves with $\Omega_k \simeq n\omega_{ci}$ will be unstable only if

$$\left(\frac{T_\parallel}{T_\perp}\right)_i = \left(\frac{\alpha_{zi}}{\alpha_{\perp i}}\right)^2 < \frac{1}{2n}. \qquad \qquad ...[3.137]$$

(If this is not true, then the positive resultant of contributions near $\Omega_k \simeq n\omega_{ci}$ is shifted to the left of $(n-1/2)\omega_{ci}$, where the negative contributions from $\Omega_k \simeq (n-1)\omega_{ci}$ predominate.) Thus, equation [3.137] is another approximate necessary condition for instability with $\Omega_k \simeq n\omega_{ci}$.

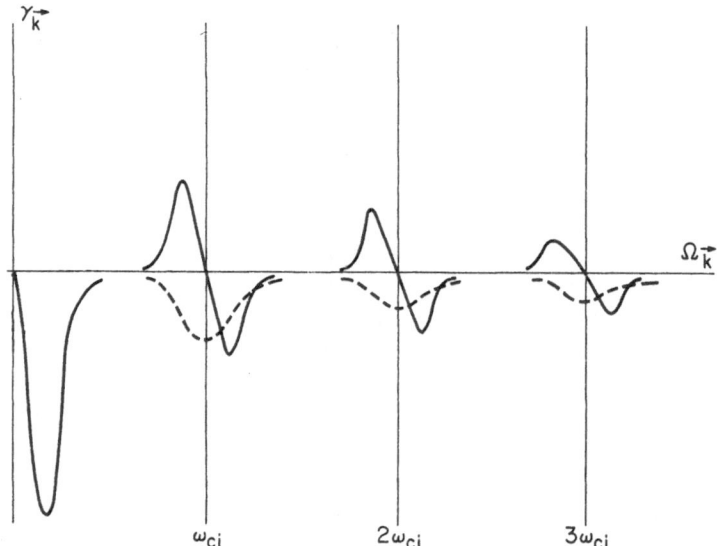

FIG. 12. Contributions of various terms of equation [131] to γ_k. Cold electrons.

If the electron temperature is increased, the negative bump near $\Omega_k = 0$ will grow, and can stabilize the oscillations near the first few harmonics of ω_{ci}. However, if the electron temperature is raised too much, these modes may become unstable again when the contributions to γ_k are as shown in Fig. 13. In this case the approximations which led to equation [3.119] are no longer valid. Instead, for the frequencies of interest

$$\langle v_z^2 \rangle_{\text{ions}} \ll \left(\frac{\Omega_k}{k_z} \right)^2 \ll \langle v_z^2 \rangle_{\text{electrons}}. \qquad \ldots[3.138]$$

Using this approximation and keeping only the $n = 0$ terms for the electrons and ions in ε_1 gives

$$\varepsilon_1(k, \omega) = 1 + \frac{1}{k^2 L_{De}^2} + \frac{\omega_{pi}^2}{\omega^2} \frac{k_z^2}{k^2}, \qquad \ldots[3.139]$$

from which

$$\Omega_k = \frac{\omega_{pi} k_z}{\sqrt{\left(k^2 + \frac{1}{L_{De}^2} \right)}} \simeq \left(\sqrt{\frac{T_e}{M}} \right) k_z. \qquad \ldots[3.140]$$

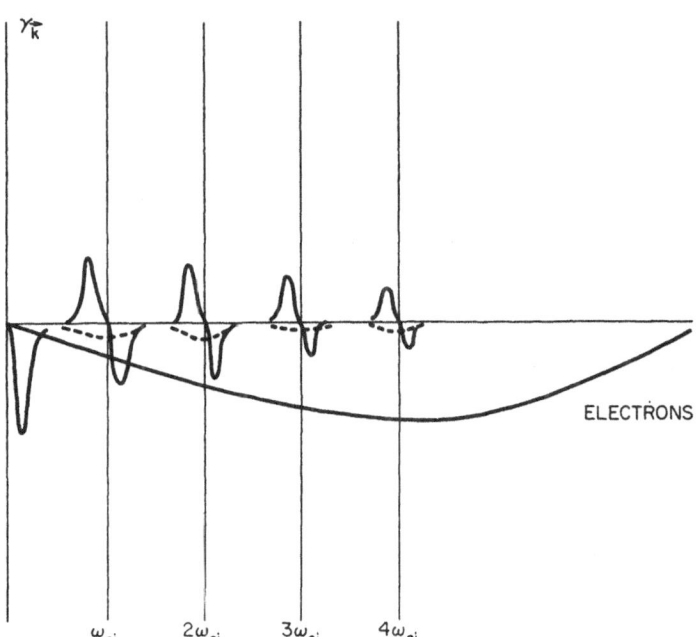

FIG. 13. Contributions to γ_k. Hot electrons.

This is an ion sound wave frequency. Its maximum value is ω_{pi}. If waves with frequency near $n\omega_{ci}$ are to be excited, it is necessary that

$$\omega_{pi} > n\omega_{ci}. \qquad\qquad ...[3.141]$$

This gives a higher density threshold by a factor M/m than equation [3.136].

We can summarize these results as follows:

(a) If the electrons are "cold", then they provide the medium through which the waves propagate. The waves are plasma oscillations with frequency given by equation [3.128].

(b) If the electrons are "hot", then the electrons, together with the cooler of the ions, provide the medium through which the waves propagate. These waves are ion sound waves with frequency given by equation [3.140].

(c) In either case, the energetic ions may emit, or absorb, these waves. The frequencies emitted or absorbed by the ions will be $n\omega_{ci}+k_z v_z$, where $k_z v_z$ is the Doppler shift due to the motion of the ion along B_0. Clearly, for a wave of frequency $n\omega_{ci}$ to be emitted or absorbed, it is necessary to have $\Omega_k = n\omega_{ci}$, which leads to equation [3.137] for the "cold" electron case and equation [3.141] for the "hot" electron case. These give instability criteria on the density of

$$n_e > n^2 \frac{B^2}{4\pi mc^2} \qquad\qquad ...[3.142]$$

for "cold" electrons and

$$n_e > n^2 \frac{B^2 m}{4\pi M^2 c^2} \qquad\qquad ...[3.143]$$

for the "hot" electron plasma.

(d) For either the "cold" or the "hot" electron plasma, equation [3.137] is necessary for instabilities with Ω_k near ω_{ci}.

(e) We have not shown this in these lectures, but the dividing lines between "cold" and "hot" electrons comes at about $T_{\|e} = 5T_{\|i}$.

What makes the instability problem so difficult to handle is the large number of parameters which must be considered. For the two-temperature M-B distribution, the dimensionless parameters of the system can be taken to be

$$\frac{\omega_{pe}}{k\alpha_{zi}}, \quad \frac{\omega_{ci}}{k_z\alpha_{zi}}, \quad \frac{T_{\|i}}{T_{\perp i}}, \quad \frac{T_{\|e}}{T_{\|i}}, \quad \tfrac{1}{2}k_z^2\rho_i^2.$$

Then the stability-instability boundary is a surface in this five-parameter space. The reader is referred to the literature for more detailed instability criteria. Unfortunately, no two writers choose the same set of parameters.

Having familiarized ourselves with the two-temperature M-B distribution, it is now rather easy to see some things that can be done to the distribution functions to produce instabilities.

First, let us note that terms with $\partial f_{s0}/\partial v_\perp$ in equation [3.129] gave rise to the terms with Z_i in equation [3.131]. These terms are negative (stabilizing), and contribute the dotted curves in Fig. 12. The sign of these terms can be changed to positive if $\partial f_{s0}/\partial v_\perp$ is positive over a range of $v_\perp$ where $J_n^2\,(k_\perp v_\perp/\omega_{cs})$ is large. We have in mind a distribution of perpendicular velocities such as the one shown in Fig. 14. For such a distribution it is possible to have

$$\int_0^\infty dv_\perp \frac{\partial f_{s0}}{\partial v_\perp} J_n^2\left(\frac{k_\perp v_\perp}{\omega_{cs}}\right)$$

greater than zero when $k_\perp$ is properly chosen. This is the "loss cone" instability of Post and Rosenbluth.[27, 28] Distributions such as that of Fig. 14 are called *loss cone* distributions. Mirror machines have loss cones because particles with small $v_\perp/v_z$ are lost through the mirrors. Therefore, distributions for which

$$f_{s0}(v_\perp, v_z) \xrightarrow[v_\perp \longrightarrow 0]{} 0$$

are developed. Distributions of the form

$$f_{s0}(v_\perp, v_z) \sim v_\perp^l\, e^{-v_\perp^2/\alpha_\perp^2 - v_z^2/\alpha_z^2} \qquad \qquad \text{...[3.144]}$$

with l a positive integer, have been extensively investigated by Guest et al.[29] Instabilities are also possible when $k_z = 0$. In this case equation [3.122] becomes

$$\varepsilon(k, \omega) = 1 + \sum_s \frac{\omega_{ps}^2}{k^2} \sum_{n=-\infty}^{+\infty} \frac{n\omega_{cs}}{(\omega - n\omega_{cs})} C_{sn}, \qquad \text{...[3.145]}$$

where

$$C_{sn} = \int d^3v \, \frac{1}{v_\perp} \frac{\partial f_{s0}}{\partial v_\perp} J_n^2\left(\frac{kv_\perp}{\omega_{cs}}\right). \qquad \text{...[3.146]}$$

For this case ε is real, and equation [3.73] cannot be used. Dory et al.[30] have shown that $\varepsilon(k, \omega) = 0$ has unstable roots for distributions of the form of equation [3.144].

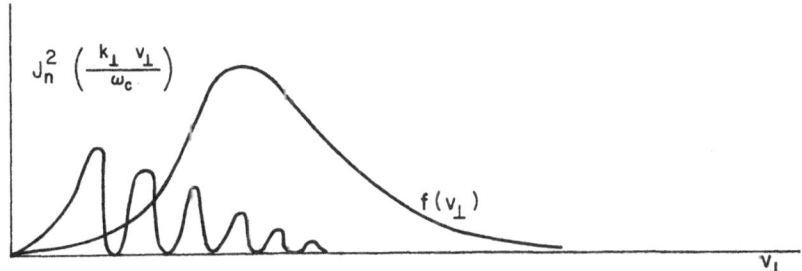

FIG. 14. A loss cone distribution of perpendicular velocities and $J_n^2\,(k_\perp v_\perp/\omega_{cs})$.

Another possible source of instability is a drift of the electrons relative to the ions.[31] To simplify the discussion of this instability, we shall assume that the distributions are single-temperature M-B distributions. The electron distribution function will be centred about the drift velocity V which is parallel to $\boldsymbol{B}_0$. Then

$$f_{0i} = \frac{n_i}{\pi^{\frac{3}{2}}\alpha_e^{\frac{3}{2}}}\, e^{-v^2/\alpha_i^2} \qquad\qquad ...[3.147]$$

$$f_{0e} = \frac{n_e}{\pi^{\frac{3}{2}}\alpha_e^{\frac{3}{2}}}\, e^{-\frac{v_\perp^2}{\alpha_e^2}-\frac{(v_z-V)^2}{\alpha_e^2}}. \qquad\qquad ...[3.148]$$

Equation [3.122] yields

$$\varepsilon(\boldsymbol{k},\omega) = 1 + \frac{1}{k^2 L_e^2}\left[1+\left(\frac{\omega-k_zV}{k_z\alpha_e}\right)C_{0e}Z\left(\frac{\omega-k_zV}{k_z\alpha_e}\right)\right]$$
$$+ \frac{1}{k^2 L_i^2}\left[1+\left(\frac{\omega}{k_z\alpha_i}\right)\sum_{n=-\infty}^{+\infty} C_{ni}Z\left(\frac{\omega-n\omega_{ci}}{k_z\alpha_i}\right)\right], \qquad ...[3.149]$$

where

$$C_{ns} = e^{-\frac{1}{2}k_\perp^2\rho_s^2}\, I_n\left(\frac{k_\perp^2\rho_s^2}{2}\right) \qquad\qquad ...[3.150]$$

and

$$L_s = \left(\frac{4\pi n_s e_s^2}{m_s}\frac{2}{\alpha_s^2}\right)^{\frac{1}{2}} \qquad\qquad ...[3.151]$$

is the Debye length of species s. We have kept only the $n = 0$ term in the electron contribution. The Fried-Conte function was defined in equation [3.135]. For small and large arguments

$$Z(x) \xrightarrow[x\to 0]{} i(\sqrt{\pi})e^{-x^2}-2x\left[1-\frac{2x^2}{3}+\ldots\right] \qquad ...[3.152]$$

$$Z(x) \xrightarrow[x\to\infty]{} -\frac{1}{x}\left[1+\frac{1}{2x^2}+\ldots\right]. \qquad\qquad ...[3.153]$$

If the electron and ion temperatures are comparable, then $\alpha_e \gg \alpha_i$. This suggests that as an approximation we use equation [3.152] for the electrons and equation [3.153] for the ions and obtain

$$\varepsilon(\boldsymbol{k},\omega) = 1 + \frac{1}{k^2 L_e^2}\left[1+i(\sqrt{\pi})\,C_{0e}\left(\frac{\omega-k_zV}{k_z\alpha_e}\right)\right]$$
$$+ \frac{1}{k^2 L_i^2}\left[1-\sum_{n=-\infty}^{+\infty} C_{ni}\frac{\omega}{\omega-n\omega_{ci}}\right]. \qquad\qquad ...[3.154]$$

We can use the relation

$$\sum_{n=-\infty}^{+\infty} C_{ni} = 1 \qquad \ldots[3.155]$$

to write the last term in equation [3.154] as

$$\sum_{n} C_{ni} \frac{\omega}{\omega - n\omega_{ci}} = 1 + \sum_{n} C_{ni} \frac{n\omega_{ci}}{\omega - n\omega_{ci}}. \qquad \ldots[3.156]$$

The real part of the frequency is determined from

$$\varepsilon_1(k, \omega) = 0 = 1 + \frac{1}{k^2 L_e^2} - \frac{1}{k^2 L_i^2} \sum_{n} C_{ni} \frac{n\omega_{ci}}{\omega - n\omega_{ci}}. \qquad \ldots[3.157]$$

A simple graphical construction shows that this has roots Ω_n lying near ω_{ci}. If all of the C_{ni} are negligible except C_{1i} then the root is

$$\Omega_1 \simeq \omega_{ci} \left[1 + \frac{C_{1i}}{k^2 L_i^2 \left(1 + \frac{1}{k^2 L_e^2} \right)} \right]. \qquad \ldots[3.158]$$

The imaginary part of the frequency is obtained from equation [3.73]. It is

$$\gamma_i \simeq - \frac{\omega_{ci}(\sqrt{\pi}) C_{1i} C_{0e}}{k^4 L_i^2 L_e^2 \left(1 + \frac{1}{k^2 L_e^2} \right)^2} \left(\frac{\Omega_1 - k_z V}{k_z \alpha_e} \right). \qquad \ldots[3.159]$$

When

$$V > \frac{\Omega_1}{k_z} \simeq \frac{\omega_{ci}}{k_z}, \qquad \ldots[3.160]$$

the wave is unstable.

3b. *Instabilities with frequencies near harmonics of the electron cyclotron frequency.* For this class of instabilities the ions may be neglected. The dielectric function is given by equation [3.122] with only the electron contribution included. Since the waves are plasma oscillations, it is necessary to have

$$\omega_{pe}^2 > n^2 \omega_{ce}^2, \qquad \ldots[3.161]$$

so

$$n_e > n^2 \frac{B^2}{4\pi mc^2} \qquad \ldots[3.162]$$

for instabilities near the nth harmonic of ω_{ce}. This gives a much higher critical density than equation [3.142] or [3.143].

If f_{e0} is a two-temperature M-B distribution, then arguments similar to those which led to equation [3.137] still apply and

$$\left(\frac{T_\parallel}{T_\perp}\right)_e < \frac{1}{2n} \qquad ...[3.163]$$

is necessary for instability of the nth harmonic. The instability criteria for the two-temperature M-B distribution has been investigated by Ozawa, Kaji and Kito[32] and by Harris.[33] This instability does not seem to have been studied for loss cone distribution, but it is obvious that there must be a loss cone instability.

3.4. Gradient-driven instabilities

We shall now investigate the effects of spatial gradients on the stability of the plasma. We shall also complicate matters somewhat by assuming a gravitational field in the y-direction. This may be thought of as crudely representing the effects of field curvature and field gradients. This gravitational field may be different for the different species. The constants of the motion in the unperturbed fields are now

$$v_z, \; v_\perp^2 - 2g_s y, \; y - v_x/\omega_{cs} = y_{cs},$$

and f_{s0} will be assumed to be a function of these.

We can find the f_{1s} without going through all of the orbit integration which was done in Section 1. Because of the gravitational field in the y-direction, there will be a drift of the guiding centre in the x-direction. Thus

$$\left(x + \frac{v_y}{\omega_{cs}}\right) = \left(x_0 + \frac{v_{y0}}{\omega_{cs}}\right) + v_d t, \qquad ...[3.164]$$

where

$$v_{ds} = \frac{g_s}{\omega_{cs}} = c\frac{m_s g_s}{e_s B}. \qquad ...[3.165]$$

The factor

$$e^{i k \cdot x_0(x, v, t) - i\omega(t - t')}$$

(see equation [3.34]) will contain as a factor

$$e^{i(\omega - k_x v_{ds} - k_z v_z)t'}. \qquad ...[3.166]$$

Therefore, where $(\omega - k_z v_z - n\omega_{cs})$ appears in equation [3.120], it must be replaced by $(\omega - k_x v_{ds} - k_z v_z - n\omega_{cs})$. There is still another difference; f_{s0} depends on v_x because it is a function of $y_{cs} = y - v_x/\omega_{cs}$ as well as because it depends on $v_\perp^2 - 2g_s y$. Going back to equation [3.11] and setting $B_1 = 0$ and $E_1 = -ik\phi$, we may write the right-hand side as

$$\frac{ie_s}{m_s}\phi k \cdot \frac{\partial f_{s0}}{\partial v} = \frac{ie_s}{m_s}\left[\frac{k_\perp v_x}{v_\perp}\frac{\partial f_{s0}}{\partial v_\perp} + k_z\frac{\partial f_{s0}}{\partial v_z} - \frac{k_\perp}{\omega_{cs}}\frac{\partial f_{s0}}{\partial y_{cs}}\right]. \qquad ...[3.167]$$

Again we are assuming $k = (k_\perp, 0, k_z)$. This gives us a term with $\partial f_{s0}/\partial y$ in the expression for f_{1s}, which now becomes

$$_{1s} = -\frac{e_s}{m_s} \phi \sum_{n=-\infty}^{+\infty} \sum_{m=-\infty}^{+\infty} \frac{J_m J_n e^{i(m-n)\alpha}}{(\omega - k_\perp v_{ds} - k_z v_z - n\omega_{cs})} \frac{n\omega_{cs} + k_\perp v_d}{v_\perp} \frac{\partial f_{s0}}{\partial v_\perp}$$

$$+ k_z \frac{\partial f_{s0}}{\partial v_\perp} - \frac{k_\perp}{\omega_{cs}} \frac{\partial f_{s0}}{\partial y_{cs}}. \qquad ...[3.168]$$

This may be integrated over velocities to get the charge density. When we did this before, the α-integration caused terms with $m \neq n$ to vanish. This is no longer the case, since now f_{s0} is a function of α, because it is a function of y_{cs}. However, we shall assume that this angular dependence is weak, so that the terms with $m \neq n$ are negligible. Another way of looking at it is to write

$$f_{s0} = f_{s0}(v_z, v_\perp^2 - 2gy, y_c)$$

$$\simeq f_{s0}(v_z, v_\perp^2 - 2gy_c, y_c), \qquad ...[3.169]$$

and hold y_c, the y-component of the position of the particle fixed during the velocity integration. This ignores the difference between the density of particles at a point and the density of guiding centres at the point. If the densities vary slowly over distances of the order of the radius of gyration, then the approximation should be good.

We shall continue to assume that

$$\phi = \phi_0 \, e^{i(k_\perp x + k_z z - \omega t)}, \qquad ...[3.170]$$

although this can no longer be strictly true. The potential must vary in the y-direction, since the plasma is non-uniform in the y-direction. The reader is referred to Rosenbluth[34] for a discussion of this approximation.

Calculating the charge density from f_{1s} and using it in Poisson's equation gives the dispersion relation $\varepsilon = 0$ with

$$\varepsilon(k, \omega) = 1 + \sum_s \frac{4\pi e_s^2}{m_s k^2} \sum_{n=-\infty}^{+\infty} \int d^3v \frac{J_n^2\left(\dfrac{k_\perp v_\perp}{\omega_{cs}}\right)}{(\omega - k_\perp v_{ds} - k_z v_z - n\omega_{cs})}$$

$$\left[\frac{n\omega_{cs} + k_\perp v_{ds}}{v_\perp} \frac{\partial f_{s0}}{\partial v_\perp} + k_z \frac{\partial f_{s0}}{\partial v_z} - \frac{k_\perp}{\omega_{cs}} \frac{\partial f_{s0}}{\partial y_{cs}}\right]. \qquad ..[3.171]$$

We shall now choose

$$f_{s0} = \frac{n_s}{\pi^{\frac{3}{2}} \alpha_s^2} \left[1 + \varepsilon_s\left(y - \frac{v_x}{\omega_{cs}}\right)\right] e^{-\frac{v_\perp^2 + v_z^2 - 2g_z y}{\alpha_z^2}}. \qquad ...[3.172]$$

Here ε_e and ε_i are small parameters introduced to give a weak dependence on y_{cs}. We will neglect ε_s^2 and higher powers. Substituting equation [3.172] into equation [3.171] and using

$$\sum_{n=-\infty}^{+\infty} J_n^2 = 1, \qquad \qquad ...[3.173]$$

we obtain

$$\varepsilon(\mathbf{k}, \omega) = 1 + \sum_s \frac{1}{k^2 L_s^2} + \sum_s \frac{e^{+\frac{2gy}{\alpha_s^2}}}{k^2 L_s} \sum_{n=-\infty}^{+\infty} \int d^3v \, \frac{[\omega + k\alpha_s^2 \varepsilon_s/2\omega_{cs}] J_n^2}{(\omega - k_\perp v_{ds} - k_z v_z - n\omega_{cs})}$$

$$= 1 + \sum_s \frac{1}{k^2 L_s^2} + \sum_s \frac{e^{+\frac{2g_s y}{\alpha_s^2}}}{k^2 L_s^2} \sum_{n=-\infty}^{+\infty} C_{sn} \frac{1}{k_z \alpha_s} [\omega + k_\perp \alpha_s^2 \varepsilon_s/2\omega_{cs}]$$

$$Z\left(\frac{\omega - k_\perp v_{ds} - n\omega_{cs}}{k_z \alpha_s}\right). \qquad ...[3.174]$$

We shall set $y = 0$ in all that follows.

4a. Finite Larmor radius stabilization of flute instabilities.[13] Now, some special cases of equation [3.174] will be examined. First, we shall let $k_z = 0$. Then from equation [3.153] we find

$$\varepsilon = 1 + \sum_s \frac{1}{k^2 L_s^2} - \sum_s \frac{1}{k^2 L_s^2} \sum_{n=-\infty}^{+\infty} C_{sn} \left[\frac{\omega + k\alpha_s^2 \varepsilon_s/2\omega_{cs}}{\omega - kv_{ds} - n\omega_{cs}}\right]. \qquad ...[3.175]$$

Now from equation [3.172] we find

$$n_{s0} = \int f_{s0} d^3v = n_s [1 + \varepsilon_s y] \, e^{+\frac{2g_s y}{\alpha_s^2}} \left(\frac{1}{n_s} \frac{dn_{s0}}{dy}\right)_{y=0} = \frac{n_0'}{n_0} = \varepsilon_s + \frac{2g_s}{\alpha_s^2}. \qquad ...[3.176]$$

We have dropped the subscript s on n_0, since charge neutrality will require that $n_{e0} = n_{i0}$. We will look for low frequency solutions of equation [3.175]. We drop the terms with $n \neq 0$, use equation [3.176] and obtain

$$\varepsilon = 1 + \sum_s \frac{1}{k^2 L_s^2} \left[1 - C_{s0} \frac{\omega - kv_{ds} + k\alpha_s^2 n_0'/2n_0\omega_{cs}}{\omega - kv_{ds}}\right]$$

$$= 1 + \sum_s \frac{1 - C_{s0}}{k^2 L_s^2} - \frac{n_0'}{kn_0} \sum_s \frac{\omega_{ps}^2}{\omega_{cs}} \frac{C_{s0}}{(\omega - kv_{ds})}. \qquad ...[3.177]$$

Now

$$C_{s0} = e^{-\frac{1}{2}k^2 \rho_s^2} I_0(\tfrac{1}{2}k^2 \rho_s^2) \xrightarrow[\rho_s \to 0]{} 1 - k^2 \rho_s^2. \qquad ...[3.178]$$

Setting $\varepsilon = 0$ in equation [3.177] and using $\rho_s^2/L_s^2 = \omega_{ps}^2/\omega_{cs}^2$ gives

$$1 + \sum \frac{\omega_{ps}^2}{\omega_{cs}^2} = \frac{n_0'}{kn_0} \sum_s \frac{\omega_{ps}^2}{\omega_{cs}} \frac{C_{s0}}{\omega - kv_{ds}}. \qquad ...[3.179]$$

This result is to be compared with equation [2.17]. The presence of C_{r0} in equation [3.179] gives a stabilization of flute instabilities due to the finite Larmor radius of the particles.

4b. *Universal instabilities*. This is a low frequency drift instability that has its origin in the diamagnetic current inherent to a confined plasma. We obtain the frequency and growth rate of this instability from equation [3.174] by neglecting all but the $n = 0$ terms and setting $v_{ds} = g/\omega_{cs} = 0$. Also we assume

$$\alpha_i \ll \frac{\omega}{k_z} \ll \alpha_e. \qquad \qquad ...[3.180]$$

Equations [3.152] and [3.153] give

$$Z\left(\frac{\omega}{k_z\alpha_e}\right) \simeq i\sqrt{\pi} \qquad \qquad ...[3.181]$$

$$Z\left(\frac{\omega}{k_z\alpha_i}\right) \simeq -\frac{k_z\alpha_i}{\omega}. \qquad \qquad ...[3.182]$$

Charge neutrality requires

$$\varepsilon_i = \varepsilon_e = \frac{1}{n_0}\frac{\partial n_0}{\partial y}. \qquad \qquad ...[3.183]$$

The diamagnetic drift velocity will be defined as

$$V = \frac{1}{n_0}\frac{\partial n_0}{\partial y}\frac{\alpha_i^2}{2\omega_{ci}} = -\frac{1}{n_0}\frac{\partial n_0}{\partial y}\frac{\alpha_e^2}{2\omega_{ce}}. \qquad \qquad ...[3.184]$$

We shall assume that

$$k_\perp\rho_e \ll 1,$$

so that

$$C_{0e} = e^{-\frac{1}{2}k_\perp^2\rho_e^2} I_0(\tfrac{1}{2}k_\perp^2\rho_e^2) \simeq 1. \qquad \qquad ...[3.185]$$

Also we assume $T_i = T_e$ and

$$k^2L_i^2 = k^2L_e^2 \ll 1. \qquad \qquad ...[3.186]$$

With these definitions and approximations, equation [3.174] becomes

$$\varepsilon(k, \omega) \simeq 2 - C_{0i}\left(1 + \frac{k_\perp V}{\omega}\right) + i\sqrt{\pi}\left(\frac{\omega - k_\perp V}{k_z\alpha_e}\right) = 0. \qquad ...[3.187]$$

Treating the imaginary part as a small correction, this may be solved for ω with the result

$$\omega = k_\perp V \frac{C_{0i}}{2 - C_{0i}}\left[1 + i\frac{2\sqrt{\pi}(1 - C_{0i})}{(2 - C_{0i})^2}\frac{k_\perp V}{k_z\alpha_e}\right]. \qquad ...[3.188]$$

Variants of this instability are obtained, if there is relative motion between ions and electrons parallel to B_0,[35] or if there is a temperature gradient.[36]

4c. *Drift cyclotron instabilities.*[37] We now look for instabilities near the ion cyclotron frequency. To this end we keep only the $n = 0$ electron term and the $n = 0, 1$ ion terms. We also assume $k_z = 0$, $T_e = T_i$, $n_e \simeq n_i$. Then equation [3.174] gives

$$\varepsilon(\boldsymbol{k}, \omega) = 1 + \frac{2}{k^2 L^2} - \frac{C_{e0}}{k^2 L^2}\left(1 - \frac{kV}{\omega}\right) - \frac{C_{i0}}{k^2 L^2}\left(1 + \frac{kV}{\omega}\right) - \frac{C_{i1}}{k^2 L^2}\frac{\omega + kV}{\omega - \omega_{ci}} = 0.$$

$$...[3.189]$$

With the further assumption that $C_{e0} \simeq 1$, equation [3.180] reduces to the quadratic equation

$$(\omega + kV)(\omega - \Omega) + a\omega(\omega - \omega_{ci}) = 0 \qquad ...[3.190]$$

where

$$\Omega = \omega_{ci}\frac{1 - C_{i0} - C_{i1}}{1 - C_{i0}} \qquad ...[3.191]$$

$$a = \frac{k^2 L^2}{1 - C_{i0} - C_{i1}}. \qquad ...[3.192]$$

If a were zero, then equation [3.190] would predict two waves. One is a drift wave with frequency $-kV$; the other is a cyclotron wave with frequency Ω. The last term in equation [3.190] provides a coupling between these waves. When

$$-\Omega kV > \frac{(\Omega + a\omega_{ci} - kV)^2}{4(1 + a)}. \qquad ...[3.193]$$

The roots are complex, indicating an instability.

4

Non-Linear Effects

The exponential growth of instabilities must eventually be limited by some non-linear effect—the subject to which we now turn our attention. There is some conceptual advantage in taking a quantum mechanical view-point and considering waves in the plasma to be composed of quasi-particles called plasmons (for the quanta of plasma oscillations), phonons (for the quanta of ion sound waves), etc. A quasi-particle of wave vector $\boldsymbol{k}$ and frequency ω will possess momentum $\hbar\boldsymbol{k}$ and energy $\hbar\omega$.

As will be seen, this gives a unified picture of many of the instabilities which we have discussed. They may be thought of as due to the emission and absorption of quasi-particles by particles. When the emission exceeds the absorption, the number of quasi-particles (and hence the amplitude of the corresponding wave) grows; this is an instability. When absorption exceeds emission the wave is damped. From this viewpoint it is easy to write down the equation describing the change of the particle distribution function due to the development of the instability.

4.1. *Quantum theory of electrostatic waves and their interaction with particles*

For the moment we shall ignore the imaginary part of the plasma dielectric constant and assume that a wave with wave vector k has the real frequency $\Omega_{k\sigma}$ which is a solution of

$$\varepsilon_1(k, \Omega_{k\sigma}) = 0. \qquad \qquad ...[4.1]$$

There may be several solutions of this equation; we distinguish between them by the index σ. We expand the electrostatic potential in a Fourier series in a large box of volume V.

$$\phi(x, t) = \sum_{k, \sigma} \left[\frac{4\pi\hbar\Omega_{k\sigma}}{Vk^2 \left[\dfrac{\partial}{\partial\omega} \omega\varepsilon_1 \right]_{\Omega_{k\sigma}}} \right]^{\frac{1}{2}}$$
$$\{B_{k\sigma} e^{i(k \cdot x - \Omega_{k\sigma}t)} + B_{k\sigma}^\dagger e^{-i(k \cdot x - \Omega_{k\sigma}t)}\}. \qquad ...[4.2]$$

The usual periodic boundary conditions are assumed. The $B_{k\sigma}$ are the Fourier coefficients. The reason for the factor in square brackets will be made clear presently.

We now calculate the energy in the electric field

$$U = \frac{1}{8\pi} \int d^3x \langle E^2(x, t) \rangle, \qquad \qquad ...[4.3]$$

where the angular brackets indicate a time average over an interval which is much longer than the periods of oscillation. We find

$$U = \sum_{k, \sigma} \frac{\hbar\Omega_{k\sigma}}{\left[\dfrac{\partial}{\partial\omega} \omega\varepsilon_1 \right]_{\Omega_{k\sigma}}} B_{k\sigma}^\dagger B_{k\sigma}. \qquad \qquad ...[4.4]$$

In addition to the electric field energy, there is also kinetic energy in the motion of the particles. It has been shown[38] that the total energy is greater than the electric field energy by the factor

$$\left[\frac{\partial}{\partial\omega} \omega\varepsilon_1(k, \omega) \right]_{\Omega_{k\sigma}}.$$

Therefore, the total energy in longitudinal waves is

$$H_L = \sum_{k, \sigma} \hbar\Omega_{k\sigma} B_{k\sigma}^\dagger B_{k\sigma}. \qquad \qquad ...[4.5]$$

The factor in square brackets in equation [4.2] was chosen so that H_L would have this simple form.

We shall now adopt a quantum mechanical viewpoint and regard H_L as the Hamiltonian for a system of quasi-particles of momentum $\hbar k$ and energy $\hbar \Omega_{k\sigma}$. We interpret $B_{k\sigma}$ as a destruction operator and $B_{k\sigma}^\dagger$ as a creation operator and assume that they obey the commutation relations

$$[B_{k\sigma}, B_{k'\sigma'}] = [B_{k\sigma}^\dagger, B_{k'\sigma'}^\dagger]_- = 0 \qquad \ldots [4.6]$$

$$[B_{k\sigma}, B_{k'\sigma'}^\dagger]_- = \delta_{kk'}\delta_{\sigma\sigma'} \qquad \ldots [4.7]$$

where

$$[A, B]_- = AB - BA \qquad \ldots [4.8]$$

It follows[39] from equations [4.6] and [4.7] that H_L has eigenvectors of the form

$$|\ldots, N_\sigma(k), \ldots, N_\sigma(k), \ldots\rangle$$

where $N_\sigma(k) = 0, 1, 2, 3, \ldots, \infty$ is the number of quasi-particles of type σ with momentum $\hbar k$. It also follows that

$$B_{k\sigma}|\ldots, N_\sigma(k), \ldots\rangle = \sqrt{[N_\sigma(k)]}|\ldots, N_\sigma(k)-1, \ldots\rangle \qquad \ldots [4.9]$$

$$B_{k\sigma}^\dagger|\ldots, N_\sigma(k), \ldots\rangle = \sqrt{[N_\sigma(k)+1]}|\ldots, N_\sigma(k)+1, \ldots\rangle \qquad \ldots [4.10]$$

$$B_{k\sigma}^\dagger B_{k\sigma}|\ldots, N_\sigma(k), \ldots\rangle = N_\sigma(k)|\ldots, N_\sigma(k), \ldots\rangle. \qquad \ldots [4.11]$$

This completes the quantum theory of the plasma waves. Next, we shall consider the particles in the plasma and then the interaction between particles and waves.

We shall assume that the single particle wave functions are known and denote these by $\chi_{sk}(x)$. For instance, if $B_0 = 0$ these will be the plane wave functions

$$\chi_{sk} = \frac{1}{\sqrt{V}} e^{ik \cdot x} \qquad \ldots [4.12]$$

with energy eigenvalues

$$E_{sk} = \hbar^2 k^2/2m_s. \qquad \ldots [4.13]$$

For present purposes we shall assume equations [4.12] and [4.13].

Following the usual procedure of the second quantization formalism[39] we write

$$\psi_s(x) = \sum_k A_{sk} \chi_{sk}(x) \qquad \ldots [4.14]$$

$$H_s = \int d^3x\, \psi_s^\dagger\left(-\frac{\hbar^2}{2m_s} \nabla^2\right) \psi_s$$

$$= \sum_k E_{sk} A_{sk}^\dagger A_{sk}. \qquad \ldots [4.15]$$

We take H_s to be the Hamiltonian for particles of species s. The operators A_{sk} and $A_{sk}^{\dagger}$ are assumed to obey the Fermion commutation relations

$$[A_{sk}, A_{sk'}]_+ = [A_{sk}^{\dagger}, A_{sk'}^{\dagger}]_+ = 0 \qquad \dots[4.16]$$

$$[A_{sk}, A_{sk'}^{\dagger}]_+ = \delta_{kk'} \qquad \dots[4.17]$$

where

$$[A, B]_+ = AB + BA. \qquad \dots[4.18]$$

This is certainly true for electrons. It may or may not be true for the ions (which may be bosons), but even if it is not true for ions, it will have no real consequences in the classical systems being discussed here.

The state vectors for the particle system will be of the form

$$| \dots, N_s(k), \dots, N_{s'}(k'), \dots \rangle$$

where $N_s(k) = 0, 1$ is the number of particles of species s with momentum $\hbar k$. This follows from equations [4.16] and [4.17]. It also follows that

$$A_{sk} | \dots, N_s(k), \dots \rangle = \pm \sqrt{[N_s(k)]} | \dots, N_s(k)-1, \dots \rangle \qquad \dots[4.19]$$

$$A_{sk}^{\dagger} | \dots, N_s(k), \dots \rangle = \pm \sqrt{[1-N_s(k)]} | \dots, 1-N_s(k), \dots \rangle \dots[4.20]$$

$$A_{sk}^{\dagger} A_{sk} | \dots N_s(k) \dots \rangle = N_s(k) | \dots N_s(k) \dots \rangle. \qquad \dots[4.21]$$

For our purposes the $\pm$ in equations [4.19] and [4.20] can be ignored.

The state vectors for the combined system of non-interacting particles and quasi-particles are of the form

$$| \dots N_s(k) \dots N_\sigma(k) \dots \rangle.$$

Next, we consider the interaction between the particles and waves. A particle of species s in an electrostatic potential $\phi(x)$ will have the potential energy $e_s \phi(x)$, so we take as the interaction Hamiltonian

$$H' = \sum_s e_s \int d^3x \, \psi^{\dagger}(x) \, \phi(x) \psi(x). \qquad \dots[4.22]$$

Using equations [4.2], [4.12] and [4.14] gives

$$H' = \sum_s \sum_\sigma H_{s\sigma}, \qquad \dots[4.23]$$

where

$$H_{s\sigma} = \sum_k \sum_q \left[\frac{4\pi e_s^2 \hbar \Omega_{k\sigma}}{Vk^2 \left[\dfrac{\partial}{\partial \omega} \omega \varepsilon \right]_{\Omega_{k\sigma}}} \right]^{\frac{1}{2}}$$
$$\{ A_{sq+k}^{\dagger} A_{sq} B_{k\sigma} + A_{sq-k}^{\dagger} A_{sq} B_{k\sigma}^{\dagger} \}. \qquad \dots[4.24]$$

This interaction Hamiltonian induces transitions between states of the non-interacting systems. The transition probability can be calculated from the well known "Fermi Golden Rule".

Transition probability per unit time

$$= \frac{2\pi}{\hbar} |M|^2 \delta(E_i - E_f), \qquad \ldots [4.25]$$

where the matrix element for the transition is given by

$$M = \langle f|H'|i\rangle + \sum_I \frac{\langle f|H'|I\rangle\langle I|H'|i\rangle}{E_i - E_I + i\eta}$$

$$+ \sum_I \sum_{II} \frac{\langle f|H'|I\rangle\langle I|H'|II\rangle\langle II|H'|i\rangle}{(E_i - E_I + i\eta)(E_i - E_{II} + i\eta)} \cdots \qquad \ldots [4.26]$$

Here $|i\rangle$ and $|f\rangle$ are initial and final states of the system, E_i and E_f are their energies, and η is a positive infinitesimal.

We now have all of the necessary machinery to discuss quasilinear theory and wave-wave coupling.

4.2. Quasi-Linear theory

For an unmagnetized plasma there are two types of longitudinal waves. There are the plasma oscillations (plasmons, denoted by λ) with frequencies

$$\Omega_{k\lambda} \simeq \omega_{pe}\left(1 + \frac{\langle k \cdot v\rangle}{\omega_{pe}} + \frac{3}{2}\frac{\langle (k \cdot v)^2\rangle}{\omega_{pe}^2}\right), \qquad \ldots [4.27]$$

and the ion sound waves (phonons, denoted by v) with frequencies

$$\Omega_{kv} \simeq \frac{\omega_{pi}kL_D}{\sqrt{(1 + k^2 L_D^2)}}. \qquad \ldots [4.28]$$

For the plasmons

$$\left(\frac{\partial \varepsilon_1}{\partial \omega}(k, \omega)\right)_{\Omega_{k\lambda}} \simeq \frac{2}{\Omega_{k\lambda}} \qquad \ldots [4.29]$$

and the interaction Hamiltonian, equation [4.24], becomes

$$H_{s\lambda} = \sum_k \sum_q \left(\frac{2\pi e_s^2 \hbar \Omega_{k\lambda}}{Vk^2}\right)^{\frac{1}{2}} \{A_{sq+k}^\dagger A_{sq} B_{k\lambda} + A_{sq-k}^\dagger A_{sq} B_{k\lambda}^\dagger\}. \qquad \ldots [4.30]$$

For the ion sound waves

$$\left(\frac{\partial \varepsilon_1}{\partial \omega}\right)_{\Omega_{kv}} = \frac{2}{\Omega_{kv}}\left(1 + \frac{1}{k^2 L_D^2}\right) \qquad \ldots [4.31]$$

and

$$H_{sv} = \sum_k \sum_q \left(\frac{2\pi e_s^2 \hbar \Omega_{kv}}{k^2 V\left(1 + \frac{1}{k^2 L_D^2}\right)}\right)^{\frac{1}{2}} \{A_{sq+k}^\dagger A_{sq} B_{kv} + A_{sq-k}^\dagger A_{sq} B_{kv}^\dagger\}. \qquad \ldots [4.32]$$

Equations [4.30] and [4.32] agree with the particle-plasmon and particle-phonon interaction Hamiltonians of Pines and Schrieffer.[40]

For simplicity, in what follows we shall consider only one species of particle (electrons) and one type of quasi-particle (plasmons).

Quasi-linear[41, 42] theory takes into consideration only first order processes of emission and absorption of plasmons by electrons. Schematic equations for the rate of change of $N_e(q)$ and $N_\lambda(k)$ can be written as in Fig. 15.

What we have done is to sum all of the processes which created a plasmon with momentum $\hbar k$ and subtract the sum of the processes which destroyed a plasmon with momentum $\hbar k$. This difference is $\partial N_\lambda(k)/\partial t$. A similar calculation gives $\partial N_e(q)/\partial t$. To obtain equations, it is only necessary to replace each diagram by the transition probability per unit time as given by equation [4.25]. Only the first term in equation [4.26] is kept. Using equations [4.9], [4.10], [4.19], [4.20] and [4.30], one immediately finds

$$\frac{\partial N_\lambda}{\partial t}(k) = \sum_q \frac{2\pi}{\hbar} \left(\frac{2\pi e^2 \hbar \Omega_{k\lambda}}{Vk^2}\right) \{N_e(q+k)[1-N_e(q)][N_\lambda(k)+1]$$

$$-[1-N_e(q+k)]N_e(q)N_\lambda(k)\} \delta\left[\frac{\hbar^2}{2m}|q+k|^2 - \frac{\hbar^2}{2m}q^2 - \hbar\Omega_{\lambda k}\right]$$

$$...[4.33]$$

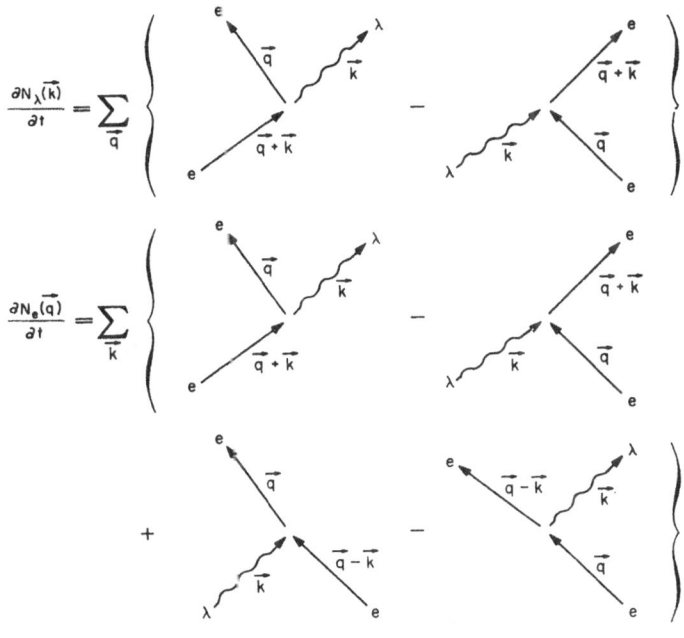

Fig. 15

$$\frac{\partial N_e}{\partial t}(q) = \sum_k \frac{2\pi}{\hbar} \left(\frac{2\pi e^2 \hbar \Omega_{k\lambda}}{Vk^2}\right)\left(\{N_e(q+k)[1-N_e(q)][N_\lambda(k)+1]\right.$$

$$-[1-N_e(q+k)]N_e(q)N_\lambda(k)\}\,\delta\left[\frac{\hbar^2}{2m}\,|\,q+k\,|^2 - \frac{\hbar^2}{2m}\,q^2 - \hbar\Omega_{\lambda k}\right]$$

$$+\{N_e(q-k)[1-N_e(q)]N_\lambda(k)-[1-N_e(q-k)]N_e(q)[N_\lambda(k)+1]\}$$

$$\left.\delta\left[\frac{\hbar^2}{2m}\,|\,q-k\,|^2 - \frac{\hbar^2}{2m}\,q^2 + \hbar\Omega_{\lambda k}\right]\right). \qquad \ldots[4.34]$$

These are the quasi-linear equations.

Some consequences of equations [4.33] and [4.34] may be seen immediately. One easily finds that

$$\frac{\partial}{\partial t}\sum_q N_e(q) = 0 \qquad\qquad \ldots[4.35]$$

$$\frac{\partial}{\partial t}\left\{\sum_q \hbar q N_e(q)+\sum_k \hbar k N_\lambda(k)\right\} = 0 \qquad\qquad \ldots[4.36]$$

$$\frac{\partial}{\partial t}\left\{\sum_q \frac{\hbar q^2}{2m} N_e(q)+\sum_k \hbar \Omega_{\lambda k} N_\lambda(k)\right\} = 0. \qquad\qquad \ldots[4.37]$$

These equations show that particles (but not quasi-particles), momentum and energy are conserved.

We can define an entropy for the electron-plasmon system as

$$S = S_e+S_\lambda$$
$$= -\sum_q \{[1-N_e(q)]\log[1-N_e(q)]+N_e(q)\log N_e(q)\}$$
$$+ \sum_k \{[N_\lambda(k)+1]\log[N_\lambda(k)+1]-N_\lambda(k)\log N_\lambda(k)\}. \qquad \ldots[4.38]$$

S_e is the appropriate definition of entropy for a system of fermions and S_λ is the appropriate definition for bosons.[43] By using equations [4.33] and [4.34] it is not difficult to show that

$$\frac{dS}{dt} \geqslant 0. \qquad\qquad \ldots[4.39]$$

Furthermore, the equality holds when

$$N_e(q) = \frac{1}{C\exp\left(\dfrac{\hbar q^2}{2mT}\right)+1} \qquad\qquad \ldots[4.40]$$

$$N_\lambda(k) = \frac{1}{\exp(\hbar\Omega_{\lambda k}/T)-1}. \qquad\qquad \ldots[4.41]$$

This seems to indicate that $N_e(q)$ approaches the Fermi-Dirac distribution, and $N_\lambda(k)$ approaches the Planck distribution. What is missing is a proof that S has only one maximum.

We now pass to the classical limit by the prescription

$$\hbar \longrightarrow 0 \qquad\qquad [4.42a]$$

$$\hbar q \longrightarrow mv \qquad\qquad \text{...}[4.42b]$$

$$N_e(q) \ll 1 \qquad\qquad \text{...}[4.42c]$$

$$N_\lambda(k) \longrightarrow \infty \qquad\qquad \text{...}[4.42d]$$

$$\hbar\Omega_{\lambda k}N_\lambda(k) = P_\lambda(k) = \text{Finite} \qquad\qquad \text{...}[4.42e]$$

$$\sum_k \longrightarrow V \int \frac{d^3k}{(2\pi)^3} \qquad\qquad \text{...}[4.42f]$$

$$\sum_q N_e(q) \longrightarrow V \int d^3v f_e(v). \qquad\qquad \text{...}[4.42g]$$

Equation [4.42c] means that the electron gas is far from degeneracy. $P_\lambda(k)$ is the energy spectrum of plasma oscillations—a classically meaningful quantity. We let the volume of the box in which the system is quantized become infinite, so sums go over into integrals; hence equations [4.42f] and [4.42g]. In the classical limits, equations [4.33] and [4.34] become

$$\frac{\partial P_\lambda}{\partial t}(k) = 2\gamma_\lambda(k)P_\lambda(k)+S_\lambda(k) \qquad\qquad ..[4.43]$$

$$\frac{\partial f_e}{\partial t}(v) = \frac{\partial}{\partial v}\cdot\left(D(v)\cdot\frac{\partial f_e}{\partial v}\right)+\frac{\partial}{\partial v}\cdot(A(v)f_e) \qquad\qquad \text{...}[4.44]$$

where

$$\gamma_\lambda(k) = \frac{2\pi^2 e^2 \Omega_{\lambda k}}{mk^2}\int d^3v\; k\cdot\frac{\partial f_e}{\partial v}\delta(k\cdot v-\Omega_{\lambda k}) \qquad\qquad \text{...}[4.45]$$

$$S_\lambda(k) = \frac{4\pi^2 e^2 \Omega_{\lambda k}^2}{k^2}\int d^3v f_e(v)\delta(k\cdot v-\Omega_{\lambda k}) \qquad\qquad \text{...}[4.46]$$

$$D(v) = \frac{4\pi^2 e^2}{m^2}\int \frac{d^3k}{(2\pi)^3}\; P_\lambda(k)\frac{kk}{k^2}\delta(k\cdot v-\Omega_{\lambda k}) \qquad\qquad ..[4.47]$$

$$A(v) = \frac{4\pi^2 e^2}{m}\int \frac{d^3k}{(2\pi)^3}\; \Omega_{\lambda k}\frac{k}{k^2}\delta(k\cdot v-\Omega_{\lambda k}). \qquad\qquad \text{...}[4.48]$$

These are the classical quasi-linear equations. The second terms on the right-hand side of equations [4.43] and [4.44] are due to spontaneous emission of plasmons by particles. These terms do not appear in some classical derivations.[41, 42] It should be noted that equation [4.45] agrees with the growth rate calculated in Section 3.2 under the assumption that $\gamma \ll \Omega$. It may easily be shown that equations [4.43] and [4.45] have the time-independent solutions.

$$P_\lambda(k) = T \text{ (Rayleigh-Jeans Law)} \qquad \text{...}[4.49]$$

$$f_e(v) = C\,e^{-mv^2/2T} \text{ (Maxwell-Boltzmann distribution)}.$$
$$\text{...}[4.50]$$

The derivation of the quasi-linear equations which has just been given may be generalized without difficulty to include other species of particle and quasi-particles. A wide class of instabilities may be described in terms of the emission of quasi-particles by particles. When the stimulated emission exceeds the absorption, then $\gamma(k)$ is positive and the plasma is unstable. Some of the instabilities which occur in a magnetized plasma have been discussed in quantum mechanical terms by Walters.[44, 45]

5

Finite Plasma Effects

It is not too difficult to discuss the instabilities of finite plasmas on the basis of the fluid equations. This was done in Chapter 2. It is much more difficult to discuss microinstabilities in finite plasmas, since for them the Vlasov equations must be used. Because of this difficulty Chapters 3 and 4 dealt only with infinite plasmas (although weak spatial gradients were discussed in some sections). In this chapter we give a brief account of some effects connected with the finiteness of the plasma.

Sturrock[46] seems to have been the first to call attention to the distinction between absolute and convective plasma instabilities. Consider a wave packet produced at the time $t = 0$. As time increases this packet will propagate, and if the plasma is unstable it will grow exponentially as it propagates. A stationary observer may observe an exponential growth; if so, we say that the instability is absolute. On the other hand the motion of the wave packet may be sufficiently rapid for a stationary observer to see a decay, even though an observer moving with the packet would see an exponential growth. If so, we say that the instability is convective. The distinction between absolute and convective instabilities can be given when the dispersion relation is known. This distinction is not simple, and we shall not consider it here; it has been adequately discussed elsewhere.[47]

If a finite plasma is absolutely unstable, a disturbance in it will grow until it is limited by some non-linear effect. A convective instability, however, may propagate out of the plasma before it has grown appreciably provided that it is not reflected at the boundaries. Studies of reflection from boundaries have been made.[48, 49] In many cases waves are strongly absorbed in the tenuous plasmas near the boundaries so negligible reflection occurs. If this is the case one may conclude that convective instabilities are relatively harmless.

One way, then, of discussing instabilities in finite plasmas is to ask whether the plasma instability is absolute or convective. If it is absolute, then one concludes that the finite plasma is unstable. If it is convective, and if there is no appreciable reflection from the boundaries, and the growth rate is not too large, or the plasma is not too big, then one concludes that the plasma is stable for all practical purposes.

It is clear that this method is less satisfactory than that of treating the plasma as finite *ab initio*. Since realistic finite plasmas are extremely difficult to treat, we shall consider a simple idealized model of a mirror machine. This model has been discussed previously by Arsenin,[50] Cotsaftis[51] and Cheng.[52, 53] We shall follow the work of Cheng.

Consider a plasma confined between planes at $z = 0$ and a as shown in Fig. 16. The planes are assumed to perfectly reflect the particles but to have no effect on the fields. They are a crude way of representing magnetic mirrors. A uniform magnetic field parallel to z is assumed. In this configuration the linearized Vlasov equation may be solved by the method of integration along the orbits which were discussed for the infinite plasma in Chapter 3. Then Poisson's equation (we only consider electrostatic waves) may be written

$$\nabla^2 \phi(x, t) = -\sum_s \frac{4\pi e_s^2}{m_s} \int d^3v \int_{-\infty}^{t} dt' \nabla \phi' \cdot \frac{\partial f_{s0}}{v'} \qquad \dots [5.1]$$

where the primes in the integrand of equation [5.1] indicate that the integration is along the unperturbed particle orbit ending at the time t at the phase

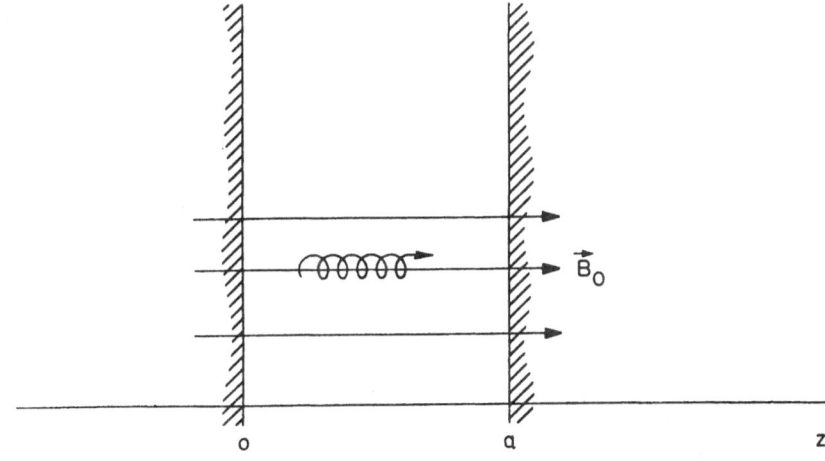

FIG. 16

space point (x, v). Equation [5.1] holds for $0 < z < a$. Outside of this region, the right-hand side of equation [5.1] should be replaced by zero. A solution may be sought in the form

$$\phi(x, t) = \phi(z) e^{i q_\perp \cdot x_\perp - i \omega t} \qquad \ldots [5.2]$$

where

$$\phi(z) = \begin{cases} A e^{+q_\perp z} & z < 0 \\ B e^{-q_\perp z} + C e^{-q_\perp(a-z)} & \\ \quad + \sum\limits_{q_z = n\pi/a} a_{q_z} \cos q_z z & 0 < z < a \\ D e^{-q_\perp(z-a)} & z > a \end{cases} \qquad \ldots [5.3]$$

The constants A, B, C and D are determined by the four boundary conditions that ϕ and $\partial\phi/\partial z$ be continuous at $z = 0$ and a. One finds coupled equations for the Fourier coefficients a_{q_z}. The condition that this set of equations have a non-trivial solution has the remarkably simple form

$$D(q_\perp, \omega) = 1 + \frac{2q_\perp}{a} \sum_{\substack{q_z = n\pi a = -\infty \\ n = \text{even, odd}}}^{+\infty} \frac{1}{(q_\perp^2 + q_z^2)\varepsilon(q, \omega)} = 0. \qquad \ldots [5.4]$$

(The reader is referred to Cheng[53] for details.) Here $\varepsilon(q, \omega)$ is the dielectric constant for an infinite plasma. The sum is over even values of n (corresponding to a perturbation which is symmetric about $z = a/2$), or over odd values of n (corresponding to a perturbation which is antisymmetric about $z = a/2$).

Although equation [5.4] looks simple, it is more difficult to analyze than the infinite plasma dispersion relation, $\varepsilon(q, \omega) = 0$. However, progress can be made by employing some of the same tricks that were effective for the infinite plasma. We write

$$D = D_1 + i D_2 \qquad \ldots [5.5a]$$

$$\varepsilon = \varepsilon_1 + i\varepsilon_2 \qquad \ldots [5.5b]$$

$$\omega = \Omega + i\gamma \qquad \ldots [5.5c]$$

and assume

$$|D_2| \ll |D_1| \qquad \ldots [5.6a]$$

$$|\varepsilon_2| \ll |\varepsilon_1| \qquad \ldots [5.6b]$$

$$|\gamma| \ll |\Omega|. \qquad \ldots [5.6c]$$

Then, we find that Ω is given approximately by

$$D_1(q_\perp, \Omega) = 0, \qquad \ldots [5.7]$$

and γ is given approximately by

$$\gamma = -\frac{D_2(q_\perp, \Omega)}{\left(\dfrac{\partial D_1}{\partial \omega}\right)_\Omega}$$

$$= \frac{\displaystyle\sum_{q_z} \frac{1}{q^2} \frac{\varepsilon_2(q, \Omega)}{\varepsilon_1^2(q, \Omega)}}{\displaystyle\sum_{q_z} \frac{1}{q^2} \frac{1}{\varepsilon_1^2(q, \Omega)} \left(\dfrac{\partial \varepsilon_1}{\partial \omega}\right)_\Omega}. \qquad \ldots [5.8]$$

The derivation is analogous to that which led to equations [3.72] and [3.73].

Before proceeding we shall call attention to a relation between equation [5.8] and the similar equation [3.73] for the growth rate in an infinite plasma. If, for some value of $q_z = n\pi/a$, $\varepsilon_1(q, \omega) \simeq 0$, then this term may give the dominant contribution to both numerator and denominator of equation [5.8] and

$$\gamma \simeq -\frac{\varepsilon_2(q, \Omega)}{\left(\dfrac{\partial \varepsilon_1}{\partial \omega}\right)_\Omega} \qquad \ldots [5.9]$$

which is the growth rate for an infinite plasma. If a, the separation of the planes becomes very large, then the spacing between values of q_z becomes very small; in this case $\varepsilon_1(q, \Omega) \simeq 0$ and equation [5.9] become better approximations.

A contour integral method[54] may be used to sum the series for $D_1(q_\perp, \omega)$. We write

$$D_1(q_\perp, \omega) = 1 + q_\perp \oint \frac{dq_z}{2\pi i} \frac{1}{q^2 \varepsilon_1(q, \omega)} \left\{ \begin{array}{c} \cot \dfrac{q_z a}{2} \\[4mm] -\tan \dfrac{q_z a}{2} \end{array} \right\} \quad \ldots [5.10]$$

where $\cot q_z a/2$ (or $-\tan q_z a/2$) is used when the sum is over even (or odd) n. The path of integration is drawn so as to enclose the poles of $\cot q_z a/2$ (or $-\tan q_z a/2$) and not to enclose any poles of $1/q^2 \varepsilon$. By distorting the path of integration, it may be shown that D_1 is given by

$$D_1(q_\perp, \omega) = 1 - q_\perp \sum \left(\text{residues at poles of } \frac{1}{q^2 \varepsilon} \right). \qquad \ldots [5.11]$$

Unfortunately, we have not been able to make this work for D_2 because ε_2 usually has branch points or essential singularities in the complex q_z-plane.

We shall now discuss some results derived from various approximations to $\varepsilon(q, \omega)$.

(a) $B_0 \neq 0$. *Cold electrons—hot ions*
We may approximate ε_1 by

$$\varepsilon_1(q, \omega) = 1 - \frac{\omega_{pe}^2}{\omega^2} \frac{q_z^2}{q_\perp^2 + q_z^2} \qquad \ldots[5.12]$$

Then

$$\frac{1}{q^2 \varepsilon(q, \omega)} = \frac{\zeta^2}{(q_z + q_\perp \zeta)(q_z - q_\perp \zeta)} \qquad \ldots[5.13]$$

where

$$\zeta^2 = 1/(\omega_{pe}^2/\omega^2 - 1) \qquad \ldots[5.14]$$

Equation [5.13] has poles at $q_z = \pm \zeta q_\perp$. After a simple calculation equation [5.11] yields

$$D_1(q_\perp, \omega) = 1 + \zeta \left\{ \begin{array}{c} \cot \dfrac{q_\perp a \zeta}{2} \\[2mm] -\tan \dfrac{q_\perp a \zeta}{2} \end{array} \right\} = 0. \qquad \ldots[5.15]$$

These are transcendental equations of ζ. There are an infinite number of roots. We shall denote the roots by ζ_n. It is easily seen that the spacing between adjacent roots is of the order of π. Solving equation [5.14] for ω gives

$$\Omega_n^2 = \omega_{pe}^2 \frac{\zeta_n^2}{\zeta_n^2 + 1} \qquad \ldots[5.16]$$

for the frequency corresponding to the root ζ_n. This should be compared with

$$\Omega^2(q_\perp, q_z) = \omega_{pe}^2 \frac{(q_z/q_\perp)^2}{(q_z/q_\perp)^2 + 1} \qquad \ldots[5.17]$$

for the frequency of a wave with wave vector $q = (q_\perp, q_z)$ in an infinite plasma. The principal effect of the finiteness of the plasma is to make the values of $q_z/q_\perp$ discrete.

Equation [5.8] may be used to calculate the γ_n corresponding to Ω_n. Thus

$$\gamma_n = \frac{1}{\left(\dfrac{\partial D_1}{\partial \omega}\right)_{\Omega_n}} \sum_{q_z = n\pi/a} \frac{1}{q_\perp^2 + q_z^2} \frac{\varepsilon_2(q, \Omega_n)}{\varepsilon_1^2(q, \Omega_n)}. \qquad \ldots[5.18]$$

The term with q_z nearest to $q_\perp \zeta_n$ will make $\varepsilon_1(q, \Omega_n) \simeq 0$. If this term dominates the sum sufficiently, then γ_n is given by equation [5.9], and the growth or damping is the same as for an infinite plasma. It may happen, of course, that equation [5.9] is not a sufficiently good approximation and the sum in equation [5.18] must be retained. Then whether γ_n is positive or negative will depend on the competition between positive and negative contributions to the sum in equation [5.18].

(b) $B_0 = 0$. *Electron plasma oscillations*
We approximate ε_1 by

$$\varepsilon_1(q, \omega) = 1 - \frac{\omega_{pe}^2}{\omega^2} - 3\frac{\omega_{pe}^4}{\omega^4} L_D^2 q^2. \qquad ...[5.19]$$

We shall omit the details, but again one finds a transcendental equation which determines the frequencies. One finds volume plasma oscillations with frequencies

$$\Omega_n^2 = \omega_{pe}^2(1 + 3L_D^2 Q^2) \qquad ...[5.20]$$

where

$$Q = \left(q_\perp, \frac{2x_n}{a}\right) \qquad ...[5.21]$$

and the x_n's are solutions of a certain transcendental equation. In addition, there are surface plasma oscillations with frequencies

$$\Omega_\pm \simeq \frac{\omega_{p:}}{\sqrt{2}}(1 \pm e^{-q_\perp a})^{\frac{1}{2}}. \qquad ...[5.22]$$

The plus sign corresponds to the mode which is symmetric about $z = a/2$, and the minus sign corresponds to the antisymmetric mode. For large a the damping of the surface waves is given by

$$\gamma(q) = -\omega_{pe}\left(\sqrt{\frac{2}{3\pi}}\right)q_\perp L_D. \qquad .. [5.23]$$

(A Maxwellian velocity distribution is assumed in calculating ε_2.) Note that the damping of surface plasma oscillations is linear in q, L_D in contrast to the damping of plasma oscillations in an infinite plasma where

$$\gamma \sim \exp(-1/q^2 L_D). \qquad ...[5.24]$$

(c) $B_0 = 0$. *Ion sound waves*
We approximate ε_1 by

$$\varepsilon_1 = 1 + \frac{1}{q^2 L_D^2} - \frac{\omega_{pi}^2}{\omega^2}. \qquad ...[5.25]$$

Again, we omit the details. We find volume ion sound waves with frequencies

$$\Omega_n^2 = \frac{\omega_{pi}^2 L_D^2 q_\perp^2 (1+\zeta_n^2)}{1+L_D^2 q_\perp^2 (1+\zeta_n^2)}. \qquad \ldots [5.26]$$

Where the ζ_n's are solutions of a certain transcendental equation. In addition, there are surface ion sound waves with frequencies

$$\Omega^2 = \frac{\omega_{pi}^2 L_D^2 q_\perp^2 (1-\beta^2)}{1+L_D^2 q_\perp^2 (1-\beta^2)} \qquad \ldots [5.27]$$

where β is a solution of a certain transcendental equation. Equations [5.26] and [5.27] may be compared with the frequency of an ion sound wave in an infinite plasma given by

$$\Omega^2 = \frac{\omega_{pi}^2 L_D^2 q_\perp^2 (1+q_z^2/q^2)}{1+L_D^2 q_\perp^2 (1+q_z^2/q^2)}. \qquad \ldots [5.28]$$

For the volume ion sound waves there is a discrete set of values of q_z/q. For the surface ion sound waves, q_z/q is imaginary.

ACKNOWLEDGMENTS

These lectures were prepared partly at the University of Tennessee and partly at the Thermonuclear Division of the Oak Ridge National Laboratory. The work done at the University of Tennessee was partially supported under contract AT-(40-1)-2598 with the U.S. Atomic Energy Commission. I have benefited greatly from conversations with colleagues at both institutions.

REFERENCES

1. B. LEHNERT. 1967. Plasma physics. *J. Nucl. Energy*, C, **9**, 301.
2. NORMAN ROSTOKER. 1966. *Plasma Physics in Theory and Application*, edited by W. Kunkel. McGraw-Hill, New York.
3. A. A. VEDENOV, E. P. VELIKHOV and R. Z. SAGDEEV. 1961. *Soviet Phys. Usp.*, **4**, 332.
4. B. B. KADOMPTSEV. 1966. *Reviews of Plasma Physics*, **2**, 153, edited by M. A. Leontovich. Consultants Bureau, New York.
5. A. JEFFREY and T. TANIUTI. 1966. *Magnetohydrodynamic Stability*. Academic Press, New York.
6. T. NORTHRUP. 1956. *Phys. Rev.*, **103**, 1150.
7. M. KRUSKAL and M. SCHWARZCHILD. 1954. *Proc. R. Soc. Lond.*, A, **223**, 348.
8. M. N. ROSENBLUTH and C. L. LONGMIRE. 1957. *Ann. Phys., N.Y.*, **1**, 120.
9. G. SCHMIDT. 1966. *Physics of High Temperature Plasmas*, Section 8–4. Academic Press, New York.
10. F. L. CURZON, A. FOLKIERSKI, R. LATHAM and J. A. NATION. 1960. *Proc. R. Soc., Lond.*, A, **257**, 386.
11. D. J. ALBARES, N. A. KRALL and C. L. OXLEY. 1961. *Physics Fluids*, **4**, 1033.
12. M. S. IOFFE. 1965. In *Plasma Physics*. International Atomic Energy Agency, Vienna

13. M. N. ROSENBLUTH, N. A. KRALL and N. ROSTOKER. 1962. *Nucl. Fusion*, Suppl. Part I, 143.
14. J. BERKOWITZ, K. O. FRIEDRICHS, H. GOERTZEL, H. GRAD, J. KILLEEN and E. RUBIN. 1958. *Proc. Second Intern. Conf., Geneva*, Vol. 31, p. 171. United Nations, New York.
15. T. OHKAWA and N. ROSTOKER. 1967 (December). *Physics today*, **20**, 48.
16. D. M. MEADE. 1966. *Phys. Rev. Lett.*, **17**, 677.
17. E. G. HARRIS. 1961. *J. Nuclear Energy*, C, **2**, 138.
18. G. BEKEFI. 1956. *Radiation Processes in Plasmas*, Section 7–2. John Wiley, New York.
19. L. D. LANDAU. 1946. *Fiz. Zh. (J. Physics, U.S.S.R.)*, **10**, 25.
20. I. B. BERNSTEIN, E. A. FRIEMAN, R. M. KULSRUD and M. N. ROSENBLUTH. 1960. *Physics Fluids*, **3**, 136.
21. E. N. PARKER. 1958. *Phys. Rev.*, **109**, 1874.
22. M. N. ROSENBLUTH and K. WILSON. 1958. *Proc. Second Intern. Conf., Geneva, 1958*, *Vol. 31, p. 89*. United Nations, New York.
23. R. Z. SAGDEEV and V. D. SHAFRONOV. 1961. *Soviet Phys. JETP*, **12**, 130.
24. D. B. CHANG. 1963. *Astrophys. J.*, **138**, 1231.
25. G. K. SOPER and E. G. HARRIS. 1965. *Physics Fluids* **8**, 984.
26. B. D. FRIED and C. CONTE. 1961. *The Plasma Dispersion Function.* Academic Press, New York.
27. M. N. ROSENBLUTH and R. F. POST. 1965. *Physics Fluids*, **8**, 547.
28. R. F. POST and M. N. ROSENBLUTH. 1966. *Physics Fluids*, **9**, 730.
29. G. E. GUEST and R. A. DORY. 1965. *Physics Fluids*, **8**, 1853.
30. R. A. DORY, G. E. GUEST and E. G. HARRIS. 1965. *Phys. Rev. Lett.*, **14**, 131.
31. W. E. DRUMMOND and M. N. ROSENBLUTH. 1962. *Physics Fluids*, **5**, 1507.
32. Y. OZAWA, I. KAJI and M. KITO. 1962. *J. Nucl. Energy*, C, **4**, 271.
33. E. G. HARRIS. *General Atomics Report GA–5581* (unpublished).
34. M. N. ROSENBLUTH. 1965. *Plasma Physics.* International Atomic Energy Agency, Vienna.
35. B. B. KADOMPTSEV. 1963. *J. Nucl. Energy*, C, **5**, 31.
36. L. L. RUDAKOV and R. Z. SAGDEEV. 1960. *Soviet Phys. JETP*, **10**, 952.
37. A. B. MIKHAILOVSKII and A. V. TIMOFEEV. 1963. *Soviet Phys. JETP*, **17**, 626.
38. L. D. LANDAU and E. M. LIFSHITZ. 1960. *Electrodynamics of Continuous Media*, p. 256. Addison-Wesley, Reading, Mass.
39. A. S. DAVYDOV. 1965. *Quantum Mechanics*, Chapters 14 and 15. Addison-Wesley, Reading, Mass.
40. D. PINES and J. R. SCHRIEFFER. 1962. *Phys. Rev.*, **125**, 804.
41. W. E. DRUMMOND and D. PINES. 1962. *Nucl. Fusion*, Suppl., 1049.
42. A. A. VEDENOV, E. P. VELIKHOV and R. Z. SAGDEEV. 1962. *Nucl. Fusion*, Suppl., 465.
43. L. D. LANDAU and E. M. LIFSCHITZ. 1958. *Statistical Physics*. Addison-Wesley, Reading, Mass.
44. G. M. WALTERS and E. G. HARRIS. 1968. *Physics Fluids*, **11**, 112.
45. G. M. WALTERS. 1967. *Quantum Mechanical Theory of Non-Linear Plasma Phenomena in a Magnetic Field*, University of Tennessee Thesis (unpublished).
46. P. A. STURROCK. 1958. *Phys. Rev.*, **112**, 1488.
47. R. J. BRIGGS. 1964. *Electron Stream Interaction With Plasmas.* M.I.T. Press, Cambridge, Mass.
48. R. E. AAMODT and D. L. BOOK. 1966. *Physics Fluids*, **9**, 143.
49. H. L. BERK, C. W. HORTON, M. N. ROSENBLUTH, D. E. BALDWIN and R. N. SUDAN. 1968. *Physics Fluids*, **11**, 365.
50. V. V. ARSENIN. 1967. *Soviet Phys. tech. Phys.*, **12**, 442.
51. M. COTSAFTIS. *Proc. Inter. Conf. on Plasma Confined in Open-Ended Geometry*, 1967. (To be published.)
52. C. C. CHENG and E. G. HARRIS. 1968. *Bull. Am. Phys. Soc.*, Ser. II, **13**, 622.
53. C. C. CHENG and E. G. HARRIS (to be published).
54. P. M. MORSE and H. FESHBACH. 1953. *Methods of Theoretical Physics*, p. 413. McGraw-Hill, New York.

5

COMPUTATIONAL PROBLEMS IN PLASMA PHYSICS AND CONTROLLED THERMONUCLEAR RESEARCH

JOHN KILLEEN

Lawrence Radiation Laboratory, University of California

1

INTRODUCTION

IN these lectures we present several areas of plasma physics research where computation has been of importance. We have taken these problems primarily from controlled thermonuclear research, because this programme has provided the impetus for most of the theoretical development of the subject and because only numerical methods could give the answers to specific questions in the design and analysis of experiments.

We consider computer experiments briefly in the last section. These methods have become popular in the last few years in testing theory in fundamental plasma physics. Some of the methods are discussed, but no results of such experiments are given. In fact, we have not presented numerical results in any of the sections, but only computational methods. The results of all the problems considered would certainly fill several books.

In Section 2 we consider numerical methods used in analysing pinch experiments. Although the title sounds quite specific, we are really discussing the solution of magnetohydrodynamic equations, so the applicability is far more general. Furthermore, the mathematics of solving initial-value problems in partial differential equations is of even more generality.

In the next section we consider resistive instability calculations. Again, these calculations are applied to pinch experiments, but these problems occur often in laboratory and space plasma research. The method of solving time-dependent perturbation equations numerically has been applied to other stability problems and should have wide applicability.

In Section 4 we consider plasma equilibria calculations. We illustrate the methods with several special two-dimensional problems.

In Section 5 we consider the numerical solution of the Fokker-Planck equation for a plasma, and in Section 6 we consider several methods of solving the Vlasov equation. Some of these methods are called computer experiments.

Computation has been of importance in the CTR programme since its beginning. As the theory developed and as the computers became faster, the range of applications increased and the models have generally been more realistic. The use of even more realistic models should be the main goal of future computer applications.

2

NUMERICAL STUDIES OF PINCH EXPERIMENTS

2.1. Infinite conductivity calculations

The use of numerical calculations in the design and analysis of pinch experiments has been of central importance. In most of these computations the equations of magnetohydrodynamics are used. The earliest problems[1] used the infinite conductivity theory in analysing the linear pinch. At Livermore, in order to analyse some experiments, we coupled the pinch dynamical equations with the external circuit equations.[2] The resulting set of differential equations were solved numerically.

As the stability theory of the pinch advanced and more complex experiments evolved, such as the diffuse "stabilized" pinch, hard-core pinch and theta pinch, more elaborate numerical computations have been made. Fortunately, at this same time the speed and capacity of computing machines have been increasing at a rapid rate.

2.2. One-dimensional, fully ionized model

Most of the above experiments have been analysed with computations on a time-dependent, one-dimensional fully ionized hydromagnetic model. Electrical resistivity and thermal conductivity of the plasma are included, and separate temperatures are assigned to the electrons and ions. The first programmes with this model were done by Hain and Roberts[3, 4] as a joint effort between A.E.R.E. Harwell and the Max-Planck-Institut, Munich. In their work the equations to be solved are given in Eulerian form. Other programmes which have been written[5, 6] use the same model, but with equations in Lagrangian form. When extending these programmes to two dimensions, it is preferable to use an Eulerian grid because of the distortion of a Lagrangian grid during the problem; so we shall discuss the Eulerian form of the equations. The model described by Hain and Roberts considers both the z and θ components of the magnetic field which is appropriate for linear and hard-core pinches. We shall consider the somewhat simpler model with a B_z only, which is appropriate for theta pinch calculations.[7, 8]

The continuity equation for the plasma is

$$\frac{\partial \rho}{\partial t} + \frac{1}{r}\frac{\partial}{\partial r}(r\rho v_r) = 0 \qquad \qquad ...[2.1]$$

where ρ is the mass density and v_r is the radial component of the velocity.

The equation of motion is

$$\frac{\partial v_r}{\partial t}+v_r\frac{\partial v_r}{\partial r} = \frac{1}{\rho}j_\theta B_z-\frac{\partial}{\partial r}(\theta_e+\theta_i)-(\theta_e+\theta_i)\frac{1}{\rho}\frac{\partial\rho}{\partial r}-\frac{1}{\rho}\frac{\partial Q}{\partial r} \quad ...[2.2]$$

where j_θ is the θ component of the current density; B_z is the z component of the magnetic field; θ_e and θ_i are the electron and ion temperatures; and Q is the von Neumann-Richtmyer artificial viscosity. The use of this term for broadening shock fronts is discussed in the book by Richtmyer and Morton.[9] The equation for the ion temperature

$$\frac{\partial\theta_i}{\partial t}+v_r\frac{\partial\theta_i}{\partial r} = -(\gamma-1)\left(\theta_i+\frac{1}{\rho}Q\right)\frac{1}{r}\frac{\partial}{\partial r}(rv_r)+\frac{1}{\rho r}\frac{\partial}{\partial r}\left(rk_i\frac{\partial\theta_i}{\partial r}\right)+\frac{\theta_e-\theta_i}{\tau_{eq}}$$

$$...[2.3]$$

where γ is the specific heat ratio, k_i the ion thermal conductivity,[10] and τ_{eq} the equipartition time.[11] The above equation includes shock heating of the ions. The equation for the electron temperature is

$$\frac{\partial\theta_e}{\partial t}+v_r\frac{\partial\theta_e}{\partial r} = -(\gamma-1)\theta_e\frac{1}{r}\frac{\partial}{\partial r}(rv_r)+\frac{1}{\rho r}\frac{\partial}{\partial r}\left(rk_e\frac{\partial\theta_e}{\partial r}\right)$$

$$-\frac{\theta_e-\theta_i}{\tau_{eq}}+\frac{\gamma-1}{\rho}\eta j_\theta^2 \quad ...[2.4]$$

where k_e is the electron thermal conductivity,[11] and η is the electrical resistivity.[10] The above equation includes ohmic heating of the electrons. The equation for the magnetic field is

$$\frac{\partial B_z}{\partial t}+v_r\frac{\partial B_z}{\partial r} = -B_z\frac{1}{r}\frac{\partial}{\partial r}(rv_r)+\frac{1}{r}\frac{\partial}{\partial r}\left(r\eta\frac{\partial B_z}{\partial r}\right). \quad ...[2.5]$$

The boundary and initial conditions for the above equations depend on the experimental situation.[3, 7, 8]

2.3. Difference methods

The above set of nonlinear partial differential equations are solved by finite-difference methods. The first two equations are of hyperbolic type,[12] and contain advective derivatives. The last three equations are of parabolic type,[12] and contain advection and diffusion terms. Different methods are appropriate for the two types of equations.

We can write a hyperbolic system[12] in the form

$$\frac{\partial U}{\partial t}+A\frac{\partial U}{\partial x}+B = 0 \quad ...[2.6]$$

where U and B are m-dimensional column vectors and A is an $m\times m$ matrix. The system [2.6] is called hyperbolic if A has all real eigenvalues and m linearly independent eigenvectors. Since we are mainly interested in the advective terms, we shall consider the simpler system

$$\frac{\partial U}{\partial t} + A\frac{\partial U}{\partial x} = 0. \qquad \qquad ...[2.7]$$

We consider a finite-difference grid with $x_j = j\Delta x$, $t_n = n\Delta t$, where j and n are integers and $U_j^n = U(x_j, t_n)$. The simplest difference approximation to equation [2.7] is

$$U_j^{n+1} = U_j^n - \frac{\Delta t}{2\Delta x} A_j^n(U_{j+1}^n - U_{j-1}^n)$$

which is unstable.[9] A simple alternative which is stable[13] is given by

$$U_j^{n+1} = \tfrac{1}{2}(U_{j+1}^n + U_{j-1}^n) - \frac{\Delta t}{2\Delta x} A_j^n(U_{j+1}^n - U_{j-1}^n). \qquad ...[2.8]$$

The stability condition is $\left|\dfrac{a\Delta t}{\Delta x}\right| < 1$ for all eigenvalues a of A. (The stability criteria given in this discussion are derived[9] assuming that the coefficients are constant, so for our non-linear equations they must be regarded as local conditions which must be tested numerically.) A better scheme is the so-called "upstream-downstream" difference equation,[9] which we give for the scalar equation

$$\frac{\partial u}{\partial t} + a\frac{\partial u}{\partial x} = 0.$$

The difference equations are

$$u_j^{n+1} = u_j^n - \frac{a_j^n \Delta t}{\Delta x} \begin{cases} u_{j+1}^n - u_j^n & \text{if } a_j^n < 0 \\[2ex] u_j^n - u_{j-1}^n & \text{if } a_j^n > 0 \end{cases}$$

The stability condition is again $\left|\dfrac{a\Delta t}{\Delta x}\right| < 1$. This is the scheme used for the advective derivatives in the Hain-Roberts[3] programme. A scheme with higher order accuracy is the "leap-frog" difference equation

$$U_j^{n+1} = U_j^{n-1} - \frac{\Delta t}{\Delta x} A_j^n(U_{j+1}^n - U_{j-1}^n) \qquad ...[2.9]$$

which has the same stability condition $\left|\dfrac{a\Delta t}{\Delta x}\right| < 1$, but has the disadvantage of being a three time-level equation. Another equation which has second-order accuracy, but uses only two time-levels is based on the expansion

$$U_j^{n+1} = U_j^n + \Delta t \left(\frac{\partial U}{\partial t}\right)_j^n + \frac{(\Delta t)^2}{2}\left(\frac{\partial^2 U}{\partial t^2}\right)_j^n.$$

If A is assumed constant, then we have the difference equation

$$U_j^{n+1} = U_j^n - \frac{\Delta t}{2\Delta x} A(U_{j+1}^n - U_{j-1}^n) + \frac{1}{2}\left(\frac{A\Delta t}{\Delta x}\right)^2 (U_{j+1}^n - 2U_j^n + U_{j-1}^n).$$

$$...[2.10]$$

When A is not constant, the above equation can become much more complicated. The condition is $\left|\dfrac{a\Delta t}{\Delta x}\right| < 1.$

It is sometimes possible to write the system [2.7] in conservation form

$$\frac{\partial U}{\partial t} + \frac{\partial F}{\partial x} = 0 \qquad ...[2.11]$$

where F is an m-dimensional column vector. The system of equations [2.1] to [2.5] can be put into this form with certain simplifying assumptions, e.g. slab geometry, equal electron and ion temperature, no heat conduction, and zero electrical resistivity. The second order difference scheme for the system [2.11] can be written in a particularly simple form, which is called the two-step Lax-Wendroff method.[14] It is particularly convenient for generalisation to two dimensions, and has the advantage of conserving the quantity U. On the first step equation [2.8] is used at the mid-points of the grid, i.e.

$$U_{j+\frac{1}{2}}^{n+\frac{1}{2}} = \frac{1}{2}(U_{j+1}^n + U_j^n) - \frac{\Delta t}{2\Delta x}(F_{j+1}^n - F_j^n). \qquad ...[2.12a]$$

The second step is the leap-frog equation

$$U_j^{n+1} = U_j^n - \frac{\Delta t}{\Delta x}(F_{j+\frac{1}{2}}^{n+\frac{1}{2}} - F_{j-\frac{1}{2}}^{n+\frac{1}{2}}). \qquad ...[2.12b]$$

This system reduces to the equation [2.10] when $\partial F/\partial x = A(\partial U/\partial x)$ and A is constant.

Another high-order scheme called the angled-derivative method and the use of a staggered mesh have been discussed by Roberts and Weiss.[15] All of the above difference approximations are explicit and have in common a stability criterion

$$(|v| + c)\frac{\Delta t}{\Delta x} < 1 \qquad ...[2.13]$$

where c is the sound speed in the fluid. For a plasma the relevant speed is the Alfvén speed, and for regions with low density and high magnetic field it can be very high. Hence the calculation could require an unacceptably small time step Δt. To remedy this, Hain and Roberts[3,4] and others[5] have devised implicit schemes which are stable when equation [2.13] is violated. There are two versions of the Hain-Roberts code; one, in which the advection is treated explicitly, using the upstream-downstream difference equation, and the other, the implicit version.

Equations [2.3]–[2.5] are of parabolic type. We can consider [2.5] in the following form

$$\frac{\partial B}{\partial t} + \frac{1}{r} \frac{\partial}{\partial r} (rvB) = \frac{1}{r} \frac{\partial}{\partial r} \left(r\eta \frac{\partial B}{\partial r} \right). \qquad \dots [2.14]$$

It is important[9] to use an implicit difference approximation for the diffusion term in [2.14]; otherwise Δt must satisfy $\eta \Delta t/(\Delta r)^2 \leqslant 1/2$, which is a very severe condition. The equation can be treated by the "splitting" technique,[9] which is also used in multi-dimensional problems. The calculation of each time-step is split into two cycles. On the first cycle the advection equation

$$\frac{\partial B}{\partial t} + \frac{1}{r} \frac{\partial}{\partial r} (rvB) = 0 \qquad \dots [2.15]$$

is solved by one of the explicit difference equations previously described. On the second cycle the diffusion equation

$$\frac{\partial B}{\partial t} = \frac{1}{r} \frac{\partial}{\partial r} \left(r\eta \frac{\partial B}{\partial r} \right) \qquad \dots [2.16]$$

is solved by an implicit difference scheme (which is stable) using the results of the first cycle. This method is used in one of the Hain-Roberts codes.[3,4]

We can also write an implicit difference approximation to the complete equation [2.14]. Let $r_j = j\Delta r$, $t_n = n\Delta t$, $B_j^n = B(r_j, t_n)$, etc. Consider the following equation

$$\begin{aligned}
B_j^{n+1} - B_j^n &= \frac{\theta \Delta t}{(\Delta r)^2 r_j} \left[r_{j+\frac{1}{2}} \eta_{j+\frac{1}{2}}^{n+1} (B_{j+1}^{n+1} - B_j^{n+1}) - r_{j-\frac{1}{2}} \eta_{j-\frac{1}{2}}^{n+1} (B_j^{n+1} - B_{j-1}^{n+1}) \right] \\
&+ \frac{(1-\theta)\Delta t}{(\Delta r)^2 r_j} \left[r_{j+\frac{1}{2}} \eta_{j+\frac{1}{2}}^{n} (B_{j+1}^{n} - B_j^{n}) - r_{j-\frac{1}{2}} \eta_{j-\frac{1}{2}}^{n} (B_j^{n} - B_{j-1}^{n}) \right] \\
&+ \frac{\theta \Delta t}{2 r_j \Delta r} \left[r_{j+1} v_{j+1}^{n+1} B_{j+1}^{n+1} - r_{j-1} v_{j-1}^{n+1} B_{j-1}^{n+1} \right] \\
&+ \frac{(1-\theta)\Delta t}{2 r_j \Delta r} \left[r_{j+1} v_{j+1}^{n} B_{j+1}^{n} - r_{j-1} v_{j-1}^{n} B_{j-1}^{n} \right] \qquad \dots [2.17]
\end{aligned}$$

where θ is a weighting constant; for stability we must have $1/2 \leqslant \theta \leqslant 1$. In the above it is assumed that v_j^{n+1}, η_j^{n+1} have already been computed. The difference approximation for [2.16] would involve only the first two terms on the right-hand side. In either case, we must solve an algebraic system of the form

$$-\alpha_j^n B_{j+1}^{n+1} + \beta_j^n B_j^{n+1} - \gamma_j^n B_{j-1}^{n+1} = \partial_j^n$$

for $j = 1, \dots, J-1$. There is a convenient algorithm[9] for solving tridiagonal matrix equations of this type. In the above equation, if advection dominates diffusion, the hydromagnetic stability condition on Δt may still have to be satisfied.

The equations for the ion and electron temperatures [2.3] and [2.4] are treated in exactly the same manner as the field equation above.

2.4. *One-dimensional, partially ionized model*

The one-dimensional model described in Section 2.2 has been generalised to include neutral gas by Roberts[16] and Duchs.[17, 18] It is called a three-fluid model, as there are three energy equations—for electron, ion, and neutral temperatures; there is an additional continuity equation and equation of motion for the neutrals. Ionization, excitation, charge exchange, and frictional heating are included. In one version[18] the effects of impurity atoms are included. (Impurity effects have also been added to the two-fluid model.[19]) The numerical methods for the three-fluid model are essentially the same as those we have given in the preceding section.

An earlier calculation[20] treated processes that occur prior to the implosion of a deuterium plasma in a linear pinch, i.e. mass motion is neglected, but the ionization, ohmic heating, and current diffusion are computed. The equations are

$$\frac{\partial j}{\partial t} = \frac{\partial^2 (\eta j)}{\partial y^2}$$

$$\frac{\partial \theta_e}{\partial t} = -\frac{\partial \theta_i}{\partial t} - \left(\frac{2\alpha}{3} + \theta_e + \theta_i\right)\frac{1}{f}\frac{\partial f}{\partial t} + A_1 \frac{1}{f}\, \eta j^2$$

$$\eta = A_2 \left(\frac{1}{f} - 1\right) \theta_e^{\frac{1}{2}} + A_3 \theta_e^{-\frac{3}{2}}$$

$$\frac{\partial f}{\partial t} = A_4 f(1-f)\, G(\theta_e)$$

$$\frac{\partial \theta_i}{\partial t} = A_5 f \theta_e^{-\frac{3}{2}} (\theta_e - \theta_i)$$

where j is the current density, η the electrical resistivity, θ_e and θ_i are the electron and ion temperatures, and f is the fraction of ionization. They are all functions of y and t; and α, A_1, A_2, A_3, A_4, and A_5 are constants. The function $G(\theta_e)$ is a fit to the ionization cross-section. The above set of differential equations was solved by finite-difference methods. The implicit scheme for the first equation above is the same as that given by equation [2.17]. It remained stable numerically, although the resistivity η changed by several orders of magnitude throughout the calculation. This programme was applied to linear pinches at Livermore and to the Zeta experiment at Harwell.

2.5. *Two-dimensional, fully ionized model*

In order to give a better description of many of the experiments in CTR programmes, a two-dimensional model is needed. We consider a cylindrical system with azimuthal symmetry. The variables are then functions of r, z, and t. Because of the success of the earlier calculations in describing certain experiments, we shall again consider the hydromagnetic model. For simplicity of discussion, we shall consider a one-fluid model (equal electron and ion temperatures) without thermal conduction. The velocity has components v_r and v_z, and the magnetic field has components B_r and B_z. This is the model described by Roberts et al.[21] in an early discussion of this problem. It is suitable for analysing theta pinches, conical plasma guns, and other cylindrical magnetic compression experiments. More recent descriptions of this model have been given at meetings.[22,23] The use of computers for two-dimensional magnetohydrodynamic calculations is well established [24] The use of other than hydromagnetic models for describing theta pinches will be discussed in a later section.

The basic equations that we shall consider are

$$\frac{\partial \rho}{\partial t} + (v \cdot \nabla)\rho = -\rho \, \text{div} \, v \qquad \qquad ...[2.18]$$

$$\rho \left(\frac{\partial v}{\partial t} + (v \cdot \nabla)v \right) = j \times B - \nabla(p+q) \qquad \qquad ...[2.19]$$

$$\frac{\partial p}{\partial t} + (v \cdot \nabla)p = -\gamma(p+q) \, \text{div} \, v + (\gamma-1)\eta j \cdot j \qquad ...[2.20]$$

$$\frac{\partial B}{\partial t} + (v \cdot \nabla)B = -B \, \text{div} \, v + (B \cdot \nabla)v - \text{curl} \, \eta j \qquad ...[2.21]$$

where p is the plasma pressure and q is the artificial viscosity. The current density is $j = (0, j_\theta, 0)$ with

$$j_\theta = \frac{\partial B_r}{\partial z} - \frac{\partial B_z}{\partial r}.$$

The nature of q in two-dimensional calculations is discussed by Schultz.[25]

The equations $[2.18]$–$[2.20]$ can be written in terms of cylindrical coordinates r and z as the following hyperbolic system

$$\frac{\partial U}{\partial t} + A\frac{\partial U}{\partial r} + B\frac{\partial U}{\partial z} + F = 0, \qquad \qquad ...[2.22]$$

where

$$U = \begin{bmatrix} \rho \\ v_r \\ v_z \\ p \end{bmatrix}; \qquad A = \begin{bmatrix} v_r & \rho & 0 & 0 \\ 0 & v_r & 0 & 1/\rho \\ 0 & 0 & v_r & 0 \\ 0 & \gamma p & 0 & v_r \end{bmatrix}$$

$$B = \begin{bmatrix} v_z & 0 & \rho & 0 \\ 0 & v_z & 0 & 0 \\ 0 & 0 & v_z & 1/\rho \\ 0 & 0 & \gamma p & v_z \end{bmatrix}; \qquad F = \begin{bmatrix} \rho(v_r/r) \\ -(B_z/\rho)j_\theta \\ (B_r/\rho)j_\theta \\ \gamma p(v_r/r)-(\gamma-1)\eta j_\theta^2 \end{bmatrix}.$$

The difference schemes discussed in Section [2.3] for one-dimensional hyperbolic systems can be generalized with some modifications to equation [2.22]. The simplest explicit scheme would be to rearrange the equations so that A and B are diagonal, with v_r and v_z as the diagonal elements, and then use the "upstream-downstream" differencing scheme. Let $z_i = i\Delta z$, $r = j\Delta r$, and $t_n = n\Delta t$; then

$$U_{i,j}^{n+1} = U_{i,j}^n - \frac{A_{i,j}'^n \Delta t}{\Delta r} \left\{ \begin{array}{ll} U_{i,j+1}^n - U_{i,j}^n & \text{if } (v_r)_{i,j}^n < 0 \\ \\ U_{i,j}^n - U_{i,j-1}^n & \text{if } (v_r)_{i,j}^n > 0 \end{array} \right.$$

$$-\Delta t F_{i,j}'^n - \frac{B_{i,j}'^n \Delta t}{\Delta z} \left\{ \begin{array}{ll} U_{i+1,j}^n - U_{i,j}^n & \text{if } (v_z)_{i,j}^n < 0 \\ \\ U_{i,j}^n - U_{i-1,j}^n & \text{if } (v_z)_{i,j}^n > 0 \end{array} \right.$$

where A' and B' are diagonal and F' is suitably modified from F. This scheme is effective for short time calculations such as the implosive phase of a theta pinch or the expansion of a plasma across a magnetic field. It has been used successfully in two-dimensional Eulerian hydrodynamic codes.[26] For longer-time integrations a higher-order scheme may be advisable, such as the angled-derivative and staggered mesh schemes discussed by Roberts and Weiss.[15, 22] Another scheme is the generalization of equation [2.10]. To make this stable, requires[27] a splitting technique, i.e. the computation is split into two cycles with r advection on one cycle and z advection on the second cycle. Since the matrices A, B are not constant, to carry this out to second order requires rather complicated difference equations.

The equations for the field components B_r and B_z in cylindrical coordinates r and z are given by

$$\frac{\partial B_r}{\partial t} + v_r \frac{\partial B_r}{\partial r} + v_z \frac{\partial B_r}{\partial z} = -B_r \left(\frac{v_r}{r} + \frac{\partial v_z}{\partial z} \right) + B_z \frac{\partial v_r}{\partial z} + \frac{\partial \eta}{\partial z} \left(\frac{\partial B_r}{\partial z} - \frac{\partial B_z}{\partial r} \right)$$
$$+ \eta \left[\frac{1}{r} \frac{\partial}{\partial r} \left(r \frac{\partial B_r}{\partial r} \right) + \frac{\partial^2 B_r}{\partial z^2} - \frac{B_r}{r^2} \right]$$

$$\frac{\partial B_z}{\partial t} + v_r \frac{\partial B_z}{\partial r} + v_z \frac{\partial B_z}{\partial z} = -B_z \left(\frac{\partial v_r}{\partial r} + \frac{v_r}{r} \right) + B_r \frac{\partial v_z}{\partial r} - \frac{\partial \eta}{\partial r} \left(\frac{\partial B_r}{\partial z} - \frac{\partial B_z}{\partial r} \right)$$
$$+ \eta \left[\frac{1}{r} \frac{\partial}{\partial r} \left(r \frac{\partial B_z}{\partial r} \right) + \frac{\partial^2 B_z}{\partial z^2} \right].$$

We see that these equations are of parabolic type. In order to avoid a severe time-step restriction, the most suitable method of solution is the alternating-direction implicit difference scheme.[9] For this purpose it is convenient to write the above system in the form

$$\frac{\partial B}{\partial t} = a\,\frac{\partial^2 B}{\partial r^2} + c\,\frac{\partial^2 B}{\partial z^2} + d\,\frac{\partial B}{\partial r} + e\,\frac{\partial B}{\partial z} + fB, \qquad \dots[2.23]$$

where

$$B = \begin{bmatrix} B_r \\ B_z \end{bmatrix}; \qquad\qquad a = c = \begin{bmatrix} \eta & 0 \\ 0 & \eta \end{bmatrix}$$

$$d = \begin{bmatrix} \left(\dfrac{\eta}{r} - v_r\right) & -\dfrac{\partial \eta}{\partial z} \\[2ex] 0 & \left(\dfrac{\eta}{r} + \dfrac{\partial \eta}{\partial r} - v_r\right) \end{bmatrix}; \qquad e = \begin{bmatrix} \left(\dfrac{\partial \eta}{\partial z} - v_z\right) & 0 \\[2ex] -\dfrac{\partial \eta}{\partial r} & -v_z \end{bmatrix}$$

$$f = \begin{bmatrix} -\left(\dfrac{v_r}{r} + \dfrac{\partial v_z}{\partial z} + \dfrac{\eta}{r^2}\right) & \dfrac{\partial v_r}{\partial z} \\[2ex] \dfrac{\partial v_z}{\partial r} & -\left(\dfrac{\partial v_r}{\partial r} + \dfrac{v_r}{r}\right) \end{bmatrix}.$$

In the computational sequence we calculate ρ, v_r, v_z, p at the new time step using equation [2.22]. Hence we know v_r, v_z, η and their derivatives at time t_{n+1} (η is a function of ρ and p), i.e. all the coefficients in equation [2.23] are known at t_n and t_{n+1}. In the ADI method we treat r derivatives implicitly on one time-step and the z derivatives implicitly on the next step. The difference equation for n even, is

$$\frac{B_{i,j}^{n+1} - B_{i,j}^{n}}{\Delta t} = a_{i,j}^{n+1}\,\frac{B_{i,j+1}^{n+1} - 2B_{i,j}^{n+1} + B_{i,j-1}^{n+1}}{(\Delta r)^2} + c_{i,j}^{n}\,\frac{B_{i+1,j}^{n} - 2B_{i,j}^{n} + B_{i-1,j}^{n}}{(\Delta z)^2}$$

$$+ d_{i,j}^{n+1}\,\frac{B_{i,j+1}^{n+1} - B_{i,j-1}^{n+1}}{2\Delta r} + e_{i,j}^{n}\,\frac{B_{i+1,j}^{n} - B_{i-1,j}^{n}}{2\Delta z}$$

$$+ \tfrac{1}{2}[f_{i,j}^{n+1}B_{i,j}^{n+1} + f_{i,j}^{n}B_{i,j}^{n}].$$

The difference equation for the next time step (n odd) is

$$\frac{B_{i,j}^{n+1} - B_{i,j}^{n}}{\Delta t} = a_{i,j}^{n}\,\frac{B_{i,j+1}^{n} - 2B_{i,j}^{n} + B_{i,j-1}^{n}}{(\Delta r)^2} + c_{i,j}^{n+1}\,\frac{B_{i+1,j}^{n+1} - 2B_{i,j}^{n+1} + B_{i-1,j}^{n+1}}{(\Delta z)^2}$$

$$+ d_{i,j}^{n}\,\frac{B_{i,j+1}^{n} - B_{i,j-1}^{n}}{2\Delta r} + e_{i,j}^{n+1}\,\frac{B_{i+1,j}^{n+1} - B_{i-1,j}^{n+1}}{2\Delta z}$$

$$+ \tfrac{1}{2}[f_{i,j}^{n+1}B_{i,j}^{n+1} + f_{i,j}^{n}B_{i,j}^{n}].$$

The above difference approximations lead to the following algebraic systems to be solved—for n even

$$-(\alpha_e)_{i,j}^{n+1}B_{i,j+1}^{n+1} + (\beta_e)_{i,j}^{n+1}B_{i,j}^{n+1} - (\gamma_e)_{i,j}^{n+1}B_{i,j-1}^{n+1} = (\delta_e)_{i,j}^{n}$$

and for n odd

$$-(\alpha_0)^{n+1}_{i,j} B^{n+1}_{i+1,j} + (\beta_0)^{n+1}_{i,j} B^{n+1}_{i,j} - (\gamma_0)^{n+1}_{i,j} B^{n+1}_{i-1,j}, = (\delta_0)^n_{i,j}.$$

In the first equation we have a tri-diagonal matrix equation to solve for each value of i, and in the second equation we have a similar problem for each j. Hence we can use the algorithm[9] mentioned earlier for one-dimensional diffusion problems, but in this case the unknowns are vectors and the coefficients are 2×2 matrices. We shall give further details and applications of this method in Sections 3, 4, and 5.

One of the main problems in using a programme of this type is the proper choice of boundary conditions on plasma and field variables. Roberts[21,22] has considered these questions for the case of the theta pinch. An important advantage of the alternating-direction implicit method for the calculation of the magnetic field is that the same set of equations can be solved in a plasma-free region surrounding the plasma. Only the coefficients are different in the two regions. We shall consider this problem again with reference to equilibria calculations in Section 4.

3

RESISTIVE INSTABILITY CALCULATIONS

3.1. *Basic equations and assumptions*

Experimental observations in pinch and stellarator research showed instabilities of configurations that the infinite-conductivity hydromagnetic theory would predict to be stable. In order to identify these instabilities, finite conductivity was included in the hydromagnetic analysis by Furth, Killeen, and Rosenbluth.[28] In that paper general equations are derived for the plane resistive current layer in the incompressible hydromagnetic approximation, and perturbations of the form

$$f_1(y) e^{i(k_x x + k_z z) + \omega t}$$

are assumed. A dispersion relation is obtained and the problem is to solve an eigenvalue problem for ω, the growth rate of the instability. In order to solve the problem the plasma is divided into two regions, a narrow inner region about the plane for which the wave vector is perpendicular to the zero-order magnetic field ($\mathbf{k} \cdot \mathbf{B}_0 = 0$) and an outer region where the infinite conductivity equations hold. Purely growing modes of the "tearing", "rippling", and gravitational types are identified.

In this treatment we assume an arbitrary time-dependence and the problem becomes an initial-value problem for the perturbed quantities. We use the same basic equations and assumptions as Ref. 28, but two regions are not used, i.e. the same equations hold throughout the plasma. The initial-value problems are solved numerically. This method of solution was developed[29] simultaneously with the analytic technique.

We assume that the hydromagnetic approximation is valid, and the ion pressure and inertia terms are neglected in Ohm's law. An isotropic resistivity is assumed, and the fluid is assumed to be incompressible. Perturbations in resistivity and the gravitational term result only from convection. The gravitational term may be interpreted as resulting from acceleration of the current layer, or the effect of field line curvature. The basic equations are

$$\frac{\partial \boldsymbol{B}}{\partial t} = \text{curl}\,(\boldsymbol{v} \times \boldsymbol{B}) - \frac{1}{4\pi}\,\text{curl}\,(\eta\,\text{curl}\,\boldsymbol{B}) \qquad \ldots[3.1]$$

$$\text{div}\,\boldsymbol{B} = 0 \qquad\qquad \text{div}\,\boldsymbol{v} = 0 \qquad \ldots[3.2]$$

$$\text{curl}\left(\rho\,\frac{d\boldsymbol{v}}{dt}\right) = \text{curl}\left(\frac{1}{4\pi}\,\text{curl}\,\boldsymbol{B} \times \boldsymbol{B} + \rho\boldsymbol{g}\right) \qquad \ldots[3.3]$$

$$\frac{\partial \eta}{\partial t} + \boldsymbol{v}\cdot\nabla\eta = 0 \qquad \ldots[3\,4]$$

$$\frac{\partial(\rho\boldsymbol{g})}{\partial t} + \boldsymbol{v}\cdot\nabla(\rho\boldsymbol{g}) = 0. \qquad \ldots[3.5]$$

The plasma equilibrium is specified by a given function $\boldsymbol{B}_0$ and $\boldsymbol{v}_0 = 0$.

3.2. First-order equations

We denote perturbed quantities by the subscript 1. We consider the linearized equations for the first order variables

$$\frac{\partial \boldsymbol{B}_1}{\partial t} = \text{curl}\,(\boldsymbol{v}_1 \times \boldsymbol{B}_0) - \frac{1}{4\pi}\,\text{curl}\,(\eta_0\,\text{curl}\,\boldsymbol{B}_1 + \eta_1\,\text{curl}\,\boldsymbol{B}_0)$$

$$= (\boldsymbol{B}_0\cdot\nabla)\boldsymbol{v}_1 - (\boldsymbol{v}_1\cdot\nabla)\boldsymbol{B}_0 + \frac{1}{4\pi}\,(\eta_0\nabla^2\boldsymbol{B}_1 + \eta_1\nabla^2\boldsymbol{B}_0$$

$$+ \text{curl}\,\boldsymbol{B}_1 \times \nabla\eta_0 + \text{curl}\,\boldsymbol{B}_0 \times \nabla\eta_1). \qquad \ldots[3.6]$$

With $\rho_0 = $ constant, the first order equation of motion is

$$-\rho_0\,\frac{\partial}{\partial t}\,(\nabla^2\boldsymbol{v}_1) = \text{curl}\,\text{curl}\left\{\frac{1}{4\pi}\,[(\boldsymbol{B}_0\cdot\nabla)\boldsymbol{B}_1 + (\boldsymbol{B}_1\cdot\nabla)\boldsymbol{B}_0] + (\rho\boldsymbol{g})_1\right\}.$$

$$\ldots[3.7]$$

These equations make use of

$$\text{div}\,\boldsymbol{B}_1 = 0 \qquad\qquad \text{div}\,\boldsymbol{v}_1 = 0. \qquad \ldots[3.8]$$

The perturbed resistivity and gravitational term are given by

$$\frac{\partial \eta_1}{\partial t} + (\boldsymbol{v}_1\cdot\nabla)\eta_0 = 0 \qquad \ldots[3.9]$$

$$\frac{\partial(\rho\boldsymbol{g})_1}{\partial t} + (\boldsymbol{v}_1\cdot\nabla)(\rho\boldsymbol{g})_0 = 0. \qquad \ldots[3.10]$$

3.3. *Sheet pinch model*

This model is equivalent to that considered in Ref. 28—a plane resistive current layer. We use cartesian coordinates and assume perturbations of the form

$$f_1(y, t)e^{i(k_x x + k_z z)}.$$

From the above equations we can separate out a pair of equations for B_{y1} and v_{y1}. The equation for B_{y1} is

$$\frac{\partial B_{y1}}{\partial t} = iv_{y1}(k_x B_{x0} + k_z B_{z0}) + \frac{\eta_0}{4\pi}\left[\frac{\partial^2 B_{y1}}{\partial y^2} - (k_x^2 + k_z^2)B_{y1}\right]$$
$$-\frac{i\eta_1}{4\pi}\frac{\partial}{\partial y}(k_x B_{x0} + k_z B_{z0}). \qquad ...[3.11]$$

The equation for v_{y1} is

$$-4\pi\rho_0 i\frac{\partial}{\partial t}\left[\frac{\partial^2 v_{y1}}{\partial y^2} - (k_x^2 + k_z^2)v_{y1}\right] = (k_x B_{x0} + k_z B_{z0})\left[\frac{\partial^2 B_{y1}}{\partial y^2} - (k_x^2 + k_z^2)B_{y1}\right]$$
$$- B_{y1}\frac{\partial}{\partial y^2}(k_x B_{x0} + k_z B_{z0}) + 4\pi i(k_x^2 + k_z^2)(\rho g)_{y1}. \qquad ...[3.12]$$

We also have

$$\frac{\partial \eta_1}{\partial t} + v_{y1}\frac{\partial \eta_0}{\partial y} = 0 \qquad ...[3.13]$$

$$\frac{\partial(\rho g)_{y1}}{\partial t} + v_{y1}\frac{\partial(\rho g)_{y0}}{\partial y} = 0. \qquad ...[3.14]$$

The plasma equilibrium is given by

$$\mathbf{B}_0 = \hat{x}B_{x0}(y) + \hat{z}B_{z0}(y); \quad \mathbf{v}_0 = 0. \qquad ...[3.15]$$

The two relevant times for the problem are the resistive diffusion time

$$\tau_R = \frac{4\pi a^2}{\langle\eta\rangle} \qquad ...[3.16]$$

and the hydromagnetic transit time,

$$\tau_H = \frac{a(4\pi\langle\rho\rangle)^{\frac{1}{2}}}{B}, \qquad ...[3.17]$$

where a is a measure of the thickness of the current layer, and B, $\langle\eta\rangle$, $\langle\rho\rangle$ are measures of the field strength, resistivity and mass density respectively. We define the following parameters

$$S = \frac{\tau_R}{\tau_H} \qquad\qquad \alpha = ka \qquad\qquad ...[2.18]$$

where $k = (k_x^2 + k_z^2)^{\frac{1}{2}}$.

We define the dimensionless variables

$$\mu = \frac{y}{a} \qquad\qquad \tau = \frac{t}{\tau_R}$$

$$\psi = \frac{B_{y1}}{B} \qquad\qquad w = -iv_{y1}k\tau_R$$

$$\theta = -\frac{\dot{\eta}_1}{\langle\eta\rangle} \qquad\qquad \gamma = ik\tau_H^2 \frac{(\rho g)_{y1}}{\langle\rho\rangle}$$

$$\rho = \frac{\rho_0}{\langle\rho\rangle} \qquad\qquad G = -\frac{\tau_H^2}{\langle\rho\rangle}\frac{\partial(\rho g)_{y0}}{\partial y}.$$

The following functions characterize the equilibrium:

$$F = \frac{1}{kB}(k_x B_{x0} + k_z B_{z0})$$

$$F' = \frac{dF}{d\mu} \qquad\qquad F'' = \frac{d^2F}{d\mu^2}$$

$$\eta = \frac{\eta_0}{\langle\eta\rangle} \qquad\qquad \eta' = \frac{d\eta}{d\mu}.$$

We can now give the four equations to be solved:

$$\frac{\partial\psi}{\partial\tau} = -Fw + \eta\left(\frac{\partial^2\psi}{\partial\mu^2} - \alpha^2\psi\right) + \alpha F'\theta \qquad\qquad ...[3.19]$$

$$\frac{\rho}{\alpha^2 S^2}\frac{\partial}{\partial\tau}\left(\frac{\partial^2 w}{\partial\mu^2} - \alpha^2 w\right) = F\left(\frac{\partial^2\psi}{\partial\mu^2} - \alpha^2\psi\right) - F''\psi + \gamma \qquad\qquad ..[3.20]$$

$$\frac{\partial\theta}{\partial\tau} = -\frac{1}{\alpha}\eta'w \qquad\qquad ...[3.21]$$

$$\frac{\partial\gamma}{\partial\tau} = -Gw. \qquad\qquad ...[3.22]$$

In the above set of equations if we set θ, η', γ, G equal to zero, we solve only [3.19] and [3.20] and can obtain[29] the pure tearing mode at $F = 0$ for $\alpha < 1$. We can have either symmetric tearing ($F = 0$ at $\mu = 0$) or asymmetric tearing ($F = 0$ at $\mu \neq 0$). This version of the programme is called RIPPLE I. If we include zero-order resistivity gradients (RIPPLE II), then we solve equations [3.19]–[3.21] and can obtain the rippling mode for $\alpha \geqslant 1$ as long as $F = 0$ at $\mu \neq 0$. If we solve the complete set of equations (RIPPLE III), we can obtain gravitational modes for $G > 0$, $\alpha \geqslant 1$.

3.4. *Difference equations for the sheet pinch*

We wish to solve the initial value problem given by equations [3.19]–[3.22]. *Note*: In the following sections, the dimensionless variables μ, τ introduced in the last section are now called y, t, also $\rho = 1$.

We consider two different models which lead to different sets of boundary conditions. If we are interested in the domain, $-\infty < y < \infty$, then for the finite difference domain we take

$$-y_{max} \leqslant y \leqslant y_{max}, \qquad y_{max} > 0,$$

where $y_{max} \gg 1/\alpha$, and we impose the following conditions:

$$\text{at} \quad y = -y_{max}: \qquad \frac{\partial \psi}{\partial y} - \alpha \psi = 0 \quad \text{and} \quad \frac{\partial w}{\partial y} - \alpha w = 0$$

$$\qquad \qquad \qquad \qquad \qquad \qquad \qquad \qquad \qquad \qquad \qquad \qquad \qquad \qquad \qquad \qquad ...[3.23]$$

$$y = y_{max}: \qquad \frac{\partial \psi}{\partial y} + \alpha \psi = 0 \quad \text{and} \quad \frac{\partial w}{\partial y} + \alpha w = 0.$$

If we are interested in a finite domain with conducting walls then

$$\text{at} \quad y = -y_{max} \qquad \psi = w = 0$$

$$\qquad \qquad \qquad \qquad \qquad \qquad \qquad \qquad \qquad \qquad \qquad \qquad \qquad ...[3.24]$$

$$\text{at} \quad y = y_{max} \qquad \psi = w = 0.$$

Consider the finite-difference mesh

$$t^n = n\Delta t \qquad \qquad n = 0, 1, 2, \ldots$$

$$y_j = j\Delta y \qquad \qquad j = -J, -J+1, \ldots, 0, \ldots, J-1, J$$

$$y_{max} = y_J = J\Delta y \qquad J > 0.$$

Let $(\delta^2 \psi)_j^n = (\psi_{j+1}^n - 2\psi_j^n + \psi_{j-1}^n)/(\Delta y)^2$. The difference equations corresponding to [3.19] and [3.20] are

$$\frac{\psi_j^{n+1} - \psi_j^n}{\Delta t} = -\tfrac{1}{2}F_j(w_j^{n+1} + w_j^n) + \alpha F_j' \theta_j^{n+\frac{1}{2}} + \tfrac{1}{2}\eta_j[(\delta^2 \psi)_j^{n+1} - \alpha^2 \psi_j^{n+1}$$

$$+ (\delta^2 \psi)_j^n - \alpha^2 \psi_j^n] \qquad ...[3.25]$$

and

$$\frac{1}{\alpha^2 S^2 \Delta t}[(\delta^2 w)_j^{n+1} - \alpha^2 w_j^{n+1} - (\delta^2 w)_j^n + \alpha^2 w_j^n]$$

$$= \tfrac{1}{2}F_j[(\delta^2 \psi)_j^{n+1} - \alpha^2 \psi_j^{n+1} + (\delta^2 \psi)_j^n - \alpha^2 \psi_j^n]$$

$$- \tfrac{1}{2}F_j''(\psi_j^{n+1} + \psi_j^n) + \gamma_j^{n+\frac{1}{2}}. \qquad ...[3.26]$$

We can write these two equations as a single vector equation for the unknown U_j^{n+1}

$$-A_j U_{j+1}^{n+1} + B_j U_j^{n+1} - C_j U_{j-1}^{n+1} = d_j^n \qquad ...[3.27]$$

where

$$U_j^n = \begin{bmatrix} \psi_j^n \\ w_j^n \end{bmatrix} \qquad\qquad d_j^n = \begin{bmatrix} d_j^n \\ k_j^n \end{bmatrix}$$

$$A_j = C_j = \begin{bmatrix} a_j & 0 \\ e_j & -1 \end{bmatrix} \qquad B_j = \begin{bmatrix} b_j & c_j \\ f_j & -h \end{bmatrix}.$$

The elements a, b, c, e, f, and h are derived by rearranging equations [3.25] and [3.26] and are known functions of y. The elements d and k are functions of ψ_j^n, w_j^n, $\theta_j^{n+\frac{1}{2}}$, $\gamma_j^{n+\frac{1}{2}}$. We also have

$$\theta_j^{n+\frac{1}{2}} = \theta_j^{n-\frac{1}{2}} - \frac{\Delta t}{\alpha}\, \eta_j' w_j^n \qquad\qquad ...[3.28]$$

$$\gamma_j^{n+\frac{1}{2}} = \gamma_j^{n-\frac{1}{2}} - \Delta t G_j w_j^n. \qquad\qquad ...[3.29]$$

In order to solve the equation [3.27] we use the algorithm

$$U_j^{n+1} = E_j U_{j+1}^{n+1} + f_j^n. \qquad\qquad ...[3.30]$$

The E_j and f_j^n are determined from the recurrence relations

$$E_j = (B_j - C_j E_{j-1})^{-1} A_j$$

$$f_j^n = (B_j - C_j E_{j-1})^{-1} (d_j^n + C_j f_{j-1}^n). \qquad\qquad ...[3.31]$$

In order to obtain E_{-J}, f_{-J}^n we use the boundary conditions at $y = y_{-J}$. With the first set of boundary conditions [3.23] we can write

$$\psi_{-J}^{n+1} = (1 + \alpha \Delta y)^{-1} \psi_{-J+1}^{n+1}$$

$$w_{-J}^{n+1} = (1 + \alpha \Delta y)^{-1} w_{-J+1}^{n+1}$$

which gives

$$E_{-J} = \begin{bmatrix} \lambda & 0 \\ 0 & \lambda \end{bmatrix}, \qquad f_{-J}^n = \begin{bmatrix} 0 \\ 0 \end{bmatrix}.$$

where $\lambda = (1 + \alpha \Delta y)^{-1}$.

With the second set of boundary conditions [3.24]

$$E_{-J} = \begin{bmatrix} 0 & 0 \\ 0 & 0 \end{bmatrix} \qquad f_{-J}^n = \begin{bmatrix} 0 \\ 0 \end{bmatrix}.$$

We now consider the first set of boundary conditions [3.23] at $y = y_J$. In difference form they are

$$\psi_J^{n+1}(1+\alpha\Delta y)-\psi_{J-1}^{n+1} = 0$$

$$w_J^{n+1}(1+\alpha\Delta y)-w_{J-1}^{n+1} = 0.$$

We have, from [3.30],

$$U_{J-1}^{n+1} = E_{J-1}U_J^{n+1}+f_{J-1}^n,$$

where E_{J-1} and f_{J-1}^n are known from [3.31]. We can write this as

$$\begin{bmatrix} \psi_{J-1}^{n+1} \\ w_{J-1}^{n+1} \end{bmatrix} = \begin{bmatrix} E_{J-1}^{11} & E_{J-1}^{12} \\ E_{J-1}^{21} & E_{J-1}^{22} \end{bmatrix} \begin{bmatrix} \lambda\psi_{J-1}^{n+1} \\ \lambda w_{J-1}^{n+1} \end{bmatrix} + \begin{bmatrix} {}^{(1)}f_{J-1}^n \\ {}^{(2)}f_{J-1}^n \end{bmatrix}$$

Hence

$$\psi_{J-1}^{n+1} = \frac{(1-\lambda E^{22})^{-1}E^{12}f^{(2)}+\lambda^{-1}f^{(1)}}{\lambda^{-1}-E^{11}-E^{12}E^{21}(1-\lambda E^{22})^{-1}\lambda}$$

$$w_{J-1}^{n+1} = \frac{\lambda E^{21}\psi_{J-1}^{n+1}+f^{(2)}}{1-\lambda E^{22}}$$

where $E^{11} = E_{J-1}^{11}$, etc. If we are using the second set of boundary conditions [3.24], we have

$$\psi_{J-1}^{n+1} = {}^{(1)}f_{J-1}^n$$

$$w_{J-1}^{n+1} = {}^{(2)}f_{J-1}^n \qquad \text{or simply} \qquad U_{J-1}^{n+1} = f_{J-1}^n.$$

In order to use this programme, we must specify the functions

$$F, F', F'', \eta, \eta'.$$

A model that is in the programme is

$$F(y) = \text{constant}+\tanh y$$

$$F'(y) = \text{sech}^2 y$$

$$F''(y) = -2 \,\text{sech}^2 y \tanh y$$

$$\eta(y) = \cosh^2 y$$

$$\eta'(y) = 2 \cosh y \sinh y.$$

This model satisfies $\eta F' = 1$ which follows from

$$\text{curl}\,(\eta_0 \,\text{curl}\,\boldsymbol{B}_0) = 0.$$

The initial conditions in the present version are $\psi = \theta = \gamma = 0$ and $w(y, 0) = -ye^{-\alpha|y|}$. By this method, time-dependent solutions are obtained for specific choices of F, η, and G. The boundary conditions can be varied, and there is no restriction on the values of α and S, i.e. there is no difference in the method for low S and high S except the choice of time step Δt.

We can exhibit the three basic modes described in Ref. 28, and we can also consider the "mixed" modes which are discussed in that paper. With a programme of this type, a great variety of possibilities can be studied. We can vary the parameters, initial conditions, and modes of calculation, i.e. with and without G, or with and without η'. In order to obtain a growth rate p from the RIPPLE III results, we run a case until it settles down to an exponential growth and then compute

$$p = \frac{(\partial \psi / \partial \tau)}{\psi}. \qquad \qquad ...[3.32]$$

This ratio is calculated at a particular value of y corresponding to ψ_{max}.

In performing these calculations it is important that the time step be chosen so that $p\Delta t \ll 1$. This is not a numerical stability requirement, but is necessary in order to give accurate values of p. In most cases the Δt should be varied to check this and also the total number of time steps needed.

3.5. *Cylindrical model*

The hydromagnetic model that has been used to study the resistive instabilities of a sheet pinch is applied in cylindrical geometry in order to study specific pinch and hard-core devices.[30]

The plasma equilibrium configuration is assumed to be known, and is given by

$$B_0 = \hat{\theta} B_{\theta 0}(r) + \hat{z} B_{z0}(r)$$

$$v_0 = 0 \qquad \eta_0 = \eta_0(r).$$

These functions are chosen to describe a particular experiment. With this programme a given zero-order magnetic field configuration can be tested for resistive instability. The stabilizing effect of the location of the conducting walls with reference to the current layer can be determined.

We assume perturbations of the form

$$f_1(r, t) e^{i(m\theta + k_z z)}.$$

From the equations for B_1 and v_1 we can find a consistent set of equations involving the components B_{r1}, $B_{\theta 1}$, v_{r1}, $v_{\theta 1}$. The equation for B_{r1} is

$$\frac{\partial B_{r1}}{\partial t} = iv_{r1} \left(\frac{m}{r} B_{\theta 0} + k_z B_{z0} \right)$$

$$+ \frac{\eta_0}{4\pi} \left[\frac{\partial^2 B_{r1}}{\partial r^2} + \frac{1}{r} \frac{\partial B_{r1}}{\partial r} - \left(\frac{m^2 + 1}{r^2} + k_z^2 \right) B_{r1} - \frac{2im}{r^2} B_{\theta 1} \right]$$

$$- \frac{i\eta_1}{4\pi} \left[\frac{m}{r^2} \frac{\partial}{\partial r} (rB_{\theta 0}) + k_z \frac{\partial B_{z0}}{\partial r} \right]. \qquad ...[3.33]$$

The equation for $B_{\theta 1}$ becomes

$$\frac{\partial B_{\theta 1}}{\partial t} = iv_{\theta 1}\left(\frac{m}{r}\,B_{\theta 0}+k_z B_{z0}\right)+v_{r1}\left(\frac{B_{\theta 0}}{r}-\frac{\partial B_{\theta 0}}{\partial r}\right)$$

$$+\frac{\eta_0}{4\pi}\left[\frac{\partial^2 B_{\theta 1}}{\partial r^2}+\frac{1}{r}\frac{\partial B_{\theta 1}}{\partial r}-\left(\frac{m^2+1}{r^2}+k_z^2\right)B_{\theta 1}+\frac{2im}{r^2}\,B_{r1}\right]$$

$$+\frac{1}{4\pi}\frac{\partial \eta_0}{\partial r}\left(\frac{\partial B_{\theta 1}}{\partial r}+\frac{B_{\theta 1}}{r}-\frac{im}{r}\,B_{r1}\right)$$

$$+\frac{\eta_1}{4\pi}\left(\frac{\partial^2 B_{\theta 0}}{\partial r^2}+\frac{1}{r}\frac{\partial B_{\theta 0}}{\partial r}-\frac{B_{\theta 0}}{r^2}\right)+\frac{1}{4\pi}\frac{\partial \eta_1}{\partial r}\left[\frac{1}{r}\frac{\partial}{\partial r}(rB_{\theta 0})\right]. \quad ...[3.34]$$

We consider only the radial component of the gravitational term. The equation for v_{r1} is

$$-4\pi\rho_0 i\,\frac{\partial}{\partial t}\left[\frac{\partial^2 v_{r1}}{\partial r^2}+\frac{1}{r}\frac{\partial v_{r1}}{\partial r}-\left(\frac{m^2+1}{r^2}+k_z^2\right)v_{r1}-\frac{2im}{r^2}\,v_{\theta 1}\right]$$

$$=\left(\frac{m}{r}\,B_{\theta 0}+k_z B_{z0}\right)\left[\frac{\partial^2 B_{r1}}{\partial r^2}+\frac{1}{r}\frac{\partial B_{r1}}{\partial r}-\left(\frac{m^2+1}{r^2}+k_z^2\right)B_{r1}-\frac{2im}{r^2}\,B_{\theta 1}\right]$$

$$+B_{r1}\left(-\frac{m}{r}\frac{\partial^2 B_{\theta 0}}{\partial r^2}-\frac{m}{r^2}\frac{\partial B_{\theta 0}}{\partial r}-\frac{m}{r^3}B_{\theta 0}-k_z\frac{\partial^2 B_{z0}}{\partial r^2}+\frac{k_z}{r}\frac{\partial B_{z0}}{\partial r}\right)$$

$$+\frac{\partial B_{r1}}{\partial r}\left(-\frac{2m}{r^2}B_{\theta 0}\right)+iB_{\theta 1}\left[-\frac{2}{r}B_{\theta 0}\left(\frac{m^2}{r^2}+k_z^2\right)\right]$$

$$+4\pi i(\rho g)_{r1}\left(\frac{m^2}{r^2}+k_z^2\right). \qquad\qquad ...[3.35]$$

The equation for $v_{\theta 1}$ is

$$-4\pi\rho_0\,\frac{\partial}{\partial t}\left[\frac{\partial^2 v_{\theta 1}}{\partial r^2}+\frac{1}{r}\frac{\partial v_{\theta 1}}{\partial r}-\left(\frac{m^2+1}{r^2}+k_z^2\right)v_{\theta 1}+\frac{2im}{r^2}\,v_{r1}\right]$$

$$=-i\left(\frac{m}{r}\,B_{\theta 0}+k_z B_{z0}\right)\left[\frac{\partial^2 B_{\theta 1}}{\partial r^2}+\frac{1}{r}\frac{\partial B_{\theta 1}}{\partial r}-\left(\frac{m^2+1}{r^2}+k_z^2\right)B_{\theta 1}+\frac{2im}{r^2}\,B_{r1}\right]$$

$$-\left[\frac{1}{r}\frac{\partial}{\partial r}(rB_{\theta 0})\right]\left[\frac{\partial^2 B_{r1}}{\partial r^2}+\frac{1}{r}\frac{\partial B_{r1}}{\partial r}-\left(\frac{m^2+1}{r^2}+k_z^2\right)B_{r1}\right]$$

$$-B_{r1}\left[\frac{\partial^3 B_{\theta 0}}{\partial r^3}+\frac{2}{r}\frac{\partial^2 B_{\theta 0}}{\partial r^2}+\left(\frac{2m^2-1}{r^2}\right)\frac{\partial B_{\theta 0}}{\partial r}+\frac{B_{\theta 0}}{r^3}+\frac{2mk_z}{r}\frac{\partial B_{z0}}{\partial r}\right]$$

$$-\frac{\partial B_{r1}}{\partial r}\left[2\left(\frac{\partial^2 B_{\theta 0}}{\partial r^2}+\frac{1}{r}\frac{\partial B_{\theta 0}}{\partial r}-\frac{B_{\theta 0}}{r^2}\right)\right]-i\frac{\partial B_{\theta 1}}{\partial r}\left[2\left(\frac{m}{r}\frac{\partial B_{\theta 0}}{\partial r}+k_z\frac{\partial B_{z0}}{\partial r}\right)\right]$$

$$+iB_{\theta 1}\left[-\frac{m}{r}\frac{\partial^2 B_{\theta 0}}{\partial r^2}-\frac{m}{r^2}\frac{\partial B_{\theta 0}}{\partial r}+\frac{3m}{r^3}B_{\theta 0}-k_z\left(\frac{\partial^2 B_{z0}}{\partial r^2}+\frac{1}{r}\frac{\partial B_{z0}}{\partial r}\right)\right]$$

$$+4\pi i\frac{m}{r}\left[\frac{\partial(\rho g)_{r1}}{\partial r}-\frac{1}{r}(\rho g)_{r1}\right]. \qquad\qquad ...[3.36]$$

We also have

$$\frac{\partial (\rho g)_{r1}}{\partial t} + v_{r1} \frac{\partial (\rho g)_{r0}}{\partial r} = 0 \qquad \text{...[3.37]}$$

$$\frac{\partial \eta_1}{\partial t} + v_{r1} \frac{\partial \eta_0}{\partial r} = 0. \qquad \text{...[3.38]}$$

It is convenient to introduce a set of dimensionless variables as has been done previously. We define the two dimensionless independent variables

$$\mu = \frac{r}{a} \qquad\qquad \tau = \frac{t}{\tau_R}.$$

Let R be the value of r such that

$$\frac{m}{r} B_{\theta 0} + k_z B_{z0} = 0.$$

This is the radius about which the instability will develop if the configuration is unstable for the mode specified by the pair of numbers m and k_z. (In the cylindrical case these two numbers are specified independently, whereas in the sheet case only one number was needed to describe a mode.)

We now define the dimensionless dependent variables

$$\psi = \frac{B_{r1}}{B} \qquad\qquad \phi = \frac{iB_{\theta 1}}{B}$$

$$w = -iv_{r1} k \tau_R \qquad\qquad U = v_{\theta 1} \tau_R k$$

$$\delta = -\frac{i\eta_1}{\langle \eta \rangle} \qquad\qquad \gamma = ik\tau_H^2 \frac{(\rho g)_{r1}}{\langle \rho \rangle}$$

where $\qquad k = \left(k_z^2 + \frac{m^2}{R^2} \right)^{\frac{1}{2}}$

and the following parameters

$$S = \frac{\tau_R}{\tau_H}, \qquad \alpha = ka, \qquad \kappa_z = k_z a \qquad \text{and} \qquad \tilde{\rho} = \frac{\rho_0}{\langle \rho \rangle}.$$

The following functions are given by the equilibrium configuration

$$F = \frac{1}{kB} \left(\frac{m}{r} B_{\theta 0} + k_z B_{z0} \right) = \frac{1}{\alpha B} \left(\frac{m}{\mu} B_{\theta 0} + \kappa_z B_{z0} \right)$$

$$F' = \frac{dF}{d\mu} \qquad\qquad F'' = \frac{d^2 F}{d\mu^2}$$

$$H = \frac{B_{\theta 0}}{B} \qquad\qquad N = \frac{1}{\mu} \frac{\partial}{\partial \mu} (\mu H)$$

$$N' = \frac{dN}{d\mu} \qquad\qquad N'' = \frac{d^2N}{d\mu^2}$$

$$\tilde{\eta} = \frac{\eta_0}{\langle \eta \rangle} \qquad\qquad \tilde{\eta}' = \frac{d\tilde{\eta}}{d\mu}.$$

We define the operator

$$L = \frac{\partial^2}{\partial\mu^2} + \frac{1}{\mu}\frac{\partial}{\partial\mu} - \left(\frac{m^2+1}{\mu^2} + \kappa_z^2\right).$$

We can now give the six equations to be solved corresponding to equations [3.33]–[3.38]

$$\frac{\partial\psi}{\partial\tau} = -Fw + \tilde{\eta}\left(L\psi - \frac{2m}{\mu^2}\phi\right) + \left(\alpha F' + \frac{2m}{\mu^2}H\right)\delta \qquad ...[3.39]$$

$$\frac{\tilde{\rho}}{\alpha^2 S^2}\frac{\partial}{\partial\tau}\left(Lw - \frac{2m}{\mu^2}U\right) = F\left(L\psi - \frac{2m}{\mu^2}\phi\right) - \left(\frac{2m}{\alpha\mu^2}H\right)\frac{\partial\psi}{\partial\mu}$$

$$-\left(F'' - \frac{1}{\mu}F' + \frac{4m}{\alpha\mu^2}N - \frac{6m}{\alpha\mu^3}H\right)\psi$$

$$+\frac{1}{\alpha^2}\left(\frac{m^2}{\mu^2} + \kappa_z^2\right)\gamma - \left[\frac{2}{\alpha\mu}H\left(\frac{m^2}{\mu^2} + \kappa_z^2\right)\right]\phi \quad ...[3.40]$$

$$\frac{\partial\phi}{\partial\tau} = -FU + \eta\left(L\phi - \frac{2m}{\mu^2}\psi\right) - \frac{1}{\alpha}\left(\frac{2H}{\mu} - N\right)w$$

$$-N'\delta - N\frac{\partial\delta}{\partial\mu} + \eta'\left(\frac{\partial\phi}{\partial\mu} + \frac{\phi}{\mu} + \frac{m}{\mu}\psi\right) \qquad ...[3.41]$$

$$\frac{\tilde{\rho}}{\alpha^2 S^2}\frac{\partial}{\partial\tau}\left(LU - \frac{2m}{\mu^2}w\right)$$

$$= F\left(L\phi - \frac{2m}{\mu^2}\psi\right) + \frac{N}{\alpha}(L\psi) + \left(\frac{2m}{\mu}F' + \frac{2m^2}{\alpha\mu^3}H + \frac{1}{\alpha}N'' + \frac{1}{\alpha\mu}N'\right)\psi$$

$$+\frac{2}{\alpha}N'\frac{\partial\psi}{\partial\mu} + \left[2\left(F' + \frac{m}{\alpha\mu^2}H\right)\right]\frac{\partial\phi}{\partial\mu}$$

$$+\left(\frac{2m}{\alpha\mu^2}N - \frac{6m}{\alpha\mu^3}H + F'' + \frac{1}{\mu}F'\right)\phi - \frac{m}{\alpha^2\mu}\left(\frac{\partial\gamma}{\partial\mu} - \frac{\gamma}{\mu}\right) \qquad ...[3.42]$$

$$\frac{\partial\delta}{\partial\tau} = -\frac{1}{\alpha}\tilde{\eta}'w \qquad\qquad ...[3.43]$$

$$\frac{\partial\gamma}{\partial\tau} = -wG \qquad\qquad ...[3.44]$$

where

$$G = -\frac{\tau_H^2}{\langle\rho\rangle}\frac{\partial(\rho g)_{r0}}{\partial r}.$$

We wish to solve the initial-value problem described by the six equations [3.39]–[3.44]. *Note*: In the following we use a slightly different notation. The dimensionless variables μ, τ are replaced by letter r, t respectively. Also $\bar{\eta}$ and $\bar{\rho}$ are simply given by η and ρ.

Hence we wish to find $\psi(r, t)$, $w(r, t)$, $\phi(r, t)$, $U(r, t)$, $\delta(r, t)$, $\gamma(r, t)$ on the domain $0 \leqslant R_i \leqslant r \leqslant R_0$, $t \geqslant 0$, where the initial distributions of the above functions are given. The boundary condition at $r = R_0$, the outer wall is $\psi = \phi = w = U = 0$.

If $R_i > 0$, the hard-core case, then $\psi = \phi = w = U = 0$ at $r = R_i$. If $R_i = 0$, the pinch case, we can use the condition that the radial derivatives vanish.

Consider the finite difference grid

$$t_n = n\Delta t \qquad\qquad n = 0, 1, 2, \ldots,$$

$$r_j = R_i + j\Delta r \qquad\qquad j = 0, 1, 2, \ldots, J$$

$$R_0 = r_J = R_i + J\Delta r.$$

Let $\psi_j^n = \psi(r_j, t_n)$, etc. We use

$$(L\psi)_j^n = \frac{\psi_{j+1}^n - 2\psi_j^n + \psi_{j-1}^n}{(\Delta r)^2} + \frac{1}{r_j}\frac{\psi_{j+1}^n - \psi_{j-1}^n}{2\Delta r} - \left(\frac{m^2+1}{r_j^2} + \kappa_z^2\right)\psi_j^n.$$

The difference equations for ψ, w, ϕ, and U are implicit, i.e. completely centred in space and time. The pair ψ and w are computed at integral time lines, and their difference equations can be written in the form

$$-a_j^{(1)}\psi_{j+1}^{n+1} + b_j^{(1)}\psi_j^{n+1} + c_j^{(1)}w_j^{n+1} - d_j^{(1)}\psi_{j-1}^{n+1} = {}^{(1)}p_j^n$$

$$-e_j^{(1)}\psi_{j+1}^{n+1} + f_j^{(1)}w_{j+1}^{n+1} + g_j^{(1)}\psi_j^{n+1} - h_j^{(1)}w_j^{n+1}$$

$$-k_j^{(1)}\psi_{j-1}^{n+1} + 1_j^{(1)}w_{j-1}^{n+1} = {}^{(1)}q_j^n,$$

$j = 1, \ldots, J-1$.

The coefficients are known functions of position and are derived from the basic difference equations as in the sheet model. These equations must be solved for ψ_j^{n+1}, w_j^{n+1} ($j = 1, \ldots, J-1$), where p_j^n and q_j^n are known functions of ψ_j^n, w_j^n, $\phi_j^{n+\frac{1}{2}}$, $U_j^{n+\frac{1}{2}}$, $\delta_j^{n+\frac{1}{2}}$, $\gamma_j^{n+\frac{1}{2}}$. We solve the pair of equations above as a single difference equation

$$-A_j^{(1)}X_{j+1}^{n+1} + B_j^{(1)}X_j^{n+1} - C_j^{(1)}X_{j-1}^{n+1} = {}^{(1)}d_j^n \qquad \ldots[3.45]$$

where

$$X_j^n = \begin{bmatrix} \psi_j^n \\ \\ w_j^n \end{bmatrix} \qquad \text{and} \qquad {}^{(1)}d_j^n = \begin{bmatrix} {}^{(1)}p_j^n \\ \\ {}^{(1)}q_j^n \end{bmatrix}$$

and $A_j^{(1)}$, $B_j^{(1)}$, $C_j^{(1)}$ are 2×2 matrices derived from the above equations.

The pair ϕ and U are computed at half-integral time lines, and their difference equations can be written

$$-a_j^{(2)}\phi_{j+1}^{n+\frac{1}{2}}+b_j^{(2)}\phi_j^{n+\frac{1}{2}}+c_j^{(2)}U_j^{n+\frac{1}{2}}-d_j^{(2)}\phi_{j-1}^{n+\frac{1}{2}} = {}^{(2)}p_j^{n-\frac{1}{2}}$$

$$-e_j^{(2)}\phi_{j+1}^{n+\frac{1}{2}}+f_j^{(2)}U_{j+1}^{n+\frac{1}{2}}+g_j^{(2)}\phi_j^{n+\frac{1}{2}}-h_j^{(2)}U_j^{n+\frac{1}{2}}-k_j^{(2)}\phi_{j-1}^{n+\frac{1}{2}}$$
$$+1_j^{(2)}U_{j-1}^{n+\frac{1}{2}} = {}^{(2)}q_j^{n-\frac{1}{2}},$$

$j = 1, \ldots, J-1$.

This pair is also solved as a single difference equation

$$-A_j^{(2)}Y_{j+1}^{n+\frac{1}{2}}+B_j^{(2)}Y_j^{n+\frac{1}{2}}-C_j^{(2)}Y_{j-1}^{n+\frac{1}{2}} = {}^{(2)}d_j^{n-\frac{1}{2}} \qquad \ldots[3.46]$$

where

$$Y_j^{n+\frac{1}{2}} = \begin{bmatrix} \phi_j^{n+\frac{1}{2}} \\ \\ U_j^{n+\frac{1}{2}} \end{bmatrix} \qquad \text{and} \qquad {}^{(2)}d_j^{n-\frac{1}{2}} = \begin{bmatrix} {}^{(2)}p_j^{n-\frac{1}{2}} \\ \\ {}^{(2)}q_j^{n-\frac{1}{2}} \end{bmatrix}$$

and $A_j^{(2)}$, $B_j^{(2)}$, $C_j^{(2)}$ are also 2×2 matrices.

The difference equations for δ and γ are simply

$$\delta_j^{n+\frac{1}{2}} = \delta_j^{n-\frac{1}{2}} - \frac{\Delta t}{\alpha}\, \eta_j' w_j^n \qquad \ldots[3.47]$$

$$\gamma_j^{n+\frac{1}{2}} = \gamma_j^{n-\frac{1}{2}} - \Delta t G_j w_j^n. \qquad \ldots[3.48]$$

We wish to solve a difference equation of the form

$$-A_j X_{j+1}^{n+1}+B_j X_j^{n+1}-C_j X_{j-1}^{n+1} = d_j^n.$$

We use the algorithm

$$X_j^{n+1} = E_j X_{j+1}^{n+1}+f_j^n; \qquad j = 0, 1, 2, \ldots, J-1.$$

Then E_j and f_j^n are determined from the recurrence relations

$$E_j = (B_j - C_j E_{j-1})^{-1}A_j$$

$$f_j^n = (B_j - C_j E_{j-1})^{-1}(d_j^n + C_j f_{j-1}^n); \quad j = 1, \ldots, J-1.$$

We obtain E_0 and f_0^n from the boundary conditions at $r = R_i$. If $\psi = w = 0$ at R_i then

$$E_0 = \begin{bmatrix} 0 & 0 \\ 0 & 0 \end{bmatrix} \qquad \text{and} \qquad f_0^n = \begin{bmatrix} 0 \\ 0 \end{bmatrix}.$$

If $\partial\psi/\partial r = 0$ at R_i, then

$$E_0 = \begin{bmatrix} 1 & 0 \\ 0 & 1 \end{bmatrix} \qquad \text{and} \qquad f_0^n = \begin{bmatrix} 0 \\ 0 \end{bmatrix}.$$

The procedure then is to compute E_j, f_j^n $(j = 1, \ldots, J-1)$. We assume that $r = R_0$ is a conducting wall, and so $\psi = w = 0$ at R_0. Then

$$X_{J-1}^{n+1} = f_{J-1}^n.$$

We can then compute the remaining X_j^{n+1}, $j = 0, 1, \ldots, J-2$, from the algorithm. The equation for Y is treated similarly.

The programme that solves the equations [3.45]–[3.48] is called RIPPLE IV. The purpose of the RIPPLE IV programme is to solve for the perturbed or first order variables. In order to do this, we must give as functions of radius the equilibrium or zero order variables. The functions that must be specified are

$$\frac{B_{\theta 0}}{B}, \frac{B_{z0}}{B}, \frac{\eta_0}{\langle \eta \rangle}$$

for a given domain $R_i \leqslant r \leqslant R_0$.

These functions then characterize the equilibrium model. To specify a particular case we give

$$a, m, k_z, S.$$

With these quantities the programme then prepares the input needed to solve the equations. The derivatives of the equilibrium functions are not computed in the programme, so actually these must be given in analytic or tabular form. A separate preliminary routine must be included which gives

$$\left(\frac{B_{\theta 0}}{B}\right), \left(\frac{B_{\theta 0}}{B}\right)', \left(\frac{B_{\theta 0}}{B}\right)'', \left(\frac{B_{\theta 0}}{B}\right)''', \left(\frac{B_{z0}}{B}\right), \left(\frac{B_{z0}}{B}\right)', \left(\frac{B_{z0}}{B}\right)'', \eta, \eta'.$$

It is advantageous to describe the equilibrium by simple analytic functions The programme has been applied to the Triax experiment,[31, 32] hard-core pinch configurations described by Bickerton et al.[33] and Rebut,[34] to the thetatron[35] and to a force-free Bessel function model.[36] All of these examples were tearing mode calculations.

4

COMPUTATION OF FINITE-BETA EQUILIBRIA

4.1. Basic method

We consider a number of two-dimensional plasma equilibria problems. In each case a confining vacuum magnetic field is given and we wish to find the distortion of this field by the presence of the plasma. We use hydromagnetic equilibria equations with either a scalar or tensor pressure. The magnetic field can be derived from a stream function $\psi(r, z)$ which satisfies a non-linear, second-order, partial differential equation. We solve the resulting boundary-value problem by finite-difference methods.[37]

We consider cases where the plasma pressure is a function of B only, or of ψ only, or of both ψ and B. (B is the magnitude of the magnetic field.) The plasma is assumed to be located within a closed constant-B surface or between closed surfaces of constant ψ. The plasma is surrounded by a plasma-free region, but the field in this region is also modified by the presence of plasma. We are in effect solving a plasma free-surface problem, because the boundary of the plasma region is determined by the solution of the field equation. We solve the equation for $\psi(r, z)$ by iteration, and can find a sequence of equilibria for increasing pressures. We increase the pressure in reasonable steps and use the equilibrium configuration obtained from the preceding case as a trial solution to begin the iteration of the next.

4.2. *Open ended, minimum-B systems*

Taylor[38] considered the equilibrium and stability of magnetic wells in the low-pressure limit, pointing out that one class of equilibrium solutions of the hydromagnetic equations could be found if the pressure tensor was a function of the magnetic field magnitude, i.e. $p = p(B)$. They would also be contained equilibria if the vacuum field into which the plasma was placed had closed B-surfaces. The combined mirror multipole field[38] was one such. Another example is the stuffed cusp,[39] which is a cusp with a current-carrying rod along the symmetry axis. Other examples having axial symmetry have been described by Furth[40] and Andreoletti.[41]

In this section we discuss the distortion of the field by the plasma, and consider the hydromagnetic equilibrium equations with the assumption that the pressure tensor is a function of B only. The relevant equations are given, and the axisymmetric case is then considered in detail by using a stream function for the magnetic field.

The equations of hydromagnetic equilibrium with a tensor pressure are

$$j \times B = \operatorname{div} p \qquad \ldots [4.1]$$

$$\operatorname{curl} B = j \qquad \ldots [4.2]$$

$$\operatorname{div} B = 0 \qquad \ldots [4.3]$$

and

$$p = p_\perp I + \frac{p_\parallel - p_\perp}{B^2} BB, \qquad \ldots [4.4]$$

where I is the identity tensor, and B is the magnitude of the magnetic field.

In general, static plasma containment is one of two types: (1) the plasma is tied to field lines and these field lines are contained within a closed volume. This means that the plasma surface is also a magnetic surface which will be of toroidal shape if B is nowhere zero. Or (2) the plasma is tied to field lines but these leave the plasma volume and containment is achieved by the mirror effect.

For $p = p(B)$ and equilibria with field lines leaving the plasma, the above equations reduce[42] to

$$\operatorname{curl} vB = 0, \qquad \ldots [4.5]$$

where

$$v = \frac{p_{\parallel}-p_{\perp}}{B^2} - 1. \qquad \ldots[4.6]$$

In the case of axial symmetry, equation [4.5] becomes the set of equations

$$\frac{\partial(vB_{\theta})}{\partial z} = 0$$

$$\frac{\partial(vB_r)}{\partial z} - \frac{\partial(vB_z)}{\partial r} = 0$$

$$\frac{1}{r}\frac{\partial}{\partial r}(rvB_{\theta}) = 0.$$

We define the stream function $\psi(r, z)$ by

$$B_r = -\frac{1}{r}\frac{\partial\psi}{\partial z} \qquad\qquad B_z = \frac{1}{r}\frac{\partial\psi}{\partial r}. \qquad \ldots[4.7]$$

We then have

$$\frac{\partial}{\partial z}\left(\frac{v}{r}\frac{\partial\psi}{\partial z}\right) + \frac{\partial}{\partial r}\left(\frac{v}{r}\frac{\partial\psi}{\partial r}\right) = 0 \qquad \ldots[4.8]$$

and

$$rvB_{\theta} = \text{constant.} \qquad \ldots[4.9]$$

It is convenient to write the above as

$$-\frac{\partial^2\psi}{\partial z^2} - r\frac{\partial}{\partial r}\left(\frac{1}{r}\frac{\partial\psi}{\partial r}\right) = \frac{1}{v}\left(\frac{\partial v}{\partial z}\frac{\partial\psi}{\partial z} + \frac{\partial v}{\partial r}\frac{\partial\psi}{\partial r}\right). \qquad \ldots[4.10]$$

We can write ψ as

$$\psi = \psi_c + \psi_p$$

where ψ_c is the stream function of the vacuum magnetic field, and ψ_p is the stream function of the magnetic field due to the plasma. We have

$$-\frac{\partial^2\psi_c}{\partial z^2} - r\frac{\partial}{\partial r}\left(\frac{1}{r}\frac{\partial\psi_c}{\partial r}\right) = 0. \qquad \ldots[4.11]$$

Hence

$$-\frac{\partial^2\psi_p}{\partial z^2} - r\frac{\partial}{\partial r}\left(\frac{1}{r}\frac{\partial\psi_p}{\partial r}\right) = \frac{1}{v}\left(\frac{\partial v}{\partial z}\frac{\partial\psi_p}{\partial z} + \frac{\partial v}{\partial r}\frac{\partial\psi_p}{\partial r}\right) + \frac{1}{v}\left(\frac{\partial v}{\partial z}\frac{\partial\psi_c}{\partial z} + \frac{\partial v}{\partial r}\frac{\partial\psi_c}{\partial r}\right).$$

The equation for ψ_p is then the non-linear partial differential equation,

$$\frac{\partial^2\psi_p}{\partial r^2} + \left(\frac{1}{v}\frac{\partial v}{\partial r} - \frac{1}{r}\right)\frac{\partial\psi_p}{\partial r} + \left(\frac{1}{v}\frac{\partial v}{\partial z}\right)\frac{\partial\psi_p}{\partial z} + \frac{\partial^2\psi_p}{\partial z^2} = -I_c, \qquad \ldots[4.12]$$

where

$$I_c = \frac{r}{\nu} \left(B_{zc} \frac{\partial \nu}{\partial r} - B_{rc} \frac{\partial \nu}{\partial z} \right). \qquad \qquad \text{...[4.13]}$$

B_{rc} and B_{zc} are the r and z components of the vacuum field. The magnetic field is then specified by

$$B_r = B_{rc} - \frac{1}{r} \frac{\partial \psi_p}{\partial z}$$

$$B_z = B_{zc} + \frac{1}{r} \frac{\partial \psi_p}{\partial r} \qquad \qquad \text{...[4.14]}$$

$$B_\theta = \frac{k}{\nu r}.$$

The magnitude of the field is then

$$B = (B_r^2 + B_\theta^2 + B_z^2)^{\frac{1}{2}}.$$

To solve the equilibrium problem in the axially symmetric case, we must solve equation [4.12] in the domain

$$-l \leqslant z \leqslant l \qquad 0 \leqslant r \leqslant R.$$

We choose l and R large enough so that $\psi_p(r, l) = \psi_p(r, -l) = \psi_p(R, z) = 0$. This is equivalent to saying the l and R are far enough away from the plasma region so that the effect on the magnetic field, due to the plasma, is negligible beyond the domain: $-l \leqslant z \leqslant l, 0 \leqslant r \leqslant R$.

We begin by specifying a vacuum field $\mathbf{B}_c = (B_{rc}, B_{\theta c}, B_{zc})$, which has a non-zero minimum and closed constant-B surfaces. We then choose given functions $p_\perp(B)$ and $p_\parallel(B)$ with the property that $p_\perp = p_\parallel = 0$ for $B > B_0$, i.e. the plasma pressure goes to zero beyond a specified constant B surface.

We wish to solve equation [4.12] by finite difference methods. In the following sections we shall use ψ for the quantity ψ_p defined earlier. The difference approximation is

$$\frac{\psi_{i,j+1} - 2\psi_{i,j} + \psi_{i,j-1}}{h^2} + \left[\left(\frac{1}{\nu} \frac{\partial \nu}{\partial r} \right)_{i,j} - \frac{1}{jh} \right] \frac{\psi_{i,j+1} - \psi_{i,j-1}}{2h}$$

$$+ \left(\frac{1}{\nu} \frac{\partial \nu}{\partial z} \right)_{i,j} \frac{\psi_{i+1,j} - \psi_{i-1,j}}{2h} + \frac{\psi_{i+1,j} - 2\psi_{i,j} + \psi_{i-1,j}}{h^2} = -(I_c)_{i,j}$$

$$\text{...[4.15]}$$

where $h = \Delta r = \Delta z$, $r_j = jh$, $z_i = ih$, and

$$\psi_{i,j} = \psi(r_j, z_i); \quad j = 0, 1, 2, \ldots, J; \quad i = -I, \ldots, 0, \ldots I; \quad R = Jh;$$
$$-l = -Ih, l = Ih.$$

The equation for $\psi(r, z)$ is non-linear because the coefficients involve $v(B)$, and B depends on ψ. We can write the full set of algebraic equations resulting from the difference approximations used.

$$a_{i,j}\psi_{i,j+1}+b_{i,j}\psi_{i,j-1}+c_{i,j}\psi_{i+1,j}+d_{i,j}\psi_{i-1,j}-4\psi_{i,j} = -\phi_{i,j} \quad ...[4.16]$$

where

$$a_{i,j} = 1-\frac{1}{2j}+\frac{v_{i,j+1}-v_{i,j-1}}{4v_{i,j}}$$

$$b_{i,j} = 1+\frac{1}{2j}-\frac{v_{i,j+1}-v_{i,j-1}}{4v_{i,j}}$$

$$c_{i,j} = 1+\frac{v_{i+1,j}-v_{i-1,j}}{4v_{i,j}}$$

$$d_{i,j} = 1-\frac{v_{i+1,j}-v_{i-1,j}}{4v_{i,j}}$$

$$\phi_{i,j} = h^2(I_c)_{i,j}$$

$$v_{i,j} = v(B_{i,j})$$

$$(I_c)_{i,j} = \frac{j}{2v_{i,j}}\left[(B_{zc})_{i,j}(v_{i,j+1}-v_{i,j-1})-(B_{rc})_{i,j}(v_{i+1,j}-v_{i-1,j})\right]$$

and

$$B_{i,j} = [(B_r)_{i,j}^2+(B_\theta)_{i,j}^2+(B_z)_{i,j}^2]^{\frac{1}{2}} \qquad ...[4.17]$$

$$(B_r)_{i,j} = (B_{rc})_{i,j}-(2jh^2)^{-1}(\psi_{i+1,j}-\psi_{i-1,j}) \qquad ...[4.18]$$

$$(B_z)_{i,j} = (B_{zc})_{i,j}+(2jh^2)^{-1}(\psi_{i,j+1}-\psi_{i,j-1}) \qquad ...[4.19]$$

$$(B_\theta)_{i,j} = \frac{k}{(v_{i,j})jh}. \qquad ...[4.20]$$

Since this set of algebraic equations is nonlinear in the unknown function $\psi_{i,j}$, we adopt an iterative technique to solve them. The method we use is based on the alternating-direction implicit scheme. We shall discuss this method in Section 4.6.

The first step in the computational procedure is to evaluate the given functions B_{rc}, B_{zc}, $B_{\theta c}$ for all values of i, j. Since we choose a vacuum field with closed constant-B surfaces, we can then compute a plasma pressure $p_\perp(B)$, $p_\parallel(B)$ for B less than some B_0. We can then proceed to solve the ψ equation by iteration and compute the modification of the magnetic field due to the plasma. We then increase the plasma pressure and repeat the procedure, thus obtaining solutions to the hydromagnetic equilibrium problem for a sequence of plasma pressures.

This programme was applied to two axially symmetric models, the "stuffed cusp"[39] and the "disc" device proposed by Furth.[40] The results of the stuffed cusp calculations are shown in Ref. 42.

The vacuum "stuffed cusp" configuration is the field produced by equal and opposite currents in two circular loops (of unit radius centred on the symmetry axis and in the planes $z = +1$ and $z = -1$), and a current in a central rod along the symmetry axis. The existence and shape of the well depends on the ratio of loop current to the current in the rod.

In both models the form of the pressure tensor we used was the one proposed by Taylor,[38] i.e.

$$\left. \begin{array}{l} p_{\parallel} = cB(B_0 - B)^m \\ p_{\perp} = cmB^2(B_0 - B)^{m-1} \end{array} \right\} \quad \text{if} \quad B \leqslant B_0$$
$$p_{\parallel} = p_{\perp} = 0 \qquad\qquad \text{if} \quad B > B_0 \qquad \text{...[4.21]}$$

where c and m are arbitrary constants. In a given calculation we fix m and B_0, and solve the equilibrium problem for an increasing sequence of values of the constant c. The parameter m varies the shape of the pressure distribution, B_0 the height to which the well is filled with plasma, and the absolute value of the plasma pressure is proportional to c.

In the general three-dimensional case we can no longer use the stream function, ψ, previously defined; however, we still have the equations

$$\text{curl } vB = 0$$
$$\text{div } \boldsymbol{B} = 0.$$

We can let $vB = \nabla\phi$, then

$$\text{div } vB = \nabla^2\phi = v \text{ div } \boldsymbol{B} + \boldsymbol{B} \cdot \nabla v.$$

Hence we have the single equation

$$\nabla^2\phi = \frac{1}{v} \nabla\phi \cdot \nabla v.$$

We can write ϕ as

$$\phi = \phi_c + \phi_p,$$

where ϕ_c is the potential function of the vacuum magnetic field, i.e. $B_c = -\nabla\phi_c$, since $v = -1$ in vacuum, and ϕ_p is the potential function of the magnetic field due to the plasma. We have $\nabla^2\phi_c = 0$; hence

$$\nabla^2\phi_p = \frac{1}{v} [\nabla v \cdot (\nabla\phi_p - \boldsymbol{B}_c)]. \qquad \text{...[4.22]}$$

We can solve equation [4.22] numerically in three dimensions, using finite difference methods, but of course the computation time is much greater than for the two-dimensional problem.

4.3. *Toroidal equilibria with scalar pressure*

We consider hydromagnetic equilibria for a toroidal plasma with a scalar pressure.[43] We assume that the system is axially-symmetric, so that we can apply this method to configurations with floating rings—multipoles or the levitron. Grad[44] has given the equation for $\psi(r, z)$ when the plasma pressure is a function of ψ.

$$-\frac{\partial^2\psi}{\partial z^2}-r\frac{\partial}{\partial r}\left(\frac{1}{r}\frac{\partial\psi}{\partial r}\right) = g'(\psi)+r^2p'(\psi) \qquad \text{...[4.23]}$$

where $p(\psi)$, the scalar pressure, and

$$g(\psi) = \tfrac{1}{2}r^2B_\theta^2 \qquad \text{...[4.24]}$$

are given functions.

As in the preceding section, we let $\psi = \psi_c+\psi_p$, where ψ_c satisfies equation [4.11]; hence we have

$$\frac{\partial^2\psi_p}{\partial z^2}+r\frac{\partial}{\partial r}\left(\frac{1}{r}\frac{\partial\psi_p}{\partial r}\right) = -I_c \qquad \text{...[4.25]}$$

where $I_c = g'(\psi)+r^2p'(\psi)$.

In the following we shall let

$$g'(\psi) = f(\psi)f'(\psi) \qquad \text{...[4.26]}$$

where $f(\psi) = rB_\theta = k/v(\psi)$, with $k = $ constant, and

$$v(\psi) = 1+bp(\psi) \qquad \text{...[4.27]}$$

with $b = $ constant. The magnetic field is then specified by equation [4.14] with v defined by equation [4.27].

The domain for equation [4.25] is $-l \leqslant z \leqslant l$, $0 \leqslant r \leqslant R$ as in the preceding case with the same boundary conditions. Since the configurations we consider here are symmetric about the $z = 0$ plane, we can use the domain $0 \leqslant z \leqslant l$ with $\partial\psi_p/\partial z = 0$ at $z = 0$.

We have applied this programme[45] to the levitron floating ring configuration. The vacuum $B_{\theta c} = k/r$ and B_{rc} and B_{zc} are computed from a system of circular coils with their centres on the major axis ($r = 0$). The plasma is confined to a region between two closed flux surfaces $\psi = \psi_1$ and $\psi = \psi_2$. We find it convenient to difference equation [4.25] with a variable mesh, because we wish to have a finer grid where the plasma is located. In writing the difference equations we let ψ_p be denoted by ψ, and ψ by $\bar{\psi}$. The difference equation for equation [4.25] is

$$\frac{2r_j}{r_{j+1}-r_{j-1}}\left[\frac{1}{r_{j+\frac{1}{2}}}\left(\frac{\psi_{i,j+1}-\psi_{i,j}}{r_{j+1}-r_j}\right)-\frac{1}{r_{j-\frac{1}{2}}}\left(\frac{\psi_{i,j}-\psi_{i,j-1}}{r_j-r_{j-1}}\right)\right]$$
$$+\frac{2}{z_{i+1}-z_{i-1}}\left[\left(\frac{\psi_{i+1,j}-\psi_{i,j}}{z_{i+1}-z_i}\right)-\left(\frac{\psi_{i,j}-\psi_{i-1,j}}{z_i-z_{i-1}}\right)\right] = -(I_c)_{i,j}$$

$$\text{...[4.28]}$$

where $r_{j+\frac{1}{2}} = \frac{1}{2}(r_j + r_{j+1})$, etc., and $j = 0, 1, 2, \ldots, J$, and $i = 0, 1, 2, \ldots, I$, with $r_J = R$ and $z_I = l$. The algebraic system corresponding to equation [4.28] is

$$a_{i,j}\psi_{i,j+1} + b_{i,j}\psi_{i,j-1} + c_{i,j}\psi_{i+1,j} + d_{i,j}\psi_{i-1,j}$$
$$-(a_{i,j} + b_{i,j} + c_{i,j} + d_{i,j})\psi_{i,j} + \phi_{i,j} = 0 \qquad \ldots[4.29]$$

with

$$a_{i,j} = \frac{4r_j}{(r_{j+1} - r_{j-1})(r_{j+1} + r_j)(r_{j+1} - r_j)}$$

$$b_{i,j} = \frac{4r_j}{(r_{j+1} - r_{j-1})(r_j + r_{j-1})(r_j - r_{j-1})}$$

$$c_{i,j} = \frac{2}{(z_{i+1} - z_{i-1})(z_{i+1} - z_i)}$$

$$d_{i,j} = \frac{2}{(z_{i+1} - z_{i-1})(z_i - z_{i-1})}$$

$$\phi_{i,j} = (I_c)_{i,j}.$$

The solution of equation [4.29] will be discussed in Section 4.6.

The pressure function that has been used[45] with this programme is

$$p(\bar{\psi}) = \begin{cases} c(\psi_2 - \bar{\psi})^m(\bar{\psi} - \psi_1)^m & \text{for } \psi_1 \leqslant \bar{\psi} \leqslant \psi_2 \\ 0 & \text{elsewhere} \end{cases} \qquad \ldots[4.30]$$

where c, ψ_1, ψ_2 are constant and m is a positive integer or zero.

4.4. Toroidal equilibria with anisotropic pressure

We assume the same axially-symmetric configuration as in the preceding case, but use the equilibria equations with a tensor pressure. Grad[44] has given the equation for $\psi(r, z)$. We have

$$-\frac{\partial^2 \psi}{\partial z^2} - r\frac{\partial}{\partial r}\left(\frac{1}{r}\frac{\partial \psi}{\partial r}\right) = \frac{1}{\sigma^2} g'(\psi) + \frac{r^2}{\sigma}\frac{\partial p_\parallel}{\partial \psi} + \frac{1}{\sigma}\left[\frac{\partial \sigma}{\partial z}\frac{\partial \psi}{\partial z} + \frac{\partial \sigma}{\partial r}\frac{\partial \psi}{\partial r}\right] \ldots[4.31]$$

where

$$\sigma(B, \psi) = 1 + \frac{p_\perp - p_\parallel}{B^2} \qquad \ldots[4.32]$$

$$g(\psi) = \frac{1}{2}\sigma^2 r^2 B_\theta^2 \qquad \ldots[4.33]$$

and $p_\perp(B, \psi)$, $p_\parallel(B, \psi)$, $g(\psi)$ are given functions. We note that equation [4.31] becomes equation [4.23] when $p_\perp = p_\parallel$. We let $\psi = \psi_c + \psi_p$, where ψ_c satisfies equation [4.11]; hence we have

$$\frac{\partial^2 \psi_p}{\partial z^2} + r \frac{\partial}{\partial r}\left(\frac{1}{r}\frac{\partial \psi_p}{\partial r}\right) + \frac{1}{\sigma}\left(\frac{\partial \sigma}{\partial r}\frac{\partial \psi_p}{\partial r} + \frac{\partial \sigma}{\partial z}\frac{\partial \psi_p}{\partial z}\right) = -I_c \qquad \dots[4.34]$$

where

$$I_c = \frac{r}{\sigma}\left(B_{zc}\frac{\partial \sigma}{\partial r} - B_{rc}\frac{\partial \sigma}{\partial z}\right) + \frac{1}{\sigma^2}g'(\psi) + \frac{r^2}{\sigma}\frac{\partial p_\parallel}{\partial \psi}.$$

The domain and boundary conditions for equation [4.34] are the same as for equation [4.25] and we use the same finite-difference scheme. The algebraic system to be solved is equation [4.29], but with

$$a_{i,j} = \frac{4r_j}{(r_{j+1}-r_{j-1})(r_{j+1}+r_j)(r_{j+1}-r_j)} + \frac{\sigma_{i,j+1}-\sigma_{i,j-1}}{\sigma_{i,j}(r_{j+1}-r_{j-1})^2}$$

$$b_{i,j} = \frac{4r_j}{(r_{j+1}-r_{j-1})(r_j+r_{j-1})(r_j-r_{j-1})} - \frac{\sigma_{i,j+1}-\sigma_{i,j-1}}{\sigma_{i,j}(r_{j+1}-r_{j-1})^2}$$

$$c_{i,j} = \frac{2}{(z_{i+1}-z_{i-1})(z_{i+1}-z_i)} + \frac{\sigma_{i+1,j}-\sigma_{i-1,j}}{\sigma_{i,j}(z_{i+1}-z_{i-1})^2}$$

$$d_{ij} = \frac{2}{(z_{i+1}-z_{i-1})(z_i-z_{i-1})} - \frac{\sigma_{i+1,j}-\sigma_{i-1,j}}{\sigma_{i,j}(z_{i+1}-z_{i-1})^2}$$

$$\phi_{i,j} = (I_c)_{i,j} = \frac{r_j}{\sigma_{i,j}}\left[(B_{zc})_{i,j}\frac{\sigma_{i,j+1}-\sigma_{i,j-1}}{r_{j+1}-r_{j-1}} - \right.$$
$$\left. -(B_{rc})_{i,j}\frac{\sigma_{i+1,j}-\sigma_{i-1,j}}{z_{i+1}-z_{i-1}}\right] + \frac{(g')_{i,j}}{\sigma_{i,j}^2} + \frac{r_j^2}{\sigma_{i,j}}\left(\frac{\partial p_\parallel}{\partial \psi}\right)_{i,j}$$

$$\sigma_{i,j} = \sigma(B_{i,j}, \bar\psi_{i,j}).$$

We apply[46] this programme to a configuration called the minimum-B levitron which has a local minimum-B region with closed constant-B surfaces. We consider the following pressure functions

$$p_\perp(B, \bar\psi) = p(\bar\psi)\left(\frac{B_0}{B}\right)^\nu\left(1-\frac{B}{B_0}\right)^{\nu+\frac{1}{2}}\left(\nu+1+\frac{B}{2B_0}\right)$$

$$p_\parallel(B, \bar\psi) = p(\bar\psi)\left(\frac{B_0}{B}\right)^\nu\left(1-\frac{B}{B_0}\right)^{\nu+\frac{3}{2}} \qquad \dots[4.35]$$

for $B \leqslant B_0$ and $p_\parallel = p_\perp = 0$ for $B > B_0$, where $p(\bar\psi)$ is given by equation [4.30] and ν is an integer.

4.5. Helical equilibria with anisotropic pressure

We consider a linear system periodic in z where the fields have helical symmetry. A helical field can be described in terms of a two-dimensional function $\psi(r, \phi)$ where

$$\phi = \theta - kz \qquad \qquad ...[4.36]$$

with k = constant. The equation for ψ has been given by Grad.[44] We have

$$\frac{1}{r^2}\frac{\partial^2\psi}{\partial\phi^2} + \frac{1}{r}\frac{\partial}{\partial r}\left(\frac{r}{1+k^2r^2}\frac{\partial\psi}{\partial r}\right) - \frac{2kf(\psi)}{\sigma(1+k^2r^2)^2} + \frac{1}{\sigma}\frac{\partial p_{\parallel}}{\partial\psi}$$
$$+ \frac{ff'}{\sigma^2(1+k^2r^2)} + \frac{1}{\sigma}\left(\frac{1}{1+k^2r^2}\frac{\partial\psi}{\partial r}\frac{\partial\sigma}{\partial r} + \frac{1}{r^2}\frac{\partial\psi}{\partial\phi}\frac{\partial\sigma}{\partial\phi}\right) = 0 \qquad ...[4.37]$$

where

$$\sigma(\psi, B) = 1 + \frac{p_\perp - p_\parallel}{B^2} \qquad \qquad ...[4.38]$$

$$f(\psi) = \sigma(B_z + krB_\theta) \qquad \qquad ...[4.39]$$

and $f(\psi)$, $p_\perp(B, \psi)$, $p_\parallel(B, \psi)$ are given functions. The magnetic field is determined by equation [4.39] and

$$B_r = \frac{1}{r}\frac{\partial\psi}{\partial\phi} \qquad\qquad B_\theta - krB_z = -\frac{\partial\psi}{\partial r}. \qquad ...[4.40]$$

We let $\psi = \psi_c + \psi_p$ as in the previous cases. The vacuum helical field[47, 48] can be normalized so that

$$B_{zc} + krB_{\theta c} = 1. \qquad \qquad ...[4.41]$$

From equation [4.39] we have

$$f(\psi) = \sigma[(B_{zc} + krB_{\theta c}) + (B_{zp} + krB_{\theta p})]$$
$$= \sigma[1 + (B_{zp} + krB_{\theta p})].$$

We define the function

$$F(\psi_p) = B_{zp} + krB_{\theta p} \qquad \qquad ...[4.42]$$

so

$$f(\psi) = \sigma(B, \psi)[1 + F(\psi_p)] \qquad \qquad ...[4.43]$$

$$B_r = B_{rc} + \frac{1}{r}\frac{\partial\psi_p}{\partial\phi}$$
$$B_\theta = B_{\theta c} + \frac{1}{1+k^2r^2}\left[krF(\psi_p) - \frac{\partial\psi_p}{\partial r}\right] \qquad ...[4.44]$$
$$B_z = B_{zc} + \frac{1}{1+k^2r^2}\left[F(\psi_p) + kr\frac{\partial\psi_p}{\partial r}\right].$$

The function $\psi_c(r, z)$ satisfies the equation

$$\frac{1}{r^2}\frac{\partial^2\psi_c}{\partial\phi^2} + \frac{1}{r}\frac{\partial}{\partial r}\left(\frac{r}{1+k^2r^2}\frac{\partial\psi_c}{\partial r}\right) - \frac{2k}{(1+k^2r^2)^2} = 0. \qquad ...[4.45]$$

Hence the equation we must solve for ψ_p is

$$\frac{1}{r^2}\frac{\partial^2\psi_p}{\partial\phi^2}+\frac{1}{r}\frac{\partial}{\partial r}\left(\frac{r}{1+k^2r^2}\frac{\partial\psi_p}{\partial r}\right)$$

$$+\frac{1}{\sigma}\left[\left(\frac{1}{1+k^2r^2}\right)\frac{\partial\psi_p}{\partial r}\frac{\partial\sigma}{\partial r}+\frac{1}{r^2}\frac{\partial\psi_p}{\partial\phi}\frac{\partial\sigma}{\partial\phi}\right]=-I, \qquad \dots[4.46]$$

where

$$I=\frac{1}{\sigma}\left[\left(\frac{1}{1+k^2r^2}\right)\frac{\partial\sigma}{\partial r}(krB_{zc}-B_{\theta c})+B_{rc}\frac{1}{r}\frac{\partial\sigma}{\partial\phi}\right]$$

$$+\frac{1}{\sigma}\frac{\partial p_\parallel}{\partial\psi}+\frac{ff'}{\sigma^2(1+k^2r^2)}-\frac{2kF(\psi_p)}{\sigma(1+k^2r^2)^2}.$$

The domain for equation [4.46] is $0\leqslant r\leqslant R$ and $\phi_a\leqslant\phi\leqslant\phi_b$ where ϕ_a and ϕ_b depend on the symmetry of the vacuum helical field.[47] The boundary conditions on ψ_p are $\psi_p(0,\phi)=0$ and $\psi_p(R,\phi)=0$ for all ϕ and $\partial\psi_p/\partial\phi=0$ at $\phi=\phi_a$ and $\phi=\phi_b$ for all r.

The difference approximation to equation [4.46] is similar to that of equation [4.25]. We obtain the algebraic system [4.29] to solve with $\psi\equiv\psi_p$, $\bar\psi=\psi$ and

$$a_{i,j}=\frac{2r_{j+\frac{1}{2}}}{r_j(1+k^2r_{j+\frac{1}{2}}^2)(r_{j+1}-r_{j-1})(r_{j+1}-r_j)}$$

$$+\frac{\sigma_{i,j+1}-\sigma_{i,j-1}}{\sigma_{i,j}(1+k^2r_j^2)(r_{j+1}-r_{j-1})^2}$$

$$b_{i,j}=\frac{2r_{j-\frac{1}{2}}}{r_j(1+k^2r_{j-\frac{1}{2}}^2)(r_{j+1}-r_{j-1})(r_j-r_{j-1})}$$

$$-\frac{\sigma_{i,j+1}-\sigma_{i,j-1}}{\sigma_{i,j}(1+k^2r_j^2)(r_{j+1}-r_{j-1})^2}$$

$$c_{i,j}=\frac{2}{r_j^2(\phi_{i+1}-\phi_{i-1})(\phi_{i+1}-\phi_i)}+\frac{\sigma_{i+1,j}-\sigma_{i-1,j}}{\sigma_{i,j}r_j^2(\phi_{i+1}-\phi_{i-1})^2}$$

$$d_{i,j}=\frac{2}{r_j^2(\phi_{i+1}-\phi_{i-1})(\phi_i-\phi_{i-1})}-\frac{\sigma_{i+1,j}-\sigma_{i-1,j}}{\sigma_{i,j}r_j^2(\phi_{i+1}-\phi_{i-1})^2}$$

$$\phi_{i,j}=I_{i,j}=\frac{1}{\sigma_{i,j}}\left[\frac{kr_j(B_{zc})_{i,j}-(B_{\theta c})_{i,j}}{1+k^2r_j^2}\left(\frac{\sigma_{i,j+1}-\sigma_{i,j-1}}{r_{j+1}-r_{j-1}}\right)\right.$$

$$\left.+\frac{(B_{rc})_{i,j}}{r_j}\left(\frac{\sigma_{i+1,j}-\sigma_{i-1,j}}{\phi_{i+1}-\phi_{i-1}}\right)\right]+\frac{1}{\sigma_{i,j}}\left(\frac{\partial p_\parallel}{\partial\psi}\right)_{i,j}$$

$$+\left(\frac{ff'}{\sigma^2}\right)_{i,j}\frac{1}{1+k^2r_j^2}-\frac{2kF_{i,j}}{\sigma_{i,j}(1+k^2r_j^2)^2}.$$

The magnetic fields are computed from the difference approximations to equation [4.44]. The pressure distributions of the preceding sections can be used for this case.

4.6. *Method of solution of the difference equations*

We wish to solve the system given by equation [4.29]. We use an iterative solution based on the alternating-direction implicit scheme introduced in Section 2 for two-dimensional diffusion problems. In this case the integer n denotes the iteration cycle number. For n even, we solve the system

$$-a^n_{i,j}\psi^{n+1}_{i,j+1}+(a^n_{i,j}+b^n_{i,j}+\rho)\psi^{n+1}_{i,j}-b^n_{i,j}\psi^{n+1}_{i,j-1} = \delta^n_{i,j} \qquad \ldots[4.47]$$

where

$$\delta^n_{i,j} = c^n_{i,j}\psi^n_{i+1,j}-(c^n_{i,j}+d^n_{i,j}-\rho)\psi^n_{i,j}+d^n_{i,j}\psi^n_{i-1,j}+\phi^n_{i,j};$$

and for n odd, we solve the system

$$-c^n_{i,j}\psi^{n+1}_{i+1,j}+(c^n_{i,j}+d^n_{i,j}+\rho)\psi^{n+1}_{i,j}-d^n_{i,j}\psi^{n+1}_{i-1,j} = \gamma^n_{i,j} \qquad \ldots[4.48]$$

where

$$\gamma^n_{i,j} = a^n_{i,j}\psi^n_{i,j+1}-(a^n_{i,j}+b^n_{i,j}-\rho)\psi^n_{i,j}+b^n_{i,j}\psi^n_{i,j-1}+\phi^n_{i,j}$$

where $\rho > 0$ is an iteration constant, which can be adjusted for speed of convergence but must be kept constant on each even-odd cycle.

There is a convenient method for solving the systems of equations given above which is a generalization of the one-dimensional method described in Section 3. For n even, we use the algorithm

$$\psi^{n+1}_{i,j} = E^n_{i,j}\psi^{n+1}_{i,j+1}+F^n_{i,j} \qquad \ldots[4.49]$$

for all $i,j = 0, 1, 2, \ldots, J-1$ where

$$E^n_{i,j} = \frac{a^n_{i,j}}{a^n_{i,j}+b^n_{i,j}+\rho-b^n_{i,j}E^n_{i,j-1}}$$

$$F^n_{i,j} = \frac{\delta^n_{i,j}+b^n_{i,j}F^n_{i,j-1}}{a^n_{i,j}+b^n_{i,j}+\rho-b^n_{i,j}E^n_{i,j-1}}$$

for all i and $j = 1, 2, 3, \ldots, J-1$. We use the boundary condition at $r = 0$ to determine $E^n_{i,0}$ and $F^n_{i,0}$; for example, if $\psi = 0$ at $r = 0$ we have

$$E^n_{i,0} = 0 \quad \text{and} \quad F^n_{i,0} = 0 \quad \text{for all } i \text{ and } n.$$

The computation consists of a double sweep along each i-line; on the first sweep the E's and F's are calculated, and on the second sweep the ψ's are calculated using the algorithm and the boundary value at $r = R$. If $\psi(R, z, t) = 0$, the $\psi^{n+1}_{i,j} = 0$ for all i and n; if $\partial\psi/\partial r = 0$ at $r = R$, then

$$\psi^{n+1}_{i,j} = \frac{F^n_{i,J-1}}{1-E^n_{i,J-1}} \qquad \text{for all } i \text{ and } n.$$

For n odd, we use the algorithm

$$\psi^{n+1}_{i,j} = E^n_{i,j}\psi^{n+1}_{i+1,j}+F^n_{i,j} \qquad \ldots[4.50]$$

for all j and $i = 0, 1, 2, 3, \ldots, I-1$, where

$$E_{i,j}^n = \frac{c_{i,j}^n}{c_{i,j}^n + d_{i,j}^n + \rho - d_{i,j}^n E_{i-1,j}^n}$$

$$F_{i,j}^n = \frac{\gamma_{i,j}^n + d_{i,j}^n F_{i-1,j}^n}{c_{i,j}^n + d_{i,j}^n + \rho - d_{i,j}^n E_{i-1,j}^n}$$

for all j and $i = 1, 2, 3, \ldots, I-1$. We use the boundary condition at $z = 0$ to determine $E_{0,j}^n$ and $F_{0,j}^n$; for example, if $\partial\psi/\partial z = 0$ we have

$$E_{0,j}^n = 1 \quad \text{and} \quad F_{0,j}^n = 0 \quad \text{for all } j \text{ and } n.$$

The computational procedure is again a double sweep using the above algorithm and the boundary value at $z = l$; e.g., if $\psi(r, l, t) = 0$ we have

$$\psi_{I,j}^{n+1} = 0 \text{ for all } j \text{ and } n.$$

5

NUMERICAL SOLUTION OF THE FOKKER-PLANCK EQUATIONS
FOR A PLASMA

5.1. *Time-dependent, two-species, isotropic velocity distributions*

In those experiments in controlled fusion research that employ the injection of energetic neutral atoms into a magnetic mirror configuration, a plasma is formed of initially hot ions and cold electrons. It is of interest to know the velocity-distribution functions of the electrons and ions as a function of time during the build-up of the plasma. The most suitable mathematical description is by means of the Fokker-Planck equations for the ion and electron distribution functions. This is because the dominant mechanism for energy transfer among the particles is by long range Coulomb interactions. The Fokker-Planck equations for the distribution functions of several species of particle, where the two-body force is an inverse-square law, have been derived in the paper of Rosenbluth, MacDonald, and Judd.[49] They use spherical polar coordinates in velocity space (v, θ, ϕ) where θ is the angle between the velocity vector and the magnetic field vector. They assume azimuthal symmetry, so the resulting distribution functions are of the form $f(v, \theta, t)$. Calculations performed in this two-dimensional velocity space[50, 51] for the ion distribution function indicate that good results can be obtained by separating the distribution function into a product of two terms. The first term is a function of v and t and the second term is a function of θ only. The equation for the function of θ is a Legendre differential equation on the domain $-\theta_c \leqslant \theta \leqslant \theta_c$, where θ_c defines the magnetic mirror loss cone. The equation for $f(v, t)$ must be solved numerically. The boundary condition on the distribution function in such a loss cone problem is $f(v, \theta_c, t) = 0$ for all v and t for each species which implies $f = 0$ at $v = 0$ in the separated solution. In those problems where we assume that the distribution functions are isotropic we take a symmetry condition at $v = 0$, i.e. $\partial f / \partial v = 0$ for all t.

Spatial dependence is not included in the model. In an earlier calculation,[52] which included the finite orbit size of the ions and the spatial dependence of the trapping process, it was found that the solutions for ion density exhibit growth rates similar to those obtained when the plasma density is assumed to be uniform.

In the equations for ions and electrons we include source terms which are appropriate for the neutral injection experiments such as ALICE[53] and PHOE-NIX.[54] We also include the loss of both species by scattering into the velocity space loss cone of the magnetic mirror configuration, and the hot ions can be lost by charge-exchange with the background gas.

A plasma potential is computed at each time step of the calculation by requiring charge neutrality. A critical velocity $v_c(t)$ is determined, such that electrons with $v < v_c$ are not lost and those with $v > v_c$ can be lost by scattering into the loss cone. At each time step the electron density is compared to the ion density and the velocity v_c modified accordingly. The plasma potential is obtained from $e\phi = \frac{1}{2}mv_c^2$.

We have coupled non-linear partial differential equations for the functions $f_e(v, t)$ and $f_i(v, t)$. We solve the equations numerically using finite difference methods. The equations are not linearized. An implicit difference scheme is used, i.e. the velocity derivatives are replaced by difference quotients taken at the new time step while the coefficients are evaluated using the distribution function of the previous time step, and extrapolated. The scheme is stable numerically in practice with no restriction on the time step. This is an essential part of the calculations, because, as the electron temperature increases, the transfer rate decreases and the time step, Δt, must be continually increased during the calculation in order to progress toward equilibrium in a sensible manner.

The equation for each species is

$$
\begin{aligned}
(4\pi\Gamma_a)^{-1} \frac{\partial f_a}{\partial t} = {} & \frac{\partial^2 f_a}{\partial v^2} \left\{ \sum_b \left[\frac{1}{3v^3} \int_0^v f_b(v', t) v'^4 dv' + \frac{1}{3} \int_v^\infty f_b(v', t) v' dv' \right] \right\} \\
& + \frac{\partial f_a}{\partial v} \left\{ \frac{1}{v} \sum_b \left[\frac{m_a}{m_b} \frac{1}{v} \int_0^v f_b(v', t) v'^2 dv' - \frac{1}{3v^3} \int_0^v f_b(v', t) v'^4 dv' \right. \right. \\
& \left. \left. + \frac{2}{3} \int_v^\infty f_b(v', t) v' dv' \right] \right\} + f_a \left(\sum_b \frac{m_a}{m_b} f_b \right) \\
& - f_a \left\{ \frac{p_a(v)}{v^3} \sum_b \left[\left(1 + \frac{m_a}{m_b} \right) \int_0^v f_b(v', t) v'^2 dv' \right. \right. \\
& \left. \left. - \frac{1}{3v^2} \int_0^v f_b(v', t) v'^4 dv' + \frac{2v}{3} \int_v^\infty f_b(v', t) v' dv' \right] \right\} + s_a(v, t).
\end{aligned}
$$

$$...[5.1]$$

The summations are taken over all the species being considered, including type a, and $p_a(v)$ is the probability that particles of type a and velocity v will be lost. The term for charge-exchange loss must be added to the above equation for ions. The term $s_a(v, t)$ represents the source of injected particles. The constant Γ_a is defined by

$$\Gamma_a = \frac{4\pi e^4}{m_a^2} \ln D$$

where D is the ratio of the Debye length to the classical distance of closest approach.

The number density of particles of type a is given by

$$n_a(t) = 4\pi \int_0^\infty f_a(v, t) v^2 dv.$$

We consider electrons and ions of $Z = 1$. We introduce the dimensionless variable $x = v/v_0$, where v_0 is a constant and is a characteristic velocity. Let $f = (4\pi v_0^3/K_e)f_e$, where K_e is determined from the equation

$$n_e(0) = K_e \int_0^\infty f(x, 0) x^2 dx;$$

i.e. the constant is determined by the initial condition with $n_e(0)$ equal to the initial electron density. Similarly, we let $g = (4\pi v_0^3/K_i)f_i$, where

$$n_i(0) = K_i \int_0^\infty g(x, 0) x^2 dx.$$

We introduce the dimensionless variable τ where $\tau = (\frac{1}{2}\Gamma_e K_e/v_0^3)t$. Let $\mu = m_e/m_i$ and $K = K_i/K_e$. We define functionals

$$M(f) = \int_x^\infty f(y, \tau) y \, dy, \qquad \qquad ...[5.2]$$

$$N(f) = \int_0^x f(y, \tau) y^2 dy, \qquad \qquad ...[5.3]$$

and

$$E(f) = \int_0^x f(y, \tau) y^4 dy. \qquad \qquad ...[5.4]$$

In terms of these new variables, the equation for the electron-distribution function becomes

$$\frac{\partial f}{\partial \tau} = A \frac{\partial^2 f}{\partial x^2} + B \frac{\partial f}{\partial x} + Cf + D \qquad \qquad ...[5.5]$$

where

$$A = \tfrac{2}{3}\left\{\left[\frac{1}{x^3}E(f)+M(f)\right]+K\left[\frac{1}{x^3}E(g)+M(g)\right]\right\}$$

$$B = \frac{4}{3x}\left\{\left[\frac{3}{2x}N(f)-\frac{1}{2x^3}E(f)+M(f)\right]+K\left[\mu\frac{3}{2x}N(g)-\frac{1}{2x^3}E(g)+M(g)\right]\right\}$$

and

$$C = 2(f+K\mu g)-p_e(x)\frac{4}{3x^2}\left\{2\left[\frac{3}{2x}N(f)-\frac{1}{2x^3}E(f)+M(f)\right]\right.$$
$$\left.+K(1+\mu)\left[\frac{3}{2x}N(g)-\frac{1}{2x^3}E(g)+M(g)\right]\right\}.$$

The term $D(x, \tau)$ describes the time-dependent source of electrons. The corresponding equation for the ion distribution is

$$\frac{\partial g}{\partial \tau} = F\frac{\partial^2 g}{\partial x^2}+G\frac{\partial g}{\partial x}+Hg+L \qquad \qquad ...[5.6]$$

where

$$F = \tfrac{2}{3}\mu^2\left\{\left[\frac{1}{x^3}E(f)+M(f)\right]+K\left[\frac{1}{x^3}E(g)+M(g)\right]\right\}$$

$$G = \frac{4}{3x}\mu^2\left\{\left[\frac{1}{\mu}\frac{3}{2x}N(f)-\frac{1}{2x^3}E(f)+M(f)\right]+K\left[\frac{3}{2x}N(g)-\frac{1}{2x^3}E(g)+M(g)\right]\right\}$$

$$H = 2\mu^2\left(\frac{1}{\mu}f+Kg\right)-H_1(x, \tau)-\mu^2 p_i(x)\frac{4}{3x^2}\left\{\tfrac{1}{2}\left(1+\frac{1}{\mu}\right)\left[\frac{3}{2x}N(f)\right.\right.$$
$$\left.\left.-\frac{1}{2x^3}E(f)+M(f)\right]+K\left[\frac{3}{2x}N(g)-\frac{1}{2x^3}E(g)+M(g)\right]\right\}.$$

The term $H_1(x, \tau)$ contains the charge-exchange-loss term, and $L(x, \tau)$ describes the time-dependent source of ions.

At any time step we can determine the number density and average energy of each type of particle. Let $I_2^-(\tau)$ and $I_4^-(\tau)$ be the second and fourth moments of the electron distribution function, i.e.

$$I_2^-(\tau) = \int_0^\infty f(x, \tau)x^2 dx$$

$$I_4^-(\tau) = \int_0^\infty f(x, \tau)x^4 dx.$$

The number of density electrons is given by

$$n_e(\tau) = K_e I_2^-(\tau) \qquad \qquad ...[5.7]$$

and the mean electron energy is given by

$$E_e(\tau) = \tfrac{3}{2}kT_e = \tfrac{1}{2}m_e v_0^2 \frac{I_4^-(\tau)}{I_2^-(\tau)}. \qquad \qquad ...[5.8]$$

The number density of ions is given by

$$n_i(\tau) = K_i I_2^+(\tau) \qquad \qquad ...[5.9]$$

and the mean ion energy is given by

$$E_i(\tau) = \tfrac{3}{2}kT_i = \tfrac{1}{2}m_i v_0^2 \frac{I_4^+(\tau)}{I_2^+(\tau)}, \qquad \qquad ...[5.10]$$

where $I_2^+(\tau)$ and $I_4^+(\tau)$ are the second and fourth moments of g.

We shall describe the form of the source terms that appear in equations [5.5] and [5.6]. The mechanisms for trapping the neutral-atom beam are the Lorentz ionization of excited hydrogen atoms and ionization of the neutral beam atoms by collisions with background gas molecules and with previously trapped ions and electrons.

We assume that the injected electrons and ions have a velocity distribution defined by $S_e(x)$ and $S_i(x)$. In the calculations we keep $n_e(\tau) = n_i(\tau)$ by adjusting the plasma potential. We have

$$D(x, \tau) = S_e(x)[l_0 + l_1 n_i(\tau)] \qquad \qquad ...[5.11]$$

$$L(x, \tau) = S_i(x)[p_0 + p_1 n_i(\tau)] \qquad \qquad ...[5.12]$$

where

$$l_0 = \frac{2v_0^3}{\Gamma_e K_e} \frac{1}{K_e N(S_e)} \frac{If^*}{V}$$

$$l_1 = \frac{2v_0^3}{\Gamma_e K_e} \frac{1}{K_e N(S_e)} \left[\frac{IL}{V} \left(\frac{\overline{\sigma_i^l v_r}}{v_0} + \frac{\overline{\sigma_i^e v_r}}{v_0} \right) + n_0 v(\sigma_i^i + \sigma_i^e) \right]$$

$$p_0 = \frac{2v_0^3}{\Gamma_e K_e} \frac{1}{K_i N(S_i)} \frac{If^*}{V}$$

$$p_1 = \frac{2v_0^3}{\Gamma_e K_e} \frac{1}{K_i N(S_i)} \left[\frac{IL}{V} \left(\frac{\overline{\sigma_i^l v_r}}{v_0} + \frac{\overline{\sigma_i^e v_r}}{v_0} \right) \right]$$

$$N(S_e) = \int_0^\infty S_e(x)x^2 dx \quad \text{and} \quad N(S_i) = \int_0^\infty S_i(x)x^2 dx.$$

The function $n_i(\tau)$ is given by equation [5.9], and I is the injected neutral beam current, V is the plasma volume, f^* is the fraction of the neutral beam ionized by the Lorentz force, L is the path length of the neutral beam through the plasma, v_r is the relative velocity between interacting particles, and v_0 is the characteristic velocity, defined earlier, which is determined by the beam velocity. The cross-sections for ionization of the beam atoms by collisions with hot ions and electrons are σ_i^i and σ_t^e. Cross-sections for ionization of the background gas by hot ions and electrons are σ_i^i and σ_i^e, and n_0 is the background gas density. In a magnetic mirror field, the cold ions (produced by charge-exchange collisions and by ionization of the background gas) may be neglected since they are rapidly scattered into the mirror escape cone and lost from the system, hence these terms are omitted from the above equation.

We can include up to ten sources of the type given by equation [5.12], corresponding to multiple ion beam injection at different energies. In the above discussion of source terms, the (σv) terms were treated as constants; however, the cross-sections have a velocity-dependence determined by experimental measurements. We have polynomial descriptions for these functions; so the terms $\sigma v n(\tau)$ can be replaced by integrals involving the distribution functions. This is illustrated in the next paragraph on charge-exchange loss.

In equation [5.6] the charge-exchange loss term is $-H(x)g(x, \tau)$, where

$$H_1(x) = (t/\tau)n_0 v \sigma_{cx}(v) = (t/\tau)n_0 v_0 x \sigma_{cx}(x), \qquad \ldots[5.13]$$

and $\sigma_{cx}(v)$ is the charge-exchange cross-section. We have fitted the experimental cross-section, σ_{cx}, with a fifth degree polynomial, so we write

$$H_1(x) = x[H_{a1} + H_{b1}x + H_{c1}x^2 + H_{d1}x^3 + H_{e1}x^4 + H_{h1}x^5] \quad \ldots[5.14]$$

where the coefficients are constants, including the constant factor, $(t/\tau)n_0 v_0$, which is an input parameter of the problem.

We denote the value of the magnetic field in the midplane by B_0, and the value at the mirror by B_{max}. We consider a plasma potential which has value ϕ in the midplane and goes to zero at the mirror. Let W be the kinetic energy of a particle, then the total energy $H = W \pm e\phi$ is a constant of the charged particle motion. We also assume that the magnetic moment $\lambda = W_\perp/B$ is a constant of the motion, where $W_\perp = \frac{1}{2}mv_\perp^2$ and $v_\perp$ is the component of velocity perpendicular to the magnetic field. The pitch angle in the midplane, α, is defined by

$$\sin^2 \alpha = \frac{W_\perp \text{ (at } B_0)}{W \text{ (at } B_0)} = \frac{\lambda B_0}{W \text{ (at } B_0)} = \frac{B_0}{W \text{ (at } B_0)} \frac{W_\perp \text{ (at } B_{max})}{B_{max}}.$$

We define the critical pitch angle by the condition that $v_\parallel = 0$ at B_{max}, and since $\phi = 0$ at B_{max} we have

$$W_\perp \text{ (at } B_{max}) = H = W \text{ (at } B_0) \pm e\phi.$$

The critical pitch angle is then given by

$$\sin^2 \alpha_c^\pm = \frac{B_0}{B_{max}} \left(1 \pm \frac{e\phi}{W \text{ (at } B_0)}\right). \qquad \ldots[5.15]$$

Electrons with energy $|W| < |e\phi|$ are not lost and electrons with $|W| > |e\phi|$ are lost with probability $p_e = 1 - \cos \alpha_c$. The plasma potential is given by

$$e\phi = \tfrac{1}{2} m_e (v_c^-)^2. \qquad \qquad \ldots [5.16]$$

The procedure for determining v_c^- is the following. At every time step, $n_e(\tau)$ and $n_i(\tau)$ are computed from equations [5.7] and [5.9], and the difference, $n_i(\tau) - n_e(\tau)$, is also computed. During the build-up of plasma, electrons tend to be lost faster than ions. Since we wish to keep $n_e(\tau) = n_i(\tau)$, the above difference is compared with a pre-assigned small number. If the difference exceeds this number, then v_c^- is increased by an amount Δv_c in order to decrease the electron loss rate and the time step is repeated. This process is repeated until $n_i(\tau) - n_e(\tau)$ is sufficiently small, and the calculation continues. As the plasma builds up and the electron energy increases, the plasma potential also increases.

The term $p_e(x)$ is

$$p_e(x) = \begin{cases} 1 - \sqrt{\left[1 - \dfrac{1}{R}\left(1 - \dfrac{x_c^{-2}}{x^2}\right)\right]} & x \geqslant x_c \\[4mm] 0 & x \leqslant x_c \end{cases} \qquad \ldots [5.17]$$

where $v_0 x_c^- = v_c^-$, and the term $p_i(x)$ which appears in equation [5.6] is then

$$p_i(x) = 1 - \sqrt{\left[1 - \dfrac{1}{R}\left(1 - \dfrac{x_c^{+2}}{x^2}\right)\right]} \qquad \qquad \ldots [5.18]$$

where $R = B_{\max}/B_0$. If the above square root becomes imaginary, then $p_i(x) = $ a given constant.

We wish to solve the two non-linear differential equations [5 5] and [5.6] on the domain $0 \leqslant x \leqslant \infty$, $\tau \geqslant 0$, with the boundary conditions $f \to 0$, $g \to 0$ as $x \to \infty$, and $\partial f/\partial x = \partial g/\partial x = 0$ at $x = 0$ for $\tau > 0$, or in the separated solution case we have $f = g = 0$ at $x = 0$. The initial distributions $f(x, 0)$ and $g(x, 0)$ are given.

For the numerical solution we choose a domain $0 \leqslant x \leqslant x_J$, where x_J is specified for each problem and is taken large enough to include the high velocity tail of the electron distribution. As the electrons increase in temperature, the distribution spreads out; thus the choice of x_J determines when the calculation must be stopped in order to preserve accuracy. At $x = x_J$, we take the boundary condition $f = g = 0$.

In the domain $0 \leqslant x \leqslant x_J$, $\tau \geqslant 0$, consider the finite-difference mesh defined by $x_j = j\Delta x$, $j = 0, 1, 2, \ldots, J$; $\tau^n = n\Delta\tau$, $n = 0, 1, 2, \ldots$. Let $f_j^n = f(x_j, \tau^n)$ and $g_j^n = g(x_j, \tau^n)$; $A_j^n = A(f_j^n, g_j^n, x_j, \tau^n)$, $B_j^n = B(f_j^n, g_j^n, x_j, \tau^n)$, etc. We define the first and second difference approximations by

$$(-\delta f)_j^n = \frac{f_{j+1}^n - f_{j-1}^n}{2\Delta x}$$

$$(\delta^2 f)_j^n = \frac{f_{j+1}^n - 2f_j^n + f_{j-1}^n}{(\Delta x)^2}.$$

We approximate equations [5.5] and [5.6] by the following implicit difference equations

$$\frac{f_j^{n+1} - f_j^n}{\Delta \tau} = \rho[A_j^{n+1}(\delta^2 f)_j^{n+1} + B_j^{n+1}(\delta f)_j^{n+1} + C_j^{n+1}f_j^{n+1} + D_j^{n+1}]$$

$$+ (1-\rho)[A_j^n(\delta^2 f)_j^n + B_j^n(\delta f)_j^n + C_j^n f_j^n + D_j^n]$$

and

$$\frac{g_j^{n+1} - g_j^n}{\Delta \tau} = \rho[F_j^{n+1}(\delta^2 g)_j^{n+1} + G_j^{n+1}(\delta g)_j^{n+1} + H_j^{n+1}g_j^{n+1} + L_j^{n+1}]$$

$$+ (1-\rho)[F_j^n(\delta^2 g)_j^n + G_j^n(\delta g)_j^n + H_j^n g_j^n + L_j^n]$$

where $\frac{1}{2} \leqslant \rho \leqslant 1$. We wish to solve these equations for the unknowns f_j^{n+1} and g_j^{n+1}; $j = 0, 1, 2, \ldots, J$.

We write the above difference equations as a set of simultaneous algebraic equations:

$$\alpha_j^{n+1}f_{j+1}^{n+1} - (1+\beta_j^{n+1})f_j^{n+1} + \gamma_j^{n+1}f_{j-1}^{n+1} = \psi_j^n \qquad \ldots[5.19]$$

$$\zeta_j^{n+1}g_{j+1}^{n+1} - (1+\eta_j^{n+1})g_j^{n+1} + \theta_j^{n+1}g_{j-1}^{n+1} = \phi_j^n \qquad \ldots[5.20]$$

for $j = 1, 2, \ldots, J-1$.

In equations [5.19] and [5.20] we have the unknowns f_j^{n+1}, g_j^{n+1}, $j = 1$, $\ldots, J-1$ on the left-hand side of the equation, and the known quantities on the right-hand side. We are interested in solving these equations for the interior points, $j = 1, \ldots, J-1$, since the boundary conditions at $x = 0$ and $x = x_J$ determine the solutions for $j = 0$ and $j = J$. Consequently, we do not have to worry about singularities at $x = 0$ in the coefficients. The system given by equations [5.19] and [5.20] is non-linear in the unknowns f_j^{n+1}, g_j^{n+1}. If we extrapolate the coefficients, α_j^{n+1}, β_j^{n+1}, etc., from their values at the previous times τ^n, τ^{n-1}, then equations [5.19] and [5.20] become a linear algebraic system in the unknowns f_j^{n+1}, g_j^{n+1}. The procedure is to extrapolate the coefficients and solve the linear system, then compute the coefficients α_j^{n+1}, β_j^{n+1}, etc. with the new values of f_j^{n+1}, g_j^{n+1}. This procedure works very well, since the coefficients change in a very smooth manner with time.

The method of solution of equations [5.19] and [5.20] is the same as that given in Section 3 for coupled diffusion equations. Further details of the difference scheme for this problem together with numerical results of sample problems have been published.[55]

5.2. *Energy and angular dependent ion distribution, Maxwellian electrons*

A computer programme has been written[56] which solves the Fokker-Planck equation for the ion distribution function $f_i(v, \theta, t)$ in two-dimensional velocity space. Electrons are also included and are described by a Maxwellian velocity distribution with the density n_e, and mean energy E_e, functions of the time. A source of ions and electrons is included as in the preceding section and the plasma potential is computed in the same manner by keeping the ion and electron densities equal to each other. The electrons are heated by the hot ions.

The ion equation is of the form[49]

$$\frac{\partial f}{\partial t} = a_1 \frac{\partial^2 f}{\partial x^2} + a_2 \frac{\partial^2 f}{\partial x \partial \theta} + a_3 \frac{\partial^2 f}{\partial \theta^2} + b_1 \frac{\partial f}{\partial x} + b_2 \frac{\partial f}{\partial \theta} + cf + d \quad ...[5.21]$$

with $x = v/v_0$. The coefficients involve moments of the ion and electron distributions as in the one-dimensional case, but here the functionals are all double integrals. This part of the programme is quite time-consuming, so they are not computed at every time step.

Equation [5.21] is solved by the alternating-direction implicit scheme which has been described in Sections 2 and 4 of these lectures. The domain for equation [5.21] is $0 \leqslant x \leqslant x_{\text{max}}$, $-\theta_c \leqslant \theta \leqslant \theta_c$, where θ_c is determined by the loss cone of the magnetic-mirror field. The boundary conditions for a loss-cone system are

$$f(0, \theta, t) = f(x_{\text{max}}, \theta, t) = 0, \quad \text{all } \theta,$$
$$f(x, -\theta_c, t) = f(x, \theta_c, t) = 0, \quad \text{all } x.$$

A z-dependent magnetic mirror field has been added to the programme[57]; so $f(z, v, \theta, t)$ is computed from the Boltzmann-Fokker-Planck equation. The plasma potential ϕ and the loss angle θ_c become functions of z. The results of these computations[57] will be published.

6

NUMERICAL SOLUTION OF THE VLASOV EQUATION

6.1. *One-dimensional models*

Consider the system of equations

$$\frac{\partial f_e}{\partial t} + v \frac{\partial f_e}{\partial x} - \frac{eE}{m} \frac{\partial f_e}{\partial v} = 0 \qquad ...[6.1]$$

$$\frac{\partial f_i}{\partial t} + v \frac{\partial f_i}{\partial x} + \frac{eE}{M} \frac{\partial f_i}{\partial v} = 0 \qquad ...[6.2]$$

$$\frac{dE}{dx} = 4\pi e \int_{-\infty}^{\infty} (f_i - f_e) \, dv \qquad ...[6.3]$$

where $f_e(x, v, t)$ and $f_i(x, v, t)$ are electron and ion distribution functions, m and M are the electron and ion masses, $-e$ and $+e$ $(Z = 1)$ are the charges, and $E(x, t)$ is the electric field.

This system has probably received more attention than any other in plasma physics, as far as computing is concerned. This is because of its straightforward, simple appearance and its importance to fundamental problems in plasma physics. This model is used in studies of instabilities such as the two-stream instability, in collisionless shocks, Landau damping and other problems. It is called the nonlinear Vlasov equation, which is rather a redundant name; it is obviously nonlinear since E is a function of f_e and f_i. However, a great deal of analytical effort has been devoted to the linearized Vlasov equation.

We shall not give a list of references for the computational work as it would be far too long. In April 1967 at Williamsburg, Virginia, a symposium was held on Computer Simulation of Plasma and Many-Body Problems. Most of the people working on the computational aspects of the Vlasov equation gave papers. The proceedings were published as a book by the National Aeronautics and Space Administration (NASA-SP-153). In addition, the book contains a bibliography prepared by C. K. Birdsall on computer experiments in electronics and plasmas.

We can write either equation [6.1] or [6.2] as

$$\frac{\partial f}{\partial t} + v \frac{\partial f}{\partial x} + F \frac{\partial f}{\partial v} = 0. \qquad \qquad ...[6.4]$$

The above equation can also be written

$$\frac{Df}{Dt} = 0 \qquad \qquad ...[6.5]$$

where

$$\frac{D}{Dt} = \frac{\partial}{\partial t} + v \frac{\partial}{\partial x} + F \frac{\partial}{\partial v}.$$

We can refer to equations [6.4] and [6.5] as the Eulerian and Lagrangian forms of the equation. Equation [6.5] implies that f remains constant along a trajectory which satisfies the equations

$$\frac{dx}{dt} = v \qquad \qquad ...[6.6]$$

$$\frac{dv}{dt} = F \qquad \qquad ...[6.7]$$

i.e. $f_j = $ constant

along a curve $x_j(t)$, $v_j(t)$ where $x_j(t)$ and $v_j(t)$ are solutions of equations [6.6] and [6.7] with $x(0) = x_j$, $v(0) = v_j$.

The various "sheet" models are based on equations [6.5]–[6.7]. In these programmes the equations of motion of a few thousand sheets are solved numerically. Generally a very simple integration scheme is used to solve equations [6.6] and [6.7] for $x_j(t)$, $v_j(t)$; $j = 1, \ldots, J$, in order to save time.

The methods of computing the integral

$$\int_{-\infty}^{\infty} f(x, v) \, dv$$

appearing in equation [6.3] vary, but generally involve counting the number of sheets in a given interval Δx or computing $\sum_k f_k \Delta v_k$ in such an interval.

There are a number of problems connected with this model such as the crossing of sheets, the number of sheets required, the mass ratio in two-species problems, and the interpretation of the results. Nevertheless it is quite popular, particularly in one-species problems. The sheet model and the corresponding two-dimensional model which we shall discuss later form the basis of what many people call "computer experiments".

Another method which is also based on equations [6.5]–[6.7] is called the "water-bag" model. In this method step-function distributions are used and the boundaries of a region in phase space of constant f are computed as a function of time by solving equations [6.6] and [6.7] for a finite set of boundary points. At each time-step a new boundary curve is constructed from the discrete points. Considerable twisting and distortion take place, so movies of the boundary in the x-v plane are useful.

A third method involves solving the Vlasov equation in the form given by equation [6.4] by using a doubly expanded representation for $f(x, v, t)$ such as

$$f(x, v, t) = \sum_{n=-\infty}^{\infty} e^{inkx} \sum_{m=0}^{\infty} e^{-v^2/2} h_m(v) Z_{mn}(t)$$

where the $h_m(v)$ are Hermite functions. Other expansions have also been used. The above series is truncated at some point and the resulting finite set of equations for $Z_{mn}(t)$ are solved numerically.

A fourth method involves solving the Eulerian form of the equation directly by finite-difference methods. We can write equations [6.1] and [6.2] as the system

$$\frac{\partial f}{\partial t} + \frac{\partial G}{\partial x} + \frac{\partial H}{\partial v} = 0 \qquad \ldots [6.8]$$

where

$$f = \begin{bmatrix} f_e \\ f_i \end{bmatrix}; \qquad G = \begin{bmatrix} v f_e \\ v f_i \end{bmatrix}; \qquad H = \begin{bmatrix} F_e f_e \\ F_i f_i \end{bmatrix} \qquad \ldots [6.9]$$

with

$$F_e = -\frac{e}{m} E \quad \text{and} \quad F_i = \frac{e}{M} E.$$

Equation [6.8] is a hyperbolic system in conservation form and is well suited to the two-step Lax-Wendroff method.[9]

Let $x_i = i\Delta x$, $v_k = k\Delta v$, and $t_n = n\Delta t$. A unit cell in the finite-difference mesh has dimensions $2\Delta x$, $2\Delta v$, and $2\Delta t$. Provisional values at t_{n+1} are obtained from

$$f_{i,k}^{n+1} = \tfrac{1}{4}(f_{i+1,k}^n + f_{i-1,k}^n + f_{i,k+1}^n + f_{i,k-1}^n)$$
$$- \frac{\Delta t}{2\Delta x}(G_{i+1,k}^n - G_{i-1,k}^n) - \frac{\Delta t}{2\Delta v}(H_{i,k+1}^n - H_{i,k-1}^n) \qquad ...[6.10]$$

then final values at t_{n+2} are obtained from

$$f_{i,k}^{n+2} = f_{i,k}^n - \frac{\Delta t}{\Delta x}(G_{i+1,k}^{n+1} - G_{i-1,k}^{n+1}) - \frac{\Delta t}{\Delta v}(H_{i,k+1}^{n+1} - H_{i,k-1}^{n+1}). \qquad ...[6.11]$$

The electric field can be obtained by integrating equation [6.3] directly. For certain boundary conditions it is convenient to solve

$$\frac{d^2\phi}{dx^2} = -\rho \qquad ...[6.12]$$

where $E = -d\phi/dx$ and ρ is the right-hand side of equation [6.3]. The approximation of equation [6.12] is

$$\phi_{i+1}^n - 2\phi_i^n + \phi_{i-1}^n = -\Delta x^2 \rho_i^n$$

which can be solved by the algorithm[9] discussed earlier in Sections 2 to 5.

6.2. Two-dimensional models

Programmes have been written to solve the Vlasov equation in two dimensions, i.e. a four-dimensional phase space. Consider the equation

$$\frac{\partial f}{\partial t} + u\frac{\partial f}{\partial x} + v\frac{\partial f}{\partial y} + F_x\frac{\partial f}{\partial u} + F_y\frac{\partial f}{\partial v} = 0 \qquad ...[6.13]$$

for both species and Poisson's equation

$$\frac{\partial^2\phi}{\partial x^2} + \frac{\partial^2\phi}{\partial y^2} = -\rho. \qquad ...[6.14]$$

We can also write

$$\frac{Df}{Dt} = 0 \qquad ...[6.15]$$

where

$$\frac{D}{Dt} = \frac{\partial}{\partial t} + u\frac{\partial}{\partial x} + v\frac{\partial}{\partial y} + F_x\frac{\partial}{\partial u} + F_y\frac{\partial}{\partial v}.$$

The characteristic equations are

$$\frac{dx}{dt} = u \qquad\qquad \frac{dy}{dt} = v$$

$$\frac{du}{dt} = F_x \qquad\qquad \frac{dv}{dt} = F_y. \qquad ...[6.16]$$

The generalization of the sheet model to two dimensions is usually called the "rod" model. The ideas are similar. In some of these programmes a time-independent magnetic field is included in the force terms. A fast method of solving Poisson's equation in difference form, which has been developed by Hockney,[58] is an important feature of these programmes.

A major difficulty in two-species problems is the large ratio of ion to electron masses. In a period of considerable electron motion the ions move very little. To avoid this difficulty some of the programmes[59] use the guiding centre equations of motion instead of equation [6.16].

Equation [6.13] can also be solved by finite-difference methods. The two-step method of equations [6.10] and [6.11] can be generalized to four dimensions. A related problem in four dimensions is the self-consistent solution of the magnetic field in Astron which has been solved by finite-difference methods.[60]

In the Lawrence Radiation Laboratory's controlled-fusion experiment, Astron, relativistic electrons are injected into a cylindrical region containing an applied magnetic field. The object is to form an E-layer—a cylindrical layer of electrons—so that the self-field exceeds the applied field. The resulting configuration is intended to be axially symmetric with no azimuthal component of the magnetic field.

The mathematical model for the build-up of the electron layer and the self-field is the time-dependent Vlasov equation coupled with Maxwell's equations. Although transient radial and axial currents exist in this model, particularly in the early stages of formation and near the injection point, these are assumed to be small compared with the azimuthal current. Field components B_r and B_z can be derived from a stream function $\psi(r, z, t)$. The canonical angular momentum, p_θ, is a constant of the motion, and we assume that all electrons are injected with the same value of p_θ. Hence, we can consider an electron-distribution function, f, defined in a four-dimensional phase space (r, z, p_r, p_z). We assume that the system is electrically neutral at every point. The ion distribution is not solved explicitly, but ions are assumed to be present, providing charge neutralization of the layer.

We specify the magnetic field by the single component of the vector potential, $A_\theta(r, z, t)$. The equation for A_θ is

$$\frac{1}{c^2}\frac{\partial^2 A_\theta}{\partial t^2} - \frac{\partial^2 A_\theta}{\partial z^2} - \frac{\partial}{\partial r}\left[\frac{1}{r}\frac{\partial}{\partial r}(rA_\theta)\right] = 4\pi j_\theta. \qquad ...[6.17]$$

The canonical angular momentum may be expressed as $p_\theta = m_0 \gamma r v_\theta + (e r A_\theta / c)$ with $\gamma = [1 - (v/c)^2]^{-\frac{1}{2}}$. It is convenient to introduce the function $\psi(r, z, t)$ defined by

$$\psi = \gamma r v_\theta / c \qquad \qquad ...[6.18]$$

so that

$$\psi = \frac{p_\theta}{m_0 c} - \frac{e}{m_0 c^2} r A_\theta. \qquad ...[6.19]$$

Since all the electrons have the same p_θ, we can use ψ in place of A_θ to determine the field. From equations [6.17] and [6.18] we have

$$\frac{1}{c^2} \frac{\partial^2 \psi}{\partial t^2} - \frac{\partial^2 \psi}{\partial z^2} - r \frac{\partial}{\partial r} \left[\frac{1}{r} \frac{\partial \psi}{\partial r} \right] = -\frac{4\pi e}{m_0 c^2} r j_\theta \qquad ...[6.20]$$

and

$$B_r = \frac{m_0 c^2}{e} \frac{1}{r} \frac{\partial \psi}{\partial z} \quad \text{and} \quad B_z = -\frac{m_0 c^2}{e} \frac{1}{r} \frac{\partial \psi}{\partial r}. \qquad ...[6.21]$$

We introduce a dimensionless velocity, $u = \gamma v/c$, so that equation [6.18] becomes $r u_\theta = \psi$. In terms of u, we have $\gamma = (1 + u^2)^{\frac{1}{2}}$ or

$$\gamma = \left[1 + u_r^2 + u_z^2 + \left(\frac{\psi}{r} \right)^2 \right]^{\frac{1}{2}}. \qquad ...[6.22]$$

Let $f(r, z, u_r, u_z, t)$ be the electron-distribution function in phase space, so that $f(r, z, u_r, u_z, t) dr dz du_r du_z$ is the number of electrons in the element $dr dz du_r du_z$ at the point (r, z, u_r, u_z) at time t. The dimensions of f are the number of electrons per square centimetre. The equation governing f is

$$\frac{\partial f}{\partial t} + \frac{\partial f}{\partial r} \frac{dr}{dt} + \frac{\partial f}{\partial z} \frac{dz}{dt} + \frac{\partial f}{\partial u_r} \frac{du_r}{dt} + \frac{\partial f}{\partial u_z} \frac{du_z}{dt} = S(r, z, u_r, u_z, t) \quad ...[6.23]$$

where S expresses the source of electrons injected into the phase space, and

$$\frac{dr}{dt} = \frac{c u_r}{\gamma} \quad \text{and} \quad \frac{dz}{dt} = \frac{c u_z}{\gamma}. \qquad ...[6.24]$$

We determine du_r/dt and du_z/dt from the relativistic equations of motion of the electrons. Using equations [6.18], [6.21] and [6.24], we obtain

$$\frac{du_r}{dt} = -\frac{c}{\gamma} \frac{\partial}{\partial r} \left(\frac{\psi^2}{2r^2} \right) \qquad ...[6.25a]$$

and

$$\frac{du_z}{dt} = -\frac{c}{\gamma} \frac{\partial}{\partial z} \left(\frac{\psi^2}{2r^2} \right). \qquad ...[6.25b]$$

The equation for f can now be written

$$\frac{\gamma}{c}\frac{\partial f}{\partial t} + u_r\frac{\partial f}{\partial r} + u_z\frac{\partial f}{\partial z} - \frac{\partial}{\partial r}\left(\frac{\psi^2}{2r^2}\right)\frac{\partial f}{\partial u_r} - \frac{\partial}{\partial z}\left(\frac{\psi^2}{2r^2}\right)\frac{\partial f}{\partial u_z} = S\frac{\gamma}{c}. \quad \text{...[6.26]}$$

The azimuthal current density, j_θ, may be written

$$j_\theta = \frac{e}{2\pi}\frac{\psi}{r^2}\int\frac{f}{\gamma}\,du_r du_z. \quad \text{..[6.27]}$$

Equation [6.20] becomes

$$\frac{1}{c^2}\frac{\partial^2\psi}{\partial t^2} - \frac{\partial^2\psi}{\partial z^2} - r\frac{\partial}{\partial r}\left(\frac{1}{r}\frac{\partial\psi}{\partial r}\right) = -2\psi\frac{r_e}{r}\int\int\frac{f}{\gamma}\,du_r du_z, \quad \text{...[6.28]}$$

where $r_e = e^2/m_0 c^2$. Equations [6.22], [6.26] and [6.28] are the self-consistent set of equations that describe the formation of the E-layer

It is useful in a numerical computation of this type to introduce dimensionless variables. In place of ψ we use the variable $\bar{\mu} \equiv m_0 c\psi/p_\theta$. In order to evaluate p_θ, we consider an equilibrium orbit in the vacuum field. At the midplane ($z = 0$) the end fields are assumed to be negligible, so that $A_\theta(r, 0) = B_0 r/2$, where B_0 is a constant determined by the injection energy and radius, and the desired pitch angle. We find $p_\theta = -(eB_0 r_0^2/2c)$, where r_0 is the radius of the equilibrium orbit at $z = 0$. From the definition of $\bar{\mu}$, it follows that

$$\bar{\mu} = -(2m_0 c^2\psi/eB_0 r_0^2). \quad \text{...[6.29]}$$

We introduce the following dimensionless quantities:

$$R = \frac{r}{r_0}, \quad Z = \frac{z}{r_0}, \quad \tau = \frac{ct}{r_0}, \quad \bar{a}_\theta = \frac{A_\theta}{B_0 r_0}, \quad \bar{b}_r = \frac{B_r}{B_0}, \quad \text{and} \quad \bar{b}_z = \frac{B_z}{B_0}.$$

From these definitions and from equations [6.19], [6.21] and [6.29], it follows that $\bar{\mu} = 1 + 2R\bar{a}_\theta$, $2R\bar{b}_r = -\partial\bar{\mu}/\partial Z$, and $2R\bar{b}_z = \partial\bar{\mu}/\partial R$.

It is convenient to let $\bar{\mu} = \mu_c + \mu$, where μ_c represents the vacuum field and μ is the contribution from the electron layer. For the electron distribution function we use the dimensionless quantity $\rho = r_e r_0 f$, introduce the parameter $C_1 \equiv -(r_e r_0 B_0/2e) = -(2 \cdot 93 \times 10^{-4})B_0 r_0$, and define the function $P(R, Z, t)$ so that

$$P = C_1^2\bar{\mu}^2/2R^2. \quad \text{...[6.30]}$$

This is the potential function for the electron motion, and the equations of motion in these variables becomes $\gamma du_r/d\tau = -\partial P/\partial R$ and $\gamma du_z/d\tau = -\partial P/\partial Z$.

We can now give the complete set of equations in dimensionless form. Equation [6.28] becomes

$$\frac{\partial^2\mu}{\partial\tau^2} - \frac{\partial^2\mu}{\partial Z^2} - R\frac{\partial}{\partial R}\left(\frac{1}{R}\frac{\partial\mu}{\partial R}\right) = -\frac{2\bar{\mu}}{R}\int\int\frac{\rho}{\gamma}\,du_r du_z \quad \text{...[6.31]}$$

and equation [6.26] becomes

$$\frac{\partial \rho}{\partial \tau} + \frac{u_r}{\gamma} \frac{\partial \rho}{\partial R} + \frac{u_z}{\gamma} \frac{\partial \rho}{\partial Z} - \frac{P_r}{\gamma} \frac{\partial \rho}{\partial u_r} - \frac{P_z}{\gamma} \frac{\partial \rho}{\partial u_z} = \sigma. \qquad \ldots [6.32]$$

where $\sigma = r_e r_0^2 S/c$, $P_r = \partial P/\partial R$, $P_z = \partial P/\partial Z$, and $\gamma = (1+u_r^2+u_z^2+2P)^{\frac{1}{2}}$.

We shall now consider the solution of equation [6.32] by finite-difference methods. To illustrate the problem, consider the one-dimensional equation

$$\frac{\partial \rho}{\partial \tau} + u \frac{\partial \rho}{\partial x} = 0 \qquad \ldots [6.33]$$

where u is a constant. The simplest two-level approximation would be

$$\rho_j^{n+1} = \rho_j^n - \tfrac{1}{2}\alpha(\rho_{j+1}^n - \rho_{j-1}^n)$$

where $\alpha = u\Delta\tau/\Delta x$. This scheme is unstable no matter how small Δt is. The simplest alternative is

$$\rho_j^{n+1} = \rho_j^n - \alpha(\rho_j^n - \rho_{j-1}^n)$$

when $u \geqslant 0$, or

$$\rho_j^{n+1} = \rho_j^n - \alpha(\rho_{j+1}^n - \rho_j^n)$$

when $u < 0$. This scheme is stable so long as $|\alpha| \leqslant 1$. The generalization of this to equation [6.32] is straightforward, and leads to the stability conditions

$$\left| \frac{kh^*\Delta\tau}{\gamma h} \right| \leqslant 1, \quad \left| \frac{lh^*\Delta\tau}{\gamma h} \right| \leqslant 1, \quad \left| \frac{P_z\Delta\tau}{\gamma h^*} \right| \leqslant 1, \text{ and } \left| \frac{P_r\Delta\tau}{\gamma h^*} \right| \leqslant 1. \qquad \ldots [6.44]$$

The first version (1962) of the LRL Layer computer code employed this scheme as well as an explicit approximation to equation [6.31]. Unfortunately, the above scheme introduces an artificial diffusion that spoils the results after a short time.

We can consider a space- and time-centred three-level approximation to equation [6.33]

$$\rho_j^{n+1} = \rho_j^{n-1} - \alpha(\rho_{j+1}^n - \rho_{j-1}^n)$$

that is stable if $|\alpha| \leqslant 1$. The extension to more than one dimension is again immediate, leading to the stability conditions in equation [6.44]. The second version (1964) of the Layer code used this scheme. Considerable computation has been done with this method and fairly satisfactory results have been obtained, but with the crude mesh employed in this problem it is still not accurate enough for extremely long running times. Furthermore, it has the disadvantage of being a three-level formula, and when the time step must be decreased for stability reasons, it is quite awkward.

The third version (1966) of Layer uses a three-point approximation to the advective term. For equation [6.33] this becomes

$$\rho_j^{n+1} = \rho_j^n - \tfrac{1}{2}\alpha(\rho_{j+1}^n - \rho_{j-1}^n) + \tfrac{1}{2}\alpha^2(\rho_{j+1}^n - 2\rho_j^n + \rho_{j-1}^n).$$

This scheme is stable if $|\alpha| \leqslant 1$. It is accurate to the second order in the quantities $u\Delta\tau$ and Δx, and has the convenience of being a two-level formula. The extension to more than one dimension is not straight-forward and can lead to an unstable method if it is not done correctly.[27] If we consider a two-dimensional equation and take the approximation given by the equation $\rho^{n+1} = (I+A+B)\rho^n$, where B is the advective difference operator in the other direction, the scheme is unstable. However, if we use the operator equation $\rho^{n+1} = (I+A)(I+B)\rho^n$, the scheme is stable. This is the scheme that is used in the third version of Layer. The difference equation can be represented by

$$\rho^{n+1} = (I+A)(I+B)(I+C)(I+D)\rho^n.$$

The computational process involved in a single time step is divided into four cycles. In the first cycle we calculate advection in the z direction; then advection in the R direction using the results of the first cycle; then advection in the u_z direction using the results of the second cycle; and, finally, advection in the u_r direction using the results of the third cycle. The mesh currently being used for this problem is 51 in z, 8 in r, 19 in u_z, and 9 in u_r, or a total of 69 768 points. We solve equation [6.31] by the alternating direction-implicit method.

A programme of this type is meant to be used for extensive parameter studies. In particular, the form and time behaviour of the vacuum fields can be varied, as can the method of injection. There is a great variety of quantities that can be printed out or plotted at any given time step. Usually, we concentrate on dependent variables, such as the current density, that are functions of the spatial coordinates r and z. This, together with the magnetic field, describes the solution to the self-consistent field problem.

The boundary conditions, other computational details, and sample results are contained in a recently published paper.[60]

Work performed under the auspices of the U.S. Atomic Energy Commission.

REFERENCES

1. M. N. ROSENBLUTH, A. ROSENBLUTH and R. GARWIN. 1954. Infinite Conductivity Theory of the Pinch. *Los Alamos Scientific Laboratory report LA–1850.*
2. J. KILLEEN and B. A. LIPPMANN. 1960. *J. appl. Phys.*, **31**, 1549–1554.
3. K. HAIN, G. HAIN, K. V. ROBERTS, S. J. ROBERTS and W. KÖPPENDÖRFER. 1960. *Z. Naturf.*, **15a**, 1039–1050.
4. K. HAIN. 1961. Pinch collapse. *AERE Harwell report AERE–R–3383.*

5. J. FLETCHER and J. KILLEEN. 1961. The magnetic compression of a fully ionized gas. In *Lawrence Radiation Laboratory Report UCRL-9969*, p. 139.
6. T. A. OLIPHANT. 1963. Numerical studies of the theta pinch. *Los Alamos Scientific Laboratory report LAMS-2944.*
7. K. HAIN and A. C. KOLB. 1962. *Nucl. Fusion Suppl.* Pt. 2, 561–569.
8. G. B. F. NIBLETT and D. L. FISHER. 1962. Numerical calculations on reversed field heating in the thetatron. *Culham Laboratory report CLM-R-19.*
9. R. D. RICHTMYER and K. W. MORTON. 1967. *Difference Methods for Initial Value Problems.* Interscience, John Wiley, New York.
10. M. N. ROSENBLUTH and A. N. KAUFMAN. 1958. *Phys. Rev.,* **109**, 1.
11. L. SPITZER. 1956. *Physics of Fully Ionized Gases.* Interscience, New York.
12. P. R. GARABEDIAN. 1964. *Partial Differential Equations.* John Wiley, New York.
13. P. D. LAX. 1954. *Communs Pure Appl. Math.,* **7**, 159–193.
14. P. D. LAX and B. WENDROFF. 1960. *Communs Pure Appl. Math.,* **13**, 217–237.
15. K. V. ROBERTS and N. O. WEISS. 1966. *Maths Comput.,* **20**, 272–299.
16. K. V. ROBERTS. 1961. (Culham Laboratory). Private communication.
17. D. DUCHS. 1963. *Proc. VI Int. Conf. on Ionisation Phenomena in Gases,* II, 567.
18. D. DUCHS and H. R. GREIM. 1966. *Physics Fluids,* **9**, 1099–1109.
19. A. C. KOLB and R. W. P. McWHIRTER. 1964. *Physics Fluids,* **7**, 519–531.
20. J. KILLEEN, G. GIBSON and S. A. COLGATE. 1960. *Physics Fluids,* **3**, 387–394.
21. K. V. ROBERTS, F. HERTWECK and S. J. ROBERTS. 1963. *Culham Laboratory report CLM-R-29.*
22. K. V. ROBERTS. 1967. *Proc. of Symposium on Computer Simulation of Plasma and Many-Body Prob.,* NASA-SP-153, p. 163.
23. K. HAIN. 1967. Numerical Solution for 1·5-Dimensional Time-Dependent Magneto-hydrodynamic Problems. *NASA-SP-153,* p. 237.
    K. HAIN. 1967. Numerical Calculations in Magnetohydrodynamics. *Proc. of APS Topical Conf. on Pulsed High-Density Plasmas,* Los Alamos, paper Fl.
24. N. O. WEISS. 1966. *Proc. R. Soc. (Lond.)* **A293**, 310–328.
25. W. SCHULTZ. 1964. *Methods in Computational Physics,* Vol. 3. Academic Press, New York.
26. W. F. NOH. 1964. *Methods in Computational Physics,* Vol. 3. Academic Press, New York.
27. C. E. LEITH. 1965. *Methods in Computational Physics,* Vol. 4, pp. 1–28. Academic Press, New York.
28. H. P. FURTH, J. KILLEEN and M. N. ROSENBLUTH. 1963. *Physics Fluids,* **6**, 459–484.
29. J. KILLEEN and H. P. FURTH. 1961. *Bull. Am. Phys. Soc.,* **6**, 309.
30. J. KILLEEN. 1964. *Bull. Am. Phys. Soc.,* **9**, 309.
31. J. KILLEEN. 1964. CTR Semiannual Report, Jan.-June 1964, *Lawrence Radiation Laboratory report UCRL-12028,* p. 85.
32. O. A. ANDERSON. 1964. CTR Semiannual Report, Jan.-June 1964, *Lawrence Radiation Laboratory report UCRL-12028,* p. 72.
33. K. L. AITKEN, R. J. BICKERTON, P. GINOT, R. A. HARDCASTLE, A. MALEIN and P. REYNOLDS. 1964. *J. Nucl. Energy,* Pt. C, **6**, 39–70.
34. P. H. REBUT. 1962. *J. Nucl. Energy,* Pt. C, **4**, 159–168.
35. H. A. B. BODIN. 1964. *Bull. Am. Phys. Soc.,* **9**, 312.
36. K. J. WHITEMAN. *J. Nucl. Energy,* Pt. C, **7**, 293–296.
37. J. KILLEEN and S. L. ROMPEL. 1966. The numerical solution of magnetic field boundary-value problems. *Lawrence Radiation Laboratory report UCRL-50127.*
38. J. B. TAYLOR. 1963. *Physics Fluids,* **6**, 1529.
39. J. BERKOWITZ, K. O. FRIEDRICHS, H. GOERTZEL, H. GRAD, J. KILLEEN and E. RUBIN. 1958. *Proc. 2nd UN Int. Conf. on the Peaceful Uses of Atomic Energy,* **31**, 171 United Nations, Geneva.
40. H. P. FURTH. 1963. *Phys. Rev. Lett.,* **11**, 308.
41. J. ANDREOLETTI. 1963. *C.R. hebd. Séanc. Acad. Sci., Paris,* **257**, 1235.
42. J. KILLEEN and K. J. WHITEMAN. 1966. *Physics Fluids,* **9**, 1846.
43. E. W. LAING, S. J. ROBERTS and R. T. P. WHIPPLE. 1959. *J. Nucl. Energy,* Pt. C, **1**, 49–54.

44. H. Grad. 1967. *Physics Fluids*, **10**, 137.
45. S. Fisher and J. Killeen. 1967. Computation of equilibrium configuration of a toroidal plasma. *Proc. Sherwood Theoretical Meeting, 1967. New York University report NYO–1480–73–MF55*, p. 16.
46. S. Fisher and J. Killeen. 1968. Proc. of Meeting on Theoretical Aspects of CTR, Berkeley, May 1968. *Lawrence Radiation Laboratory report PA–002*, p. 17.
47. H. Grad. 1967. *Phys. Rev. Lett.*, **18**, 585.
48. Y. Sone. 1968. (New York University). Private communication.
49. M. N. Rosenbluth, W. M. MacDonald and D. L. Judd. 1957. *Phys. Rev.*, **107**, 1.
50. G. Bing and J. Roberts. 1961. *Physics Fluids*, **4**, 1039.
51. D. J. Ben Daniel and W. P. Allis. 1962. *J. Nucl. Energy*, Pt. C, **4**, 31.
52. A. H. Futch, Jr., W. Heckrotte, C. C. Damm, J. Killeen and L. E. Mish. 1962. *Physics Fluids*, **5**, 1277.
53. A. H. Futch, Jr., C. C. Damm, J. H. Foote, R. Freis, F. J. Gordon, A. H. Hunt, J. Killeen, K. G. Moses, R. F. Post and J. F. Steinhaus. 1966. *Plasma Physics and Controlled Nuclear Fusion Research II (1966)*, 3–22. IAEA, Vienna.
54. W. Bernstein, V. V. Chechkin, L. G. Kuo, E. G. Murphy, M. Petravic, A C. Riviere and D. R. Sweetman. 1966. *Plasma Physics and Controlled Fusion Research II (1966)*, 23–44. IAEA, Vienna.
55. J. Killeen and A. H. Futch., Jr. 1968. *J. Comput. Phys.*, **2**, 235–254.
56. K. D. Marx. 1968. *Bull. Am. Phys. Soc.*, **13**, 297.
57. K. D. Marx. 1968. A numerical model for a plasma in a magnetic mirror machine. Ph.D. Thesis, University of California, Davis.
58. R. W. Hockney. 1966. *Physics Fluids*, **9**, 1826.
59. C. G. Smith. 1968. Computer simulation of plasma confinement in a stellarator. *Proc. Meeting on Theoretical Aspects of CTR, Berkeley, May 1968. Lawrence Radiation Laboratory report PA–002*, p. 15.
60. J. Killeen and S. L. Rompel. 1966. *J. Comput. Phys.*, **1**, 29–50.

6

TURBULENCE

M. G. Rusbridge

Physics Department,
University of Manchester Institute of Science and Technology

1

INTRODUCTION

IT is striking to compare the predictions of plasma stability theory with experiment. In theory, almost every confined distribution of plasma is unstable to the exponential growth of a wide spectrum of modes of oscillation; in practice, most plasma physics experiments show a spectrum of random fluctuations which does not grow exponentially, and is indeed almost constant over the time scale suggested by the theoretical growth rates—rather they grow and decay on the much longer time scale of the existence of the plasma. It is reasonable to believe that these fluctuations represent the visible evidence of the theoretically predicted instability, but then we must suppose that the fluctuations themselves alter the properties of the plasma sufficiently to suppress the predicted growth. (For similar reasons, the identification of any observed mode of plasma oscillation with a predicted unstable mode requires rather more careful justification than is usually given.)

Let us classify the observations of naturally occurring oscillations in a plasma as follows:

(1) Stability—no oscillations above the thermal level.
(2) Oscillation in a single mode with harmonics, at a constant amplitude.
(3) Statistically stationary random fluctuations constituting a broad spectral band of oscillations, with phase relations between them weak or absent.
(4) Uncontrolled build-up of oscillations leading to rapid destruction of the plasma.

Examples of all of these are known, although the natural selection of plasma physics experiments tends to eliminate those of class (4), while those in class (1) form a very small set at the moment! Class (3) comprises those situations which we commonly call turbulent, and which will be discussed in these lectures.

The study of plasma turbulence is interesting, both for its own sake and

256

because its presence can alter the properties of the plasma, both by suppressing the visible growth of unstable modes, as we have already seen, and in many other ways. For example, a high rate of diffusion of plasma across a magnetic field is often attributed to turbulence, as are anomalously high rates of plasma heating and the acceleration of a few suprathermal particles to very high energies. Because of its effect on diffusion and plasma loss, the study of turbulence is often confined to the attempt to find means of suppressing it; but we ought to consider the possibility that its consequences may be useful. For example, in a hypothetical thermonuclear reactor the nuclear reactions occur almost exclusively among the ions with high velocities in the tail of the Maxwellian distribution. If we could use a suitable form of turbulence to produce many more fast particles than this distribution allows, we could obtain an adequate reaction rate with a correspondingly lower ion temperature.

To give some initial guidance in the study of plasma turbulence, we go to the ideas of hydrodynamic turbulence in search of a feeling for the right questions to ask. Here we have a well-established[1] picture of energy being injected into eddies of size comparable with the mean flow by virtue of the instability of this flow, and being transferred to ever smaller and smaller eddies by non-linear interactions, until it finally reaches such small eddies that viscous damping takes over and removes it. Overall, the damping rate just balances the growth, and the level of turbulence remains constant. We cannot take this picture over for a plasma in anything like detail, but we can note that the direction and rate of energy flow among the possible modes of oscillation form the most important part of the description of any turbulent state.

In fact, the theory of plasma turbulence, as it has been developed so far, is more nearly analogous to that of the turbulence composed of waves on the ocean surface, as developed, for example, by Hasselmann.[2] Here the individual waves interact with each other only rather weakly, and it comes rather naturally out of the analysis to regard the waves as 'quasi-particles', with their interactions as analogous to particle collisions. Even this, however, does not completely describe the plasma case, which presents features with no analogues in other fields.

On the other hand, it must be admitted that there has been distressingly little detailed experimental testing of this theory, and in fact in experiments on turbulence in Zeta[3, 4] we have found that the theory of hydrodynamic turbulence provides a more useful analogy. It seems quite likely that in many cases the appropriate description lies somewhere between the two extremes —the most difficult case to deal with.

Nevertheless, the basic ideas of the plasma theory will probably survive in some form, in an improved version, and are interesting in themselves, so that it is worth while to consider them even if no exact application of the theory as it stands now can be found.

Accordingly I shall discuss the theory of weak turbulence in a collisionless plasma, described by the Vlasov equation. It is useful to start by discussing the problem of stochastic acceleration of particles moving in a random electric field. This is relevant to the theory of turbulence, because energy gained by particles must be lost by the field, and thus the same mechanism represents damping from the point of view of turbulence; we have already seen that an understanding of damping mechanisms is fundamental to the theory of turbulence.

My principal aim will be to describe the basic physical processes included in the theory and to show how these can be combined to lead to a kinetic equation for the turbulent fluctuations which in principle can describe a stationary state. I do not propose to give details of the formal derivation, for which the references should be consulted.

I shall then give a brief description of the experimental situation and try to indicate what in my opinion are the important points for future experiments to concentrate on.

2

STOCHASTIC ACCELERATION

Consider a charged particle moving in a uniform electric field which varies randomly with time. When such a situation arises, there is not usually any interest in following the behaviour of a particular particle in a particular form of the variation of electric field with time—although this problem could be solved easily enough. Rather we are interested in the statistical behaviour of a large number of particles liberated randomly in different versions of the electric field; and we consider, therefore, an ensemble of random functions of time representing the random electric field, and we enquire how the statistical behaviour of the velocity of the particles as a function of time is related to the statistical properties of the ensemble. We use brackets $\langle \rangle$ to denote averages over the ensemble, and we assume the ensemble is stationary, i.e. such averages are independent of time. In particular we assume $\langle E \rangle = 0$, and the ensemble is characterized sufficiently for our purpose by the mean square field and the auto-correlation function $R(\tau)$ defined by

$$\langle E(t)E(t) \rangle = \langle E^2 \rangle$$
$$\langle E(t)E(t+\tau) \rangle = \langle E^2 \rangle R(\tau). \qquad \ldots[1]$$

A general discussion of the properties of stationary random functions is given by Bendat.[5] Now a charged particle released in the field with zero velocity at time $t = 0$ subsequently moves according to

$$m \frac{dv}{dt} = eE(t)$$
$$v(t) = e/m \int_0^t E(t')dt'. \qquad \ldots[2]$$

The velocity is also a random function of time, but not, of course, stationary. Clearly $\langle v \rangle = 0$, but

$$\langle [v(t)]^2 \rangle = e^2/m^2 \int_0^t \int_0^t \langle E(t')E(t'') \rangle \, dt' \, dt'', \qquad \ldots[3]$$

assuming that the operations of integrating and taking the average can be interchanged. Now from above

$$\langle E(t')E(t'') \rangle = \langle E^2 \rangle R(t''-t'),$$

and by changing the variables in equation [3] to

$$\tau = t'' - t'$$
$$\theta = t'' + t',$$

we find that the integrand is independent of θ, and the integration over θ can be carried out to give

$$\langle [v(t)]^2 \rangle = 2(e^2/m^2)\langle E^2 \rangle \int_0^t (t-\tau) R(\tau) \, d\tau. \qquad \ldots[4]$$

Now normally both $\int_0^\infty R \, d\tau$ and $\int_0^\infty \tau R \, d\tau$ are finite, and provided this is so, in the limit $t \to \infty$ we find approximately

$$\langle [v(t)]^2 \rangle = 2(e^2/m^2)\langle E^2 \rangle t \int_0^\infty R(\tau) \, d\tau, \qquad \ldots[5]$$

that is the mean square velocity increases linearly with time; the result is identical in form to the equation describing the mean square displacement in space of a particle following a random walk, and we say that the particle undergoes a random walk in velocity space. By the same sort of analogy we can define a diffusion coefficient in velocity space by setting

$$v^2 = 2D_v t$$

so that

$$D_v = (e^2/m^2)\langle E^2 \rangle \int_0^\infty R(\tau) \, d\tau. \qquad \ldots[6]$$

The quantity $\tau_c = \int_0^\infty R(\tau) \, d\tau$ is a characteristic time scale of the random function, and is known as the *correlation time*. The limit $t \to \infty$ can be taken to mean $t \gg \tau_c$.

Now, according to the theory of random functions,[5] the power spectrum of the electric field is the Fourier cosine transform of the correlation function (Wiener-Khinchin theorem)

$$G(\omega) = \frac{1}{\pi} \langle E^2 \rangle \int_0^\infty R(\tau) \cos \omega \tau \, d\tau,$$

and thus

$$D_v = \pi \frac{e^2}{m^2} G(0), \qquad \qquad ...[7]$$

which says that over sufficiently long times the particle simply responds to the component of the electric field at zero frequency—in other words diffusion is a resonance process. The value of this statement is that it remains true in more complex situations, provided we interpret "zero frequency" as meaning zero frequency as seen following the particle in its orbit. Thus, if the electric field varies in space and time, there should be a contribution to the diffusion coefficient from any Fourier component with angular frequency ω and wave number k if

$$\omega - kv = 0$$

where v is the particle velocity, and in place of equation [7] we can write

$$D_v = \pi(e^2/m^2) \iint G(\omega, k)\delta(\omega - kv)d\omega dk, \qquad ...[8]$$

where the power spectrum G now depends explicitly on k as well as ω. This is an important and fundamental result; a rigorous derivation is given by Macmahon and Drummond,[6] who also exhibit the conditions for its validity —mainly that the velocity shall change only slightly in one correlation time of the fluctuations as seen by the particle. This is because the velocity is taken to be unperturbed by the fluctuations in evaluating equation [8].

So far we have implicitly limited ourselves to motion in one dimension in the absence of a magnetic field. When these limitations are removed, the algebra is more complicated; but we can still recognize the results as having the same form and significance as equation [8]. Thus Hall and Sturrock[7] obtain detailed solutions for the case of fluctuations of electric and magnetic fields in three dimensions and with a non-zero mean magnetic field. I shall quote the results in the limit of small Larmor radius, and for electric field fluctuations only. Parallel to the mean field, the velocity space diffusion coefficient is (in our notation)

$$D_{v, \parallel} = \pi(e^2/m^2) \iint G_\parallel(\omega, k)\delta(\omega - k_\parallel v_\parallel)d\omega dk, \qquad ...[9]$$

where $G_\parallel$ is the power spectrum of the parallel component of the electric field. The perpendicular diffusion coefficient on the other hand is

$$D_{v, \perp} = \pi(e^2/m^2) \iint G_\perp(\omega, k)\delta(\omega - k_\parallel v_\parallel - \Omega)d\omega dk, \qquad ...[10]$$

where Ω is the cyclotron frequency—in other words the particle resonates at its cyclotron frequency, exactly as we should expect.

The same analysis gives an expression for the diffusion coefficient in space across the magnetic field. The transverse velocity of a particle is simply E/B, where B is the mean magnetic field; and an exactly parallel analysis gives

$$D_{x, \perp} = (\pi/B^2) \iint G_\perp(\omega, k)\delta(\omega - k_\parallel v_\parallel)d\omega dk. \qquad ...[11]$$

Results are also given[7] for the case when magnetic field fluctuations dominate, as is the case, for example, for cosmic rays in the interplanetary field. Here[7,8] it turns out that velocity space diffusion occurs only in pitch angle, while the magnitude of the particle velocity does not change.

So far we have placed no restrictions on the form of $G(\omega, k)$, but have simply taken it as given. The next extension is to limit the form of $G(\omega, k)$ to contain only those modes consistent with the dispersion relation for waves in a plasma. Then $G(\omega, k)$ is of the form

$$G(\omega, k) = G_1(k)\delta(\omega - \omega_k), \qquad \qquad ...[12]$$

where ω_k is the frequency given by the dispersion relation. The particular case of drift waves has been studied by Stix[9]; these represent a dangerous class of unstable modes in a non-uniform plasma. For these waves, the parallel and perpendicular electric fields are related in such a way that the transverse spatial diffusion coefficient given by equation [11] is proportional to the parallel velocity diffusion coefficient equation [9]; the particle diffuses along a line in phase space, and stochastic acceleration is closely linked with spatial diffusion.

The application of these ideas directly to a plasma, however, is subject to two difficulties. The condition for resonance $\omega - k_{\parallel} v_{\parallel} = 0$, applied to the ions, is precisely the condition for ion Landau damping—indeed we can now recognize the process of stochastic acceleration as simply Landau damping, looked at from the point of view of the particles. For any particle within the main body of the ion distribution with $v_{\parallel} \lesssim v_{\theta i}$, the ion thermal speed, we expect the ion Landau damping to be so strong that the power spectrum for the corresponding modes will be strongly attenuated, and perhaps reduced to the thermal level. Thus these simple ideas apply only to "suprathermal" particles with $v \gg v_{\theta i}$. The other difficulty is that equation [12] cannot be exactly correct for any finite amplitude of waves in a plasma; as we shall see later, there must always be some spread in frequency around the linear dispersion relation. Thus the infinitely sharp plasma resonance $\delta(\omega - \omega_k)$ should be replaced by some broader resonance shape. Bass, Fainberg, and Shapiro[10] attempted to do this in a rather simple way; but any interesting effects that they found depended on the shape of the far wings of the resonance, which their method could not describe accurately. Strictly speaking, we cannot find the form of the power spectrum without solving the problem of turbulence.

Macmahon and Drummond[6] have discussed the situation in the limit of small amplitudes, where perturbation theory can be used. They show that two non-linear effects contribute to the acceleration of particles in the body of the ion distribution. The first comes from a more careful consideration of the particle orbits. The response of non-resonant particles to a wave ω_k is simply an oscillation of the particle velocity at the frequency $\omega' = \omega_k - k_{\parallel} v_{\parallel}$. Now we find that a new resonance is possible with a different wave ω_q, for which the resonance condition is

$$\omega' = \omega_q - q_{\parallel} v_{\parallel}$$

or

$$(\omega_k - \omega_q) - (k_{\parallel} - q_{\parallel}) v_{\parallel} = 0, \qquad \qquad ...[13]$$

and it is possible to satisfy this, even though neither of the individual waves concerned can resonate with particles; that is to say, we can have

$$\left| \frac{\omega_k - \omega_q}{k_{\parallel} - q_{\parallel}} \right| \lesssim v_{\theta i},$$

even though

$$\left| \frac{\omega_k}{k_{\parallel}} \right|, \left| \frac{\omega_q}{q_{\parallel}} \right| \gg v_{\theta i}.$$

The second effect is that the non-linear interaction of two waves ω_k and ω_q gives rise to a virtual wave with frequency $\omega_k - \omega_q$ and wave number $k - q$; such a wave will not in general satisfy the dispersion relation, and has no existence apart from the two exciting waves—but it can resonate with particles, and the resonance condition is also given by equation [13].

These two processes almost exactly cancel for electrons (at least in the case considered[6]), but for ions the second is dominant and will form the principal source of diffusion of ions. The nature of these processes will become clearer when we consider the weak turbulence theory.

These results are obtained by perturbation theory and are valid only for small amplitudes. For large amplitudes the problem is that the perturbation of the particle orbit may become large in one field correlation time. Dupree[11] has pointed out that, if the perturbations are large enough and random, the electric field correlation along the particle orbit will depend primarily on the orbit and not on the correlation properties of the fields themselves. On this basis he shows that the particle-wave resonance broadens out, so that all particles interact with all waves. The result is very similar to that of Bass, Fainberg and Shapiro,[10] and subject to some of the same difficulties—again the results depend crucially on the shape of the resonance far away from the central peak, and again the theory cannot adequately predict this shape.

Another change occurs if the fluctuating field amplitudes become large enough: a single peak of the field may be large enough to stop a particle and reverse its motion. Then the acceleration process is of a totally different nature—the particles are scattered by field irregularities with which they collide, and, if the irregularities are moving, the particles will absorb energy in trying to come into equilibrium with them. This is precisely the Fermi[12] process, originally suggested as the origin of cosmic rays; and we see that the resonance process and the Fermi process are complementary: if a particle approaches a moving field irregularity it can either pass through it and gain energy by resonance, or be reflected and gain energy by scattering.

The acceleration of particles to high energy is a widespread phenomenon in cosmic and laboratory plasmas,[13] and the resonance and Fermi processes are the only mechanisms yet suggested as its origin. In no case, however, has a completely satisfactory detailed description of the acceleration process been given. As an example, consider the case of Zeta.[3] Here most of the ions have a Maxwellian distribution, although with an anomalously high temperature[14]; but there is a high-energy tail (orders of magnitude above that expected for a Maxwellian) which produces a detectable yield from nuclear reactions.[15] The observed field fluctuations are too small for the Fermi process to be important. The power spectrum $G(\omega, k)$ is not well known experimentally,[4] but it has not been possible to find any plausible extrapolation of the existing results which would produce a diffusion coefficient able to account for the observed acceleration. (An additional complication here is that collisions between the accelerating ions and electrons are not negligible, and can strongly reduce the effectiveness of any accelerating mechanism.)

There is, therefore, as yet no clear experimental confirmation of these theoretically predicted processes; only indirect evidence from the well-established existence[13] of high-energy particles.

3

WEAK TURBULENCE

We have discussed the stochastic acceleration problem as representing, from the point of view of turbulence, the principal damping mechanism in a collisionless plasma. The growth mechanism is the instability predicted by linearized theory, which is discussed in Chapter 4 of this volume by Harris. To generate a picture of turbulence, therefore, we must couple growth and damping by considering the non-linear interactions between waves.

We have to remember, however, that the growth of turbulence may act to suppress the linear growth mechanism. The first attempt to discuss the effect of instability beyond the linear stage of growth showed such an effect; this is the "quasi-linear" theory of Drummond and Pines.[16] Consider a uniform plasma without a magnetic field. The Vlasov equation in one dimension is

$$\frac{\partial f}{\partial t}+v\,\frac{\partial f}{\partial x}+\frac{e}{m}\,E\,\frac{\partial f}{\partial v}=0. \qquad\qquad ...[14]$$

Now we write
$$E = \sum_k E_k \exp\left[i(\omega_k t - kx) + \gamma_k t\right]$$
$$f = f_0 + \sum_k f_k \exp\left[i(\omega_k t - kx) + \gamma_k t\right],$$

with $E_k^* = E_{-k}, f_k^* = f_{-k}$ since E and f are real. Here f_0 is the mean distribution function, and ω_k and γ_k are the frequency and growth (or damping) rate of the mode with wave number k. We now allow f_0 to vary slowly in time, and try to find the effect of the fluctuations E_k, f_k on it. To this end we substitute for E and f in equation [14], to give separate equations for f_0 and for each f_k. We neglect the non-linear terms in the f_k equations, giving a relation between f_k and E_k, but keep them in the equation for f_0. The result takes the form of a diffusion equation in velocity space:

$$\frac{\partial f_0}{\partial t} = \frac{\partial}{\partial v}\left(D_v \frac{\partial f_0}{\partial v}\right) \qquad \qquad ...[15]$$

where the diffusion coefficient is given by

$$D_v = \sum_k \left(\frac{e}{m}\right)^2 |E_k|^2 \, \mathscr{R}\left(\frac{1}{i(\omega_k - kv) + \gamma_k}\right).$$

We consider the case $\gamma_k \ll \omega_k$. Then the diffusion coefficient takes on two forms. For resonant particles we have

$$D_v = \sum_k \pi(e/m)^2 |E_k|^2 \delta(\omega_k - kv),$$

exactly the same form as appeared in the discussion of stochastic acceleration. For non-resonant particles we have

$$D_v = \sum_k \left(\frac{e}{m}\right)^2 \frac{\gamma_k |E_k|^2}{(\omega_k - kv)^2}$$

which in view of the significance of γ_k can be rewritten

$$D_v = \tfrac{1}{2} \sum_k \left(\frac{e}{m}\right)^2 \frac{1}{(\omega_k - kv)^2} \frac{d}{dt} |E_k|^2.$$

For these non-resonant particles with this form of D_v, equation [15] is not really a diffusion equation at all, but merely gives the change in f_0 necessarily arising from setting up the fluctuating field; this change is perfectly reversible, and if a set of fluctuations grew and then decayed, there would be no net effect on f_0. For the resonant particles, on the other hand, the equation is a genuine irreversible diffusion equation.

Now in general terms, diffusion acts to smooth out any non-uniformity, and since non-uniformity promotes instability, we see the tendency for insta-

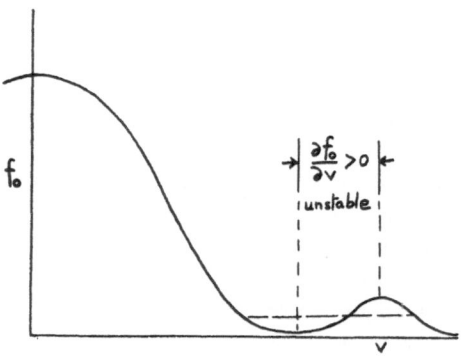

FIG. 1

A distribution function unstable to waves with phase velocities in the range indicated. The dashed line shows the ultimate distribution according to quasi-linear theory.

bility to be self-limiting. The specific illustration given in[16] is of an electron distribution f_0 consisting of a Maxwellian with a small high-energy bump (Fig. 1). Such a distribution may be unstable for waves whose phase velocities lie within the region shown where $df_0/dv > 0$. Then the resonant particle diffusion acts to smooth out the bump, producing something like the distribution shown dashed, for which $df_0/dv \leqslant 0$ everywhere and the instability driving mechanism has disappeared. Thus the unstable modes would grow only up to this point, and thereafter remain stationary or decay slowly due to slower processes not considered in this model.

Nevertheless, this mechanism of self-limiting is not strong enough to account for the observations; for in many cases we see that the level of fluctuations is stationary even when the instability driving force still exists. Thus drift modes that are driven by a density gradient and are observed in alkali-metal plasmas[17] reach a stationary state long before the density gradient is eliminated—which would, of course, require the disappearance of the plasma. Evidently we must consider the neglected non-linear terms in equation [14].

Before attempting to do this in any formal sense, it is worthwhile to consider physically what the possible processes are. Suppose we set up in a plasma a certain spectrum of natural oscillations $E(\omega, k)$ with $\omega = \omega_k$ the natural frequency given by the dispersion relation. Now because of the non-linearity any two modes with wave numbers k and q will interact to try to drive the plasma to oscillate in modes $k' = k \pm q$ with frequencies $\omega' = \omega_k \pm \omega_q$ respectively, and the plasma responds to these driving terms as it would to an externally applied electric field. If no combination frequency ω' is a natural frequency of the plasma for the corresponding k', the response is that of a forced oscillation; it builds up to a steady level proportional to $[\varepsilon(\omega', k')]^{-1}$ where ε is the dielectric constant of the plasma. Now the plasma dispersion relation is $\varepsilon(\omega, k) = 0$, and if $\varepsilon(\omega', k') \to 0$, then we have a forced oscillation at a natural frequency of the plasma. As in the corresponding situation in mechanics, the response of the plasma increases linearly with time.

Thus we see that the mode coupling produced by the non-linear terms can be either resonant or non-resonant. In the former case the spectrum evolves with time as energy is exchanged between different natural modes of the plasma; in the latter the combination mode is no more than an adjustment of the plasma to the simultaneous presence of two other modes. Suppose we introduce two modes k and q into the plasma for a limited time only; resonant mode coupling will leave the plasma oscillating at a combination frequency, while non-resonant modes can exist only in the presence of exciting modes, and disappear when these are removed. (Notice the parallel with the resonant and non-resonant interactions between waves and particles.) For this reason non-resonant modes are also called *virtual waves*.

Even so, we cannot neglect the virtual waves. We have already seen that virtual waves are real enough to interact with particles and suffer Landau damping; in addition it may interact with a third mode l to produce a fourth real mode. This gives in effect resonant coupling of three modes instead of two—and evidently we can imagine a succession of resonant interactions involving higher and higher numbers of different modes. In principle, we ought to take all of these into account, which implies summing an infinite series. However, the amplitude of a virtual wave is evidently proportional to the product of the amplitudes of the exciting waves; and therefore, if the amplitudes are small, we can consider only terms involving the smallest possible number of distinct modes. This is the solution obtained by perturbation theory.

In the case of resonant mode coupling the growth rate of the mode ω', k' is proportional to the product of the amplitudes of modes k and q, and so also is the non-linear damping rate of these modes. Now, in any resonance the width is determined by the damping rate; and thus, as the amplitude rises, the width of the plasma resonance increases.

Thus, when the turbulence level is small (weak turbulence), the fluctuations consist almost entirely of modes very closely obeying the linear dispersion relation, and the level of non-resonant modes or virtual waves is very small. As the level of turbulence increases, the plasma resonance becomes smeared out and simultaneously the relative level of virtual waves rises. Eventually the distinction between resonant and non-resonant modes disappears and we have strong turbulence. The various stages are shown in Fig. 2.

Let us now restrict ourselves to weak turbulence and summarize the various types of interaction between modes which we have discussed so far. We have:

(1) Resonant mode coupling between three modes k, q, l. This requires that we simultaneously satisfy

$$k+q = l$$
$$\omega_k+\omega_q = \omega_l. \qquad\qquad ...[16]$$

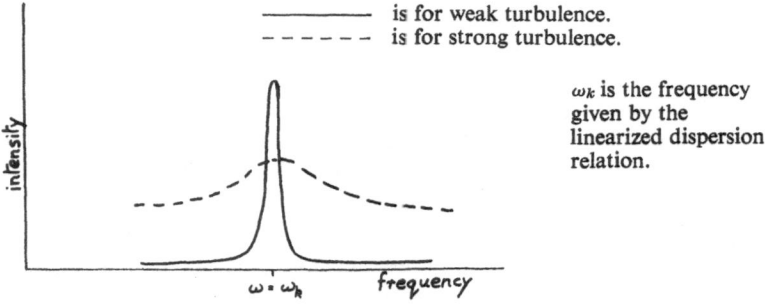

———— is for weak turbulence.
– – – – – – is for strong turbulence.

ω_k is the frequency given by the linearized dispersion relation.

FIG. 2. The power spectrum in the turbulent state for a particular mode k.

(2) Non-linear Landau damping. The condition is either

$$\omega_k + \omega_q - (k_{\parallel} + q_{\parallel})v_{\parallel} = 0 \qquad \qquad ...[17]$$

or

$$\omega_k - \omega_q - (k_{\parallel} - q_{\parallel})v_{\parallel} = 0. \qquad \qquad ...[18]$$

(3) Resonant mode coupling between four modes, k, q, l, p. There are two possible types: in the first, we have

$$k + q + l = p$$
$$\omega_k + \omega_q + \omega_l = \omega_p \qquad \qquad ...[19]$$

and in the second

$$k + q = l + p$$
$$\omega_k + \omega_q = \omega_l + \omega_p. \qquad \qquad ...[20]$$

There are two general remarks to be made about these relations. It is not always possible to satisfy equations [16] simultaneously; for this to be possible a fairly stringent condition on the dispersion relation is required. In an isotropic plasma the condition is simple—the dispersion relation plotted with ω as a function of k must be concave upwards[18] (Fig. 3). For the aniso-tropic plasma case no very simple condition can be given, but in fact resonant mode coupling is impossible over the greater part of the plasma dispersion relation. The same applies to four-wave interactions of the type of equation [19], although those of the form of equation [20] are always possible. In fact, for waves on a liquid surface (for which, of course, equations [17] and [18] do not apply), equation [20] defines the dominant non-linear interaction.

The second remark is that equations [16]–[20] have a wider significance; they govern the flow of momentum and energy during non-linear interactions, in exactly the way that their quantum mechanical analogues would. Thus, in particular, the relative sign of the energy transfer to a particular mode is the same as the relative sign in these equations. In equation [16], energy may

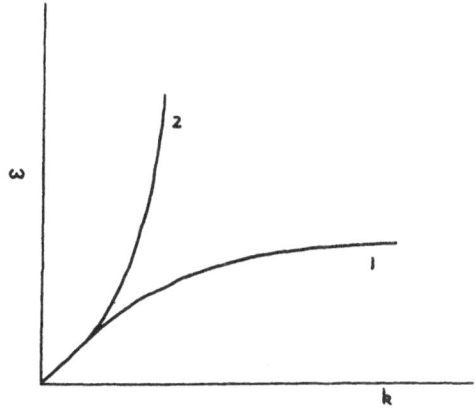

FIG. 3

Possible forms for isotropic dispersion relations.

Curve 1: resonant mode coupling forbidden.

Curve 2: resonant mode coupling allowed.

be lost by both modes k and q as mode l builds up, or mode l may decay into the other two modes (a decay instability[18, 19]); but it is not possible for mode k to gain and mode q to lose energy in the course of this particular interaction. In non-linear Landau damping governed by equation [17], both modes k and q lose energy; but in equation [18], if mode k loses energy, this energy is shared between mode q and the particles. These effects can be established by detailed calculation from the formal theory but are true in general of all non-linear interactions of waves.[20]

Let us now indicate very briefly the formal development of these ideas. We return to equation [14], but now substitute

$$E = \sum_k E_k e^{-ik \cdot x}, \quad f = f_0 + \sum_k f_k e^{-ik \cdot x},$$

where E_k, f_k are now functions of time. In the resulting equations, in which we retain the non-linear terms, we eliminate f_k in terms of E_k using Poisson's equation. The resulting equation for E_k can be written

$$\frac{d}{dt} E_k(t) = i\omega_k E_k(t) + \sum_{q \neq 0} L(k, q) E_{k-q} E_q + O(E^3), \qquad \dots [21]$$

where $L(k, q)$ is an integral operator. Now suppose we initially have some spectrum of waves which are uncorrelated in phase, and we form from equation [21] an equation for the time variation of $|E_k^2| \equiv \langle E_k(t) E_{-k}(t) \rangle$; we find

$$\frac{d}{dt} |E_k^2| = 2\gamma_k |E_k^2| + \sum_{q \neq 0} M(k, q)\langle E_{k-q} E_q E_{-k} \rangle + O(E^4), \qquad \dots [22]$$

where M is another operator, γ_k is the linear growth rate, and the triple correlation embodies the non-linear terms. Initially, since the phases are random, the triple correlation vanishes; but the non-linearity must induce a finite value. We therefore proceed in the same way to find an equation for $\langle E_{k-q} E_q E_{-k} \rangle$ and find that this involves a fourth-order correlation of the form $\langle E_{k-q} E_{q-l} E_l E_{-k} \rangle$. Again using the random phase assumption, we find that this vanishes initially unless $k - q + l = 0$, when we have

$$\begin{aligned}
\langle E_{k-q} E_{q-l} E_l E_{-k} \rangle &= \langle E_{k-q} E_{-k+q} \rangle \langle E_k E_{-k} \rangle \\
&+ \langle E_l E_{-l} \rangle \langle E_k E_{-k} \rangle \\
&+ \langle E_l E_{-l} \rangle \langle E_{q-l} E_{-q+l} \rangle. \qquad \dots [23]
\end{aligned}$$

Since this shows that the triple correlation is actually of order E^4 the $O(E^4)$ terms in equation [22] must be considered explicitly, although many of them do not contribute to non-linear energy transfer but only to a shift in the wave frequency. The full expressions are given by Aamodt and Drummond.[21] The resulting kinetic equation derived from equation [22] is given in a useful and simple form by Al'tshul' and Karpman,[22] who show that it can be expressed in terms of the action n_k defined by setting $n_k \omega_k$ equal to the energy of the mode k. The result is of the form

$$\frac{dn_k}{dt} = 2\gamma_k n_k + S\{n\} + R^+\{n\} + R^-\{n\} + O(n^3) \qquad \ldots[24]$$

where S, R^+, R^- are terms quadratic in the n_k representing the processes defined by equations [16], [17] and [18] respectively. It is a simple step to use the quantum mechanical analogy to regard n_k as the number of quanta of the mode k, in the classical limit.

The form of $S\{n\}$, derived from equation [23], is simple and can be understood intuitively. We have

$$S\{n\} = \frac{1}{16\pi} \sum_{q+l=k} [M_1(q, l)n_q n_l - 2M_2(q, l)n_k n_q]\delta(\omega_k - \omega_q - \omega_l).$$

The summation and δ-function embody the conditions of equation [16]; the first term in square brackets describes the increase in n_k by resonant interactions between other modes, while the second describes loss due to interactions between the mode k itself and all others.

The derivation of equation [24] depended essentially on the assumption of initially random phases which led to equation [23]. Thus, strictly equation [24] can be valid only for the initial development of the turbulence from the assumed state with random phase. However, if the intensity of turbulence remains sufficiently small, equation [23] remains approximately valid, and so also does equation [24]. This is the "random phase approximation". It is precisely this approximation that breaks down in attempts to treat strong turbulence by perturbation theory. When equation [23] is not valid, we must seek an equation for the fourth-order correlation—but this will involve correlations of the fifth order, and so on; we develop an infinite hierarchy of equations, with no known method of reducing them to a finite set.

Let us now assume that the turbulence is weak and equation [23] is approximately satisfied. Then equation [24] remains valid for all time and is the kinetic equation we seek. It could in principle describe the approach to a stationary solution in which $\frac{dn_k}{dt} \to 0$ (although, as we shall see, it may not do so). One of the most important applications of such an equation is to obtain general conservation laws; and indeed an instructive conservation law does emerge from equation [24]. As we have already seen, there are cases where $S\{n\}$ vanishes because of the shape of the dispersion relation. In addition, equation [17] cannot be satisfied unless the modes k, q individually have phase velocities comparable to the ion thermal speed; and if such modes are all heavily Landau damped, then $R^+\{n\}$ also vanishes. The remaining quadratic non-linear term $R^-\{n\}$ conserves the total "number of quanta", i.e.

$$\sum_k R^-\{n\} = 0,$$

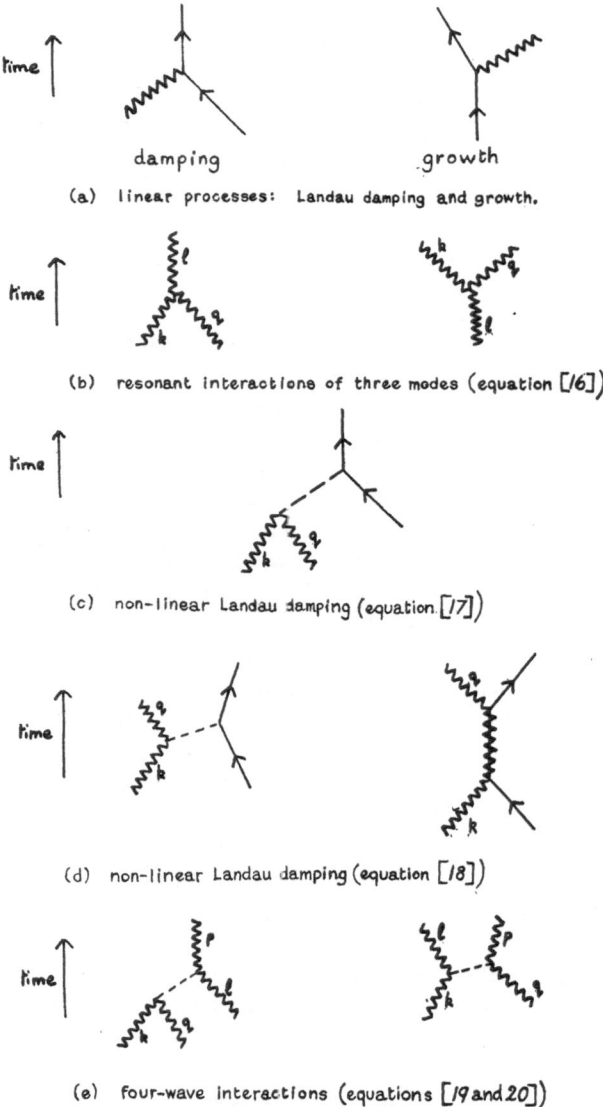

(a)   linear processes: Landau damping and growth.

(b)   resonant interactions of three modes (equation [16])

(c)   non-linear Landau damping (equation [17])

(d)   non-linear Landau damping (equation [18])

(e)   four-wave interactions (equations [19 and 20])

FIG. 4. Representation of interactions discussed in the text.

so that

$$\frac{d}{dt} \sum_k n_k = 2 \sum_k \gamma_k n_k.$$

We can extend this to four wave processes (included in $O(n^3)$ in equation [24]). If three-wave resonant coupling is impossible, so is four-wave coupling of the type of equation [19]; but that of the type of equation [20], the transformation of two modes into two other modes, is always possible; and it is perhaps now intuitively obvious that this interaction will also conserve $\sum_k n_k$.

Under a wide range of circumstances, therefore, the non-linear terms cannot completely balance the linear growth terms, but must act simply to transfer energy between regions of growth and damping in phase space; and an ultimate steady state obeying these equations must satisfy

$$\sum_k \gamma_k n_k = 0.$$

The various non-linear processes are conveniently represented by diagrams, of which those corresponding to equations [16]–[20] are shown in Fig. 4. This method of presentation is very commonly used, and indeed becomes almost essential in treating higher-order terms. The two diagrams corresponding to equation [18] represent the two interfering processes discussed above as contributing to non-linear stochastic acceleration. Such diagrams make clear, for example, which terms conserve the total number of quanta and which do not.

As we have seen, the kinetic equation could in principle lead to a stationary solution. In fact, however, it is not clear that a solution exists; when written out in full with $dn_k/dt = 0$, equation [24] takes the form of an integral equation of a type which does not necessarily have a solution.[18] Kadomtsev[18] has obtained a solution for ion acoustic waves by artificially limiting the directions of propagation to within a narrow cone; Akhiezer[23] has removed this restriction, but his solution has very strange properties, and it is not clear how realistic it can be. I have found no other solution in the literature; and the possibility that a solution does not exist must be considered. In this case, the level of turbulence must continue to rise until higher-order terms become important; but in this case, all higher-order terms become of the same order, and perturbation theory has broken down; we have strong turbulence. It is, of course, conceivable that not even then will a stationery solution exist, and presumably such a case will lead to the catastrophic destruction of the plasma.

What is the theoretical position for strong turbulence? It is still possible to write down the perturbation theory and obtain a formal analogue of equation [24], containing an infinite series which must be summed, but there is no guarantee of convergence, and attempts at partial summation lead nowhere.[24] The situation is exactly the same as that for hydrodynamic turbulence; there too, much ingenuity has been used in attempts to sum the perturbation theory expansion, but again with no useful success. The most successful approach here has been the dimensional analysis based on physical ideas given by Kolmogorov[25]; but it is hard to find a situation in plasma physics simple enough for dimensional analysis, although Chen[26] has given an argument for drift waves, and Zakharov[27] for plasma waves.

Clearly there is scope for approximate theories. An interesting example is that of Dupree,[11] already referred to, in which he seeks to consider the actual particle orbit in the fluctuating field, and in effect to solve for the orbit and the field simultaneously, rather than starting with the unperturbed orbit. We consider the non-linear interaction between a wave and a particle moving under the influence not just of one other wave, as in equation [18] and Fig. 4, but of all other waves simultaneously. The result is to broaden the Landau resonance for each mode until it can contain all the particles, leading to non-linear damping. (Dupree calls the process "trapping". It must be distinguished from the more usual use of the word to describe the physical trapping of a particle between two crests of a single wave.) A somewhat similar starting point is taken by Galeev,[28] while Orszag and Kraichnan[24] hope to apply Kraichnan's[29] theory of hydrodynamic turbulence in suitably modified form.

4

The Experimental Situation

I shall conclude by discussing briefly the experimental evidence on plasma turbulence, and outline some of the measurements suggested by the theory. The first point is that there are numerous observations of statistically steady random fluctuations, so that we can conclude that non-linear terms can in fact produce a steady state. In no case, however, is it clear that the level of turbulence is low enough for the theory of weak turbulence to be a good description. Usually, the observed fluctuations are identified with some one of the modes predicted by linear instability theory, which implicitly assumes the weakly turbulent description. This, clearly, is a necessary preliminary to any attempt to investigate the internal dynamics of the turbulence—you must know what modes you have to play with. A typical case is that of Young,[30] who has quite successfully analysed the fluctuations in the Model C Stellerator. (Our measurements of turbulence in Zeta,[4] and to be published, are in a slightly different category; for we know from the start that the turbulence must be strong, and we have made no attempt to identify the modes with any linear theory.) It is often a difficult experimental problem to measure the properties of fluctuations sufficiently accurately to make comparison with theory possible. For example, in Zeta,[4] we knew that the wavelength along the lines of force was long, but we could not measure it, because we could not find any practicable method of accurately tracing the paths of lines of force. A further difficulty lies in the fact that, while a frequency spectrum of a fluctuating quantity is easily measured, the wave number spectrum, requiring Fourier analysis in space, is much more difficult to obtain with any degree of resolution.

Specific measurements on the non-linear interactions of waves have been made by Cano et al.,[31] Fedorchenko et al.,[32] and by Stern and Tzoar,[33] who have demonstrated the occurrence of resonant mode coupling, while both resonant and non-resonant coupling have been observed by Ellis and Porkolab.[34] I have not found any experiment demonstrating non-linear Landau damping.

The experiment of Malmberg and Wharton[35] is one of the few in which the unstable mode is actually observed while growing, so that it can be followed throughout the exponential phase of growth, and the growth rate compared with linear theory; further experiments are expected to be concerned with the non-linear state of saturation which can be seen in their results.

Several experiments[36] are concerned with attempts to set up very strong turbulence and thus heat the plasma rapidly. So far, however, these have not reached the point of investigating the actual turbulent state in any detail.

(I have not discussed at any length experiments on turbulent diffusion, although a good many experiments are concerned solely with this. I believe that we do not yet know the spectrum of turbulence in sufficient detail to have any confidence in attempts to calculate diffusion coefficients; we simply note, therefore, that experimentally one of the effects of turbulence is enhanced diffusion.)

This does not add up to anything like a complete picture of turbulence. What experimental programme can be suggested which might contribute to such a picture? First of all, we need good measurements of frequency and wave-number spectra. For a given wave-number we need to be able to find the frequency spectrum and compare it with those shown in Fig. 2; this will be the decisive test of the ideas of weak turbulence—if these are applicable, a plasma resonance must be visible. Second, we need to measure the triple correlations which represent the non-linear transport of energy. It should be possible to calculate specific differences in the form of this correlation, depending on which of the various non-linear processes is dominant. The ability to detect and measure fluctuations in the ion distribution function would help enormously. Third, we need to set up further experiments to demonstrate and investigate quantitatively the predicted non-linear interactions, using externally excited modes of oscillation.

5

CONCLUSION

In these lectures I have said nothing about the application of the theory of a turbulent plasma to explain the "anomalous" effects observed in plasma experiments. Enough is known to make it clear that these effects could be accounted for in principle as arising from turbulence in the plasma; but I do not believe that a completely satisfactory description is available for any of them; and in my view no such description will be wholly convincing until the basic theory is more securely related to experiment. I have put this forward, therefore, as the primary aim of present work on plasma turbulence.

REFERENCES

1. G. K. BATCHELOR. 1956. *The Theory of Homogeneous Turbulence.* C.U.P.
2. K. HASSELMANN. 1962. *J. Fluid Mech.*, **12**, 481. 1963. *J. Fluid Mech.*, **15**, 273, 385.
3. W. M. BURTON *et al.* 1962. *Nucl. Fusion Supplement*, p. 903.
4. D. C. ROBINSON and M. G. RUSBRIDGE. 1964. *Proceedings of the International Symposium on Diffusion of Plasma across a Magnetic Field, Feldafing*, p. 162.
5. J. S. BENDAT. 1958. *Principles and Applications of Random Noise Theory.* Chapman and Hall, London.
6. A. B. MACMAHON and W. E. DRUMMOND. 1967. *Physics Fluids*, **10**, 1714.
7. D. E. HALL and P. A. STURROCK. 1967. *Physics Fluids*, **10**, 2260.
8. J. R. JOKIPII. 1966. *Astrophys. J.*, **146**, 480; 1968. *Astrophys. J.*, **152**, 671.
9. T. H. STIX. 1967. *Physics Fluids*, **10**, 1601.
10. F. G. BASS, YA. B. FAINBERG and V. D. SHAPIRO. 1966. *JETP*, **22**, 230.
11. T. H. DUPREE. 1967. *Physics Fluids*, **10**, 1049.
12. E. FERMI. 1949. *Phys. Rev.*, **75**, 1169.
13. V. N. TSYTOVICH. 1966. *Sov. Phys. Uspekhi*, **9**, 370.
14. B. B. JONES and R. WILSON. 1962. *Nucl. Fusion Supplement*, p. 889, and R. WILSON, private communication.
15. R. A. COOMBE and B. A. WARD. 1963. *Plasma Physics (J. Nucl. Energy C)*, **5**, 273.
16. W. E. DRUMMOND and D. PINES. 1962. *Nucl. Fusion Supplement*, p. 1049.
    A. A. VEDENOV, E. P. VELIKHOV and R. Z. SAGDEEV. 1962. *Nucl. Fusion Supplement*, p. 465.
17. H. W. HENDEL, B. COPPI, F. PERKINS and P. A. POLITZER. 1967. *Phys. Rev. Lett.*, **18**, 438. N. S. BUCHEL'NIKOVA, A. M. KUDRYAVTSEV, R. A. SALIMOV. 1965. *Sov. Phys. Tech. Phys.*, **10**, 53.
18. B. B. KADOMTSEV. 1965. *Plasma Turbulence.* Academic Press, London.
19. J. D. JUKES. 1965. *Physics Fluids*, **8**, 1531.
20. P. A. STURROCK. 1964. *Proc. Int. School of Physics "Enrico Fermi", Course 25.* Academic Press. 1960. *Ann. Phys.*, **9**, 422.
21. R. E. AAMODT and W. E. DRUMMOND. 1964. *Physics Fluids*, **7**, 1816.
22. L. M. AL'TSHUL' and V. I. KARPMAN. 1965. *JETP*, **20**, 1043.
23. I. A. AKHIEZER. 1965. *JETP*, **20**, 1519.
24. S. A. ORSZAG and R. H. KRAICHNAN. 1967. *Physics Fluids*, **10**, 1720.
25. A. N. KOLMOGOROV. 1941. *Compt. rend. Acad. sci. U.R.S.S.*, **30**, 301; and **32**, 16.
26. F. F. CHEN. 1965. *Phys. Rev. Lett.*, **15**, 381.
27. V. E. ZAKHAROV. 1967. *JETP*, **24**, 455.
28. A. A. GALEEV. 1967. *Physics Fluids*, **10**, 1041.
29. R. H. KRAICHNAN. 1965. *Physics Fluids*, **8**, 575.
30. K. M. YOUNG. 1967. *Physics Fluids*, **10**, 213.
31. R. CANO *et al.* 1967. *Physics Fluids*, **10**, 2260.
32. V. D. FEDORCHENKO, V. I. MURATOV and B. N. RUTKEVICH. 1966. *Sov. Phys. Tech. Phys.*, **10**, 1549.
33. R. A. STERN and N. TZOAR. 1966. *Phys. Rev. Lett.*, **17**, 903.
34. R. A. ELLIS, JR., and M. PORKOLAB. 1968. *Phys. Rev. Lett.*, **21**, 529.

35. J. H. MALMBERG and C. B. WHARTON. 1967. *Bull. Am. Phys. Soc. II*, **12**, 742.
36. S. M. HAMBERGER, A. MALEIN, J. H. ADLAM and M. FRIEDMAN. 1967. *Phys. Rev. Lett.*, **19**, 350. T. H. JENSEN and F. R. SCOTT. 1967. *Phys. Rev. Lett.*, **19**, 1100. E. K. ZAVOISKII and L. I. RUDAKOV. 1967. *At. Energ. (U.S.S.R.)*, **23**, 417.

7

COLLISIONLESS SHOCKS

H. Völk

Max-Planck-Institut für Extraterrestrische Physik,
Garching bei München

1

INTRODUCTION

INTEREST in collisionless shock-waves has grown considerably in the last few years. If we try to find the motive for the intensified research in this field, we can single out three major developments which mutually stimulate each other:

First of all, it has only recently been possible in the laboratory to generate shocks whose thickness is smaller than the shortest mean free path for two-body interactions, like ionizing collisions, charge exchange or Coulomb collisions. Only these are what one should call collisionless shocks.

Measurements by satellites and interplanetary probes have proven the existence of a bow-shock in the solar wind standing in front of the magnetosphere. An enormous amount of data has already been accumulated.

The wealth of details regarding the presence of waves in the shock area has strengthened the belief that turbulent dissipation as well as dispersion plays a major role in the shock structure. But only through the rapid development of plasma turbulence theory has it become possible to construct models of turbulent shocks on an at least semi-quantitative basis.

Let us first turn briefly to some of the experimental and observational results. Most of the experiments are done in θ- and z-pinch configurations, such that the shock propagates perpendicular to the magnetic field. In a θ-pinch shock, for example, the filling gas in the cylindrical discharge tube is usually preionized either by the UV-radiation from auxiliary z-pinches at both ends or through discharging a primary θ-pinch. This plasma is then compressed by a slowly rising B-field, which penetrates into the plasma and provides the bias field. At some chosen value of the bias field, typically of the order of some 100 Gauss, a fast and strong θ-pinch is triggered, with a few kG maximum field amplitude. This causes the plasma to implode towards the axis and serves as a piston. Ahead of this piston a cylindrically converging shock-wave propagates.[1, 2, 3, 4] In a z-pinch with an axial bias field, the fast rising azimuthal field provides the piston.[5, 6]

It is important to obtain a sufficiently homogeneous and reproducible plasma ahead of the shock. In addition, one must require that the shock separates clearly from the piston and becomes stationary before it gets very near to the discharge axis; otherwise cylindrical effects become important and the results are difficult to interpret, as far as the comparison with theory goes, which usually assumes an idealized steady state and plane shock. Because of these difficulties the results are not entirely unambiguous. Nevertheless, there are a number of very interesting pertinent features.

For small $\beta (\beta \lesssim 0.1;\ \beta = p \cdot 8\pi/B^2,\ p =$ plasma pressure upstream) there is characteristically a different magnetic structure of the shock-front for Mach numbers below and those above a critical value of about three. The relevant Mach number M is the ratio of the shock-velocity u to the upstream magnetosonic speed $c_A = [B^2/(4\pi\rho) + c_s^2]^{\frac{1}{2}}$; ρ and c_s are the upstream mass-density of the plasma and the speed of sound, respectively; B is the strength of the bias field. For $M \lesssim 3$ the shock-thickness is a few (typically ten) times c/ω_{pe}. For Mach numbers above the critical value, the shock front broadens to scales of the order of c/ω_{pi} ($\omega_{pi,e}$ = ion, electron plasma frequency) and sometimes[1, 2, 5, 6] develops a double structure, consisting of a broad, probably non-stationary,[2] precursor ramp ($\simeq c/\omega_{pi}$) followed by a narrow transition, which is still of dimensions $\simeq c/\omega_{pe}$. This double structure seems to disappear at still higher values of $M \gtrsim 6$.[5]

For a small β upstream plasma and $M = 1.3$, the shock appears to have a rather laminar structure of damped oscillations behind the front; while for higher M, the back of the shock is turbulent.[2]

The average values of the dynamical quantities agree fairly well with those deduced from the MHD conservation laws across the shock (Rankine-Hugoniot-relations[7, 8]). This agreement with the Rankine-Hugoniot-relations has also been confirmed for the solar wind simulation experiment at AVCO,[9] where a pulsed beam from a powerful plasma gun is blown axially on a magnetic dipole. Thus the shock is curved and in general oblique (the shock-"plane" is not parallel to B). The measured shock widths are a few times c/ω_{pi}.

Another, and a very interesting, example of an oblique shock wave experiment has been reported in a θ-pinch type geometry.[10] Here the piston field is created by the rapidly rising current in a single turn loop wrapped around the discharge tube in its *midplane*. The loop produces a curved magnetic piston driving radially and symmetrically inwards into the plasma, generating a shock that propagates ahead of the piston in the manner of a bow shock wave. Considering, as usual, a small piece of the shock-front as planar, the shock structure can be investigated for a wide range of angles θ

between the shock plane and the axial bias field. Preliminary results for $M \simeq 4$ show shock-thicknesses of several c/ω_{pi}, increasing with θ. Also an oscillatory *azimuthal* **B**-field is created in the shock, which is interpreted as a precursor in the whistler mode. Such modes have also been seen in other experiments.[2,9]

If in all these experiments the shock-thickness is generally only a few times smaller than the shortest mean free path, this problem is totally absent in the earth's bow shock. The mean free path in the solar wind near the earth is of order 1 AU (= earth-sun distance). The mean shock stand-off distance

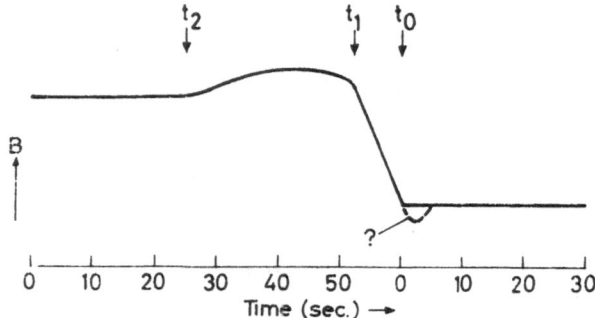

FIG. 1. $|\mathbf{B}|$ is the r.m.s. modulus of the magnetic field through the shock on an outbound passage; time is measured in the satellite. The cartesian components of $\mathbf{B}$ show a similar behaviour. The initial dip, dotted, with a question mark, is sometimes observed, but is not as statistically certain as the other features.

from the earth is about 14 earth radii (R_E) and about 4 R_E from the magnetospheric boundary, measured along the earth-sun line. The shock usually exhibits an irregular oscillatory motion[11,12,13] with a r.m.s. speed of the order of 10 km/sec, which is considerably larger than typical satellite speeds (signified by the occurrence of multiple shock-crossings along a satellite trajectory). Still, however, the shock-thickness can be estimated to be of the same order of magnitude (50–100 km) as the *characteristic* ion Larmor radius c/ω_{pi}, where the ion velocity is taken as the Alfvén velocity.[11] Typical values of the parameters in the solar wind near the earth are‡: $n \simeq 5 \text{ cm}^{-3}$, $v_{\text{wind}} \simeq 400$ km/sec, $T_i \simeq 5$ eV, $T_e/T_i \simeq 10$, $B \simeq 5 \times 10^{-5}$ Gauss, $c_A \simeq 50$ km/sec, $M = v_W/c_A \simeq 8$, $\beta_i \lesssim 1$. Actually the bow shock has a wealth of detailed structure, which varies with time and depends in a complex way on solar and geomagnetic activity, as do the stand-off distance and the shock speed. Yet it seems possible to determine some salient typical properties

‡ These numbers vary considerably with solar perturbations, and should serve as merely a guideline; especially, T_e/T_i is very poorly known. For a review, see for instance Refs. 14 and 15.

of the magnetic structure, for example, which are summarized in Fig. 1, a schematic drawing by Heppner et al.[11]

While it is already difficult to convert satellite times into spatial distances, because of our scant knowledge of the motion of the shock, there are three "thicknesses", given by the initial rise time of B, $t_0 \pm 2$ sec; $t_0 - t_1$; and $t_0 - t_2$. The initial rise time corresponds to a length of the order of c/ω_{pi}, the thickness mentioned above. The other two scales are correspondingly larger. Structures similar to that given in Fig. 1 have also been seen in the laboratory[2] for high β ($\simeq 0.6$) and moderate Mach numbers $M \simeq 4.5$. It should, however, be stressed that this is a statistical average. Also much narrower shock transitions seem to have been observed. For a discussion, see Tidman.[16, 17]

Superimposed on the above average structure are coherent as well as incoherent fluctuations with frequencies f in the range 0.5 Hz $< f < 1.5$ Hz and $f > 3$ Hz respectively, as seen in the satellite frame of reference.[11] At least the coherent fluctuations are circularly polarized. They are typically damped out after three or four cycles, and are also observed somewhat upstream, therefore propagating at least as fast as the shock itself. Thus they should be whistler modes (compare Refs. 2, 9, 10).

Downstream, until far into the transition region between the shock and the magnetosphere, large-amplitude, long-wavelength magnetic structures have been observed. In general they are not completely turbulent and there is much more power in the fluctuations of the components of the field vector than in those of its magnitude. Because of the long wavelengths and the persistence in amplitude, these could be torsional Alfvén waves, generated in the shock front, as are probably most of the other fluctuations. However, a mere compression of upstream Alfvén waves swept through the shock might be an additional contribution.[14]

Very recent measurements by the OGO-5 satellite with electrostatic probes[18] have in many cases shown *electrostatic* turbulence within the shockfront. The r.m.s. electric fields measured by 15 per cent bandwidth filters at 200, 560 and 1300 Hz are often found to correspond to peaks of the order of 20 to 100 mV/m, with most of the amplitudes nearer 5 to 10 mV/m compared with background levels of 0.2 mV/m. A broadband analysis of these turbulent electrostatic fields shows significant energy from 200 up to 3000 Hz. The regions of electrostatic turbulence are spatially limited, and the strongest turbulence is always associated with regions of rapid change in the mean B-field, suggesting that the currents responsible for magnetic compression also drive the electrostatic instability, which may consist of ion sound waves. The frequencies observed do not rule out the possibility of the presence of electron cyclotron waves. At present, the OGO-5 measurements are being studied carefully in an attempt to identify unambiguously the modes involved in the electrostatic turbulence.

This summary of some observational results is a crude and incomplete bird's-eye view, which because of its brevity puts the results in a much more clearcut form than they are. The literature and material about the bow-shock, for instance, could fill a small library. Instrumental limitations usually allow only a quick and often controversial glimpse at the exciting details of the phenomena associated with the shock. Thus, from the observations it already seems clear that no single unified shock *theory* can be expected. Indeed, theory itself has to demonstrate in the first place how a shock can come about in a plasma through collective interactions only. We shall see that dispersion and turbulent dissipation are the two concurrent mechanisms to limit the shock steepness.

The overall structure of the shock entails an intrinsically non-linear problem which is, in a description by moment equations, already very difficult mathematically, let alone in the Vlasov-description. We shall in this regard, therefore, adopt the fluid approach, although this is often hard to justify. The turbulent dissipation, on the other hand, is generally due to resonant instabilities, which are excited in the strongly inhomogeneous medium of the shock transition. These can only be described by a kinetic theory, because they depend on the details of the electron and ion distribution functions. The big problem then is to combine the two parallel approaches into a consistent picture. The usual way is to obtain somehow an effective (turbulent) collision frequency and to put this function into the macroscopic equations for the overall structure.

In these lectures we shall concentrate first on the basic question of shock-formation and then go in some detail through two arbitrarily chosen, but representative, models to demonstrate the essential features of the two known aspects of the limitation of shock-front steepness. In this way we shall be able to point out some of the connections with the experimental findings.

2

SHOCK FORMATION

2.1. *Continuum flow*

An ordinary gas, where the interaction between particles is furnished by short range two-body collisions, can be described by the equations of gas dynamics. This is true, as long as the mean free path is small compared with all macroscopic dimensions of the system. For instance, the flow around an obstacle of large enough linear dimensions is adequately described in terms of a few moments of the particle-distribution function, like density, mean velocity, pressure, etc. However, once the body becomes so small that its size is smaller than a mean free path, the fluid nature of the flow is lost and the particles interact with the body individually. The flow then becomes Newtonian, and we have to use the equations of motion of a single particle in order to determine the overall physical situation.

In a plasma in a magnetic field, however, we have a new length scale: the gyroradius r_i of a typical ion, where $r_i = v[m_i c/(q_i B)]$; (v = ion velocity; q_i, m_i = ion charge and mass, respectively; B = magnetic field; c = speed of light).

Now the gyro-radius can take the role of a mean free path (at least for plasma motions perpendicular to the magnetic field) and this quantity may be much smaller than the two-body collisional mean free path. In the solar wind near the earth, for instance, r_i is of the order of 100 km, while the scale of the magnetosphere is about 10 R_E. Thus we certainly can describe the earth's bow shock wave by continuum equations. Furthermore, the particle interaction is collective in a hot and dilute plasma, in the sense that a particle overwhelmingly interacts with all others simultaneously rather than with its individual neighbours. This introduces another scale, the Debye length $\lambda_D = v_{th}/\omega_{pe}$ (v_{th} = electron thermal speed). Thus the term "collisionless" does not at all mean that there are no interactions at all; it merely means the absence of entropy-producing two-body collisions.

Such a plasma is best described in the terms of the Vlasov-Maxwell equations. It can support many modes of oscillation, which we can call a shock wave in the large amplitude, non-linear case, if it constitutes a transition between two different homogeneous asymptotic states of the plasma and travels faster than the fastest propagating linear, information-carrying wave. Under certain circumstances, within the overall structure of such a non-linear wave, other modes can be destabilized. These modes will interact in a turbulent fashion with themselves, or with individual particles, thus introducing other "collisionless" mean free path-lengths, which are of the order of the wavelength of a representative unstable mode and can play the same role as r_i or λ_D in the previous discussion.

2.2. Shock-formation and steepening

Any perturbation in an ordinary gas, which would in the linear approximation correspond to a sound pulse, steepens. Because of the non-linearity of the Navier-Stokes equations, higher and higher harmonics are generated, which travel slightly faster than the long wavelength-components within the pulse. This process is finally stopped when viscous dissipation sets in at small scales. The result is a shock, where the production of short waves is balanced by their viscous dissipation. A remarkable fact is that the time of formation of the shock is generally very short compared with the timescale of its asymptotic decay due to viscosity. Thus the ultimate fate of any sound pulse is a shock wave (if it is not of such small amplitude that it gets damped out before significant steepening occurs). It then constitutes a quantity in its own right.

In a more general system, like a plasma, the necessary conditions for steepening of a pulse, consisting of the Fourier components of one type of linear oscillation, is a linear dependence of frequency ω on wave-number k

for small $k^{19, 20}$ (Fig. 2). We shall call this a "fluid-like" dispersion relation.

Consider for example the gas dynamic momentum equation

$$\frac{\partial}{\partial t} \boldsymbol{v} + (\boldsymbol{v} \cdot \boldsymbol{\nabla}) \boldsymbol{v} = -\frac{1}{\rho} \boldsymbol{\nabla} p \qquad \qquad ...[2.1]$$

The term $(\boldsymbol{v} \cdot \boldsymbol{\nabla}) \boldsymbol{v}$ is the essential non-linearity.

If we try to solve this equation (and the other gas-dynamic relations) by a perturbation scheme, assuming $\boldsymbol{v}$ (but not the other variables) to be zero in lowest approximation, we write: $\boldsymbol{v} = \boldsymbol{v}_0 + \boldsymbol{v}_1 + \boldsymbol{v}_2 + \ldots$; etc. Then to lowest order we obtain, as $\boldsymbol{v}_0 = 0$,

$$D\boldsymbol{v}_1 = 0 \qquad \qquad ...[2.2]$$

$$D\boldsymbol{v}_2 = \mathscr{L}\{\boldsymbol{v}_1, \boldsymbol{v}_1\}. \qquad \qquad ...[2.3]$$

D is a linear operator, which describes sound waves ($\omega/k = \text{constant} = \text{speed of sound} = a$). $\mathscr{L}$ contains quadratic terms.

If we assume $\boldsymbol{v}_1 \sim \exp\{i(\boldsymbol{k} \cdot \boldsymbol{x} - \omega t)\}$, then $\mathscr{L}\{\boldsymbol{v}_1, \boldsymbol{v}_1\} \sim \exp\{2i(\boldsymbol{k} \cdot \boldsymbol{x} - \omega t)\}$. Now we can view equation [2.3] as the equation of a linear oscillator $\boldsymbol{v}_2$ with $\omega/k = \text{const} = a$, driven by the "external force" $\mathscr{L}\{\boldsymbol{v}_1, \boldsymbol{v}_1\}$.

Since $\omega(2k) = 2\omega(k)$, resonant energy transfer will occur from $\boldsymbol{v}_1$ to $\boldsymbol{v}_2$. In this way higher and higher multiples of k are produced in a cascade.

Steepening occurs, if in addition these smaller scales try to overtake longer ones.

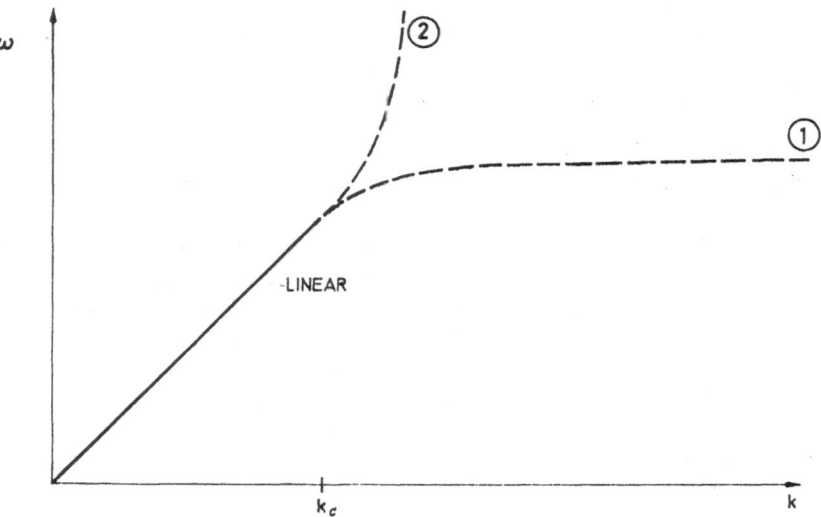

FIG. 2. ω is a linear function of k up to k_c, where dispersion sets in in some way, indicated by the dotted curves. For a dispersion relation of the type (1), $\omega/k \to 0$ for large k, while for type (2), $\omega/k \to \infty$.

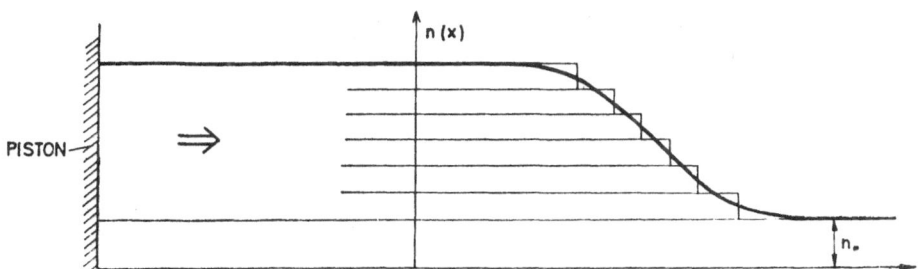

FIG. 3. The piston, moving to the right, generates a pulse in the same direction, which can be decomposed into a series of small steps.

Imagine, for example, a density perturbation of finite amplitude[21] (see Fig. 3). We can decompose this pulse into a series of small steps, each riding on top of its predecessor (in the $+x$-direction). Consider a single step. It does not change its form, because of the linearity of the dispersion relation; but it propagates in a medium, with a density $> n_0$ and which is already in a forward motion.

Thus the rate at which a step catches up with the one ahead of it is $(\rho/a)\delta(a+u)/\delta\rho$, where δ means the change across a step. (ρ = density; u = mean velocity of the step.) A compressional pulse ($\delta\rho > 0$) steepens, if

$$\frac{\rho}{a}\frac{\delta(a+u)}{\delta\rho} > 0. \qquad ...[2.4]$$

The validity of the above arguments is not confined to gas dynamics. If we describe a plasma by moment equations, then the steepening term $(v\cdot\nabla)v$ which corresponds to the term $(\rho/a)(\delta u/\delta\rho)$ in $[2.4]$ also occurs in the plasma momentum equation. If the dispersion relation is "fluid-like", i.e. linear for at least small k, then a pulse will steepen if $[2.4]$ is fulfilled.‡

Obviously, then, a fluid-like dispersion relation is not a sufficient criterion for the formation of a shock wave. In a plasma this can be inhibited because of polarization properties of a particular mode. For the intermediate Alfvén-mode, for instance, which is purely transverse, $(v\cdot\nabla)v$ vanishes. There is also no change in density, and steady state sinusoidal, large-amplitude waves are possible. Mode coupling thus does not occur for those waves (see Section 4).

Apart from dissipation, steepening can ultimately be checked by *dispersion* for large values of k (cf. Fig. 2). Then the short wavelength waves no longer travel at the same phase velocity as the long waves and cannot stay within the shock front. Thus the largest k-component in a non-linear wave

‡ Incidentally it seems that "fluid-like" plasma waves can to a sufficient degree of accuracy and with proper interpretation, always be described by moment equations. This is not entirely trivial and justifies to some extent the proposition made in the introduction that one could use a moment approach for the analytical description of the overall shock structure. Nevertheless, there are quite a few interesting shock theories starting from the outset with the kinetic equation.[22] We shall not discuss them in these lectures.

is k_c. In ordinary gases dissipation usually takes over, long before dispersive effects due to the excitation of new internal degrees of freedom begin to play a role. In a plasma the effect of turbulent dissipation is not so easily determined and dispersion is most important. Thus in general both mechanisms will be competitive.

Let us consider an example. In a cold plasma, magnetosonic waves propagate perpendicular to B with a phase velocity $\omega/k = B/(4\pi m_i n)^{\frac{1}{2}}$ ($n =$ particle number density). This is true for small k. For large k we have to retain electron inertia in Ohm's law[23]:

$$\frac{4\pi}{\omega_{pe}^2}\frac{\partial j}{\partial t} = E + \frac{1}{c}(v \times B). \qquad ...[2.5]$$

In order to see at which k the current inertia term becomes comparable in order of magnitude with the Lorentz term on the right-hand side, we put $(\omega/\omega_{pe}^2)|j| = O[(1/c)|v||B|]$. From the momentum equation, however, we have $\omega m_i n|v| = O[(1/c)|j||B|]$. Thus $\omega^2 = O(\omega_e \omega_i)$, where $\omega_{e,i}$ are the electron and ion gyro frequencies, and dispersion sets in for $\omega^2 \gtrsim \omega_i \omega_e$ (the lower hybrid frequency for $\omega_e^2 \ll \omega_{pe}^2$). Assuming that for this ω we still have approximately the above ω/k, we get $k_c^2 = O(\omega_{pe}^2/c^2)$.

Thus for the non-linear version of such a wave we would expect from dispersion arguments alone scales not smaller than c/ω_{pe}.

With Ohm's law in the form [2.5], the dispersion relation has the form shown as type (1) in Fig. 2.

There is an interesting analogy to water waves in a shallow channel where there is practically no dissipation. The corresponding dispersion relation is

$$(\omega/k)^2 = (g/k)\tanh(kh_0) \qquad ...[2.7]$$

($g =$ gravitational acceleration, $h_0 =$ channel depth). In such channels, stationary non-linear *solitary* waves have been observed, which travel faster than the fastest linear waves, which are the long ones, of course. (Compare Fig. 4.) The scale length in the solitary wave is of order h_0, where ω/k starts to deviate from its constant value $\sqrt{(gh_0)}$.

From this analogy we should expect that such solitary waves exist in the plasma example discussed above. This will be shown in the next section. Yet there is another point: at some critical Mach number, $u/\sqrt{(gh_0)}$, the solitary wave breaks and goes over into a hydraulic bore, which is highly turbulent and has a considerably broader scale-length. Breaking means that the flow becomes triple-valued in the shock-front. In a water wave the one-dimensional description implied in Fig. 4 breaks down, and the flow velocity now also depends on the elevation of a fluid element. An unstable tangential discontinuity develops.

Thus we may tentatively conclude that dispersion alone is in general insufficient to prevent a wave from breaking for higher Mach numbers. Even the inclusion of turbulent resistivity in a collisionless plasma shock cannot prevent breaking, at least in many cases treated up to date. Formally this means that a particular shock-solution cannot be extended beyond a critical Mach number. The experimental results (Section 1) indicate that this happens for $M \simeq 3$ for perpendicular shocks.

We can then think of essentially three types of shock theories: first of all, models where dispersion is the basic effect and turbulence in addition determines the shock-structure; such shocks are conventionally called *laminar*, although the transition to the second, a shock-theory where turbulent dissipation operates almost exclusively, is not clearly determined. Common to both is the property that one can still approximately distinguish between a gross, overall structure and, as far as dissipation is concerned, a turbulent sub-structure (described by weak turbulence theory). Such models will eventually be limited by the breaking phenomenon. The third class of model should be independent of a Mach number-limitation and be able to describe the transition across a critical Mach number and what happens afterwards. Naturally, this is the most difficult area and is not yet very well explored analytically[24, 25]; but some very interesting computer calculations have been performed,[26] which seem to confirm the c/ω_{pe} behaviour below $M \approx 3$ and the possibly non-stationary c/ω_{pi} structure suggested by the experiments for $M \gtrsim 3$.

As an example of the first type of theory, we treat in the next section low Mach number solitary waves and shocks propagating perpendicular to the B-field in a cold plasma, and we add some remarks concerning oblique shocks. In Section 3 we will discuss a model[14] of a parallel propagating high β shock, where dispersion effects are neglected from the outset. Whether this model is Mach number-limited is yet to be determined. We will have little to say about shock theories which are not Mach number-limited, or which are purely electrostatic,[16, 17] because they are, at least for the author, beset with conceptual difficulties.

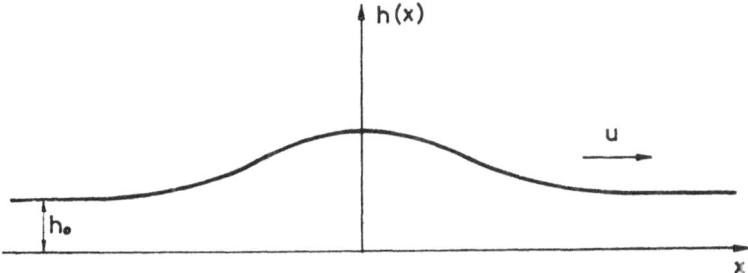

FIG. 4. A solitary wave in water. $h(x)$ is the surface elevation of the water. The wave is entirely symmetrical with respect to x. u is the propagation speed.

3

LAMINAR SHOCKS

3.1. *Low β shocks propagating perpendicular to B*

Let us look for plane, stationary shock solutions in a cold, quasi-neutral plasma. In this subsection we will assume that the plane of the shock is parallel to B.[27, 28, 29] Thus the shock propagates perpendicular to B; this is called a perpendicular shock. The assumption of a cold plasma means $\beta = nT8\pi/B^2 \ll 1$ (T = plasma temperature), while quasi-neutrality implies $\omega_e^2 \ll \omega_{pe}^2$, i.e. $B^2/(8\pi) \ll nm_ec^2$; consequently we can neglect the displacement current in Maxwell's equations and we will neglect the thermal motion of the particles. Let the shock propagate in x-direction. Thus $\partial/\partial t = \partial/\partial y = \partial/\partial z = 0$; also $B_x = B_y = 0$, and we have the geometry indicated in Fig. 5.

We shall include a frictional force between ions and electrons, which may arise from Ohmic dissipation, or, according to the theme of these lectures, from dissipation by the excitation of instabilities in the shock front. The attendant heating effect, however, will be neglected. Obviously this implies small friction.

Define $v_{i,e} = (u_{i,e}, w_{i,e}, 0)$; $B = (0, 0, B)$, where i, e denote ions and electrons respectively. Then we get the following systems of equations in the shock frame:

$$n_i(x) = n_e(x) = n(x) \quad \text{(quasi-neutrality)} \qquad \ldots[3.1]$$

$$(d/dx)(nu_i) = (d/dx)(nu_e) = 0 \qquad \ldots[3.2]$$

$$m_iu_i(dv_i/dx) = q_i[E+(1/c)(v_i \times B)]+vm_e(v_e-v_i) \qquad \ldots[3.3]$$

$$m_eu_e(dv_e/dx) = q_e[E+(1/c)(v_e \times B)]-vm_e(v_e-v_i) \qquad \ldots[3.4]$$

where v is the effective collision frequency for momentum transfer.

$$\text{curl } E = 0 \qquad \ldots[3.5]$$

$$\text{curl } B = (4\pi/c)(q_iv_i+q_ev_e) \qquad \ldots[3.6]$$

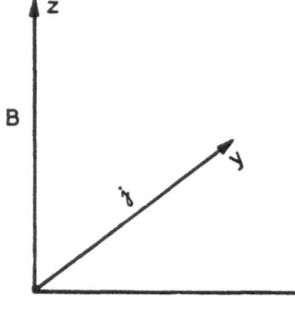

FIG. 5. Geometry of a perpendicular shock. B is always in the z-direction, while the current flows in the y-direction. The shock propagates parallel to the x-axis. There is no z-component of the mean velocity or the electric field. The shock propagates to the right.

$$\text{div } \boldsymbol{B} = 0 \qquad \qquad ...[3.7]$$

We shall assume singly ionized ions, i.e. $q_i = -q_e = q > 0$. From $[3.6]$ we get

$$u_i = u_e = u(x). \qquad \qquad ...[3.8]$$

Thus ions and electrons travel together in the x-direction. From $[3.5]$:

$$E_y = \text{const,} \ E_z = 0, \text{ where we choose } E_z = 0. \qquad ...[3.9]$$

Adding $[3.3]$ and $[3.4]$ results in

$$(d/dx)\{(m_i+m_e)nu^2 + B^2/(8\pi)\} = 0 \quad \text{and} \qquad ...[3.10]$$

$$(d/dx)\{m_i w_i + m_e w_e\} = 0. \qquad \qquad ...[3.11]$$

The mass flow in the y-direction can always be made to vanish, by a suitably chosen Galilean transformation. Thus we may choose $m_i w_i + m_e v_e = 0$; which means that the electrons carry almost all the current.

Since we look for shock-like solutions, we require that all quantities reach a constant value for $x \to +\infty$. The asymptotic values in this upstream region will have a subscript $_0$.

From $[3.3]$ and $[3.4]$ we get:

$$-q_i \int_\infty^x dx' E_x(x') = -\tfrac{1}{2}(m_i - m_e)(u^2 - u_e^2);$$

which says that the x-motion of the ions is determined almost entirely by the electric field.

Neglecting from now on terms of the order m_e/m_i compared with 1, we get the final system of equations

$$(d/dx)(nu) = 0, \text{ i.e. } nu = n_0 u_0. \qquad \qquad ..[3.12]$$

$$(d/dx)[m_i nu^2 + B^2/(8\pi)] = 0, \text{ i.e. } u = u_0 - \frac{B^2 - B_0^2}{8\pi m_i n_0 u_0} \qquad ...[3.13]$$

$$m_e nu(d/dx) w_e = -nq[E_y - (u/c)B] - m_e nv w_e \qquad ...[3.14]$$

$$(d/dx)B = (4\pi/c)nqw_e. \qquad \qquad ...[3.15]$$

In the upstream region no current shall flow. This leads to

$$E = (u_0/c)B_0. \qquad \qquad ...[3.16]$$

In principle, we could now eliminate n, u, w_e from the above equations and thereby obtain an equation for B alone. The equations are, however, easier to interpret[27] if we introduce the time of flight of a particle $\tau = \int_\infty^x dx'/[v(x')]$, i.e. $d/d\tau = u(d/dx)$. Then we obtain the following equation[28]:

$$\frac{d}{d\tau}\left\{\tfrac{1}{2}\frac{c^2}{\omega_{pe}^2}\left[\frac{d}{d\tau}\left(\frac{B}{B_0}\right)\right]^2 + \psi(M, B)\right\} = -v\,\frac{c^2}{\omega_{pe}^2}\left[\frac{d}{d\tau}\left(\frac{B}{B_0}\right)\right]^2$$

$$\psi(M, B) = \tfrac{1}{2}c_A^2[(B/B_0)-1]^2\{\tfrac{1}{4}[(B/B_0)+1]^2 - M^2\}; \qquad \dots[3.17]$$

$$c_A^2 = B^2/(4\pi n_0 m_i), \quad M = u_0/c_A.$$

If we omit for the moment the dissipative term on the right, [3.17] can be interpreted as the motion of a particle with coordinate B/B_0 in the potential ψ. The relevant time-scale of this problem is obviously $c/(c_A\omega_{pe}) = (\omega_i\omega_e)^{-\frac{1}{2}}$. For moderate Mach numbers M, $u_0 \simeq c_A$, and this time-scale corresponds to a length scale $\simeq c/\omega_{pe}$, which agrees with the order of magnitude estimate given in the last section, where we considered the same problem in the context of dispersive shock steepness limitation. For $M > 1$, $\psi(M, B)$ qualitatively has the form given in Fig. 6, where

$$B_{\max} = B_0(2M-1)$$
$$B_\pm = \tfrac{1}{2}B_0\{-1+\sqrt{(1+8M^2)}\}.$$

Obviously, u, given by equation [3.13], has to be positive for all x, i.e. all τ; otherwise we would get particles crossing. Then our "cold" approximation breaks down; the wave breaks. The maximum allowed value for B is then $B_0\sqrt{(1+2M^2)}$, from [3.13].

For $v = 0$ we can integrate equation [3.17] over τ. This gives us a constant of integration, which corresponds to the "energy" of the particle. To obtain a solution that looks like a shock, however, we have to equate this constant to zero. A particle starting at B_0, then takes an infinite amount of time (corresponding to the uniform upstream conditions of the flow) to roll

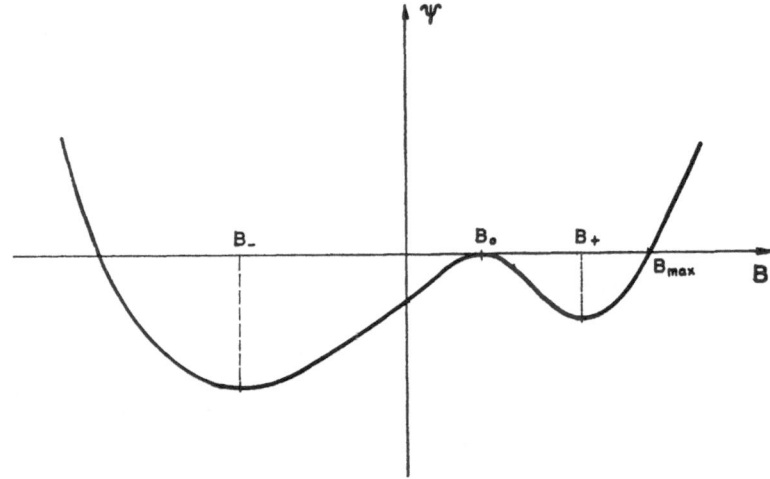

FIG. 6. The form of the effective potential ψ for $M > 1$.

down the potential well, say to the right in Fig. 6, reach a maximum at B_{max} and return to B_0 in a symmetrical way. Actually, the left branch of ψ is not accessible, because B_- is larger than the maximum allowed value for B. Since u is monotonically decreasing with B, we obtain by the requirement $u(B_{max}) > 0$ a limitation on the Mach number:

$$M < 2. \qquad\qquad ...[3.18]$$

The resulting profile for $B(x)$ is a solitary wave of the type shown in Fig. 1. Here we have to replace $h(x) \rightarrow B(x)$. (Since τ is a monotonic function of x, we may write again $B(x)$.) For $M \rightarrow 2$ the solitary wave becomes a cusp. For M slightly larger than 2, most (for $T = 0$, actually *all*) of the ions in the shock front[29,19] would be reflected back upstream. This would lead to an asymmetry, i.e. to a true shock.

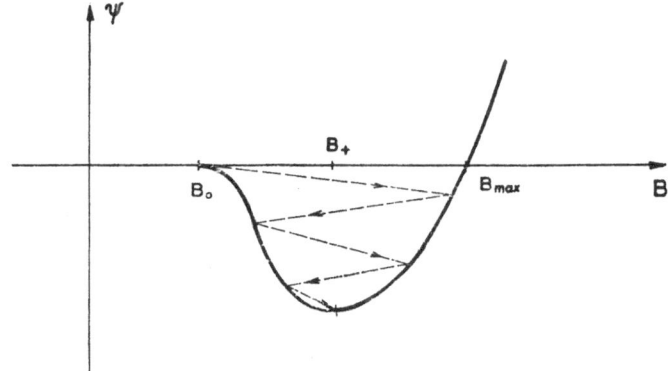

FIG. 7. The motion of the equivalent particle B in the potential well ψ with small friction.

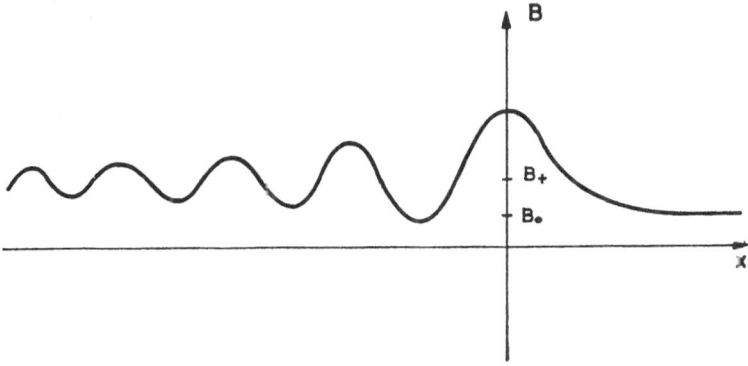

FIG. 8. A laminar shock with small dissipation. After the initial rise, B executes damped oscillations around B_+.

Within our model, for $M < 2$, however, the retention of the dissipative term in [3.17] also leads to an asymmetry. The motion of the "particle" then gets damped (see Fig. 7). B does not quite reach the value B_{max} any longer; moreover, it cannot get back to B_0 and so oscillates in the potential well, until it finally ends up at B_+. The x-dependence of B is shown in Fig. 8.

For small v, the characteristic scale L_e of the leading edge of the wave is[28]

$$L_e \simeq [M/\sqrt{(M^2 - 1)}](c/\omega_{pe}) \qquad \ldots[3.19]$$

while the damping length L_D of the downstream oscillation is

$$L_D \simeq \frac{1 + \sqrt{(1 + 8M^2)}}{2M} \cdot \frac{|\omega_i \omega_e|}{v} \cdot \frac{c}{\omega_{pe}} \sim \frac{1}{v}. \qquad \ldots[3.20]$$

If v becomes very large, however, then, as can be seen from Fig. 7, B monotonically rises from B_0 to B_+, thus exhibiting a shock transition in the conventional sense. In this case, however, the cold approximation certainly is inapplicable, because large heating will occur.

Actually the dissipation may not at all be due to collisions, but due to turbulent friction. Obviously, to create the gradient of the B-field in the leading edge, we must have a large current density in the y-direction which increases with M from [3.19]. Thus, locally within the front a two-stream instability could arise, because the electrons stream in the opposite direction to the ions. The critical relative speed between electrons and ions is $w_{ec} \simeq \sqrt{(T/m_e)}$, the electron thermal velocity v_{th}^e. (In these considerations, we have to retain the finite temperature.) Calculating the relative speed from the solitary wave profile, one obtains a critical Mach number M_c, above which the instability sets in.[19]

$$M_c \simeq 1 + \tfrac{3}{8}\left(\frac{8\pi nT}{B^2}\right)^{\frac{1}{4}}. \qquad \ldots[3.21]$$

Because of the low β, M_c is still small compared with the breaking value, $M = 2$.

For $M > M_c$, ordered kinetic energy gets fed into unstable plasma oscillations, which preferentially heat the electrons. Once $T_e \gg T_i$, ion sound modes also can be destabilized (compare Ref. 18). Here the threshold (relative) velocity is about $\sqrt{(T_e/m_i)}$ which will in general be lower than the threshold for the two-stream instability discussed above, and strong dissipation will set in. To calculate the dissipative effect of these resonant instabilities, one has to solve the coupled system of the equations of weak turbulence (in the ion sound mode) together with the equations for the average distribution functions $\langle f_{i,e} \rangle$ which describe the overall shock-structure. The corresponding effective collision frequency has been estimated[20] to be of the order of $v_{eff} \simeq \omega_{pi}(u_0/v_{th}^e)(T_e/T_i)$, but unless M is not extremely small, so that the heating may still be neglected, the overall structure of the shock will not even approximately be described by equation [3.17], which does not contain a pressure term; we, therefore, will not pursue this argument much further here (for an excellent review, see Ref. 20). There have been a number of attempts to improve the overall description of the shock-structure by including a pressure term in the equations[30] and thereby taking into account the resistive heating in some heuristic way.[31-35] Usually also these models break down for a Mach number $M \simeq 3$. This problem seems to be related to the phenomenon of the "isomagnetic jump" in conventional MHD-shock theory.[36]

There exists a one-dimensional numerical model of an originally completely cold plasma,[26] where particle crossing can be taken into account and the heating behind the shock front can be calculated. Above $M = 2$ the rather laminar wavelike structure starts to become highly turbulent and non-stationary. The width of the shock then becomes of the order of c/ω_{pi} with a fine structure $\simeq c/\omega_{pe}$, which compares favourably with the experiments.[1-6] The broadened shock structure has been phenomenologically explained by R. Z. Sagdeev,[19] to whom the theory of collisionless shocks owes most of its exciting aspects: in a similar fashion as in the breaking process of a water wave, where the high velocity parts of the bore are prevented from over-shooting the upstream quiet fluid through the instability of the resulting tangential discontinuity, the reflected ions in a breaking laminar plasma shock are turned around by the upstream B-field in about a characteristic Larmor radius c/ω_{pi}. This scale then should characterize the shock structure.

3.2. Low β oblique shocks

If the angle θ between the plane of the shock and B is not zero, we call such a shock oblique. Although this is obviously the general case, we shall confine ourselves to a few remarks. One reason for this is that most experiments have been done for strictly perpendicular shocks. Another one is that the construction of laminar wave-trains proceeds in a similar spirit as in the perpendicular case. Yet there are some very important differences.

First of all for $(m_e/m_i)^{\frac{1}{2}} \ll \theta$ the dispersion properties for small amplitude waves change from a linear behaviour of ω for small k to one characterized by the curve type (2) in Fig. 2. Dispersion sets in for $\omega \gtrsim \omega_i$ and therefore electron inertia, which sets in at $\omega \gtrsim (\omega_i\omega_e)^{\frac{1}{2}}$ is not important. For θ also small compared to 1 the corresponding characteristic scale is $(c/\omega_{pi})\theta$.[37] Now short waves travel faster than long ones, and without dissipation one obtains a solitary rarefaction wave instead of a compression wave. Even the slightest dissipation converts this solitary wave into a laminar wave-train. However, this time the oscillations do not trail the "shock-front" but constitute a leading wave train. The main reason for this difference is that magnetic field lines cross the shock plane, and therefore the highly mobile electrons can short out electric fields across the shock in contrast to a strictly perpendicular shock.

The two-stream instability need not necessarily develop because the projection of the gradient of $|B|$ onto the shock normal may be too small, if θ is not small, although another instability due to O. Buneman (see Ref. 20), may develop with frequencies $\omega_i \ll \omega \ll \omega_e$.

The main point we should make, however, is the instability of the leading nonlinear wave-train to "decay".[20,38,39] This is another type of instability, where a non-linear, approximately sinusoidal wave of frequency ω_0, and wave number k_0, can decay into two others ω_1, k_1; ω_2, k_2, if the linear dispersion relation allows the "decay"-relations to be fulfilled:

$$k_0 = k_1 + k_2 \quad \text{and} \quad \omega(k_0) = \omega(k_1) + \omega(k_2) \qquad ...[3.22]$$

This is possible within one branch of $\omega(k)$ for a dispersion relation of the type (2) in Fig. 2, but not for type (1). Actually the non-linear version of a wave of type (1) can also be unstable, but the resulting decay products cannot both be waves of the same type. Using the kinetic wave equation, and equating the change $(\partial\omega/\partial k + u)\partial N_k/\partial x$ of the number N_k of waves with wave-vector k, per volume element, with their rate of destruction by turbulent decay $2\nu N_k$ (ν = growth rate of decay instability) one can estimate the decay length, for small M, to be[39]:

$$L \simeq 4(c/\omega_{pi}) \tan \theta (M-1)^{-\frac{3}{4}} \qquad \text{for } \tan \theta \simeq 1 \qquad ...[3.23]$$

as also experiments seem to indicate,[10] and

$$L \simeq (2c/\omega_{pi}) \pi^{-1} (M-1)^{-\frac{3}{4}} \qquad \text{for } \tan \theta \simeq 0.$$

Thus the appearance of the oscillatory structure leads to its own decay. If one considers the shock-thickness to be given by the equivalent to L_D (see equation [3.20]) in both cases, it is obvious that the decay "collision" frequency reduces this number considerably, although in the perpendicular case the appearance of the much faster ion sound instability will mask this effect.

3.3. High β perpendicular shocks

Hitherto we have considered only low β laminar shocks. However, as has been argued, the eventual strong dissipation in a perpendicular shock can heat the plasma considerably. Thus it is reasonable to consider such shocks, when the condition of $\beta \ll 1$ is somewhat relaxed. As can be seen, the thermal ion Larmor radius already becomes comparable with the scale of the leading edge $L_e \simeq c/\omega_{pe}$ for $\beta \gtrsim m_e/m_i$, which is extremely small. Thus, besides the pressure terms, the inclusion of finite Larmor radius corrections into the basic shock-equations is very important. For a model which proceeds again with the construction of a (leading) wave train and its ensuing destruction by turbulent decay, we simply refer to Ref. 40.

All the models discussed up to now more or less start with a quasi-laminar solution determined by dispersion and then, so to say on top of that, discuss its modification by turbulent dissipation, although in the case of strong dissipation the distinction from an exclusively dissipative shock lies more in the guiding philosophy rather than in the physical content. In the next section shocks will be discussed which are turbulent from the outset.

4

TURBULENT SHOCKS

4.1. *High Mach number shocks*

A shock-theory which is not Mach number-limited will in principle have to start directly from the Vlasov equation. Only within this framework will it be possible to include effects like selective reflection of particles (with respect to their place in the distribution function) in the shock front, or resonance effects at Doppler-shifted multiples of the gyrofrequencies, due to the finite temperature at least in the back of the shock.

An analytical theory will have to start with the equations of weak turbulence and, assuming these to be applicable also for high Mach numbers, proceed to solve the equations for the averaged distribution function $\langle f_{i,e} \rangle$ together with the kinetic wave equation (including, in particular, quasi-linear effects) for the fluctuations. Indeed the kinetic wave equation has been postulated[24, 25] and later rigorously derived[39] within the framework of shock-theory. Although such a programme looks hopeless analytically, in almost all cases it at least serves as a guide to bracket the kind of waves which probably constitute the dissipative ingredient, and accordingly allows shock thicknesses to be estimated. In the case of the propagation perpendicular to B in a cold plasma, whistler waves (which can stay in the shock front because their phase velocity is larger than c_A) can provide the necessary random fluctuations[24, 25, 39]. Experiments[2, 9, 10] and observations[11] in the bow shock seem to agree with this idea, as well as the estimated shock thicknesses of a few c/ω_{pi}, although these measurements are done in plasmas of moderate β or even $\beta \gtrsim 1$.

For very high Mach numbers the Rankine-Hugoniot relations indicate[8] that the magnetic field should somewhat lose its importance for the shock-structure and, therefore, purely electrostatic shocks have been suggested. As far as these electrostatic shock models are not laminar,[41] the logical chain of first a non-linear pulse, which then steepens and becomes limited by dispersion and dissipation, has not yet been explored. The alternative is then to consider two counterstreaming plasmas, which, in the process of their mutual interpenetration, build up a shock-like interface through the excitation of electrostatic instabilities[16, 17]. This problem has been considered in an interesting numerical model.[42]

4.2. *Turbulent high β parallel shocks*

We now turn to a particular model,[14] where we can go through the programme described in the beginning of this section, in at least a qualitative fashion. It describes a truly turbulent shock, in the sense that dispersive effects are neglected from the outset. On the other hand, the picture of a steepening wave is still retained.

The discussion in Ref. 14 which we shall follow in principle, without however, giving all the detailed arguments, is in terms of the Vlasov equation. Only at the end a system of moment equations derived from the quasi-linear equations is used to qualitatively obtain a shock-solution. It seems, however, that the shock from the outset can be described entirely in terms of a set of two-fluid equations for small Mach numbers, although this approach is still under active investigation. The reason for this is that the attendant (firehose or garden-hose) turbulence[43, 14] is mainly generated by non-resonantly unstable Alfvén waves, which do not require a kinetic description.[44]

We shall assume $\beta_i \gg 1$, where $\beta_i = 8\pi p_i/B^2$. In addition we assume $T_e/T_i \gg 1$ and $c \gg c_A$, i.e. the Debye-length is small compared with the thermal ion Larmor radius‡ (*i, e* again refer to ions and electrons). It is well known that quasi-undamped ion sound-waves exist in a plasma under these conditions. Since $\beta > 1$, we have $c_A^2 < (v_{th}^i)^2 < c_s^2$, where $c_s^2 = T_e/m_i$, the ion sound phase velocity. Thus the ion sound wave is the fastest fluid-like mode propagating along B for $\beta \gg 1$, and, in the following, planar ion sound shock-solutions will be considered with the shock-normal parallel to B (a parallel shock).

As a matter of fact, ion sound waves have a dispersion relation corresponding to type (1) in Fig. 2; k_c^{-1} in this case equals the Debye length $\lambda_D = v_{th}^e/\omega_{pe}$. The corresponding critical frequency is ω_{pi}. In other words, ion sound oscillations become dispersive for $k > \lambda_D^{-1}$. This is also the wavenumber above which significant departure from quasineutrality sets in.

Thinking more in terms of laminar shocks, we now could start to construct a solitary wave with scale-lengths of the order of λ_D and with the inclusion of some dissipation obtain a shock solution.[45] It will be shown, however, that ion sound waves, propagating parallel to B, will destabilize Alfvén waves with k-vectors parallel to B and in magnitude less than a reciprocal ion gyroradius. The ensuing garden-hose turbulence will limit the shock thickness to scales of the order of an ion gyroradius—long before dispersion sets in at λ_D. Therefore, we can neglect dispersive effects altogether. The following discussion assumes that garden-hose turbulence is the only mechanism determining the shock-structure. Let us first show that ion sound waves steepen and can couple to unstable Alfvén waves.

‡ Although these conditions may in part be only marginally fulfilled in the solar wind, they certainly approximate the conditions in the bow shock much better than any cold plasma theory. Consequently this model should hopefully be relevant for the earth's bow shock (see also the model in 3.3 in this regard).

Since the electrons are very hot, their thermal velocity is much larger than c_s and they see essentially a static electric field. Therefore they will approximately obey a Boltzmann-distribution:

$$n_e = n_0 \cdot \exp\{e\phi/T_e\}; \qquad \qquad ...[4.1]$$

$T_e = $ const., $-e = $ electronic charge, and $\phi(z)$ is the electrostatic potential. Since both torsional Alfvén waves and ion sound waves are quasi-neutral we shall assume

$$n_i = n_e = n. \qquad \qquad ...[4.2]$$

From now on, quantities without an index refer to the ions. If electron quantities occur, they will have the superscript ^{e}.

The ions are practically cold, i.e. $v_{\mathrm{th}}^i \ll c_s$. Thus we may take moments of the ion Vlasov equation and neglect the ion heat current, but allow for a pressure-anisotropy:

$$\partial n/\partial t + \mathrm{div}\,(nv) = 0 \qquad \qquad ...[4.3]$$

$$n\left[\left(\frac{\partial}{\partial t}\right) + v \cdot \nabla\right]v = \frac{e}{m}\,nE + \frac{en}{mc}\,[v \times B] - \frac{1}{m}\frac{\partial}{\partial x_l}\,p_{hl} \qquad ...[4.4]$$

$$\left[\left(\frac{\partial}{\partial t}\right) + v_l\left(\frac{\partial}{\partial x_l}\right)\right]p_{mn} = -p_{mn}\frac{\partial v_l}{\partial x_l} - p_{nl}\frac{\partial v_m}{\partial x_l} - p_{lm}\frac{\partial v_n}{\partial x_l}$$
$$+ \varepsilon_{nrs}p_{mr}\Omega_s + \varepsilon_{mrs}p_{nr}\Omega_s. \qquad \qquad ...[4.5]$$

Here we have used a mixed vector and index notation. Summation over like indices is implied; $\Omega_r = eB_r/(mc)$; ε_{mrs} is the totally antisymmetric unit tensor.

In addition we have Maxwell's equations

$$\mathrm{curl}\,E = -(1/c)(\partial B/\partial t); \quad \mathrm{curl}\,B = (4\pi/c)en(v - v^e) + (1/c)dE/dt;$$
$$\mathrm{div}\,B = 0; \quad E_z = -\partial\phi/\partial z. \qquad \qquad ...[4.6]$$

As for the electron current $-env^e$, we assume the electrons to be isothermal with an isotropic pressure $p^e = nT^e$. Neglecting their inertia, their transverse velocity is simply given by

$$v_\perp^e = (B_\perp/|B_z|)v_z^e + c(E_\perp \times B_z)/|B_z|^2; \; B_z = (0, 0, B_z),$$

while v_z^e follows from their continuity equation which is analogous to equation [4.3].

To describe parallel ion sound waves the system [4.1]–[4.6] is linearized about a static homogeneous equilibrium with an isotropic pressure:

$$n = n + \delta n; \; v = \delta v; \; \delta v \times B = 0, \text{ i.e. } \delta v = (0, 0, \delta v_\parallel)$$
$$p_{mn} = p\delta_{mn} + \delta p_{mn}; \; B = (0, 0, B); \; \delta B = 0 \qquad \qquad ...[4.7]$$
$$E = -\nabla\delta\phi.$$

δn, δv, δp_{mn}, $\delta \phi$ are the perturbations, which will be taken $\propto \exp \{-i\omega t + i k \cdot x\}$; $k = (0, 0, k)$.

Since the shock-structure is anticipated to be limited to scales of a few ion gyroradii, we assume $\omega \ll \Omega$, from which it follows that δp_{mn} is diagonal and isotropic around B:

$$\delta p_{mn} = \begin{pmatrix} \delta p_\perp & 0 & 0 \\ 0 & \delta p_\perp & 0 \\ 0 & 0 & \delta p_\parallel \end{pmatrix}.$$

From the linearized equations the ion sound dispersion relation follows:

$$(\omega/k)^2 = c_s^2 \simeq T^e/m, \quad \text{for} \quad T^e/T \gg 1. \qquad \ldots[4.8]$$

The density perturbation is connected with $\delta v_\parallel$ by

$$\delta n = \delta v_\parallel (n/c_s) \qquad \ldots[4.9]$$

while the pressure tensor fulfils the following relations:

$$\delta p_\parallel = 3p\delta v_\parallel (1/c_s); \quad \delta p_\perp = p\delta v_\parallel (1/c_s). \qquad \ldots[4.10]$$

Inserting [4.9] into the steepening criterion [2.4] and observing that c_s is density independent, we obtain $(n/c_s)\delta(c_s+v_\parallel)/\delta n = 1$. Thus compressional ion sound waves steepen.

From equations [4.10] and [4.9]: $(\delta p_\parallel - \delta p_\perp)/p = 2\delta n/n > 0$, for $\delta n > 0$. A compressive ion sound pulse generates an anisotropy, such that $p_\parallel > p_\perp$.

Now, low frequency transverse Alfvén waves constitute just another branch of oscillations for the system [4.1]–[4.6], besides ion sound waves. These transverse modes become unstable, if (electrons included)

$$(p_\parallel + p_\parallel^e) - (p_\perp + p_\perp^e) > B^2/4\pi. \qquad \ldots[4.11]$$

From what has been said above, then, a sufficiently strong primary ion sound compression can locally excite the garden-hose instability. If we assume an ion sound pulse of the form indicated in Fig. 3, then to the left the turbulence operates, while within the decreasing part of $n(z)$ there will be a point z_0, such that for $z > z_0$ [4.11] is not yet satisfied.

At this point an important remark has to be made: garden-hose turbulence saturates, at least in the quasi-linear approximation.[43] There is a tendency to restore the pressure isotropy (viscous effect), while generating a stationary spectrum of magnetic fluctuations. Thus in the back of the ion sound pulse the firehose instability will be saturated while in the front, where n is increasing, the instability gets excited in course of time. Its attendant viscous effect on the overall ion sound structure can then be expected to limit the steepening of the shock, at least for a weak shock.

In the stability criterion [4.11] and also in the equation governing the turbulence, only the sum of electron and ion terms ever occur.[44] Since the electrons are not much influenced by the ion sound wave (they have a high heat conductivity and do *not* undergo an adiabatic compression like the ions), we may neglect $p_\parallel^e - p_\perp^e$ altogether in [4.11]. This is compatible with the Ansatz [4.1] and [4.6]. Defining the quantity

$$\Delta = \frac{p_\parallel - p_\perp}{p} - \frac{2}{\beta} \qquad ...[4.12]$$

we have from [4.11] the instability criterion: $\Delta > 0$. At the same time this form of the criterion shows that a high β-plasma can be easily destabilized.‡

From now on we treat the system of equations [4.1]–[4.6] in an approximate fashion. All quantities are assumed to depend only on z and t. This can be done, because both ion sound and unstable Alfvén waves propagate parallel to B. At first we split every quantity into an ensemble average plus a fluctuation (assuming a random phase approximation to be valid for the turbulent substructure). The averages will describe the overall structure of a steepening ion sound pulse, while the fluctuations are due to firehose amplification.

If mode coupling is negligible we can then confine ourselves to the lowest approximation, which is the quasilinear one.[43, 14, 44] Indeed, for simple torsional Alfvén waves propagating in an isotropic medium, in the zero Larmor radius approximation mode coupling does not occur (see Section 2.2). The garden-hose instability, however, anticipates an anisotropy of the background and can only be described correctly in a finite Larmor radius approximation. Denoting by $\omega(k)$ and $\gamma(k)$ the real and imaginary parts of ω respectively and by R the ion Larmor radius, the instability is described by the following relations, for $0 < \Delta \ll 1$ (see Fig. 9 taken from Ref. 14):

$$\left.\begin{array}{l} \omega(k) \simeq \Omega\tfrac{1}{2}(kR)^2 \quad (= 0 \text{ in zero Larmor radius approximation}) \\ \gamma(k) \simeq k(p_\parallel/nm)^{\frac{1}{2}}[\Delta - \tfrac{1}{4}(kR)^2]^{\frac{1}{2}} \\ \gamma_{max} \simeq \Omega\Delta = \omega(k) \quad \text{for} \quad k = (2\Delta)^{\frac{1}{2}}/R \ll 1/R \\ \gamma(k^*) = 0 \quad \text{for} \quad k^* = 2\Delta^{\frac{1}{2}}/R \ll 1/R. \end{array}\right\} ...[4.13]$$

Thus a mode-coupling effect can occur only to the extent that the fluctuation amplitude is not small (because it is a non-linear effect) and the deviation from isotropy and the zero Larmor radius approximation are appreciable. Thus to neglect mode coupling we require from [4.13]

$$|h_k|^2 = O(\Delta) = O[(p_\parallel - p_\perp)/p] = O(2/\beta) \ll 1, \qquad ...[4.14]$$

where $$|h_k|^2 = |\delta B_k|^2/(2B^2).$$

In the following, the ordering [4.14] is assumed to hold, which probably implies a restriction to weak shocks. Furthermore it will be anticipated that the wavelength of the most unstable wave is small compared to the scale-length in the shock-transition. Then we can neglect the fact that the instability is excited in the inhomogeneous medium of the shock and use a

‡ This is even more true if, as in the solar wind, the upstream (interplanetary) plasma is already anisotropic.[15, 14]

WKB-approximation in z and t. Since also the group velocity $\partial\omega/\partial k$ is small compared with c_s, because of [4.14], and since the unstable spectrum is not strongly peaked, so that both $(\partial\omega/\partial k)\,1/c_s$ and the wave packet distortion along z can be neglected, we arrive at the following equations for the shock structure in the quasi-linear approximation for a steady shock:

$$\frac{\partial}{\partial z}(\rho u) = 0 \qquad\qquad\qquad ...[4.15]$$

$$\frac{\partial}{\partial z}(\rho u^3 + 3up_\| + 2\rho u\psi) = 4(p_\perp - p_\| \cdot 2)u\frac{\partial}{\partial z}\sum_k h_k(z) \qquad ...[4.16]$$

$$\frac{\partial}{\partial z}(up_\perp) = 2p_\| u\frac{\partial}{\partial z}\sum_k h_k(z) \qquad\qquad ...[4.17]$$

$$\rho = \rho_0 \exp\{\psi(z)/c_s^2\} \qquad\qquad ...[4.18]$$

$$u\frac{\partial}{\partial z}h_k \simeq 2\gamma_k(z)h_k(z). \qquad\qquad ...[4.19]$$

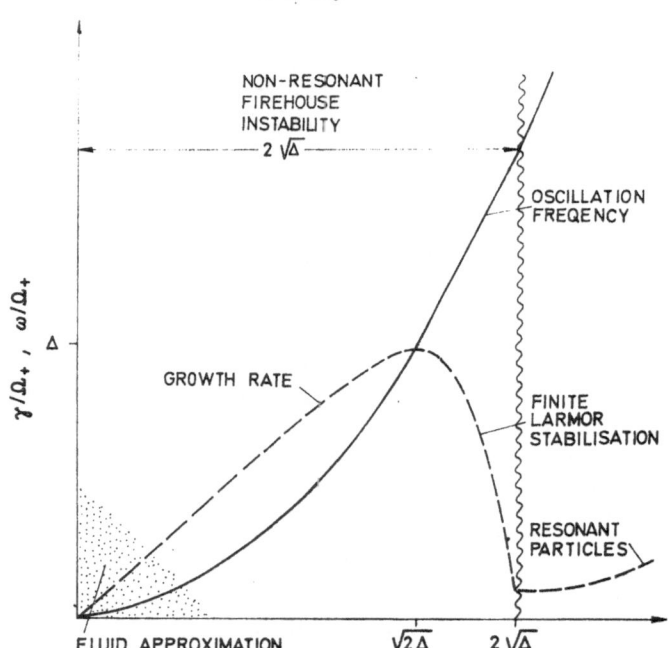

FIG. 9. A plot of the garden hose instability. The subscript $_+$ denotes ion quantities. At $kR_+ = 2\sqrt{\Delta}$ the growth rate goes to zero and resonant effects could occur. Such effects are not considered in the text and are hopefully small (Kennel and Sagdeev, 1967).

Here
$$u = \langle v_z \rangle, \; \rho = \langle n \rangle m, \; \psi = e \langle \phi \rangle / m$$
$$p_\perp = \langle p_- \rangle, \; p_\parallel = \langle p_\parallel \rangle, \; \langle v_x \rangle = \langle v_y \rangle = 0.$$

The ensemble averaged quantities describe the non-linear ion sound shock modified by turbulent dissipation as represented by the right-hand side of equations [4.15]–[4.17]. The spatial evolution of the fluctuations is given by [4.19], where $\gamma_k(z)$ should be taken from [4.13], in which latter equation all quantities now have to be identified by their ensemble averages. This system of nonlinear integro-differential equations can, of course, not be solved exactly. It is probably even more complicated than the original set [4.1]–[4.6], but its physical content is more obvious.

To indicate that shock solutions of this system seem to exist, we give a few remarks and point to Ref. 14 for a more detailed discussion.

Very roughly, the asymptotic solutions (at $z = \pm\infty$) should be given by $\gamma_{\tilde{k}}(\pm\infty) = 0$, where $\tilde{k}$ is a representative unstable wave-number. We give a subscript $_0$ to the upstream ($z = +\infty$) values of all quantities. Thus the Mach number $M = u_0/c_s$.

Assuming a weak shock,

$$M^2 - 1 = O[(u - u_0)/u_0] = O(T/T^e) = O(h - h_0) \ll 1,$$

where $h = \sum_k |h_k|^2$ is the energy density of the fluctuations, leads to:

$$(p_\parallel - p_\perp)/p \simeq -2\{\chi + 3(h - h_0)\}, \text{ where } \chi = (u - u_0)/u_0$$

Thus a compressional pulse ($\chi < 0$) favours fire-hose turbulence ($(p_\parallel - p_\perp)/p > 0$), while a build-up of h (i.e. $h - h_0 > 0$) diminishes that tendency, as we already know. Substituting L_s^{-1} for (d/dz), and putting into the formula [4.13] for the growth rate γ a typical anisotropy (neglecting finite Larmor-radius effects), leads to *two* solutions for χ. The first one is, within the assumptions, equal to zero (upstream state). The other one is:

$$-\chi \simeq \tfrac{3}{2}(M^2 - 1) + (T/T^e) \quad \text{(downstream state)}.$$

Both solutions correspond to a steady turbulent level, i.e. to the asymptotic solutions on either side of the shock transition. Then an estimate for the shock-thickness follows:

$$L_s \simeq R/(M^2 - 1)^2 \gg R;$$

similarly:

$$h - h_0 \simeq (M^2 - 1).$$

Also $\tilde{k} L_s \simeq (M^2 - 1)^{-1}$, which indicates that the WKB-approximation is also justified.

Obviously this approximate solution should not be taken as more than an argument that the whole approach is reasonable. Besides its internal consistency, its major physical feature is the conclusion that large amplitude torsional Alfvén waves will be generated. Since mode-coupling is small for these waves, they should persist far downstream.

As far as the earth's bow shock is concerned, this persistence seems to be indicated also by the observations (see Section 1), although the bow-shock is curved, and this model of a parallel shock can, therefore, certainly contain only one facet of the complex overall situation.

REFERENCES

1. G. C. GOLDENBAUM. 1967. *Physics Fluids*, 10, 1897.
2. E. HINTZ. 1968. Report to the Novosibirsk Conference on Plasma Physics and Controlled Nuclear Fusion Research.
3. R. CHODURA, M. KEILHACKER, M. KORNHERR and H. NIEDERMEYER. 1968. *Ibid.*
4. R. K. KURTMULLAEV, YU. E. NESTERIKHIN, V. I. PIL'SKII and R. Z. SAGDEEV. 1966. *Plasma Physics and Controlled Nuclear Fusion Research*, Vol. 2, p. 367. IAEA, Vienna.
5. J. W. M. PAUL, L. S. HOLMES, M. J. PARKINSON and J. SHEFFIELD. 1965. *Nature*, 208, 133.
6. J. W. M. PAUL *et al.* 1967. *Nature*, 216, 363.
7. F. DE HOFFMANN and E. TELLER. 1950. *Phys. Rev.*, 80, 692.
8. R. LÜST. 1953. *Zeitschr. Naturf.*, 8a, 279.
9. R. M. PATRICK and E. R. PUGH. 1967. *AVCO-Everett Res. Rep.*, No. 280.
10. A. ROBSON, J. SHEFFIELD and R. J. BICKERTON. 1968. ORO-3647-1, *Progress Rep. No. 1, Center for Plasma Physics and Thermonuclear Research, Univ. of Texas, Austin.*
11. J. P. HEPPNER, M. SUGIURA, T. L. SKILLMAN, B. G. LEDLEY and M. CAMPBELL. 1967. *NASA-Goddard Document X-612-67-150.*
12. R. E. HOLZER, M. G. McLEOD and E. J. SMITH. 1966. *J.G.R.*, 71, 1481.
13. J. H. WOLFE, R. W. SILVA and M. A. MYERS. 1966. *J.G.R.*, 71, 1319.
14. C. F. KENNEL and R. Z. SAGDEEV. 1967. *J.G.R.*, 72, 3303.
15. R. LÜST. 1967. *Solar-Terrestrial Physics.* Academic Press, New York and London.
16. D. A. TIDMAN. 1967. *Physics Fluids*, 10, 547.
17. D. A. TIDMAN. 1967. *J. Geophys. Res.*, 72, 1799.
18. R. FREDERICKS. Private communication.
19. R. Z. SAGDEEV. 1966. *Reviews of Plasma Physics*, Vol. 4 (ed. by M. A. LEONTOVICH). Consultants Bureau, New York.
20. R. Z. SAGDEEV and A. A. GALEEV. 1966. *International Center for Theoretical Physics, IC/66/64.* Trieste.
21. A. R. KANTROWITZ and H. E. PETSCHEK. 1966. *Plasma Physics in Theory and Application* (ed. by W. B. Kunkel), pp. 148–206. McGraw-Hill, New York.
22. C. K. CHU and R. A. GROSS. 1968. *Plasma Laboratory, Columbia University New York —Report No. 41.*
23. R. LÜST. 1959. *Fortschritte Physik*, 7, 503.
24. F. J. FISHMAN, A. R. KANTROWITZ and H. E. PETSCHEK. 1960. *Rev. Mod. Phys.*, 32, 959.
25. M. CAMAC, A. R. KANTROWITZ, M. M. LITVAK, R. M. PATRICK and H. E. PETSCHEK. 1962. *Nucl. Fusion Suppl.*, part 2, 423.
26. P. L. AUER, H. HURWITZ and R. W. KILB. 1962. *Physics Fluids*, 5, 298.
27. L. R. DAVIS, R. LÜST and A. SCHLÜTER. 1958. *Zeitschr. Naturforschung*, 13a, 916.
28. J. H. ADLAM and J. E. ALLEN. 1958. *Phil. Mag.*, 3, 448.
29. C. S. GARDNER, H. GOERTZEL, H. GRAD, C. S. MORAWETZ, M. H. ROSE and H. RUBINS. 1958. *Proc. 2nd U.N. Intern. Conf. on Peaceful Uses of Atomic Energy*, 31, 230.

30. K. HAIN, R. LÜST and A. SCHLÜTER. 1960. *Rev. Mod. Phys.*, **32**, 967.
31. A. CAVALIERE and F. ENGELMANN. 1967. *Nucl. Fusion*, **7**, 137.
32. P. J. KELLOG. 1964. *Dept. of Physics, Univ. of Minnesota, Tech. Rep. CR–69.*
33. K. I. GOLDEN, H. K. SEN and Y. M. TREVE. 1961. *Proc. 5th Int. Conf. on Ionization Phenomena in Gases, Munich 1961*, Vol. 2, p. 2109.
34. G. BARDOTTI, A. CAVALIERE and F. ENGELMANN. 1966. *Nucl. Fusion*, **6**, 46.
35. R. CHODURA. 1968. *Physics Fluids*, **11**, 400.
36. W. MARSHALL. 1956. *Proc. Roy. Soc.*, A, **233**, 367.
37. V. I. KARPMAN and R. Z. SAGDEEV. 1964. *Sov. Phys.-Tech. Phys.*, **8**, 606.
38. V. N. ORAEVSKII and R. Z. SAGDEEV. 1963. *Sov. Phys.-Tech. Phys.*, **8**, 955.
39. A. A. GALEEV and V. I. KARPMAN. 1963. *Sov. Phys. JETP*, **17**, 403.
40. C. F. KENNEL and R. Z. SAGDEEV. 1967. *J. Geophys. Res.*, **72**, 3327.
41. D. MONTGOMERY and G. JOYCE. 1968. *Univ. of Iowa Rep.*, 68–24.
42. S. A. COLGATE and C. W. HARTMAN. 1967. *Physics Fluids*, **10**, 1288.
43. V. D. SHAPIRO and V. I. SHEVCHENKO. 1964. *Sov. Phys. JETP*, **18**, 1109.
44. R. C. DAVIDSON and H. J. VÖLK. 1968. *Physics Fluids*, **11**, 2259.
45. S. S. MOISEEV and R. Z. SAGDEEV. 1963. *Journ. Nucl. Energy C*, **5**, 43.

8

COLLISIONLESS SHOCK WAVES

J. W. M. PAUL

U.K.A.E.A. Culham Laboratory

PART I

A GENERAL REVIEW

1

RELEVANCE OF SHOCK STUDIES

1.1. *Occurrence of shocks*

THE first upsurge of interest in plasma shock waves was stimulated by the possibility of heating ions rapidly to thermonuclear temperatures. Under these conditions the plasma and the shock would necessarily be free from classical binary collisions, that is collisionless. More recently, however, interest has centred on geophysical and astrophysical phenomena, in which the same collisionless conditions prevail.

A shock is produced by the rapid compression of a plasma and it necessarily results in energy dissipation, that is in irreversible plasma heating. The non-classical heating mechanism involved in collisionless shocks has attracted much theoretical interest.

In the laboratory collisionless shocks have been observed in various experiments.[1-13] The pinch effect is frequently used to compress an initial plasma, and consequently to generate a cylindrically imploding shock. Non-cylindrical pinches, such as occur in very short pinch tubes,[13] toroidal pinches,[14] plasma guns[15] and plasma focus devices[16] also produce shocks. These various pinch devices are used in fusion research and as sources of electromagnetic and neutron radiation. Simulation of geophysical shocks has been performed in a plasma wind tunnel driven by a special arc source.[8] Some early plasma shock studies were performed using electromagnetic shock tubes of both "T"[17] and annular geometry.[18]

In the field of geophysics, satellite observations have shown that the earth's magnetosphere is embedded in a supersonic flow of plasma from the sun, but protected from it by a collisionless bow shock,[19] as shown in Fig. 1. Such observations have also demonstrated the correlation between terrestrial magnetic storms and interplanetary shock waves[20] travelling through the

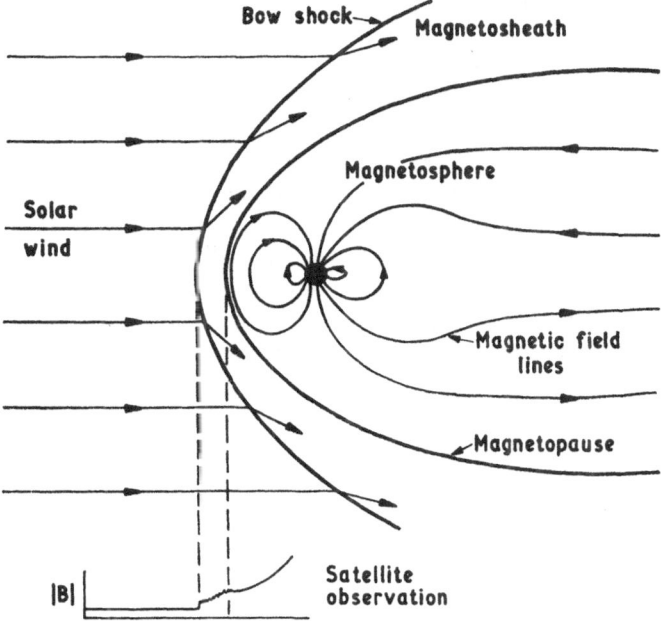

FIG. 1. Schematic of magnetospheric bow shock.

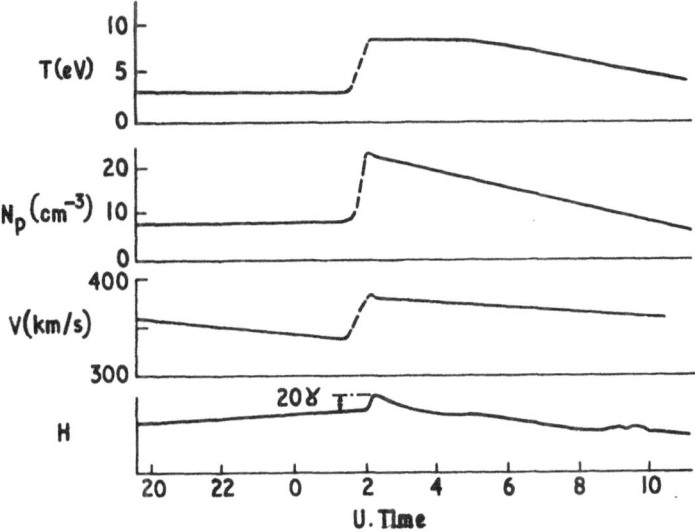

FIG. 2. Interplanetary shock observed by VELA 3A satellite and at Guam
on the earth, January 19-20th 1966.[20]

solar wind (Fig. 2). These satellite observations have involved a considerable expenditure of scientific effort and money and have given rise to a large number of theoretical papers on collisionless shock phenomena.[21-26]

In addition to the above observations of collisionless shocks, there are many theories, or speculations, which involve collisionless shock waves in solar and astrophysical contexts. Three types of shock are currently thought to occur in connection with solar flares: (i) the flow pattern of plasma and magnetic field near the magnetic neutral point is thought to involve a slow shock front[27]; (ii) the explosive release of energy in the flare gives rise to a shock wave travelling over the surface of the sun which has been observed[28] through the radio emission from the hot electrons; and (iii) another shock[29] travels out into the solar wind along the open magnetic field lines emanating from the flare (Fig. 3).

In the more distant and speculative areas of astrophysics, there are various examples of explosive plasma events which are thought to involve shock waves. Some models of radio sources,[30] quasars[30] and pulsars[31] have invoked shock phenomena (e.g. Fig. 4). The best known explosive event, the supernova, is generally accepted as involving some sort of expanding shock wave.

1.2. *Theoretical significance*

Until recently, most theoretical plasma physics has been concerned with understanding the great variety of small amplitude (linear) waves and their related linear instabilities. Attention is now moving on to waves of appreciable amplitude and the consequent modifications to the instabilities. A whole new range of non-linear phenomena occur, such as limitation of instability and turbulence. Shock waves physics can be regarded as an extension of the existing wave studies into the non-linear regime, where a wealth of new plasma physics is to be found.

The most important type of shock, the collisionless shock, is dependent on plasma turbulence, rather than collisions, for its heating mechanism. This shock provides a convenient opportunity to study plasma turbulence, which is in itself of importance in many branches of plasma physics. In particular, some laboratory collisionless shocks appear to be dominated by microturbulence driven by a universal instability.[32] This type of instability is of importance in plasma containment experiments. Thus shock wave experiments may provide a testing ground for non-linear and turbulent plasma theories.

2

NATURE OF THE SHOCK TRANSITION

2.1. *Gas-dynamic shock*

Before considering the problem of the plasma shock, it is useful to con-

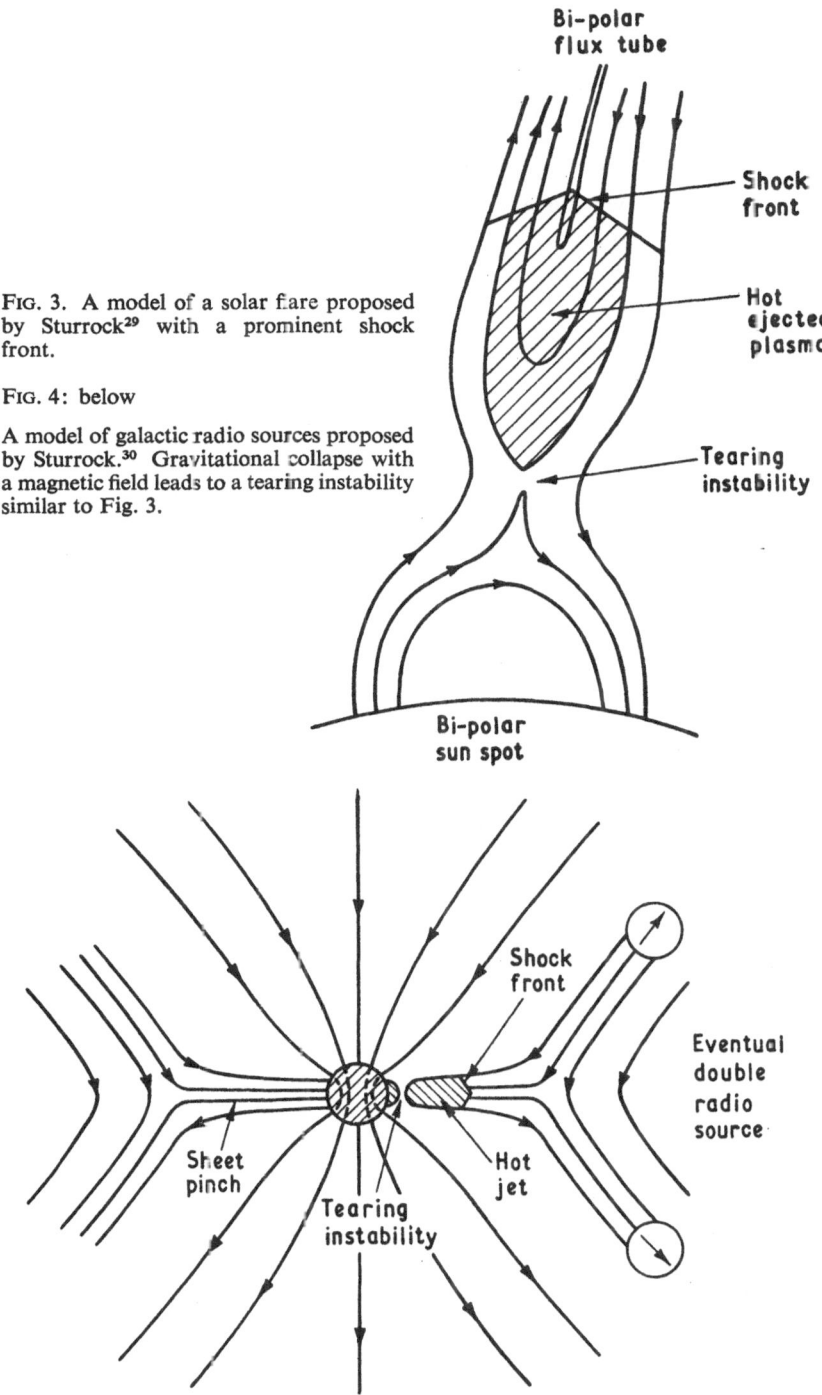

**Bi-polar
flux tube**

**Shock
front**

FIG. 3. A model of a solar flare proposed by Sturrock[29] with a prominent shock front.

**Hot
ejected
plasma**

FIG. 4: below

A model of galactic radio sources proposed by Sturrock.[30] Gravitational collapse with a magnetic field leads to a tearing instability similar to Fig. 3.

**Tearing
instability**

**Bi-polar
sun spot**

**Shock
front**

**Eventual
double
radio
source**

**Sheet
pinch**

**Hot
jet**

**Tearing
instability**

sider the much simpler gas-dynamic shock. Compression of a gas results in the propagation of a sound wave. This wave propagates by reason of the binary collisions in the gas. The velocity of these waves increases with compression. Consequently, for a finite amplitude, successive parts of the compression overtake each other, resulting in a steepening of the front. However, unless the compression is sufficiently rapid, e.g. its velocity is greater than the sound velocity, the wave is damped before it steepens.

This steepening process can be seen in terms of the simple dispersion curve of frequency ω against wave number $k = 2\pi/\lambda$. This is a straight line for a gas. The non-linearity, resulting from the finite amplitude, produces harmonics ($\sin^2 \omega t \rightarrow \cos 2\omega t$), which results in higher k, i.e. shorter wavelength and steepening. This steepening process is limited by viscous dissipation (irreversible heating) in the finite velocity gradient of the front. This dissipation is the non-linear consequence of the collisions which caused the wave to propagate in the first instance. The diffusive aspect of the viscous transport process balances the non-linear steepening. This balance occurs, and a steady state shock results, when the transport process provides sufficient dissipation to satisfy the steady conservation of mass, momentum and energy across the shock front. The conservation relations, called the Rankine-Hugoniot jump conditions,[33] are independent of the mechanism of the shock front, which can be treated as a discontinuity.

The detailed structure of the shock front can be considered in terms of the fluid equation, using the Navier-Stokes approximation. Such calculations give a width for strong shocks which is of the order of the collision mean free path. This implies that the fluid equations are invalid, and the full kinetic equations should be used. Attempts have been made to construct kinetic solutions, but no completely satisfactory answer has been obtained. However, there is general agreement between both fluid and kinetic treatments and experimental results.

In gas dynamics, if there are no collisions there are no sound waves, and consequently no shock waves; the particles just interstream. Plasma physics has always leaned very heavily on gas dynamics for its models and concepts. Consequently, it was at one time thought that it would be impossible to create a plasma shock wave in the absence of classical binary collisions.

2.2. Collisions in a plasma

Before discussing collisionless shocks, it is essential to have a clear concept of "collisions" in a plasma. Plasma physics essentially involves solutions of the many-body problem which results from the long range Coulomb forces. There are few truly binary collisions in a plasma. Particles are deflected stochastically in the fluctuating electric fields of many other particles. This problem has usually been tackled by reducing the many-body interactions to an effective binary one. This allows the use of normal gas-dynamic fluid and kinetic concepts in plasma physics.

In a plasma the range of the interactions is limited by Debye screening, and this facilitates an approximation to binary oscillations. On the classical plasma model, a collision occurs when a series of small angle deflections, calculated assuming a succession of binary collisions between un-correlated centres of a Debye screened Coulomb field, results in a 90° deflection. Because of the screening term, this is equivalent to considering only collisions with impact parameters less than the Debye distance (λ_D).

This classical collision frequency can also be derived from the stochastic deflection of particles in the random fluctuating electric field of the thermal plasma. Fluctuating fields with $\lambda < \lambda_D$ originate in the incoherent thermal motions of single particles, while fluctuations with $\lambda > \lambda_D$ arise from the collective screening effects of many particles. The latter can be analysed conveniently into coherent waves. The classical collision approximation corresponds to neglecting these collective fluctuations. For a plasma in thermal equilibrium the ratio of the collective to the classical collision frequency is about ($0.4/\ln \Lambda$), where Λ is the Coulomb logarithm. This ratio is usually negligibly small, so that the classical approximation is valid.

However, real plasmas are rarely in thermal equilibrium. There is usually free energy present in the various gradients (density, magnetic field, velocity, temperature), currents, differences of temperature between particles or non-Maxwellian distribution functions. Some aspects of this non-equilibrium character can frequently interact with the low level of thermal waves in the plasma. This interaction is usually a resonance plasma instability which causes these waves to grow to a non-negligible level. The non-linear phase of this growth usually results in plasma turbulence. These collective fluctuating electric fields with $\lambda > \lambda_D$ produce stochastic particle deflections and lead to an effective binary collision frequency and transport coefficients, just as did the fluctuations with $\lambda < \lambda_D$. Such collective transport processes are usually called anomalous, turbulent or supra-thermal. Anomalous transport coefficients are usually invoked to explain experiments rather than derived theoretically. The latter requires sufficient knowledge of the non-linear phase of instability for the spectrum $\langle E^2(\omega, k) \rangle$ to be predicted.

The wide range of possible plasma instabilities within collisionless shocks is beyond this lecture, but some of the more important, e.g. the two-stream[32, 34] and ion-acoustic[34-36] instabilities, will be discussed later.

This generalization of transport coefficients to include collective effects extends the useful range of the fluid model of a plasma, but it has one serious limitation. The classical binary approximation automatically gives rise to particle heating. The collective processes cause the growth of waves. The heating results finally from the damping of these waves. Thus there is an additional energy term in the system, corresponding to the energy in these waves. If this energy is small compared with other energies in the system, as is often the case, then the fluid model is a reasonable approximation. The complete process described above results, as would be expected, in the conversion of the available free energy into random thermal energy.

2.3. *Plasma shock*

Compression of a plasma can produce longitudinal waves (plasma sound waves), but in the presence of a magnetic field transverse non-compressional and mixed longitudinal and transverse waves are also possible. These waves occur even in the absence of classical binary collisions and are the result of the collective plasma effects. As in the case of the gas-dynamic shock the dispersion usually starts linearly, and such waves can steepen as the amplitude is increased forming the usual shock discontinuity. In a plasma this steepening can be limited by two different types of process, dissipation and dispersion.

(a) *Dissipative limitations.* Dissipation occurs as in gas-dynamics, but now involves more transport coefficients because there are two types of particles and a magnetic field. Also both classical and collective transport are possible.

For a steady-state shock the conservation relations and Maxwell's equations must be satisfied. The resulting jump conditions are called the de Hoffmann-Teller relations,[37] and these specify the required dissipation within the shock. Consider a simple case of a shock propagating perpendicular to a magnetic field. The jump conditions $(1 \to 2)$ in the shock frame are the same as the Rankine-Hugoniot condition but with magnetic terms added to the pressure and internal energy:

$$p^* = p + B^2/(2\mu_0)$$
$$E^* = E + B^2/(2\mu_0 \rho). \quad \text{(M.K.S. units)}$$

Conservation and other equations are listed below:

Mass: $\qquad \rho_1 v_1 = \rho_2 v_2$

Momentum: $\qquad (v_1 - v_2)^2 = (p_2^* - p_1^*)(1/\rho_1 - 1/\rho_2)$

Energy: $\qquad E_2^* - E_1^* = \left(\dfrac{1}{\rho_1} - \dfrac{1}{\rho_2}\right)\left(\dfrac{p_1^* + p_2^*}{2}\right)$

State: $\qquad E = p/(\gamma - 1)$

Maxwell: $\qquad v_1 B_1 = v_2 B_2$

Further simplifying to the case of a negligible ratio of initial particle to magnetic pressure ($\beta_1 \ll 1$), these reduce to a quadratic relation between shock strength $S = p_2^*/p_1^*$ and compression $F = \rho_2/\rho_1$

$$\left(\frac{\gamma-2}{\gamma-1}\right) F^2 - \left(\frac{S+3}{2}\right) F + \frac{(\gamma+1)S + (\gamma-1)}{2(\gamma-1)} = 0.$$

All other parameters can be obtained from F and S, for example the Alfvén Mach number,

$$M_A = \sqrt{\frac{F(S-1)}{2(F-1)}}$$

and the temperature behind the shock is given by

$$kT_2 = \frac{B_1^2}{2\mu_0 n_1}\left(\frac{S-F^2}{F}\right).$$

The more general cases are treated in Ref. 38.

(b) *Dispersive limitation.* Two characteristic forms of plasma dispersion curves are shown in Fig. 5. If the steepening is limited by dissipation before the change of gradients of the dispersion curve, then a normal dissipative shock is formed. In the absence of such dissipation, steepening of the shock is limited by the nature of the dispersion curves. At the higher harmonics the phase velocity changes.

In case (a) the phase velocity of the wave decreases as the wave steepens. The higher harmonics, shorter wavelengths, are left behind, forming a set of compression oscillations behind the main front. This type of dispersion occurs for (i) wave propagation perpendicular to a magnetic field in a low β plasma as a result of electron inertial resonance at $\omega \sim \sqrt{(\omega_{ce}\omega_{ci})}$ and $(1/k) \sim (c/\omega_{pe})$ and (ii) an unmagnetized plasma at $\omega \sim \omega_{pi}$ and $(1/k) \sim \lambda_D$ from the ion plasma resonance.

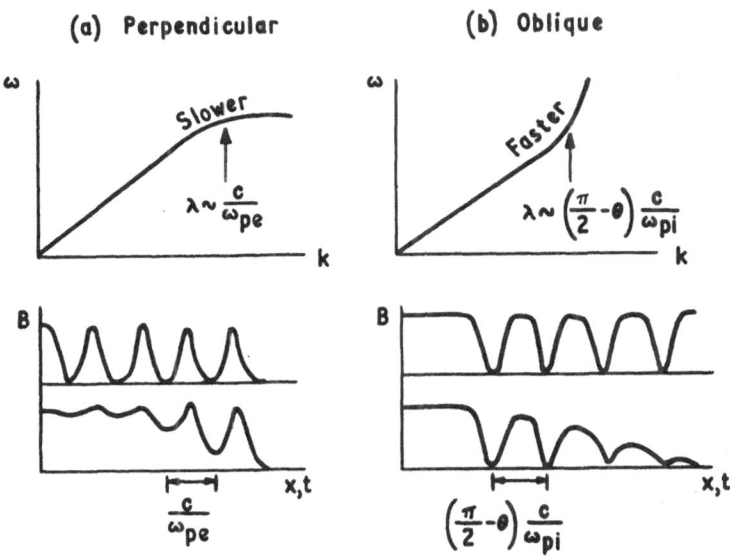

Fig. 5. Dispersion effects: (a) Backward wave train, e.g. perpendicular propagation and electron inertia waves.
(b) Forward wave train, e.g. oblique propagation and whistler waves.

In case (*b*) the same processes occur, but the phase velocity is now increasing and consequently the high frequency short wavelengths move in front of the main shock, forming a set of rarefaction oscillations. This second type of dispersion occurs for oblique propagation at an angle θ to the magnetic field in a low β plasma as a result of ion gyration in the whistler mode at $\omega \sim \omega_{ci}$, $(1/k) \sim \cos \theta \, (c/\omega_{pi})$.

Neither of these cases results in the formation of a shock. It is possible to add the energy of these oscillations to the energy equation and thereby satisfy the shock jump relations. But there is no real dissipation. These oscillations can be damped, either by classical or collective effects, to produce a damped wave train either behind or in front of the main jump, as shown in Fig. 5. Clearly the forward wave train could not exist without such damping. Sometimes such damped oscillatory shock structures are called collisionless, because collisionless dispersion dominates, irrespective of the nature of the damping.

3

MHD CLASSIFICATION OF SHOCKS

Shocks have been classified on the one fluid MHD model by the nature of the jump conditions without reference to the structure.[38] The conservation equations of normal and transverse momentum, together with the energy equation (all including magnetic field effects), form three intercepting surfaces in a three-dimensional diagram with axes of temperature, transverse magnetic field, and specific volume. In general these three surfaces intersect at four points, labelled 1, 2, 3, 4 in order of increasing compression. These intercepts are possible end points for a shock jump, which must occur from a lower to a higher number for entropy increase. These points are also divided in the middle by the initial Alfvén velocity appropriate to the magnetic field component in the direction of shock propagation. This leads to the following classification of shocks:

> (*a*) *fast* or super-Alfvénic shocks $1 \to 2$, which are observed in the laboratory and are known theoretically to be stable;
>
> (*b*) *intermediate* shocks $1 \to 3$, $1 \to 4$, $2 \to 3$, $2 \to 4$, which have not been observed and are thought theoretically to be unstable;
>
> (*c*) *slow shocks* $3 \to 4$, which have not been observed but are thought to be stable;
>
> (*d*) the transition points between fast and intermediate, and between intermediate and slow, shocks correspond to *Switch on* and *Switch off* shocks respectively; and these have uncertain stability.

All subsequent discussion will be limited to fast shocks.

The existence of end points for a shock jump does not imply the existence of a continuous solution of the equations from the starting point to the final point. The jump conditions are independent of the mechanism, but the construction of a solution depends on the existence of suitable dissipation mechanisms to join the two points. This problem has been treated by means of the two fluid MHD equations. Fortunately, it is not necessary to solve the structure problem in order to prove the existence of the solution, because this is dependent only on the nature of the equations at the end-point singularities. Linear expansion of the equations about the singular points allows them to be classified as nodes, foci or saddles (Fig. 6). This classification then determines whether solutions exist[39, 40] between the two points (no entry into saddle), and if so, whether they are monotonic (node) or oscillatory (focus). Finally, after proving the existence of a solution, it is necessary to demonstrate the stability of this solution to small amplitude wave disturbances. This analysis[38] has shown that all fast shocks which exist, to which this discussion is limited, are stable.

4

MHD Shock Structures

The shock structure problem requires the solution of the two-fluid equations of mass flow, momentum, energy, and state of the plasma together with Maxwell's equations through the shock. Complete analytical solutions are available only for very weak shocks propagating perpendicular to a magnetic field.[41] Strong shock solutions can be obtained analytically for certain simple cases[39, 42-45] with a reduced number of transport processes (e.g. no thermoelectric effect). Computed solutions[46] tend to be for specific cases and lack detailed interpretation. There is much computational activity at present which should find its way into the literature soon.

To illustrate the equations involved consider a simple case of a perpendicular shock in which the electrons and ions are shock-heated only by resistivity and viscosity respectively. Thermal conductivity, thermoelectric effects, electron inertia, etc. are ignored. In the frame of the shock, $v \equiv (v_x, 0, 0)$, and the equations are as follows:

(a) Continuity: $\qquad\qquad nv = \text{const.}$

Node   **Focus**   **Saddle**

Fig. 6. Nature of end-point singularities of shock jump.

(b) Momentum of ions or fluid:

$$Mnv\frac{dv}{dx} = J_yB_z - \frac{d(p_i+p_e)}{dx} + \frac{d}{dx}\left(\tfrac{4}{3}\mu\frac{dv}{dx}\right) \quad \text{(x-component)}$$

(c) Momentum of electrons or generalized Ohm's Law:

$$neE_x = J_yB_z - \frac{dp_e}{dx} \quad \text{(x-component)}$$

$$\eta J_y = (E_y - vB_z) - \frac{J_yB_z}{ne} \quad \text{(y-component)}$$

(d) Energy:

(i) ions: $\tfrac{3}{2}nvk\dfrac{dT_i}{dx} = -p_i\dfrac{dv}{dx} + \tfrac{4}{3}\mu_i\left(\dfrac{dv}{dx}\right)^2$ (adiabatic and viscous heating)

(ii) electrons: $\tfrac{3}{2}nvk\dfrac{dT_e}{dx} = -p_e\dfrac{dv}{dx} + \eta J_y^2$ (adiabatic and resistive heating)

(e) Equation of state: $p = nkT$

(f) Maxwell's equations: $E_y = vB_z = v_1B_{z1}$

$$J_y = -\frac{1}{\mu_0}\frac{dB_z}{dx}.$$

In the shock problem the two important dissipative coefficients in the momentum equations are viscosity, associated with gradient of velocity; and resistivity, associated with the gradient of the magnetic field. Shocks can be classified as viscous or resistive, depending on which of these mechanisms dominates. Viscosity (μ) arises from interstreaming collisions, and for collision frequency v is given by

$$\mu = nkT/v \quad (\propto T^{\frac{5}{2}} \text{ for classical collisions}).$$

The various combinations of inter-particle collisions (e.g. ion-ion, electron-electron, ion-electron) each give rise to different viscosities (μ_{ii}, μ_{ee}, μ_{ie}). Resistance (η) arises from the relative motion of electrons and ions involved in a current flow and is given by

$$\eta = \left(\frac{m_e}{ne^2}\right)v \quad (\propto T^{-\frac{3}{2}} \text{ for classical collisions}).$$

One of the main differences in these two classes of shocks is the different dependence on v. As the shock-heating depends on the product of the transport coefficient with the square of the appropriate gradient, there is a tendency for the resistive shock to steepen and for the viscous shock to broaden as v decreases; that is, with increasing temperature for classical collisions.

Thermal conductivity, another important transport coefficient, does not appear directly in the momentum equation and cannot by itself support a strong shock but can modify the structure through the energy equation. Other phenomena, like the Hall effect and thermo-electric effect, can similarly modify the structure of the shock.

The most acceptable *definition of the so-called "collisionless" shock* is that the irreversible shock heating in a "collisionless" shock cannot be accounted for by using the classical transport coefficients in the fluid model.

The above fluid considerations apply to collective transport coefficients, provided always that these form a reasonable approximation. A shock in which such collective coefficients dominated would be "collisionless" although fluid-like.

5

CRITICAL MACH NUMBER FOR RESISTIVE SHOCKS

The previous discussion (Section 3) of the existence of continuous solutions between existing end points is high-lighted by the case of the resistive fast shock. Solutions joining the end points exist only below a certain critical Alfvén Mach number. There is no such limit if viscosity is present.

It is possible to produce a resistive shock in which viscous processes are negligible. For example, if the classical ion-ion mean free path is much longer than the characteristic compression length, or more simply the shock width, and no collective viscosity is present, then a purely resistive shock results. Above a critical Alfvén Mach number, resistivity alone cannot satisfy the conservation relations. As a result, a second steepening occurs within the resistive structure and this must be limited by a viscous process.

This critical Mach number is related to the existence of two Mach numbers in a magnetized plasma.

(a) The Alfvén Mach number ($M_A = V_S/V_A$), which relates the shock velocity (V_S) to the Alfvén velocity [$V_A = B/(\mu_0 nM)^{\frac{1}{2}}$],‡ is relevant when the magnetic field is coupled to the compression.

(b) The acoustic Mach number ($M_s = V_S/C_s$), which relates V_S to the plasma sound speed (C_s), is relevant when there is no magnetic field or the field is uncoupled from the compression. In this case, if $M_s > 1$, steepening will lead to the formation of a viscous shock.

The value of C_s depends on the combination of adiabatic or isothermal sound speeds, $a = (\gamma kT/M)^{\frac{1}{2}}$ and $s = (kT/M)^{\frac{1}{2}}$ respectively, of the electrons and ions. For the collisionless resistive shock, the ions are adiabatic and the electrons isothermal, so that $C_s^2 = a_i^2 + s_e^2$.

On a scale-length short compared with the resistive shock-width, the magnetic field and plasma are uncoupled. Consequently, if the local flow becomes acoustically supersonic, $M_s > 1$, acoustic waves will steepen form-

‡ Strictly, Alfvén waves do not exist for propagation perpendicular to *B*, but V_A is still used.

ing a viscous shock as a steep jump near the rear of the resistive structure.

Above the critical Mach number, if there is no viscous process to limit the second steepening, the fluid equations break down by becoming multi-valued. This is usually interpreted to mean that the shock "breaks" or "overturns", producing interstreaming plasmas. This interstreaming, and the resulting interstreaming instability, is thought to produce the required collective viscosity for these shocks to exist.

The critical Alfvén Mach number M_A^* for the appearance of this double structure, or for "breaking", is obtained from the condition that the flow changes from acoustically super- to sub-sonic in passing through the shock (i.e. local $M_s = 1$). Thus M_A^* will depend on whether the electrons or ions are heated and whether they are adiabatic or isothermal. Woods[40] gives a consistent set of assumptions and values of M_A^* (Table I) which apply for collective as well as classical transport coefficients.

TABLE I

Critical Alfvén Mach numbers $(\beta_1 \ll 1)$
(k = thermal conductivity, η = resistivity, μ = viscosity)

Non-zero coefficient	$\gamma = 5/3$		$\gamma = 2$	
	$T_i = 0$	$T_i = T_e$	$T_i = 0$	$T_i = T_e$
$k_e, k_i, k_e + k_i$	1	1	1	1
η	2·76	2·76	2·95	2·95
$\eta + k_i$	2·76	3·01	2·48	2·80
$\eta + k_e$	3·46	3·01	3·56	2·80
$\eta + k_i + k_e$	3·46	3·46	3·56	3·56
$\mu + \ldots$	∞	∞	∞	∞

6

NON-FLUID MODELS

6.1. *Vlasov treatment*

The Vlasov equation considers changes of the particle distribution function produced by smoothed out electric and magnetic fields. These fields can clearly be analysed into plasma waves. Both analytical[35] and computational work[47, 48] has been done in this field, but the description will be given in terms of the computational method. If the initial system is not in equilibrium (there is free energy), the initial smoothed out fields will change the initial distribution function of the particles in phase space. The resulting new average $(n_i - n_e)$ and current will then give the modified smoothed out fields, and so the calculation proceeds by steps. This method follows the resonant interactions of particles and waves.

In effect the computation is as follows:

(*a*) *Instability*: the interaction of particles with waves;

(b) *Non-linear effects*: wave decay, interaction of waves with waves, and scattering of waves on particles;

(c) *Damping* of waves and consequent heating of plasma.

The complete cycle transfers the free energy of the particles into thermal particle energy. The computational treatment is usually limited by the use of only one spatial dimension, a limited number of mesh points, and un-realistic m/M and λ_D. Up to the present, this method has not been applied to shock structures as a whole.

A similar, but more restricted, computation[21, 23] has been performed using one-dimensional fluid sheets. The electron and ion fluids are coupled by the space charge field. Distribution functions as such are not considered directly. Above M_A^* these sheets interpenetrate, but with certain assumptions the computation can be continued and interpreted. These computations have been performed for shocks in the realistic geometry of a theta-pinch.

6.2. *Wave kinetics*

The recently developed wave kinetic model[49-51] provides an alternative approach which leans heavily on analogies with quantum mechanics. Insta-bility, that is inverse Landau damping, is replaced by Cherenkov emission of a plasma wave quantum (phonon for ion wave and plasmon for electron wave) by the particle. The wave quanta make various quantum transitions; decay ($\omega_1 \to \omega_2 + \omega_3$), mutual wave-wave interactions ($\omega_1 + \omega_2 \to \omega_3 + \omega_4$), and interactions with particles (absorbing and scattering quanta, $\omega_1 + p_1 \to p'_1$, $\omega_1 + p_1 \to p'_1 + \omega_2$). A formal quantum mechanical treatment is possible, but its use is limited by the difficulty of evaluating the transition probabilities.

7

PARAMETERS FOR THE CLASSIFICATION OF SHOCKS

It is difficult to make a clear classification of fast shocks. In this section the various parameters which, according to theory or experiment, affect the formation and nature of the structure will be listed with brief comments.

7.1. *State of the initial plasma*

(a) Degree of ionization: assume effectively fully ionized, (also state of ionization if $Z \neq 1$).

(b) Uniformity (e.g. n_e, B): affects dynamics of compression (e.g. axial gradients of n and B in θ-pinch tend to give non-cylindrical implosion).

(c) Quiescence/turbulence: macroscopic or microscopic (e.g. the solar wind appears to be turbulent).

(d) Mass or current flow: (e.g. compression of two interstreaming non-interacting plasmas).

7.2. Initial plasma parameters

(a) Presence or absence of magnetic field and ratio of particle to magnetic pressure.

$$\beta_e = \frac{n_e k T_e}{B^2/(2\mu_0)} = \left(\frac{s_e}{V_A}\right)^2; \quad \beta_i = \frac{n_i k T_i}{B^2/(2\mu_0)} = \left(\frac{s_i}{V_A}\right)^2$$

(b) Densities n_e, n_i and mass ratio m/M.

(c) Temperatures T_e, T_i and ratios T_e/T_i, $T_\parallel/T_\perp$. Distribution functions f_e, f_i if non-Maxwellian.

(d) The ratio of the Alfvén velocity (V_A) to the velocity of light (c) is important in that it is directly related to the ratios of characteristic frequencies

$$\frac{\omega_{ci}}{\omega_{pi}} = \frac{V_A}{c}; \quad \frac{\omega_{ce}}{\omega_{pi}} = \frac{M}{m}\left(\frac{V_A}{c}\right)$$

$$\frac{\omega_{ce}}{\omega_{pe}} = \left(\frac{M}{m}\right)^{\frac{1}{2}}\frac{V_A}{c} = \left(\frac{B^2}{\mu_0 n m c^2}\right)^{\frac{1}{2}} = \alpha; \quad \frac{\lambda_D}{r_{ce}} \approx \alpha.$$

In particular, if $\alpha > 1$, charge neutrality is violated over the electron orbit ($r_{ce} < \lambda_D$), which is then non-adiabatic. Also, if classical and collective collisions are negligible, the electron inertia waves of Section 2.3(b) have relativistic electron drift velocities.

7.3. Piston and compression

(a) Geometry of piston in relation to magnetic field and any gradients determines geometry of shock.

(b) Stability and "porosity" of the piston influence formation of separated shock.

(c) Relation of acceleration and steady phases to useful dimensions of apparatus influences formation of steady shock.

(d) Piston velocity influences Mach number.

7.4. Shock conditions

(a) Clearly the values of the parameters listed in Section 7.2 can be important within the shock.

(b) A steady state, approximately plane shock front is desirable but not always achieved in experiments.

(c) The angle (θ) which the normal to the plane of the shock front makes with the magnetic field. Perpendicular ($\theta = \pi/2$), oblique ($0 < \theta < \pi/2$) or parallel ($\theta = 0$) shock.

(d) Mach numbers, Alfvén ($M_A = V_S/V_A$), acoustic ($M_s = V_S/C_s$) and magnetosonic [$M_M = V_S/(V_A^2 + C_s^2)^{\frac{1}{2}} \simeq M_A/(1 + \frac{1}{2}\gamma\beta)^{\frac{1}{2}}$] and relation to critical values.

(e) Is the observed shock structure (E, B, n)

    (i) macroscopically steady and reproducible or turbulent?

    (ii) monotonic or oscillatory?

    (iii) single or double?

(f) The ratio of shock width and rise time to characteristic lengths and times such as those of classical collisions λ_{ii}, τ_{ei}, gyro radii r_c, and λ_D, (c/ω_{pe}), (c/ω_{pi}), ω_{ce}^{-1}, ω_{ci}^{-1}, $(\omega_{ce}\omega_{ci})^{-\frac{1}{2}}$, ..., etc.

(g) The conservation relations should be satisfied across the transition.

(h) Irreversible shock heating: strong or weak shock, electrons and/or ions heated.

(j) The ratio (R) of the observed to the calculated classical heating measures the importance of collective effects, that is, to what extent the shock is "collisionless". If $R > 1$, what type of instability is present and is there micro-turbulence $(\lambda \sim \lambda_D)$ within the shock?

(k) Is the shock heated plasma thermal or turbulent?

8

REVIEW OF MAIN EXPERIMENTS AND RESULTS

The main laboratory experiments, summarized in Table II, will be reviewed briefly in relation to the present state of understanding. The various parameters measured and the diagnostic techniques used are summarized in Table III. The particular experiments with which the present author has been concerned are discussed in Part II of this chapter.

8.1. *Perpendicular shocks with low β_1 and $M_A < M_A^*$*

In the absence of classical or collective collisions, such a shock would form a backward wave train of electron inertia oscillations, provided that $\alpha < 1$. A few heavily damped oscillations of this type with $\lambda \sim (c/\omega_{pe})$ have probably been observed.[9, 10]

For $\alpha > 1$ the electron current drift velocity (v_d) should, theoretically, become relativistic

$$v_d \sim \frac{1}{\mu_0 n e} \frac{B}{L_S}; \quad L_S \sim \frac{c}{\omega_{pe}}; \quad \frac{v_d}{c} \sim \frac{B\omega_{pe}}{\mu_0 n e c^2} = \alpha.$$

Thus, if $\alpha > 1$, the shock width L_S must be greater than (c/ω_{pe}) and is given by

$$v_d \sim c; \quad L_S \sim \frac{B}{\mu_0 n e c} = \frac{V_A}{\omega_{pi}}.$$

Shocks of this type have probably been observed.[9]

The above structures are possible only if instabilities are absent. The most likely instability is the two-stream instability of the current.[34] This occurs if $v_d > v_{eth}$, the electron thermal velocity, and has a growth rate $\gamma \sim \omega_{pi}$. There

Table II

Summary of principal experiments

Description	L_S
(I) In U.S.S.R.:	
1. Novosibirsk (Nesterikhin, Kurtmullaev, Alikhanov); ref. 3, 4, 9	
(i) θ-pinch; $r = 8$ cm, $\beta_1 < 1$, various M_i	
(a) Perp.; $M_A < M_A^*$, below critical n_e^* oscillatory behind, $\alpha > 1$	V_A/ω_{pi} (30 mm)
$\alpha < 1$	c/ω_{pe} (7 mm)
above n_e^* monotonic, anomalous η	$10 \, c/\omega_{pe}$ (15 mm)
$M_A > M_A^*$, double $\rightarrow$ mono, ion heating observed	c/ω_{pi} (30 mm)
(b) Oblique; $M_A < M_A^*$, oscillatory forward	$(\frac{1}{2}\pi - \theta)(c/\omega_{pi})$
$M_A > M_A^*$, double $\rightarrow$ mono	
No details of initial plasma, steadiness, B jump, separation or convergence.	
(ii) No B expt., "VOLNA"; $L = 100$ cm; electron beam ionization, electrostatic piston	
(a) Weak, oscillatory	λ_D (20 mm)
(b) Strong	$10\lambda_D$
2. Moscow (Smolkin); θ-pinch UV-1; $r = 3\cdot5$ cm; $\beta_1 < 1$; perp. shock; ref. 52, 53	
$M_A < M_A^*$	$10c/\omega_{pe}$
$M_A > M_A^*$	c/ω_{pi}
(II) In U.S.A.:	
1. Austin (Robson, Sheffield); Loop θ-pinch; $r = 22$ cm, $\beta_1 < 1$; ref. 13	
(a) Perp. on mid-plane, $M_A \gtrsim M_A^*$, as TARANTULA	$\dfrac{2\pi \cos\theta}{(M_A^2 - 1)^{\frac{1}{2}}} \left(\dfrac{c}{\omega_{pi}} \right)$
(b) Oblique, $M_A \gtrsim M_A^*$, oscillatory	
Detailed study of oblique steady state shock.	
2. Maryland (Hintz, de Silva, Goldenbaum)	
(i) Small θ-pinch; $r = 4\cdot5$ cm, $\beta < 1$; ref. 6, 54, 55	
Non-steady perp. shock, T_e measurement gives $R \sim 1$ so may be classical collisions.	
(ii) Large θ-pinch; $r = 23$ cm, $\beta < 1$; ref. 12	
$M_A < M_A^*$, Non-steady perp. shock, turbulent diffusion of piston	$3\cdot5 \, (c/\omega_{pe})$

3. AVCO (Patrick, Pugh); "TERRELLA" dipole expt; $L \sim 180$ cm, $\beta < 1$, $M_A < M_A^*$; ref. 8 Plasma wind tunnel from arc, oblique not oscillatory, detailed study of fluctuation behind shock.

(III) In Germany:
1. Julich (Hintz); Reverse field θ-pinch; $r = 9$ cm, $\beta_1 \gtrsim 1$, perp. shock; ref. 10
  (a) $\beta_1 < 1$, $M_A < M_A^*$; $M_A \sim 1.3$, unsteady oscillation — $10(c/\omega_{pe})$
    $M_A \sim 2.4$, steady monotonic — $10(c/\omega_{pe})$
  (b) $\beta_1 \sim 0.5$; fluctuation behind shock ($\sim 10\omega_{ci}$);
    $M_A < M_A^*$ — $10(c/\omega_{pe})$
    $M_A > M_A^*$; no double structure for $M_A < 6.5$ — $0.3(c/\omega_{pi})$
  (c) $\beta_1 \sim 1$; fluctuation behind shock;
    $M_A \sim 8$ — c/ω_{pi}

2. Garching (Keilhacker); θ-pinch; $r = 7$ cm, $\beta_1 \gtrsim 1$, perp. shock; ref. 11
  (a) $\beta_1 \ll 1$; $M < M^*$, unsteady — $10c/\omega_{pe}$ (18 mm)
  (b) $\beta_1 \sim 5$; $M_A > M_A^*$, stationary; $M_M \sim 2$ — $12c/\omega_{pe}$
    $M_M \sim 3$ — c/ω_{pi}

(IV) In Italy:
1. Frascati (Ascoli-Bartoli, Martone); θ-pinch "CARIDDI"; $r = 9$ cm, $\beta_1 < 1$; ref. 5, 58
  (a) Perp. steady; — $20c/\omega_{pe}$ (3 mm)
  (b) Oblique $M_A < M_A^*$; whistler oscillation — (15 mm)

(V) In U.K.:
1. Culham (Paul); z-pinch "TARANTULA"; $r = 25$ cm, $\beta_1 < 1$, perp. shock; ref. 1, 2, 7, 57
  (a) Low, $M_A < M_A^*$; steady, monotonic, conservation T_e — $7c/\omega_{pe}$ (1.4 mm)
    ion acoustic turb. observed
  (b) Intermediate, $M_A^* < M_A < 6$; steady double structure — $10c/\omega_{pe}$, $2c/\omega_{pi}$ (15 mm)
  (c) High, $M_A \sim 6$; unsteady oscillation — $3c/\omega_{pi}$
  Non-classical electron heating demonstrated for all M_A.

$\left(\dfrac{c}{\omega_{pi}}\right)$

is sufficient time[24] for this instability to grow in an electron inertia wave if

$$M_A > M_A' = 1 + \tfrac{3}{8}(\beta_1)^{\ddagger}, \qquad (\beta < 1).$$

In the experiments mentioned above, these low Alfvén Mach numbers were obtained by using low densities and high magnetic fields. Above M_A', instability can be expected to damp out the oscillations, producing a monotonic structure. This change of structure has been observed experimentally.[9]

The monotonic structure observed for $2 < M_A < 3$ is well documented on many different experiments. Universally $L_S \sim 10 \ (c/\omega_{pe})$. Experiments show that the electrons, and not the ions, are heated.[7] Also this observed heating cannot be explained by classical transport coefficients[7]; these are inadequate by two orders of magnitude. The shocks are collisionless. The change to a double structure at $M_A^* \sim 3$ suggests that resistivity (collective) is dominant. An effective collision frequency ν^*, has been obtained from L_S and from the observed electron heating. Forward scattering experiments[57] have demonstrated the presence within the shock front of a micro-instability

TABLE III

Diagnostics

Process	Parameters	Technique
Initial Plasma	n (relative), T_e	Double Langmuir probe
	$\int n_e \, dl$	Interferometry
		(visible, I.R., microwave)
	$\int n_n \, dl$	Two λ interferometry
	n_n	Magnetosonic waves
	n_e	Local microwave interferometer
	n_e, T_e	Spectroscopy
		(Stark, line ratios)
Collapse	Dynamics (B, n)	Magnetic probes, interferometry
	(*i*) piston and shock	
	velocity $\rightarrow M_A$	
	(*ii*) Separation, steady	
	(*iii*) Conservation jumps	
Heating	T_e, T_i	Laser scattering, particle analysis, spectroscopy
	(*i*) Species heated	
	(*ii*) Conservation jump	
	(*iii*) Collisionless	
	(*iv*) ν^*	
Macro-structure	B, n, E	Electric and magnetic probes, interferometry
	mono, double oscill., L_S	
	L_S, ν^*	
Micro-structure	Turbulence, $\langle n^2 \rangle$, $\langle E^2 \rangle$	Laser scattering, emission radiation, probes

with frequency nearer ω_{pi} than ω_{pe}. An experiment on this type of shock will be described in detail in Part II.

These monotonic shocks are interpreted as resistive shocks dominated by a collective resistivity from micro-turbulence driven by the current. Until recently the only theory of these streaming instabilities was for a homogeneous plasma without a magnetic field.[34] On this model, for an isothermal initial plasma, compression would tend to develop electron inertia waves. This development could be limited by either of two mechanisms which heat the electrons;

   (a) $v_d > v_{eth}$ would drive the two-stream instability with frequency $\omega \sim \omega_{pe}$ and heat the electrons giving $T_e > T_i$;

   (b) classical collisional resistivity at the front of the shock giving $T_e > T_i$. Once $T_e > T_i$, the ion acoustic instability with $\omega \leqslant \omega_{pi}$ can be driven if $v_d > C_s$.

Kadomtsev[35] has calculated the spectrum of turbulence $\langle E^2(\omega, k) \rangle$ resulting from the non-linear phase of the ion acoustic instability. Sagdeev[36] has used this spectrum to derive stochastically an effective collision frequency ν^*. The observed collision frequencies[7] agree well with those predicted by Sagdeev.

More recently Krall[32] has considered the inhomogeneous problem in the presence of a magnetic field. He finds that the appropriate instability is an electron drift instability (universal) driven by a combination of ∇B, ∇n, and $E \times B$ drifts.

The next important steps towards understanding these shocks will be, on the theoretical side, to combine the universal instability mechanism with the non-linear treatment of turbulence and, on the experimental side, to measure the turbulent spectrum and correlate it with the observed heating and with theories.

8.2. Perpendicular shocks with low β and $M_A > M_A^*$

On the fluid model, the resistive shock is expected to develop a viscous sub-shock at the rear for $M_A > M_A^*$. There is no experimental evidence for this happening. On the contrary, experimentally[1,2] at $M_A > M_A^*$, the resistive shock develops a broad foot ($L \sim c/\omega_{pi} \sim r_{ci}$) at the front. There are a number of possible explanations for this discrepancy, including high β effects; but the most attractive one involves the reflection of ions from the resistive shock.[24]

On the single particle model, which is more appropriate when discussing the collisionless ions, ions are reflected from the longitudinal electric field within the shock front. The number of ions reflected depends on the spread of the distribution function (i.e. the temperature), because only ions with certain velocities are reflected. Robson and MacMahon[68] have suggested that a viscous sub-shock at the rear of the resistive shock could produce appreciable ion heating and consequently spread the ion distribution function so that more ions are reflected forwards. These reflected ions then gyrate in the magnetic field acquiring transverse energy before finally passing through the shock. During this forward gyration the requirement of charged neutrality will cause electrons to follow the ions. The magnetic field will follow the electrons, giving rise to the observed foot. The interstreaming of ions in this foot could also give rise to interstreaming instabilities and consequent plasma heating. Experimentally there is anomalous electron heating[7] and some evidence from charge exchange neutrals that ions are counterstreaming in the foot.[9]

The interpretation of these high Mach number shocks is not clear either experimentally or theoretically.

8.3. *Perpendicular high β shocks*

Experimental data has only recently become available[10, 11] and the interpretation is not yet clear. There is experimental evidence for appreciable turbulence behind these shocks, and this could be related to the turbulence observed behind the magnetospheric bow shock.

8.4. *Oblique low β shocks*

There is clear detailed evidence for a train of forward whistler oscillations in oblique shocks.[9, 13, 58] In one experiment[13] these shocks have been shown to require collective resistivity for $M_A < M_A^*$ and collective resistivity and viscosity for $M_A > M_A^*$. For $M_A > M_A^*$ the Austin[13] and Novosibirsk[9] results disagree. The latter say that the oblique shock becomes broad and monotonic like a perpendicular shock, while the former still get oscillations. No forward oscillations are observed in the flow experiment[8] but appreciable turbulence occurs behind the shock.

8.5. *Shocks without magnetic field*

In the absence of a magnetic field, the ion acoustic dispersion relations should give a backward train of oscillations with $\lambda \sim \lambda_D$ at the ion plasma resonance $\omega \sim \omega_{pi}$. Such oscillations have been observed experimentally[9] but at high amplitude the structure broadens to $10 \, \lambda_D$, presumably because of some interstreaming instability.

PART II

THE TARANTULA EXPERIMENT

1

INTRODUCTION

THE Tarantula experimental programme[1, 2, 7, 57] has covered the formation, propagation and structure of collisionless shocks which propagate perpen-

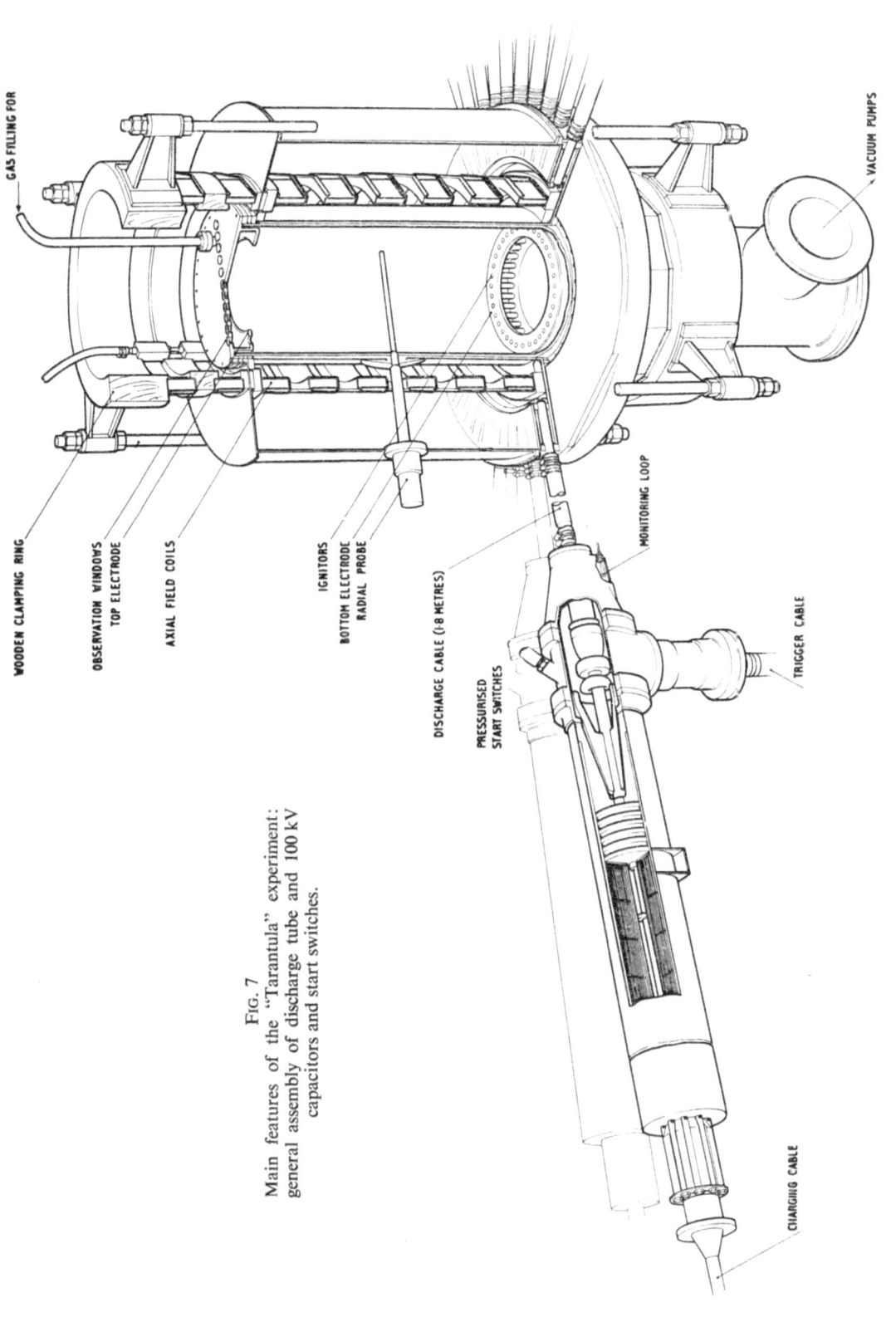

GAS FILLING FOR

WOODEN CLAMPING RING

OBSERVATION WINDOWS
TOP ELECTRODE

AXIAL FIELD COILS

IGNITORS
BOTTOM ELECTRODE
RADIAL PROBE

DISCHARGE CABLE (1·8 METRES)

PRESSURISED
START SWITCHES

MONITORING LOOP

TRIGGER CABLE

CHARGING CABLE

VACUUM PUMPS

Fig. 7

Main features of the "Tarantula" experiment:
general assembly of discharge tube and 100 kV
capacitors and start switches.

dicular to a magnetic field through a highly ionized initial plasma of low β (i.e. $\beta < 25\%$) and $\omega_{ce}/\omega_{pe} \ll 1$. The aim is to elucidate the mechanism of energy dissipation in the shock when classical transport processes are ineffective; that is, when the shocks are "collisionless".

The shock wave is produced by the cylindrical compression of an initial plasma by a fast linear z-pinch in a quartz tube of large diameter, 0·5 m, and length, 1·0 m. The apparatus, called Tarantula[59], is illustrated in Fig. 7. The use of z-pinch, rather than θ-pinch, geometry, and the large diameter distinguish this study from most others on collisionless shocks.

The study will be described in five sections: (*i*) initial plasma, (*ii*) dynamics of compression, (*iii*) macro-structure of the shock, (*iv*) shock heating, (*v*) micro-structure of the shock.

2

INITIAL PLASMA

2.1. *Axial discharge*

The discharge tube is normally immersed in a uniform steady axial magnetic field $(0 < B_z < 2\mathrm{kG})$ and filled with hydrogen or deuterium in the range 1–40 mtorr pressure.

The initial plasma is produced by an oscillatory axial current $(I \sim 100\,\mathrm{kA})$. At the first current reversal the plasma becomes turbulent and rapidly fills the whole tube. At the third current peak $(t = 170\,\mu\mathrm{s})$ the driving voltage is removed and a quiescent reproducible plasma develops as the current decays. The shock is produced in this afterglow, when the current is effectively zero $(t \sim 300\,\mu\mathrm{s})$.

2.2. *Experimental methods*

The initial conditions for the shock experiment have been carefully documented by three experimental methods.

(1) Double Langmuir probes provide local measurements of number density $n_e(r, z, t)$ and electron temperature $T_e(r, z, t)$.

(2) Infra-red laser interferometry provides an absolute measure of $\int n_e(r, z, t)\,dz$.

(3) The radial propagation of small amplitude magnetosonic waves has been used to determine the density of neutral hydrogen (n_n). Below a critical frequency $(f_c \sim \frac{1}{2}n_i\sigma_{cx}v_i)$, charge exchange couples both ions and neutrals into the wave motion. The measured wave velocity is then related to the total density $(n_n + n_i)$.

2.3. *Results*

The standard conditions, $p_1 = 20$ mtorr hydrogen and $B_{z1} = 1\cdot2$ kG, have been diagnosed in detail. The plasma is axially uniform for 98 cm, and radially uniform for the central 20 cm dia. (Fig. 8) with parameters,

85% ionized; $n_e = 7 \times 10^{20}$ m^{-3}, $T_e = 1.4$ eV, $\beta_1 = 0.04$.

The plasma is reasonably quiescent and reproducible. Experiments have been performed over the ranges 10^{20} m$^{-3} < n_{e1} < 10^{21}$ m^{-3}; $0 < B_{z1} < 1.5$ kG; $0.04 < \beta_1 < \infty$.

The initial plasma appears to be contained axially by a narrow (~ 1 cm) neutral layer on the electrode and radially by the axial magnetic field and a more diffuse neutral layer.

FIG. 8. Density distribution in the initial plasma (350 μs after start).

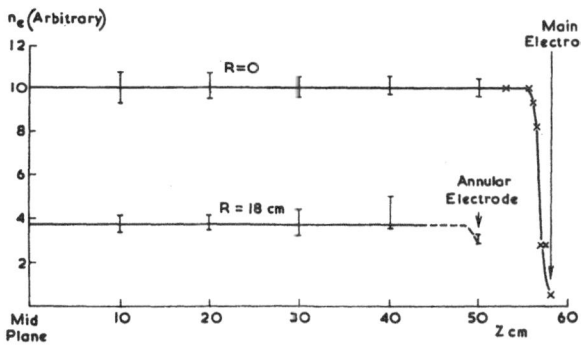

... i) Axial variation of electron density (double Langmuir probe).

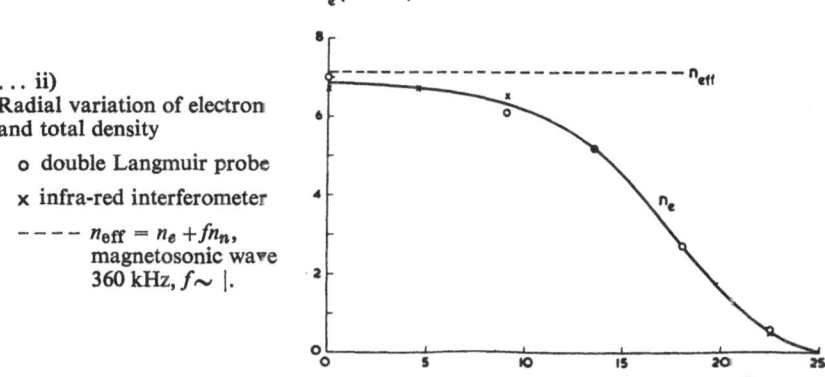

... ii)
Radial variation of electron and total density

  o double Langmuir probe

  x infra-red interferometer

  - - - - $n_{\text{eff}} = n_e + fn_n$, magnetosonic wave 360 kHz, $f \sim 1$.

3

DYNAMICS OF COMPRESSION

3.1. *Pinch device*[59] *(see Fig. 7)*

The fast pinch is produced by discharging a low inductance capacitor bank (20 μF) through 40 spark gap switches (source inductance 14 nH) into the initial plasma (load inductance 30 nH). The capacitor bank is inductively coupled to the discharge electrodes, so that the plasma remains at earth potential and probles can be inserted without high voltage insulation. The inductive coupling also provides an approximately square current pulse. For standard initial conditions the measured parameters are: $V_c = 75$ or 50 kV; $E_z = 62, 32$ kV/m, $(dI/dt)_0 = 1\cdot8, 1\cdot1$ TA/s and $I_p = 420, 320$ kA respectively. Standard pinch conditions refer to $V_c = 50$ kV.

3.2. *Magnetic field measurements of piston and shock*

The axial (B_z) and azimuthal (B_θ) magnetic fields have been measured during the compression by using multiple magnetic probes (7 and 3·5 mm dia.) and single probes down to 0·9 mm outside diameter. The shot to shot and year to year reproducibility is within about $\pm15\%$. The piston collapse velocity $(V_P \sim 110$ km/s) is greater than the Alfvén velocity and thus drives a cylindrically imploding shock.

The shock front is observed as a very sharp change of B_z which moves with velocity $V_S \sim 200$–500 km/s in front of, and clearly separated from, the piston. For standard conditions there are three phases of the shock implosion as shown in Fig. 9.

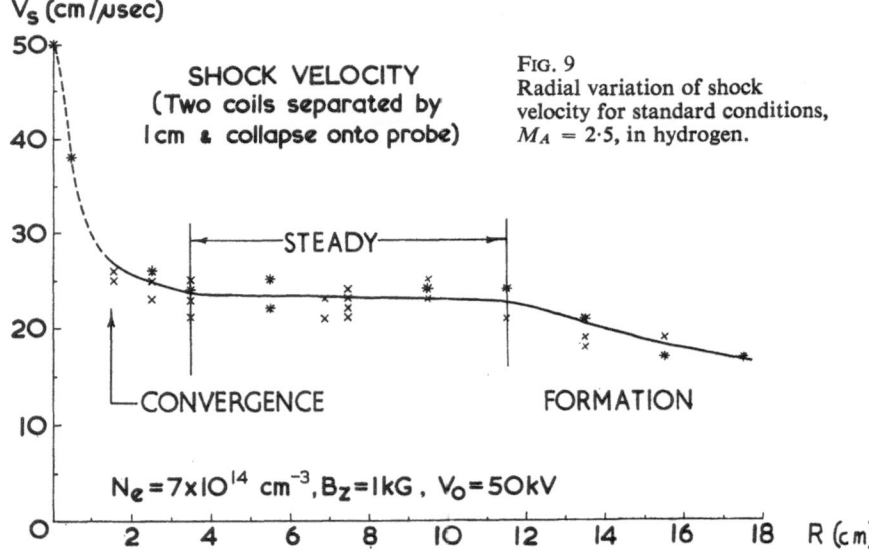

FIG. 9
Radial variation of shock velocity for standard conditions, $M_A = 2\cdot5$, in hydrogen.

The overall dynamics of the collapse for standard conditions is illustrated with the aid of a space-time diagram in Fig. 10. The detailed study of shock structure has been made at $r = 9$ cm, where the shock is steady and in a highly ionized uniform plasma. The shock is well separated in space from the piston and in time from the reflected shock, while still approximately planar.

The angle (θ) between the normal to the shock front and the axial magnetic field has been estimated from simultaneous measurements of the time of arrival of the shock at $r = 9$ cm for different axial positions. For standard conditions, the normal to the plane of the shock is perpendicular to B_{z1} near the centre and $\theta \leqslant 2°$ for ±25 cm away axially. Oblique effects occur only if $\theta \geqslant M_A(m_e/m_i)^{\frac{1}{2}} \sim 3.5°$ and can therefore be neglected.

3.3. *Comparison with MHD computation*

The dynamics of the collapse, as measured by magnetic probes, has been

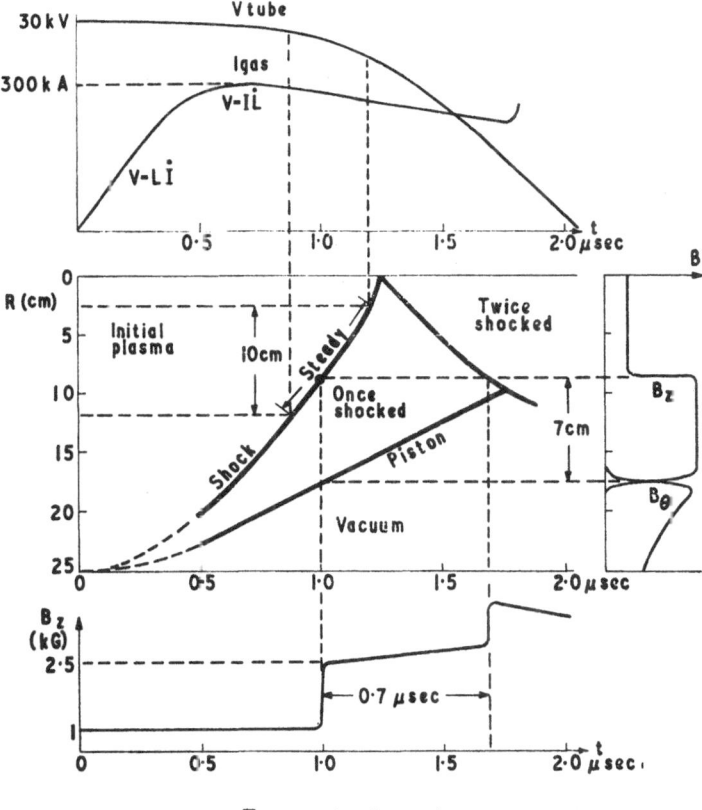

Tarantula dynamics

FIG. 10. Schematic of dynamics for standard conditions, $M_A = 2.5$.

compared with the predictions of a two-fluid MHD computer calculation,[60] based on the known initial conditions of the circuit and plasma. With the normal, binary collision transport coefficients the computed shock structure broadens, becoming comparable with the plasma radius. If viscosity is omitted, the structure then steepens beyond the limit of the computation (cf. Section I.4). These shock structure effects, which disagree with the experimental results, have been removed by introducing into the computation an artificially fixed shock width. The dynamics and energetics of the computation are unaffected by this artifice. The total shock heating $(T_i + T_e)$ can be predicted but not the ratio T_i/T_e.

Provided the measured initial total density $(n_i + n_n)$ is used, rather than the electron density, good agreement is obtained between experimental and computational magnetic profiles and streak diagrams (Fig. 11) for $M_A = 2.5$ (and 3.7).

4

MACRO-STRUCTURE OF SHOCK

The macro-structure of electric and magnetic field has been measured by imploding the shock onto small magnetic (0.9 mm) and coaxial electric probes situated at $r = 9$ cm. The latter measures the radial potential (V_r).

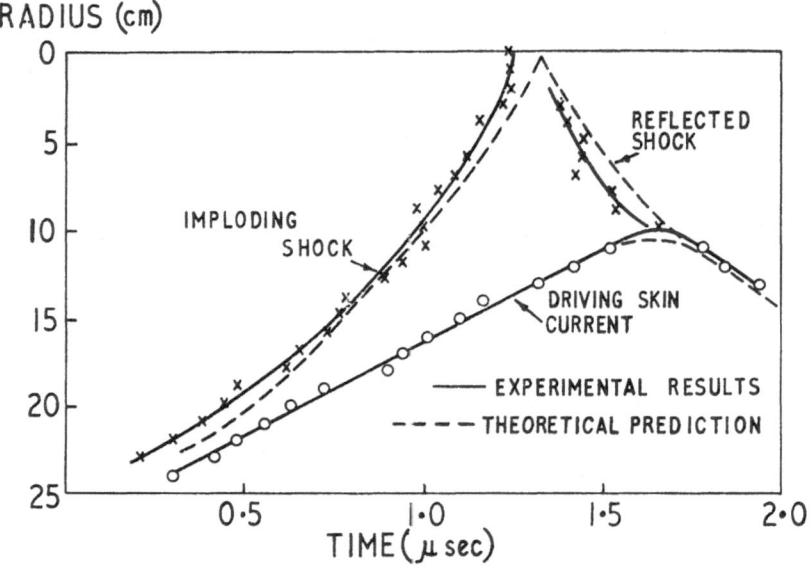

FIG. 11. Comparison of experimental and theoretical dynamics for standard conditions, $M_A = 2.5$.

TABLE IV

Macro-structure classification

Alfvén Mach number	Low 2·5	Intermediate 3·7	High 6·3
B_{z1} (kG)	1·2	0·75	0·43
n_{e1} (10^{20} m^{-3})	6·2	5·9	4·0
β_1	0·04	0·10	0·20
Structure	Single, steady	Double, steady	Oscillations, unsteady
L_S (mm)	1·4	1·4 15·0	3·0
L_S	$7c/\omega_{pe}$	$7c/\omega_{pe}$ $2c/\omega_{pi}$	$3c/\omega_{pi}$

Three specific cases of low, intermediate and high M_A are listed in Table IV and profiles are shown in Figs. 12–14. The dependence of shock width on (c/ω_{pe}) for low M_A has been checked over a range of parameters, as has the change from low to intermediate M_A structure at $M_A \sim 3$.

For low and intermediate M_A the change of radial potential across the shock accounts for the radial ion motion through the shock, indicating the absence of appreciable ion heating.

The importance of classical collisions can be measured crudely by the ratio of collision time to shock rise time (τ_S). For low M_A, if the ions were heated $\tau_{ii} \gg \tau_S \sim 10$ ns; but if, as described later, the electrons are heated, τ_{ee} is only marginally greater than τ_S for standard conditions. At lower pressures $\tau_{ee} > \tau_S$.

It is possible to solve simplified two-fluid equations of motion through the shock by using the experimental profiles of B_z and V_r. In this calculation it is assumed that the electrons are heated by an effective resistivity, the ions are adiabatic and there is no azimuthal momentum. The computation gives predicted profiles of n_e, v_i, T_e, η^*. The final T_e agrees well with the measurements described in the next section.

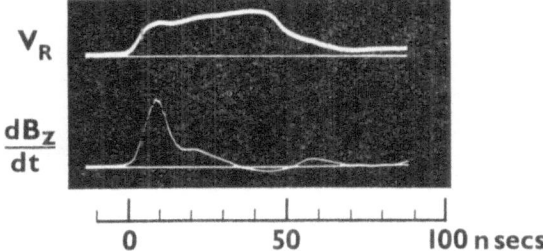

FIG. 12. Oscillograms of radial potential, V_R (1 cm separation coaxial probe) and dB_z/dt for standard conditions, $M_A = 2·5$.

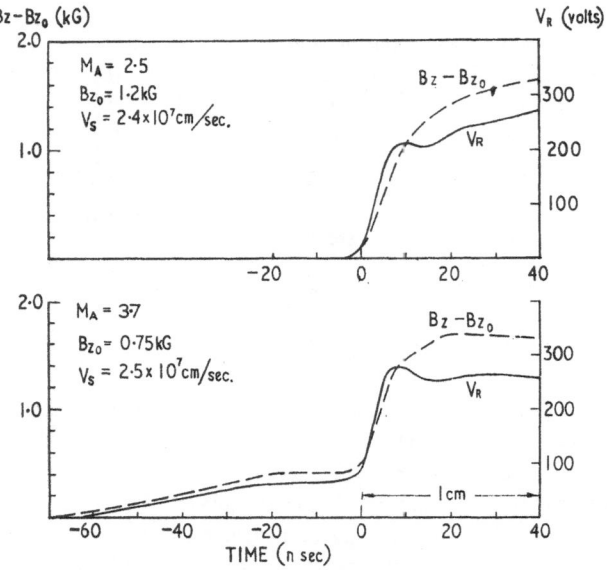

FIG. 13. Structure of V_R and B_z for low and intermediate M_A.

FIG. 14. Structure of B_z for low, intermediate and high M_A.

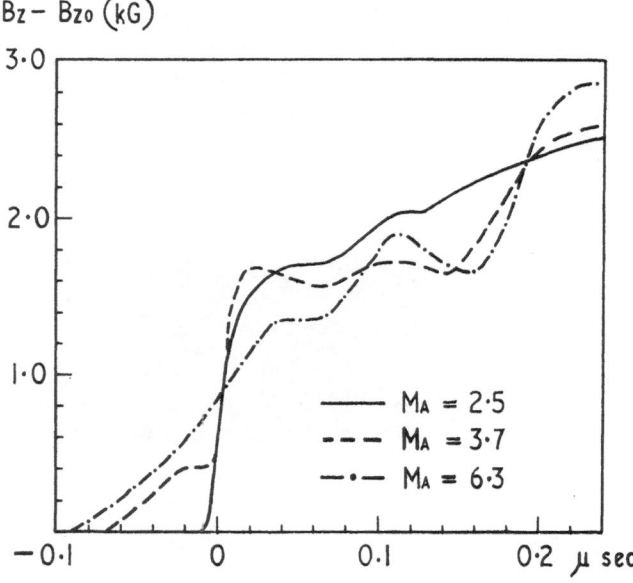

5

SHOCK HEATING

5.1. *Thomson scattering of laser light*

Heating of the electrons has been measured by Thomson scattering.[61] A 400 MW 20 ns light pulse from a ruby laser is passed across the diameter of the discharge tube at the midplane. The laser pulse is timed relative to the shock structure using an electric probe moved azimuthally out of the optical paths. The light, Thomson scattered into the axial direction from a 1 cm length of the laser beam centred at $r = 9$ cm, is detected by a photomultiplier (Fig. 15). The spectral profile of the scattered light is obtained by using interference filters with narrow pass-band (3 to 35 Å) and high rejection ratios ($\sim 10^{-4}$).

This spectral profile, which results from Doppler broadening by the electrons ($k\lambda_D \gg 1$), fits a Gaussian to within $\pm 15\%$ for over a half width of the profile, and yields an electron temperature with standard deviation of $\pm 10\%$ (Fig. 16). Systematic errors increase the possible error to $\pm 15\%$. Calculations, based on these measured temperatures, show that there should be sufficient time for the electrons to thermalize before the measurements.

5.2. *Measured electron temperatures and comparison with computations*

The measured electron temperatures (T_e) for $M_A = 2\cdot5$ and $3\cdot7$ are presented in Fig. 17 as a function of time (τ) relative to the sharp transition in the shock structure recorded by the electric probe. For $M_A = 3\cdot7$ the change in magnetic field (ΔB_z) is plotted as well to show the measurement of T_e within the broad forward transition in the shock structure.

The electron temperatures in Fig. 17 are compared with the total shock heating ($T_e + T_i$) derived (assuming $\gamma = 5/3$) from (i) plane geometry conservation relations, using the measured V_S, n_{e1} and B_{z1}, and (ii) the cylindrical collapse MHD computations[60] based on the initial conditions. This collapse computation has been shown above to give good agreement with the measured dynamics of the experiments and consequently should provide a reasonable estimate of ($T_i + T_e$). This comparison is summarized in Table V.

TABLE V

Comparison of temperatures (eV)

M_A	2·5	3·7	6·3
T_e measured (peak)	44	56	44
T_e extrapolated to $\tau = 0$	46	66	—
($T_e + T_i$) conservation	42	72	140
($T_e + T_i$) computation	49	84	145

There is good agreement between the measured T_e and the predicted ($T_e + T_i$) immediately behind the shock for $M_A = 2\cdot5$, and consequently the

(a)

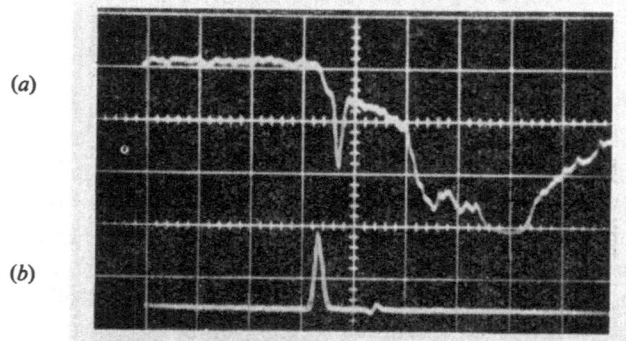

(b)

FIG. 15

(a) Scattered light signal with 15 Å wide optical filter set at 6824 Å ($\Delta\lambda \simeq$ 120 Å). The sharp rise of background light just before the scattered signal corresponds to the arrival of the shock at the point. (b) Photodiode monitor of input laser power (350 MW). (c) Accurate timing of measurements relative to electric probe signal. The delay of 60 ns corresponds to 1·5 cm behind the shock.

(c)

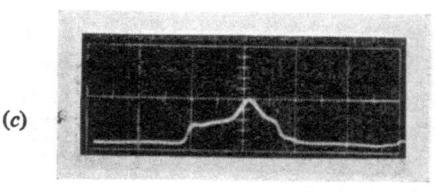

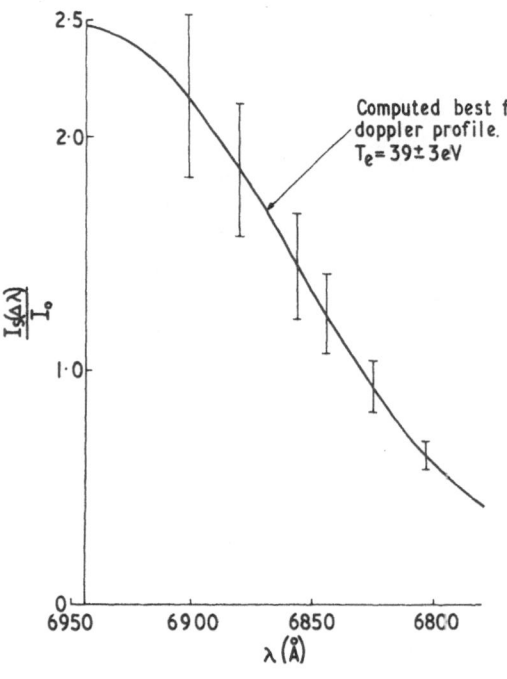

FIG. 16

Plot of ratio of scattered to incident power against wavelength with best fit Doppler profile.

ions cannot be appreciably heated in this case. However, for $M_A = 3.7$ the measured T_e is less than the predicted $(T_e + T_i)$, by some 28%, allowing the possibility of some ion heating.

The possibility of temperature limitation by end effects or impurities was investigated for $M_A = 2.5$ in an experiment performed in deuterium at half the number density. The dynamics was the same as for hydrogen ($M_A = 2.5$), and so the temperature should be doubled. The measured temperature, $T_e = 91$ eV for $\tau = 50$ ns confirmed this simple scaling and made temperature limitation unlikely.

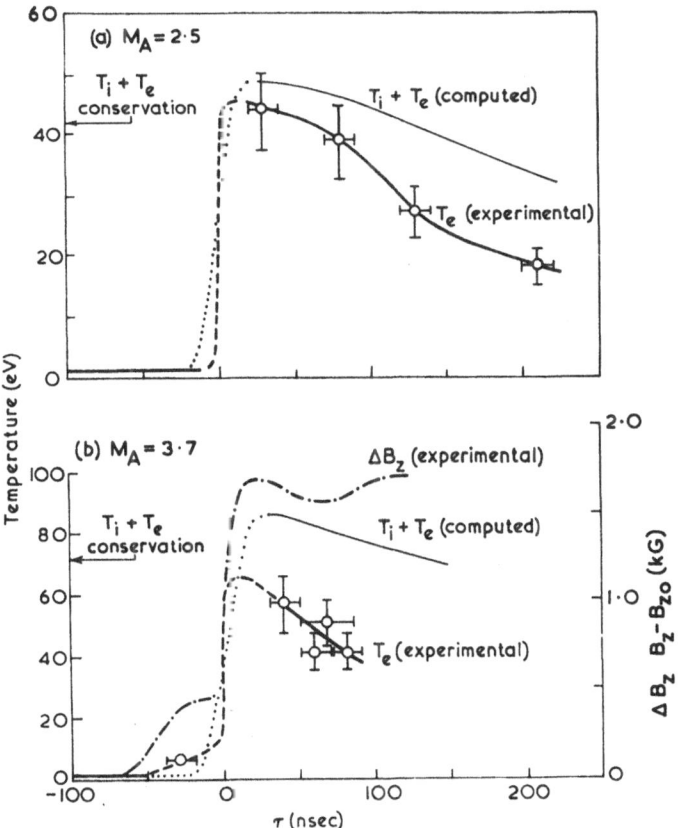

FIG. 17. Temperatures for low and intermediate M_A shocks in hydrogen.

6

COLLISIONLESS SHOCK

6.1. *Inadequacy of classical transport coefficients*

The electron heating to be expected from the classical transport coefficients can be calculated from the measured fields. In order to calculate the adiabatic and viscous heating it is necessary to calculate self consistent profiles of $n(x)$ and $v(x)$ through the shock. However, the calculation shows that these effects are secondary to the main resistive heating.

The electron energy equation has the form

$$\tfrac{3}{2}nkv\frac{dT}{dx} = -A+V+R,$$

(*i*) in which the adiabatic part $\quad A = nkT_e\frac{dv_x}{dx}$,

(*ii*) the viscous part $\quad V = \mu_e\left[C_1\left(\frac{dv_x}{dx}\right)^2 + C_2\left(\frac{dv_y}{dx}\right)^2\right]$,

where μ_e = electron viscosity[62] $v_y = -1/(\mu_0 ne)\, dB/dx$ and C_1 and C_2 are functions of $\omega_e\tau_e$ given by Braginski,[62] and

(*iii*) the resistive part $\quad R = \frac{\eta}{\mu_0^2}\left(\frac{dB}{dx}\right)^2$

where η is the Spitzer-Härm resistivity. Electron inertial effects have been shown to be negligible because $L_S > (c/\omega_{pe})$. The heat flux vector, which can redistribute but not generate heat, has been omitted. The equation was integrated "step-wise" on a computer using the observed B_z and V_r profiles.

For $M_A = 2\cdot5$ this calculation results in a temperature of only 7·5 eV which is a factor of six down on that observed. Viscous heating is negligible and adiabatic heating some 20% of the total, so that resistivity and the B profile dominate. If the classical resistivity is increased by a factor of about 100 throughout the structure, the observed heating can be obtained. Thus classical resistivity is inadequate, by about two orders of magnitude, to explain the observed electron heating and a "collisionless" mechanism is necessary.

Similarly, for $M_A = 3\cdot7$ and 6·3 classical heating is inadequate. The ratio (R) of observed to classical heating for the broad and sharp features at $M_A = 3\cdot7$ are $R = 6$ and 8 respectively; while for $M_A = 6\cdot3$, $R = 10$.

By integrating the simple electron energy equation across the shock, an effective mean resistivity and corresponding collision frequency (v_M^*) can be obtained from (T_2-T_1), (B_2-B_1), V_S and L_S. For both hydrogen (20 mτ) and deuterium (10 mτ) shocks with $M_A = 2\cdot5$,

$$v_M^* \approx 0\cdot5 \times 10^{10} \text{ s}^{-1}.$$

6.2. *Collisionless mechanism for low M_A* (see Section 8.1)

The most probable mechanism for these shocks is the excitation of electrostatic plasma waves by the drift of electrons through ions in the high current density of the shock transition. This electron motion results from a combination of $\nabla_r B$, $\nabla_r n$ and $E_r \times B_z$ drifts. The instability has the nature of a high frequency universal drift mode.[32]

For lack of better theoretical guidance, this problem has often been discussed in terms of the "drift" instability of a homogeneous plasma carrying a current in the *absence* of a magnetic field.[34] Two types of instability are considered:

(i) *ion acoustic instability* arising from a resonance of electrons with ion acoustic waves ($\omega < \omega_{pi}$) for $v_d > C_s$ and $T_e > T_i$,

(ii) *two-stream instability* arising from resonance of ions with electron plasma waves ($\omega = \omega_{pe}$) for $v_d > v_{eth}$.

The occurrence of these instabilities depends on two ratios, (i) $R_v = v_d/v_{eth}$ and (ii) $R_T = T_i/T_e$, as shown in Fig. 18. In the shock problem $(1/\omega_{ce}) \ll \tau_S \ll (1/\omega_{ci})$ so that the magnetic field is more likely to affect the

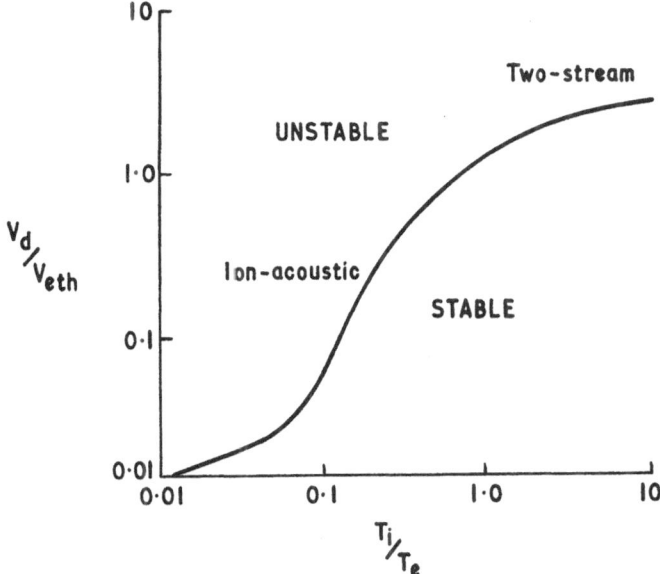

FIG. 18. Current driven streaming instability, after Stringer.[34]

resonant electrons of the ion acoustic than the ions of the two stream instability.

The measured magnetic field gradient corresponds to a peak azimuthal current density of 4 kA/cm². If it is assumed that the mean density $\frac{1}{2}(n_1 + n_2)$ and mean electron temperature $\frac{1}{2}(T_{e1} + T_{e2})$ occurs at this peak, the resulting ratio of electron drift to thermal velocity

$$R_v = \frac{v_d}{v_{\mathrm{eth}}} \approx 10\%.$$

Assuming in addition adiabatic compression of the ions with $\gamma = 2$,

$$R_T = \frac{T_i}{T_e} \approx 10\%.$$

Such conditions, according to Stringer,[34] are unstable to the ion acoustic but not to the two stream instability.

Nevertheless the two stream instability could be present, if (i) the electrons are non-Maxwellian with a high velocity component, or (ii) there is a small unresolved region of high drift velocity.

The ion-acoustic instability by itself cannot provide a self-consistent model because it requires $T_e > T_i$, which is not satisfied at the start of the shock structure. A "trigger" mechanism is therefore required to raise T_e so that the ion acoustic instability can proceed. Both ordinary resistivity and the two-stream instability have been suggested as trigger mechanisms operating at the front of shock.

The non-linear phase of the ion acoustic instability in a homogeneous unmagnetized plasma has been studied by Kadomtsev.[35] He assumes a constant growth rate at resonance from linear theory and non-linear "diffusion" in (ω, k) space produced by scattering of waves on ions. Thermalization of the waves is not considered. From this study he derives the spectrum, $\langle E^2(k) \rangle$, of the resulting turbulent electric field.

The interaction of electrons with this turbulent electric field gives rise to an effective collision frequency (ν^*) and corresponding turbulent resistivity (η^*), which have been calculated by Sagdeev[36]:

$$\nu^* = 10^{-2} \frac{T_e}{T_i} \left(\frac{M_i}{kT_e} \right)^{\frac{1}{4}} v_d \omega_{pi} = 10^{-2} \frac{T_e}{T_i} \frac{v_d}{v_{\mathrm{eth}}} \omega_{pe}.$$

(Note that this ν^* is 100 times smaller than that given by Sagdeev and Galeev.[63]‡) Sagdeev also predicts

$$\frac{dT_e}{dT_i} = 43 \frac{v_d}{v_{\mathrm{eth}}},$$

‡ It appears from recent discussions[70] that the factor 10^{-2} given by Sagdeev[36] was introduced to give better agreement with experiment!

which, for experimental values, agrees with the observed dominance of electron heating. Substitution of mean shock parameters into the equation yielded

$$v^* = 1\cdot4 \times 10^{10} \text{ s}^{-1},$$

in reasonable agreement with the value derived from experiment.

The scaling experiment, using hydrogen and deuterium at constant $M_A = 2\cdot5$, has $n_H = 2n_D$, $L_{SH} = (1/2)L_{SD}$, $T_{eH} = (1/2)T_{eD}$, $v_H^* = v_D^*$ and corresponds to the Sagdeev formula.

The stepwise computation of the classical plasma heating through the shock (Section 6.1) showed sufficient electron heating for ion acoustic instability to develop at the front of the shock at the start of the steep gradient. The computation was then continued with the addition of the above turbulent resistivity with its functional dependence. The instability switches off at the end of the steep gradient. The calculated and the observed electron heating agreed when η^* was reduced by a factor 0·6 from that given by Sagdeev. The computed effective collision frequency averaged over the shock width was $v^* = 0\cdot8 \times 10^{10} \text{ s}^{-1}$.

While there is circumstantial agreement between the Kadomtsev-Sagdeev theory and experiment, it must be remembered that the assumed conditions do not correspond with experiment.

Recently Krall[32] has developed a linear theory for local stability within a shock front with ∇B, ∇n, $E_r \times B_z$ drifts. Also recently Drummond and Sloan[64] have criticized Kadomtsev's assumption that ion scattering dominates the non-linear phase.

7

MICRO-STRUCTURE OF SHOCK FROM FORWARD SCATTERING

7.1. Description

Recently direct evidence for the existence of micro-instability within the shock has been obtained for $M_A = 2\cdot5$. Forward scattering of ruby laser light from plasma density fluctuations[61] with wavelength greater than the Debye length (see also Chapter 11 of this book by Ramsden), shows an appreciable enhancement[57] over the stable thermal level, of these fluctuations within the shock front. The enhancement occurs for relatively small frequency shifts from the laser line corresponding to the range of low frequency plasma waves (ion feature). This result reinforces previous suggestions that ion acoustic rather than two stream turbulence dominates in this shock.

A 50 MW laser pulse of half height full width 35 ns was timed and positioned to hit the shock front in the mid-plane of the discharge tube when the

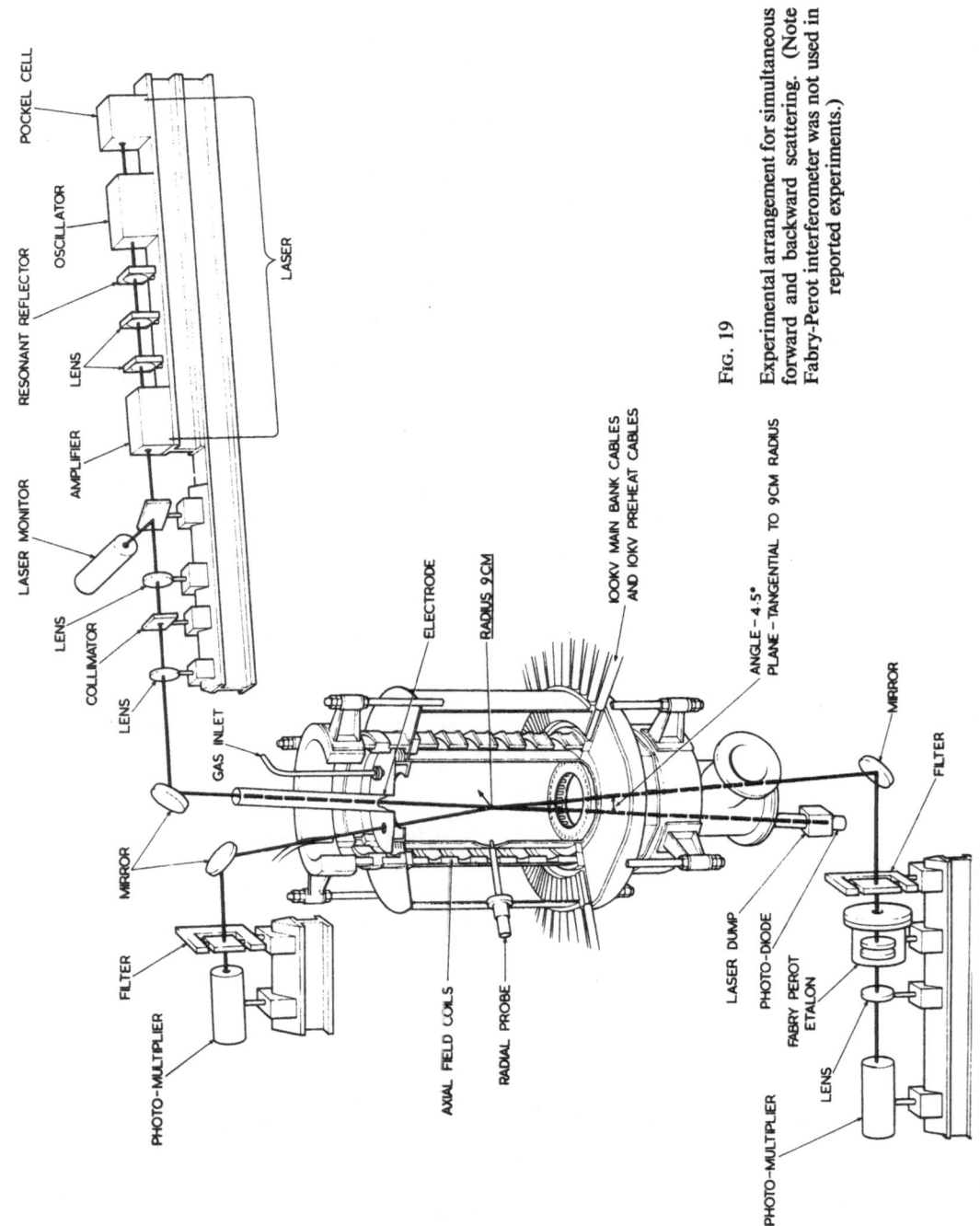

FIG. 19

Experimental arrangement for simultaneous forward and backward scattering. (Note Fabry-Perot interferometer was not used in reported experiments.)

shock is at 9 cm radius. The relative timing of the laser pulse to the shock structure was obtained from an electric probe at the same 9 cm radius but moved azimuthally out of the field of view.

The laser beam is in a plane which is tangential to the cylindrical shock front (electric vector perpendicular to this plane) and within this plane is at 2·25° to the axial direction. The forward scattered light is detected in the same plane at an angle of 4·5° to the laser beam (Fig. 19). The plasma fluctuations which scatter light into this detector have wave vector k colinear with the azimuthal current in the shock and $|k| \leqslant 1/\lambda_D$ (i.e. collective scattering). A back-scattered signal is detected in a slightly different plane at an angle of 170° to the laser beam. This back-scattered light signal arises from fluctuations with $|k| \gg 1/\lambda_D$ (i.e. random thermal fluctuations) and monitors the electron heating. Both detecting systems are focused onto the 2 mm diameter laser beam at the mid-point of the discharge tube.

Spectral resolution of the forward-scattered light was obtained by using optical filters in front of the photomultiplier. A narrow pass-band (3 Å) filter was used to accept light scattered from the low frequency ($\Delta\omega \sim \omega_{pi} \equiv$ 0·1 Å) fluctuations (ion feature), while rejecting most of the light scattered from the high frequency ($\Delta\omega \sim \omega_{pe} \equiv 7$ Å) fluctuations (electron feature). A wider pass-band (35 Å) filter was used to accept light scattered from both features. Spectral resolution of the back-scattered signal was obtained by using a wide pass-band (30 Å) filter set about 50 Å off the ruby line so that a signal appears only when the electrons are appreciably heated.

7.2. Results

A pulse of forward-scattered power is observed (Fig. 20) which is much shorter than the laser emission. This pulse corresponds in time and duration with the passage of the shock front through the laser beam. The rise time of the pulse (5 ns) is just within the limit of the photomultiplier response. Clearly there is no resolution in space or time within the shock front.

The time variation of the ratio of scattered to incident power, hereafter called normalized scattered power, has been derived from Fig. 20 and plotted in Fig. 21. This illustrates the enhancement of scattering within the shock over that from the pre- and post-shock plasma. With more gain in the detection system the pre- and post-shock scattering can be seen, as in Fig. 20, together with the enhancement within the shock seen in the wings of the laser pulse.

The back-scattered power, also shown in Fig. 20, is insignificant from the cold initial plasma ($T_e \sim 1·4$ eV) but appreciable from the hot post shock plasma ($T_e \sim 45$ eV).

The normalized scattered power from the plasma is not very reproducible. The average of about ten measurements from either the pre-, on, and post-shock regions, gave a standard deviation of about 30%. The average spurious signal (no plasma) was more reproducible, with standard deviation of about 5%. These average measurements of the normalized scattered power, corrected for spurious signal, and relative to pre-shock plasma, are given in Table VI[57], for both 3 Å and 35 Å filters.

Within experimental error, the ratio of peak normalized scattered power from the shock to that from the initial plasma, is the same for both filters, yielding a ratio $R = 16$. This result demonstrates that most of the enhanced

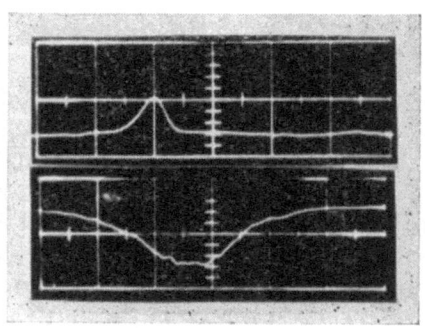

1. On-shock

2. Pre-shock   3. Post-shock

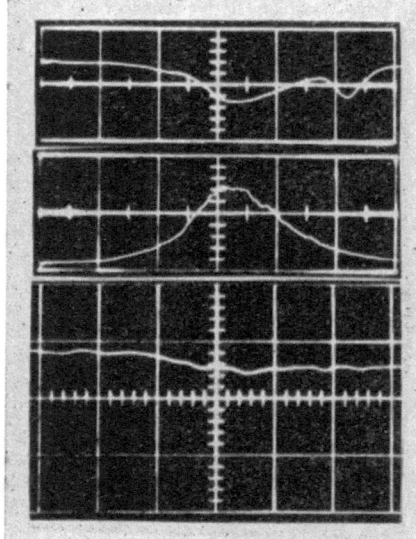   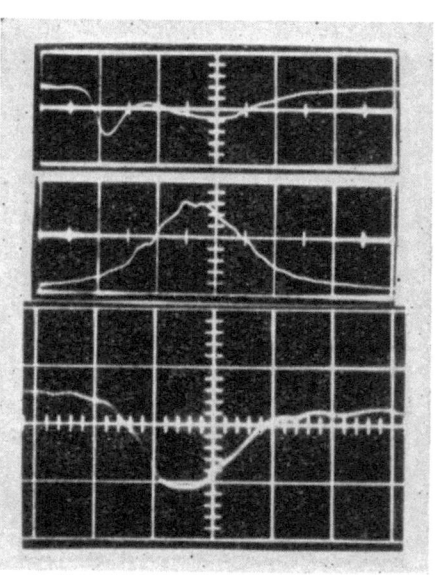

FIG. 20. Enhanced forward scattering from shock front (20 ns/cm).

1. On-shock only:	(a) Laser power	(b) Scattered power.
2. With pre-shock:	(a) Forward scattered.	
	(b) Laser power.	
	(c) Back scattered.	
3. With post-shock:	(a), (b), (c) as for 2.	

scattering from the shock occurs within the ion feature rather than the electron feature.

TABLE VI

	35 Å	3 Å
Pre-shock	1	1
On shock	16	16
Post-shock	2	0·7
Spurious	1·5	1·0

The observed ratio, R, should be corrected for the ratio of scattering volumes. The shock does not occupy the whole cross-sectional area of the laser beam, whereas the pre-shock plasma does. For the assumed dimensions and geometry, effective shock thickness 1·4 mm and laser beam width ~2 mm, this correction raises the ratio to $R = 37$.

7.3. Local enhancement

The measured ratio, R, is not the true local enhancement relative to what the scattered power in the ion feature would be from a stable plasma with the same parameters. The power in the ion feature from such a stable plasma, but with no current, depends on $\alpha = 1/(k\lambda_D)$ and T_e/T_i.[61] Both of these parameters vary through the shock, $4·4 > \alpha > 1·2$ and $1 < T_e/T_i < 12$ from pre- to post-shock plasma. Unfortunately the theoretical dependence of the scattering on these parameters for $T_e/T_i > 3$ is not available without compu-

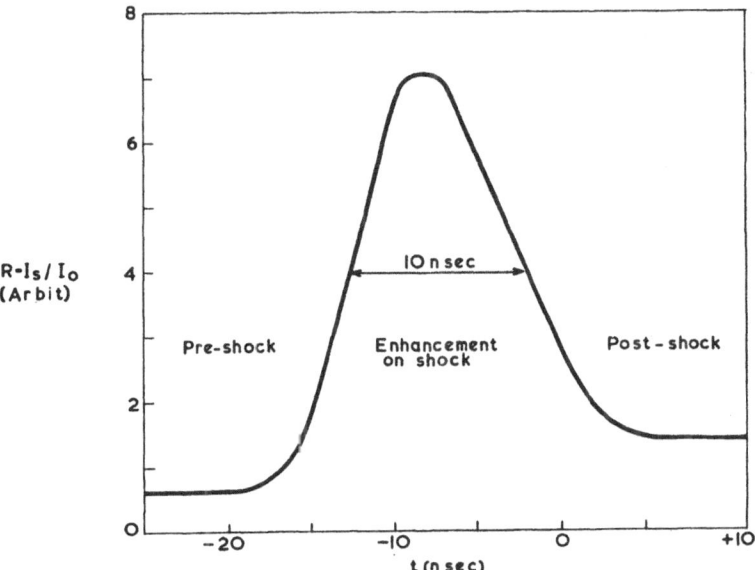

FIG. 21. Normalized scattered power derived from the oscillogram Fig. 20(a).

tation. Such a computation[65] shows that the total (ion and electron feature) scattering cross section is almost constant for the experimental variation of α and T_e/T_i. However, as the contribution of the electron feature to this total cross section increases through the shock, that of the ion feature must decrease. Thus the local enhancement of the ion feature is greater than the quoted ratio, R, relative to the initial plasma.

The above estimate of the scattering from a similar stable plasma took no account of the electron current drift velocity, the source of instability. According to Rostoker and Rosenbluth,[61] the presence of an electron current within a stable plasma produces only a small ($< \times 2$) enhancement of the ion feature for up to 90% of the critical current for instability.

The firm conclusion from this discussion of the experimental results is that the observed peak in the forward scattered power cannot be explained without assuming the presence of a microinstability within the shock front.

7.4. Inferred effective collision frequency

It is possible to extend the interpretation of these results onto less certain ground. The measured density fluctuations can be converted into electric field fluctuation, which can then be used to derive an effective electron collision frequency. This latter step involves assuming the k-spectrum of the turbulence, and this has not been measured. Nevertheless, it is interesting to consider the consequences of combining theoretical spectra with the observed level of fluctuation at one k value.

The ratio of the scattered signal from the shock (subscripted S) to that from the initial plasma for a given k and within the ion feature is simply the ratio of the Fourier spectral densities of the corresponding plasma density fluctuations within the frequency range of the ion feature:

$$R(k) = \frac{\langle n_e^2(k)\rangle_S}{\langle n_e^2(k)\rangle_1}.$$

The electric field fluctuation, $\langle E^2(k)\rangle$, can be derived from $\langle n^2(k)\rangle$, by assuming the low frequency dispersion relations for ion acoustic waves and an effective mean temperature for the shock.

$$\langle E^2(k)\rangle_S = (\alpha_1/\alpha_S)^4 \, R(k) \, \langle E^2(k)\rangle_1.$$

If the initial plasma is thermal, the Rayleigh-Jeans law and the dispersion relations yield

$$\varepsilon_0 \langle E^2(k)\rangle_1 = kT_{e1} \, F(\alpha_1) \quad \text{(M.K.S. units)}$$

where

$$F(\alpha) = \alpha^2/[(1+\alpha^2)(1+2\alpha^2)] = F(k).$$

However, recently the scattering from the initial plasma has been calibrated absolutely by Rayleigh scattering from nitrogen gas. This has shown that the scattering is a factor $A \approx 40$ above the thermal level.‡ Then

$$\varepsilon_0 \langle E^2(k) \rangle_1 = A \, kT_{e1} \, F(\alpha_1).$$

A test particle with initial velocity u_0 experiences stochastic deflections in a turbulent electric field $\langle E^2(\omega, k) \rangle$. With the usual definition of an effective collision frequency

$$v^* = \frac{1}{u_0^2} \frac{\partial}{\partial t} \langle u_\perp^2 \rangle \bigg|_{t=0} = \frac{1}{u_0^2} D_\perp.$$

The Fokker-Planck perpendicular diffusion coefficient in velocity-space has been evaluated[66, 67] in the form

$$D_\perp = \frac{1}{(2\pi)^3} \left(\frac{e}{m} \right)^2 \int \left(1 - \frac{(k \cdot u_0)^2}{k^2 u_0^2} \right) \langle E^2(\omega, k) \rangle \delta(\omega + k \cdot v) d^3 k d\omega$$

where

$$\langle E^2(\omega, k) \rangle = \int \langle E^2(r, t) \rangle \exp\{-i(k \cdot r + \omega t)\} dr \, dt.$$

The form of the collision frequency can be simplified[69] by assuming that the test particle is an electron with the mean thermal velocity (v_{eth}) which is much greater than the ion acoustic wave phase velocity

$$u_0 = v_{eth} \gg (\omega/k)$$

and that the turbulence is isotropic. Then

$$v^* = \frac{1}{6\pi^2} \left(\frac{e}{m} \right)^2 \frac{1}{(v_{eth})^3} \int_0^{k\lambda_D = 1} \langle E^2(k) \rangle k \, dk. \qquad \ldots[4]$$

The limits correspond to the cut-off of collective wave effects above $k = 1/\lambda_D$.

Substitution of various spectral dependencies of $\langle E^2 \rangle$ on k and plasma parameters between the initial and final states, results in a wide range of v^*, $10^8 – 3 \times 10^{10}$ s^{-1}, which overlaps the value derived from experiment. Two types of spectrum have been used:

(a) *power law*; the highest v^* is given by the $1/k$ dependence, which corresponds to the Kadomtsev spectrum,[35] and

(b) *thermal spectral dependence* $F(\alpha)$; this gives the highest v^* for the initial temperature.

These results on the interpretation of the forward-scattering experiments are very preliminary and are included as an indication of the importance of

---

‡ The scattered signals from the pre- and post-shock plasma have been shown to have a spurious origin. Consequently they do not imply supra-thermal fluctuations. This spurious signal does not affect the scattered signal from the shock nor the conclusions of this section.

the technique. When more detailed measurements of the spectrum are made and the various theoretical assumptions checked, it should be possible to construct a self-consistent model of the turbulent dissipation within the shock.‡

REFERENCES

1. J. W. M. PAUL, M. J. PARKINSON, J. SHEFFIELD and L. S. HOLMES. 1965. *7th Int. Conf. Phen. in Ionized Gases (Belgrade) II*, 819.
2. J. W. M. PAUL, L. S. HOLMES, M. J. PARKINSON and J. SHEFFIELD. 1965. *Nature Lond.*, **208**, 133.
3. A. M. ISKOL'DSKII, R. KH. KURTMULLAEV, YU. E. NESTERIKHIN and A. G. PONOMARENKO. 1965. *JETP*, **20**, 517.
4. R. KH. KURTMULLAEV, YU. E. NESTERIKHIN, V. I. PILSKI and R. Z. SAGDEEV. 1965. *I.A.E.A. Culham Conf. II*, 367.
5. U. ASCOLI-BARTOLI, S. MARTELLUCI and M. MARTONE. 1965. *I.A.E.A. Culham Conf. II*, p. 275.
6. G. C. GOLDENBAUM and E. HINTZ. 1965. *Physics Fluids*, **8**, 2111.
7. J. W. M. PAUL, G. C. GOLDENBAUM, A. IIYOSHI, L. S. HOLMES and R. A. HARDCASTLE. 1967. *Nature Lond.*, **216**, 363.
8. E. PUGH and R. PATRICK. 1967. *Physics Fluids*, **12**, 2579.
9. S. G. ALIKHANOV, N. I. ALINOVSKI, G. G. DOLGOV-SAVELEV, B. G. ESELEVICH, R. KH. KURTMULLAEV, V. K. MALINOVSKII, R. YU. NESTERIKHIN, V. I. PILSKII, R. Z. SAGDEEV and V. N. SEMENOV. 1968. *I.A.E.A. Novosibirsk Conf.*, CN–24/A–1.
10. E. HINTZ. 1968. *I.A.E.A. Novosibirsk Conf.*, CN–24/A–2.
11. R. CHODURA, M. KEILHACKER, M. KORNHERR and H. NIEDERMEYER. 1968. *Novosibirsk Conf.*, CN–24/A3.
12. A. W. DE SILVA, D. F. DUCHS, G. C. GOLDENBAUM, H. R. GRIEM, E. HINTZ, A. C. KOLB, H. J. KUNZE and I. M. VITKOVITSKY. 1968. *I.A.E.A. Novosibirsk Conf.*, CN–24/A–8.
13. A. E. ROBSON and J. SHEFFIELD. 1968. *I.A.E.A. Novosibirsk Conf.*, CN–24/A–6.
14. D. E. T. F. ASHBY and J. W. M. PAUL. 1959. *4th Intern. Conf. on Ionization Phen. in Gases (Uppsala)*, IVA, 961.
15. R. G. CRUDDACE. 1967. Thesis, Oxford (R. G. CRUDDACE and M. HILL. *Culham Laboratory Report* CLM–M52, 1966).
16. YU. A. KOLESNIKOV, N. V. FILIPPOV and T. I. FILIPPOV. 1966. *Kurchatov Report* 18/904 (Culham CTO–8).
17. A. C. KOLB. 1959. *4th Intern. Conf. on Ionization Phen. in Gases (Uppsala)* II, IVc, 1021.
18. F. J. FISHMAN, A. R. KANTROWITZ and H. E. PETSCHEK. 1960. *Rev. Mod. Phys.* **32**, 959 (collisionless claim later withdrawn).
19. N. F. NESS, C. S. SCEARCE and J. B. SEEK. 1964. *J. Geophys. Res.*, **69**, 3531.
20. J. T. GOSLING, J. R. ASBRIDGE, S. J. BAME, A. J. HUNDHAUSEN and I. B. STRONG. 1968. *J. Geophys. Res.*, **73**, 43.
21. P. L. AUER, H. HURWITZ and R. W. KILB. 1961-2. *Physics Fluids*, **4**, 1105; and **5**, 298.
22. K. W. MORTON. 1964. *Physics Fluids*, **7**, 1800.
23. V. J. ROSSOW. 1965. *Physics Fluids*, **8**, 358.
24. R. Z. SAGDEEV. 1966. *Rev. of Plasma Physics*, **4**, 23.
25. C. F. KENNEL and R. Z. SAGDEEV. 1967. *J. Geophys. Res.*, **72**, 3303 and 3327.
26. D. A. TIDMAN. 1967. *Physics Fluids*, **10**, 547.
27. H. E. PETSCHEK. 1963. *N.A.S.A. Conference on Physics of Solar Flares, Goddard*.
28. J. P. WILD. 1968. *Conf. on Plasma Instabilities in Astrophysics, Asilomar* (private communication). And with K. V. SHERIDAN and K. KAI. *Nature Lond.*, **218**, 536.

‡ Such a self-consistent model has now been presented.[71, 72]

29. P. A. STURROCK. 1966. *Plasma Astrophysics Summer School at Varenna*, p. 168; and *Nature Lond.*, **211**, 695.
30. P. A. STURROCK. 1966. *Plasma Astrophysics Summer School at Varenna*, p. 338; and *Nature Lond.*, **211**, 697.
31. A. HEWISH, S. J. BELL, J. D. H. PILKINGTON, P. F. SCOTT and R. A. COLLINS. 1968. *Nature Lond.*, **217**, 709.
32. N. KRALL. Private communication.
33. R. COURANT and K. O. FRIEDRICKS. 1948. *Supersonic Flow and Shock Waves*. Interscience.
34. T. E. STRINGER. 1964. *J. Nucl. Energy, C*, **6**, 267.
35. B. KADOMTSEV. 1965. *Plasma Turbulence*. Academic Press.
36. R. Z. SAGDEEV. 1967. *Proc. Symp. in Applied Maths, XVIII*, 281.
37. F. DE HOFFMAN and E. TELLER. 1950. *Phys. Rev.*, **80**, 692.
38. J. E. ANDERSON. 1963. *MHD Shock Waves*. M.I.T.
39. K. I. GOLDEN, H. K. SEN and Y. M. TREVE. 1961. *Fifth Conf. Ionization Phen. in Gases*, **2**, 2109.
40. L. C. WOODS. 1967. *Culham Report CLM–P153*. 1969. *Plasma Physics*, **11**, 25.
41. P. N. HU. 1966. *Physics Fluids*, **9**, 89. (Later paper by GRAD and HU.)
42. W. MARSHALL. 1955. *Proc. Roy. Soc. A*, **233**, 367.
43. J. D. JUKES. 1957. *J. Fluid Mech.*, **3**, 275.
44. V. D. SHAFRANOV. 1957. *JETP*, **5**, 1183.
45. W. GEIGER, H. J. KAEPPELER and B. MAYSER. 1962. *Nuclear Fusion, Supplement*, **2**, 403.
46. R. J. BICKERTON, A. E. ROBSON and L. C. WOODS. Private communications.
47. O. BUNEMAN. 1964. *Phys. Fluids, Suppl.*, p. S4.
48. C. SMITH and J. DAWSON. 1963. *Princeton Report MATT–151*.
49. E. G. HARRIS. 1968. *Lectures at this Summer School* (chapter 4).
50. D. W. ROSS. *Univ. Texas Report ORO–3458–12*.
51. G. M. WALTERS and E. G. HARRIS. 1968. *Physics Fluids*, **11**, 112.
52. S. P. ZAGORODNIKOV, L. I. RUDAKOV, G. E. SMOLKIN and G. V. SHOLIN. 1965. *7th Intern. Conf. Phen. in Ionized Gases (Belgrade)* II, 791; and *Kurchatov Report I.A.E. 909* (Culham CTO 355, 1967).
53. S. P. ZAGORODNIKOV, G. E. SMOLKIN and G. V. SHOLIN. 1966–7. *Kurchatov Report I.A.E. 1263 (1966)* (*Culham CTO 356, 1967*).
54. G. C. GOLDENBAUM. 1967. *Physics Fluids*, **10**, 1897.
55. A. W. DE SILVA and J. A. STAMPER. 1967. *Phys. Rev. Lett.*, **19**, 1027.
57. J. W. M. PAUL, L. S. HOLMES, R. A. HARDCASTLE and C. C. DAUGHNEY. 1968. *I.P.P.S. Conf. on Plasma Diagnostics, Culham*.
58. M. MARTONE. 1966. *Phys. Letters*, **22**, 73.
59. W. R. BELL, A. E. BISHOP, H. J. CRAWLEY, G. D. EDMONDS, J. W. M. PAUL and J. SHEFFIELD. 1966. *Proc. I.E.E. (London)*, **113**, 2099.
60. K. HAIN, K. V. ROBERTS and D. L. FISHER. Private communication.
61. M. N. ROSENBLUTH and N. ROSTOKER. 1962. *Physics Fluids*, **5**, 776.
62. S. I. BRAGINSKI. 1963. *Rev. Plasma Physics*, Vol. I. Consultants Bureau.
63. R. Z. SAGDEEV and A. A. GALEEV. 1966. *Trieste Lecture on Non-Linear Theory*.
64. W. E. DRUMMOND and M. L. SLOAN. Private communication.
65. D. R. MOORCROFT. 1963. *J. Geophys. Res.*, **68**, 4870.
66. W. B. THOMPSON and J. HUBBARD. 1960. *Rev. Mod. Phys.*, **32**, 714.
67. A. G. SITENKO. 1967. *Electromagnetic Fluctuation in a Plasma*. Academic Press.
68. A. E. ROBSON and A. B. MACMAHON. Private communication.
69. R. J. BICKERTON and I. COOK. Private communication.
70. R. Z. SAGDEEV. *E.S.R.I.N. Conference on Collision-Free Shock Waves in the Laboratory and in Space, Frascati, 1969*. To be published.
71. J. W. M. PAUL. *E.S.R.I.N. Conference on Collision-Free Shock Waves in the Laboratory and in Space, Frascati, 1969*. To be published.
72. J. W. M. PAUL, C. C. DAUGHNEY and L. S. HOLMES. 1969. *Nature Lond.*, **223**, 822. Also *Culham Laboratory pre-print CLM–P201*.

9

LASER PRODUCED PLASMAS

S. A. RAMSDEN

Department of Applied Physics, University of Hull

1

INTRODUCTION

OBSERVATION of laser-induced breakdown in a gas was first reported[1] at the 3rd International Conference on Quantum Electronics held in Paris in 1963 as a by-product of some experiments on non-linear optical effects in the focused beam from a Q-spoiled ruby laser. Since then, over two hundred papers have been published dealing with experimental and theoretical studies of the breakdown mechanism, the properties of the plasma itself, experimental and theoretical investigation of plasmas produced by the irradiation of solid targets and single particles and the possibility of heating the plasmas so produced to extremely high temperatures of thermonuclear interest.

In the short space of two lectures it is obviously impossible, and certainly undesirable, to cover the whole field. Instead, attention will be confined to the essential features of the phenomena observed and, where possible, their most probable interpretation.

Experimental and theoretical work on gas breakdown is considered first. This is followed by an account of the experimental work on, and theoretical interpretation of, the properties of the plasma so formed. Finally, an account is given of work on laser-produced plasmas from solid targets and single particles.

2

GAS BREAKDOWN

Gas breakdown can be observed at pressures of the order of atmospheric and above by focusing the output of a Q-spoiled ruby or neodymium glass laser with a short focal length lens. The phenomenon has a relatively well defined threshold corresponding to a flux density of the order of 10^{11} W/cm^2 ($\sim 10^{30}$ photons/cm^2 s) and an electric field strength in the light wave of $10^6 - 10^7$ V/cm, depending on the type and pressure of gas used.

Direct photo-ionization of the gas is impossible. For example, the ionisation potential of argon is 15·8 eV whereas the energy of a ruby laser quantum is only 1·78 eV. Further, it is quite evident that cosmic radiation cannot act as a source of initiating electrons. The free electron density in gases due to cosmic radiation is too low to provide a probability of close to unity that at least one electron exists within the small volume of $\sim 10^{-6}$ cm^3 at the focal spot, as required by the experimental data of Tomlinson[2] and of Meyerand and Haught,[3] which show only small statistical variations in the time of breakdown; also, an applied electric field and external ionization sources do not influence the breakdown.

From the calculations of Gold and Bebb,[4] Keldysh[5] and others, it appears that the most probable source of the first few electrons is multiphoton ionization either of the gas itself or of any low ionization potential trace impurities. Experiments by Voronov and Delone[6] on the ionization of low pressure gases have provided direct experimental evidence for this process, and, for Kr and Xe, give values for the ionization probability in order of magnitude agreement with the calculations of Keldysh.

Multiphoton ionization cannot, however, explain the rapid growth of the electron density within the focal volume except, perhaps, as will be seen later, for very short pulses of the order of 10^{-11} s in duration. For normal Q-switched pulses of the order of 10^{-8} s duration the growth mechanism seems to be due to cascade ionization. The electron absorbs light quanta by colliding with the atoms, in a process which is the inverse of bremsstrahlung, and is thereby accelerated. After accumulating an energy sufficient for ionization, the electron ionizes an atom, so that one electron is replaced by two slower ones, and the process repeated. It is also possible for an electron to excite an atom to an intermediate state from which it can then be ionized by the radiation. The growth of electron density is given by $n(t) = n(0)2^k$, where k is the number of generations in time t and $n(0)$ is the initial electron density. The value of k is very insensitive to the values of the initial and final density. If, for example, the cascade begins with one electron within the focal volume of $\simeq 10^{-6}$ cm^3, after 43 generations the electron density is 10^{19} cm^{-3}.

Although there has been much controversy over the matter, the process can be treated classically, as in the case of microwave breakdown at lower frequencies. Under conditions of microwave breakdown the energy of oscillation ($\simeq 10^{-3}$ eV) is greater than the quantum energy of the radiation (10^{-4}–10^{-5} eV), and the classical treatment is obviously valid. For breakdown at optical frequencies, the position is reversed in that the quantum energy (1·78 eV for ruby) is much greater than the energy of oscillation (10^{-2} eV) and, as a result, some authors have maintained that the process is quantum-like in character and that classical theory is therefore inapplicable. Others have used classical theory without any stipulation as to its validity. At frequencies much greater than the collision frequency it can, however, be shown that microwave breakdown theory and inverse bremsstrahlung are the classical and quantum description of one and the same process and that classical theory is in fact valid.

In the classical theory, a free isolated electron oscillates in the alternating electric field E of a wave of angular frequency ω with a mean energy of $e^2 E^2/(2m\omega^2)$, where e and m are the charge and mass of the electron respectively. The electron acquires energy from the wave only when it collides with an atom, when the abrupt change in the velocity causes the energy of oscillation to be converted into translational energy. The rate of growth of energy, ε, is given by

$$\frac{d\varepsilon}{dt} = \frac{e^2 E^2}{2m\omega^2} \cdot v_c \cdot \frac{\omega^2}{\omega^2 + v_c^2}, \qquad \ldots[2.1]$$

where $v_c = N_a v \sigma$ is the effective collision frequency between electrons and atoms, N_a the number of atoms per unit volume, v the electron velocity and σ the cross-section for the process.

For $\omega^2 \gg v_c^2$ equation [2.1] reduces to

$$\frac{d\varepsilon}{dt} = \frac{e^2 E^2}{2m\omega^2} \cdot v_c, \qquad \ldots[2.2]$$

i.e.

$$\frac{d\varepsilon}{dt} = v_c \cdot \frac{\hbar^2 \omega^2}{\varepsilon} \cdot \frac{I(t)}{I_0} \qquad \ldots[2.3]$$

where $I(t) = E^2 c/(8\pi)$ is the intensity of the radiation and

$$I_0 = mc\hbar^2 \omega^4/(4\pi e^2 \varepsilon).$$

The total gain in energy

$$\int_0^\tau \frac{d\varepsilon}{dt}\, dt = \frac{k}{\alpha}\chi \qquad \ldots[2.4]$$

where k is the number of generations during the time τ of the laser pulse, χ is the ionization energy and α is a factor introduced to take into account the loss of energy due to other processes

i.e.

$$\frac{v_c \hbar^2 \omega^2}{\varepsilon I_0} \int_0^\tau I(t)\, dt = \frac{k\chi}{\alpha}. \qquad \ldots[2.5]$$

The time averaged threshold intensity for breakdown is thus

$$\bar{I}_{\text{thr}} = \frac{1}{\tau}\int_0^\tau I(t)\, dt = \frac{I_0 k \varepsilon \chi}{\alpha \hbar^2 \omega^2 \tau v_c} \qquad \ldots[2.6]$$

i.e.

$$\bar{I}_{\text{thr}} = \frac{mck\omega^2\chi}{4\pi e^2 \sigma v N_a \tau}. \qquad \ldots[2.7]$$

Equation [2.7] shows the influence of the different factors, such as frequency, ionization potential, gas density and laser pulse length, on the threshold intensity for breakdown. We have, of course, considered a very idealized case, and have neglected factors such as diffusion losses, radiation losses, recombination, energy level scheme, influence of excited states and broadening of the excited levels in a strong electric field. The theory does, however, give the correct dependence and is a good approximation for the order of magnitude of the breakdown threshold. Thus, it is observed (Fig. 1) that the threshold decreases with increasing pressure up to 100 atmospheres or more. Above this Gill and Dougal[7] have observed a minimum, as is found in microwave breakdown. Measurements at $1{\cdot}06\,\mu$ (Nd glass laser), the second harmonic at 5300 Å, and at 6943 Å (ruby laser), show that the threshold increases with increasing frequency. Measurements by Buscher et al.[8] show a subsequent decrease in threshold at the second harmonic (3471 Å) of the ruby laser wavelength and although there is no existing theory to explain this, departures from the simple classical theory might be expected as the photon energy approaches the excitation energy of the gas. Smith and Haught[9] have considered energy loss processes, and on the basis of experiments with a Penning mixture of Ne and Ar come to the conclusion that radiative transport of excitation energy may be an important loss mechanism. Mitsuk et al.[10] have shown that diffusion losses may also be important at pressures lower than atmospheric. A more comprehensive theory including some of these factors has been given by Phelps.[11]

In the above theory we have assumed that the radiative flux and electric field in the focal spot are uniformly distributed, whereas in most lasers they fluctuate rapidly in both space and time, due to changes in the mode of

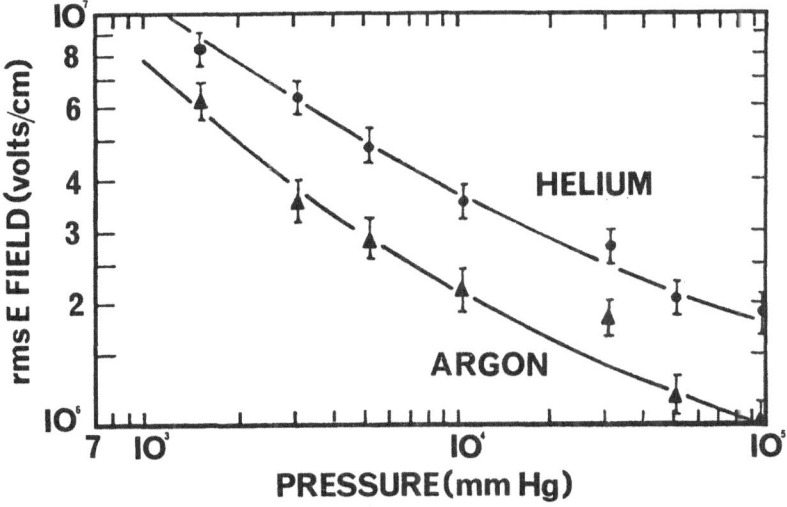

FIG. 1. Breakdown field strength in He and Ar as a function of pressure.[3]

structure, and it might be expected that this would greatly affect the breakdown threshold. The measurements of Smith and Tomlinson[12] with a single mode ruby laser and a mode locked neodymium laser seem, however, to disprove this, the thresholds obtained being in substantial agreement with previous data for conventional multiple-mode laser radiation.

Now let us look at what happens if we decrease the duration of the laser pulse. At higher rates of the avalanche process the influence of diffusion, radiative, and recombination losses become less, and the idealized picture considered above should be a better approximation to reality. The above treatment, however, is valid only if $I \leqslant I_0$, and this requirement is violated if $\tau < \tau_0 = k\varepsilon\chi/(\alpha\hbar^2\omega^2 v_c)$. For nitrogen at a pressure of an atmosphere and the ruby laser frequency $\tau_0 = 3 \times 10^{-10}$ s.

As has been shown by Bunkin and Fedorov,[13] for $I > I_0$ and thus $\tau < \tau_0$, the role of multiphoton bremsstrahlung absorption is greatly enhanced. The dependence of I_{thr} on pulse duration becomes less, as it is necessary to take into account strong non-linear dependence of the rate of energy transfer, $d\varepsilon/dt$, on I. With further increase of I the rate of energy transfer becomes slower and reaches a maximum of $d\varepsilon/dt \simeq v_c\varepsilon$ when $I/I_0 \simeq [\varepsilon/(\hbar\omega)]^2$, and afterwards diminishes as $(I/I_0)^{\frac{1}{4}}$. The existence of a maximum corresponds to the situation where the colliding electrons have the highest probability to absorb simultaneously $n = \varepsilon/\hbar\omega$ photons.

The existence of a maximum rate of energy gain by the electron means that the rate of ionization has a maximum and the time interval for one generation of electrons has a minimum. Since it is necessary, for avalanche breakdown, to obtain a definite number of generations, we can conclude that there exists a minimum laser pulse duration τ_{min} for which the avalanche process is possible. The order of magnitude of τ_{min} is given by $\tau_{\text{min}} \simeq k\chi/(\alpha\varepsilon v_c)$. The corresponding threshold intensity is $I_{\text{thr}} \simeq I_0[\varepsilon/(\hbar\omega)]^2$. For pulse duration $\tau < \tau_{\text{min}}$, breakdown occurs as a result of multiphoton ionization only. The value of τ_{min} is independent of laser frequency and for nitrogen at a pressure of one atmosphere is $\sim 10^{-11}$ s. The corresponding threshold intensity for the neodymium glass laser is $\sim 3 \times 10^{13}$ W/cm^2.

Preliminary investigations of breakdown with ultra-short optical pulses have been reported by Kaitmasov et al.[14] and by Alcock and Richardson.[15] In the latter experiment a single mode-locked laser pulse was coupled out of the cavity by means of a Pockels cell-polarizer combination. Two-photon fluorescence studies indicated that the pulse length was $\sim 10^{-11}$ s. Breakdown was observed when the laser output was focused with a 2 cm focal length lens. Streak photographs revealed that the development of the plasma was significantly different from that observed with conventional Q-switched lasers. Studies of the pressure dependence of the breakdown threshold seemed to indicate, however, that the breakdown process could still be described in terms of cascade ionization.

3

PROPERTIES OF LASER PRODUCED PLASMAS IN GASES

When the laser power noticeably exceeds the threshold for breakdown, the gas becomes highly ionized and the resultant plasma absorbs the laser beam strongly (Plate 1) as a result of the relatively high cross-section for free-free transitions of electrons in the field of the ions. A large amount of energy, of the order of a joule or more, is absorbed in a relatively small volume and the plasma is heated to a high temperature. Further, as was first observed by Ramsden and Davies,[16] during the time of the laser pulse the plasma expands rapidly, at a velocity of the order of 10^7 cm/s, in a direction opposite to that of the laser beam (see Plate 2). The light is absorbed, not in a single volume with dimensions characteristic of the radius of the focal spot, but in a relatively thin layer, of the order of the mean free path of the incident radiation. As soon as the degree of ionization ahead of the layer absorbing at a given instant reaches, for one reason or another, a sufficiently high value, a new layer becomes highly absorbing, and thus the absorption region is continuously displaced in the direction towards the lens in the manner of an "absorption and heating" wave. This effect prevents the entire pulse energy from being released in a very small volume and thus limits the attainment of very high temperatures.

Three mechanisms have been proposed for the existence of such a heating and absorption wave.

(i) *Detonation wave mechanism.*[17, 18] The heated gas in the absorbing layer expands and transmits a shock wave in all directions, including the direction opposite to that of the beam. In the shock wave the gas is heated and ionized, so creating new regions in which absorption and heating can take place. This mechanism is similar in many ways to a detonation wave in an explosive, energy being fed into the gas by absorption from the laser beam rather than by the release of chemical energy.

(ii) *Breakdown wave mechanism.*[18] If the light flux at the focus is appreciably larger than the threshold for breakdown, then it also exceeds the threshold in other parts of the light channel. Breakdown therefore takes place in these places also, but with a delay with respect to the narrowest region; the larger the cross-section of the channel and the smaller the flux density the larger the delay. Thus, a "breakdown" wave moves into the beam.

(iii) *Radiation wave mechanism.*[18] The gas in front of the absorbing layer becomes ionized and absorbing because of absorption of thermal radiation from the highly ionized and strongly heated region to the rear.

The detonation wave mechanism has been considered by Ramsden and Savic,[17] Raizer,[18] and others. The mechanism is analysed in terms of the

Chapman-Jouget detonation theory in reacting gases, which assumes that the velocity of the wave relative to the heated matter behind the wave coincides exactly with the local velocity of sound.

The velocity of the wave is given by

$$V = \left[\frac{2(\gamma^2 - 1)W}{\Delta t \pi r^2 \rho_0} \right]^{\frac{1}{3}} \qquad \ldots [3.1]$$

where W is the energy absorbed from the laser beam during time Δt, r is the radius of the focal spot, ρ_0 is the density of the gas and γ is the adiabatic index. Thus in the case of the original experiments of Ramsden and Davies[16] on a laser produced spark in air, using the measured values $W = 0.2$ joules, $r = 0.004$ cm, $\Delta t = 10^{-8}$ s we obtain a value of $V = 2 \times 10^7$ cm/s, which is in good agreement with the velocity observed.

Ramsden and Savic[17] also deduced the relationship between the distance, x, travelled by the front and the time, t, after breakdown. The dependence obtained, $x \propto t^{0.6}$, was in good agreement with the observed distance-time relationship for the duration of the laser pulse. Good agreement of distance-time relationships with the detonation wave model has also been obtained by a number of other workers.[19]

According to equation [3.1] the velocity should be proportional to the cube root of the laser power, and such a dependence has been observed by Mandelshtam et al.[20]

The specific internal energy, U, of the gas acquired as a result of absorption of the beam is given by $U = V^2 \gamma / [(\gamma^2 - 1)(\gamma + 1)]$ and from the measured velocity we can thus obtain a value for the temperature. Thus, Mandelshtam et al.[20] observed a velocity of 133 km/s, corresponding to an equilibrium temperature of $\sim 10^6$ °K, which was in good agreement with a value for the temperature determined, using absorbing foils, from the absolute value of the soft X-ray emission from the plasma. More refined temperature measurements[21] using the ratio of the soft X-ray fluxes transmitted by foils of different thickness have also given values of the temperature of $\sim 10^6$ °K. Spectroscopic studies[22] of a laser-produced spark in Ar have revealed lines due to ArIX, ArX and ArXI which also point towards the existence of a high temperature plasma.

Interferometric measurements[23] using either the same, or a second, laser as a light source for the interferometer have shown that shortly after the end of the laser pulse the gas is almost completely ionized, corresponding to an electron density of $> 10^{19}$ cm^{-3}, and have also shown the existence of a shock wave.

It would appear that in many cases the detonation wave model provides a reasonably good description of the expansion and properties of the plasma. It is certainly the most likely mechanism for low powers and very short focal length lenses, where the breakdown wave mechanism gives a value for the velocity much lower than the measured values. On the other hand, for higher

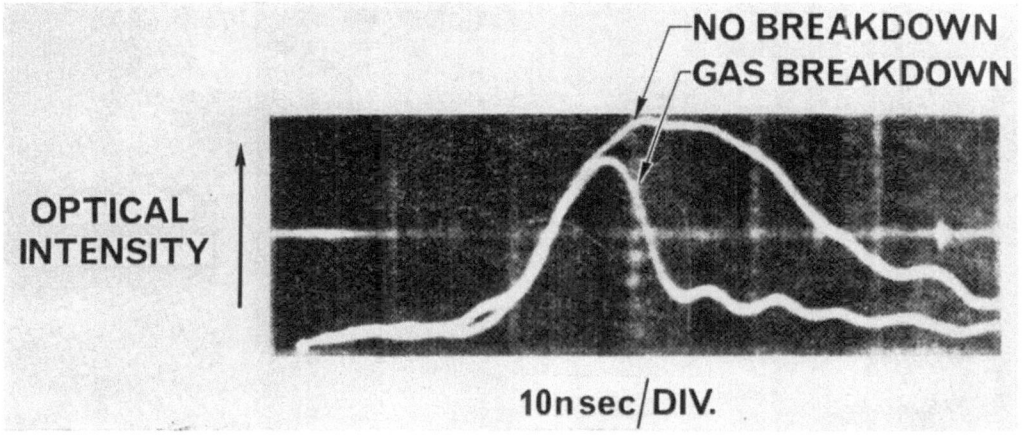

PLATE 1. Attenuation of the transmitted laser beam by the plasma.

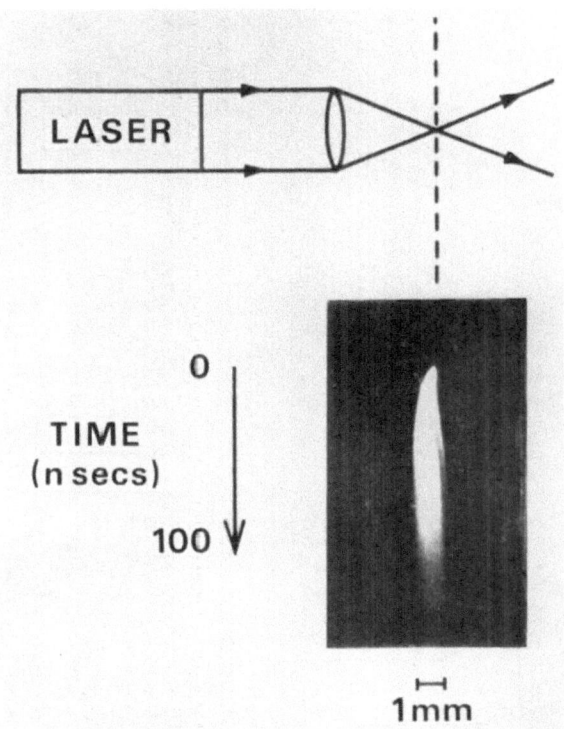

PLATE 2. Streak photograph of a laser produced spark in air.

PLATE 3

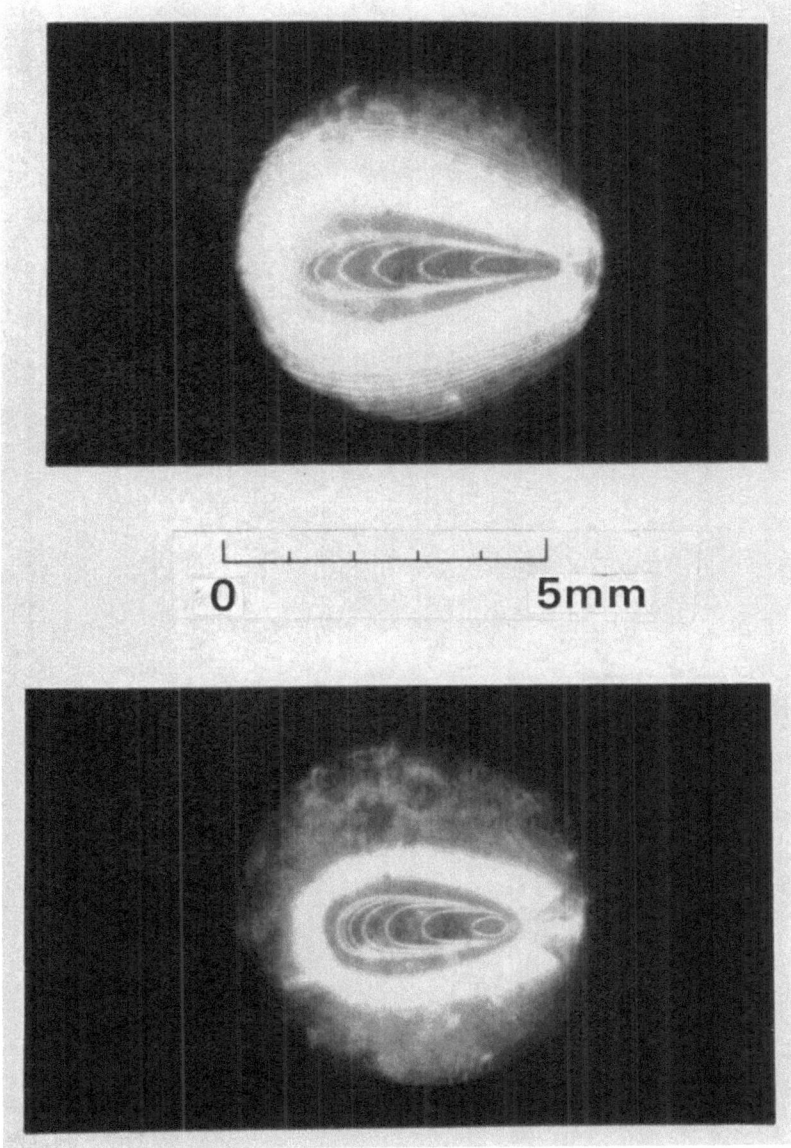

Schlieren photographs of a laser produced spark in air taken using a train of mode-locked pulses of separation 5·5 ns and duration 5 ps, produced by a Nd glass laser. The ruby laser creating the spark is incident from the left.

PLATE 4

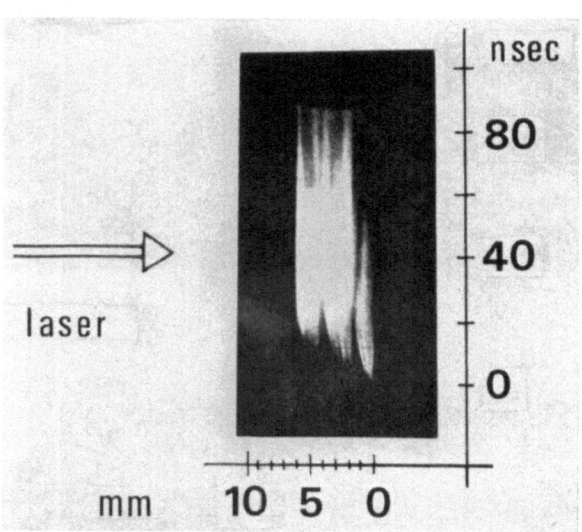

Streak photograph of multiple breakdown.

powers and longer focal length lenses, all three mechanisms give nearly equal velocities in rough agreement with the experimental values. It is not certain which is the dominant mechanism and it is not improbable that all three operate either simultaneously or during different phases of the expansion. Thus, the radiation wave may serve as a "tongue", which heats the gas ahead of the shock front as is the case in other shock waves of very large amplitude. For high power lasers and long focal length lenses, which produce light channels with small convergence in the focal region, it is most probable that the initial expansion is in the form of a breakdown wave, although after some time the hydrodynamic or radiative mechanisms must also begin to play a role. Thus, recent measurements by Alcock et al.[24] of the initial phase of the expansion of a laser spark in air produced using a ruby laser of power up to 300 MW give values for the velocity which are more in agreement with a modified breakdown wave mechanism than the detonation wave theory. Schlieren effect photographs (Plate 3) were taken of the expanding plasma using a train of mode-locked pulses produced by a Nd-glass laser, and measurements made of both the axial and radial velocities of the plasma. Because of the short duration of the mode-locked pulses, excellent time resolution was obtained in the first few nanoseconds of the expansion. Measurements at different pressures showed that the velocity varied as the square root of the pressure, whereas the detonation wave predicts an inverse velocity relationship with pressure. Assuming that electrons are produced ahead of the breakdown wave, possibly by precursor radiation, and that the time to breakdown is a constant, they were able to obtain an excellent fit to a wide range of experimental data.

Evidence for a breakdown wave has also been obtained by Bobin et al.[25] in experiments with a high energy, high power, short pulse length (10 joules in 5 ns) Nd glass laser. In air the plasma was observed to expand initially with a velocity consistent with the detonation wave model, but after the first few nanoseconds the speed of propagation increased suddenly, by as much as a factor ten, in the manner of a breakdown wave. Contrary to the radiation supported detonation regime, the luminous front speed in the second stage increased with increasing pressure. Transition to the breakdown wave regime was also observed in experiments with a 20 joule 30 ns ruby laser in deuterium gas as the pressure was increased above 25 atmospheres.

At high peak powers breakdown occurs not only at one but at several points along the axis (Plate 4) and the expansion is no longer smooth. Whilst this is no doubt due in some cases to spatial inhomogeneities in the light channel, Korobkin et al.[26] explain the phenomenon on the basis of precursor X-ray radiation from the heated plasma. At temperatures of the order of 10^6 °K the mean free path of quanta with energy $\sim KT$ radiated by the heated gas is considerably longer than the mean free path of the light quanta, the width of the wave, or even the characteristic dimensions of the focal spot.

Due to the presence of a gas layer with considerable ionization ahead of the hot plasma, breakdown starts at the front of this layer also and the region of strong light absorption jumps for a distance of the order of the X-ray mean free path in the cold gas. The initial hot plasma is now shielded from the supporting radiation by a new absorbing layer and the whole process repeated. The process is obviously dependent on gas pressure and recent experiments[27] with a single mode laser give regular reproducible jumps with good correlation between jumps and the mean free path at different pressures. Between jumps the expansion follows the detonation wave model. The formation of multiple sparks with a high power (1 GW) Nd glass laser and a long (2·5 m) focal length lens has been explained by Basov et al.[28] as due to temporal and spatial variation of the mode structure across the face of the laser rod.

After the end of the laser pulse the plasma is observed to expand in the manner of a blast wave. Schlieren effect studies by Buchl et al.[29] have been compared with Sakurai's blast wave theory, and Panarella and Savic,[30] taking into account the fact that the blast wave originates in an asymmetric source, have obtained excellent agreement with the interferometric observations of Alcock et al.[23] Schlieren effect[31] and holographic[32] studies of this later stage of the plasma yield values of electron density of $\sim 10^{19}$ cm^{-3}. Spectroscopic investigations[33-37] show that the spectrum consists essentially of a continuum with superimposed lines of singly and multiply ionized atoms corresponding to electron densities of up to 10^{19} cm^{-3} and electron temperatures of the order of a few eV. Askaryan et al.[38] have observed a microwave reflection signal lasting for several hundred microseconds, indicating that during this time the electron density remains above the critical value of 10^{13} cm^{-3}. This same group has used the magnetic effects[39] of a spark produced in a magnetic field to follow the development of the plasma, and Korobkin and Serov[40] have investigated the magnetic field of the plasma itself. The results of the latter investigation indicate that a magnetic moment exists in the plasma in a direction perpendicular to the direction of propagation, although the reason for this is not clear.

A magnetic field might also be expected to confine the plasma to a greater or less extent depending on the strength of the field. This would be useful in that, if it limited the expansion of the plasma, more energy could be absorbed in a smaller volume with consequent greater heating of the plasma. Unfortunately, however, the energy density of the plasma is so high that fields of the order of a megagauss or more would be required to confine the plasma in its initial stages. A diminution in expansion velocity by almost a factor two has, however, been observed by Chan et al.[41] in the case of a He spark produced in a pulsed magnetic field of ~ 200 kG. No appreciable effect was observed for sparks in air and butane; and whether this is due to more rapid diffusion across the magnetic field in the case of the heavier gases or simply due to the lower absorption, and hence lower kinetic pressure, of a spark in He is not clear. No decrease in breakdown threshold was observed, in contrast to the results obtained by Vardzigulova et al.[42] who observed a significant decrease in the breakdown threshold for an air spark in a magnetic field of 200 kG, due, apparently, to the slowing down of the diffusion of electrons from the focal region. The direction of the laser beam with respect to the axis of the magnetic field, however, was different in the two experiments.

4

LASER PRODUCED PLASMAS USING SOLID TARGETS AND SINGLE PARTICLES

Plasmas may also be produced by laser beam irradiation of solids or liquids. There is no threshold as in the case of a gas, the light being absorbed at any intensity. A plasma is formed, however, only when the flux density is high and comparable with that necessary for gas breakdown.

Many experiments have been carried out with solid targets and thin foils.[43-57] Ion energies of > 1 keV have been observed; spectroscopic observations have shown the existence of highly ionized (up to NiXVIII) species; interferometric measurements have indicated plasma densities of $> 10^{17}$ cm^{-3}; absorption and other measurements densities ranging from 10^{21} cm^{-3} in the nucleus of the flare to 10^{11} cm^{-3} near the periphery. Temperatures estimated from the emission spectrum are of the order of 10^5–10^6 °K.

Of great interest is whether a laser beam can be used to heat a small pellet of hydrogen or deuterium in a vacuum to temperatures of thermonuclear interest. In this way a specific mass of the substance may be heated and used to fill a magnetic trap, for example, in thermonuclear fusion research.

As stated earlier, the plasma absorbs energy as a result of free-free transitions of the electrons in the field of the ions. If the radiation is to penetrate the plasma, the density must be less than the critical value for reflection (i.e. the frequency must be less than the plasma frequency) and in the case of the ruby laser must be less than 3×10^{21} cm^{-3}. At a temperature of 10^7 °K and a density of 3×10^{21} cm^{-3} the mean free path of the light quanta is $\sim 10^{-2}$ cm, which is of the order of the dimension of the focal spot, and the electron-ion relaxation time ($\sim 10^{-10}$ s) is sufficiently short to allow time for the ions to be heated. There will be 10^{17} ions within a focal volume of dimension $\sim 10^{-2}$ cm, and to heat these to 10^7 °K will require an energy of $\simeq 10$ joules.

The heating is limited, however, by gas dynamical expansion of the plasma, which not only causes the density to drop but also moves the plasma out of the focal volume. For effective heating, it is necessary that the plasma be unable to expand appreciably during the time of the laser pulse. The gas dynamic expansion velocity is of the order of $(KT/M)^{\frac{1}{2}}$, where T is the temperature, M the atomic mass and K is Boltzmann's constant; for hydrogen at a temperature of 10^7 °K. this is of the order of 10^7 cm/s. For a focal spot of radius 10^{-2} cm the duration of the laser pulse should thus be < 1 ns, implying a laser of power > 10 GW.

It would thus appear possible to heat a small pellet of hydrogen or deuterium of $\sim 10^{-2}$ cm in size and containing $\sim 10^{17}$ atoms to a temperature of 10^7 °K with a laser capable of giving 10 joules in 1 ns and of peak power 10 GW. Whilst at the time the original papers of Basov and Krokhin[58] and Dawson[59] were written such a laser seemed rather remote, now, only a few years later, it is well within the realm of possibility and more than one such laser system has already been built. Apart from the laser itself, however, there are the technical problems of producing a small pellet of hydrogen or deuterium in a vacuum and of hitting it with a laser beam and, whilst there has recently been considerable progress in this direction, much of the experimental work to date has been carried out using extruded filaments of hydrogen and deuterium and small targets and single particles of other substances.

A considerable amount of work[60-64] has been done using small charged particles of LiH suspended in a vacuum by means of a three phase electric field applied to opposite pairs of a cubic electrode array.[65] The particles are stored in an injector located within the vacuum system and are mechanically projected up into the suspension region by means of a wire spring (Fig. 2). Electrons emitted from a hot filament charge the particles as they enter the suspension region. Particles with the proper charge-to-mass ratio are captured within the suspension volume. By adjustment of the a.c. and d.c. potentials applied to the six electrodes all but one of the particles are ejected

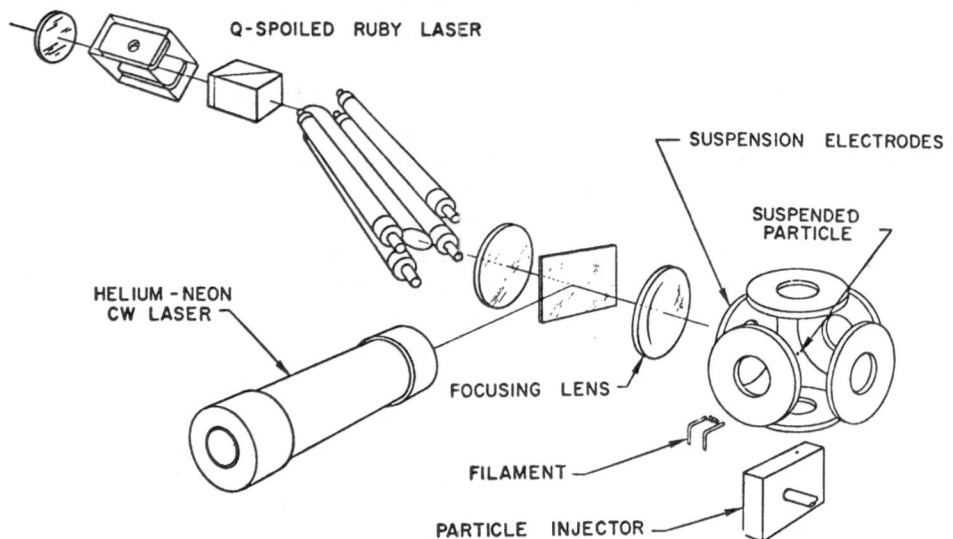

FIG. 2. Apparatus for producing a plasma from a suspended LiH particle.[60]

from the suspension region and the orbit of the remaining particle is reduced to a minimum of the order of the size of the particle itself. The particle size is usually approximately an order of magnitude less than the diameter of the focused laser beam to allow the plasma to expand and the density to drop to the value at which absorption can take place. Experiments have been carried out with particles 10–60 μ in size using ruby and Nd glass lasers of power up to 1 GW and pulse lengths 4–35 ns. Charge collection measurements have shown that the smaller particles are fully ionized and that the plasma expands symmetrically with a velocity of $\sim 10^7$ cm/s corresponding to ion energies at the boundary of ~ 1 keV and mean plasma energies of up to 200 eV, in good agreement with fluid dynamic (Fig. 3) and similarity solution calculations of the expansion of an initially cold plasma sphere into a vacuum. Streak photographs of the expanding plasma also show a symmetrical expansion with initial velocity of $\sim 10^7$ cm/s, and soft X-ray measurements yield an electron temperature of between 50 and 100 eV. Experiments with mirror and minimum-B magnetic fields of up to 20 kG indicate confinement at

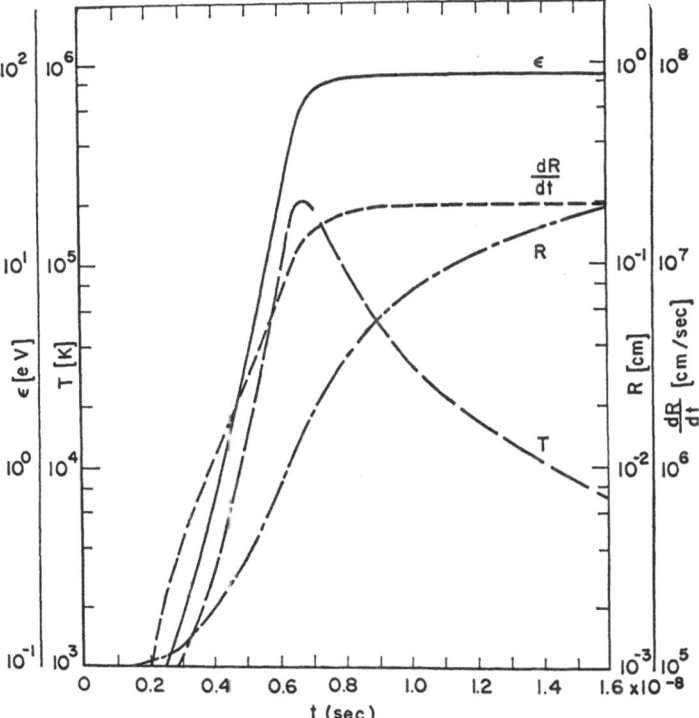

FIG. 3. Calculated time history[60] of the plasma produced by irradiating a LiH particle with a ruby laser beam. The particle is 10μ in diameter, the diameter of the focal spot is 100 μ and the laser is of peak power 20 MW.

electron densities of $\sim 10^{13}$ cm^{-3} for times limited only by scattering into the loss cones of the magnetic field.

Confinement studies in magnetic mirror geometry, using plasmas produced by focusing a Q-switched ruby laser on small aluminium discs, foils and spheres, have been carried out by Sucov et al.[66] High speed photographs show that in the absence of a confining magnetic field the plasma generated from ball targets is roughly symmetrical, whereas the plasma generated from flat targets shows asymmetries which are accentuated in the presence of a magnetic field and lead to a rapid escape of a large fraction of the plasma across the magnetic field lines. Although the expansion velocities are not appreciably reduced by the magnetic field, confinement of a small fraction of the total number of target atoms is observed for times of 10–40 μs, the average rate of plasma loss being roughly consistent with scattering by low temperature ions and electrons into the loss cones of the magnetic field. This same group has also developed a system[67] for suspending a particle in a vacuum for a brief period of time. The particle is placed in a minute cup at the end of a fine needle which is suddenly withdrawn just before the laser is fired. Experiments[68] have been carried out with small aluminium balls 50–150 μ diameter. Streak photographs show the existence of two distinct luminous regions—an inner spherically symmetric core and an outer shell expanding more rapidly and slightly asymmetrically back towards the laser. The behaviour of the outer shell, which contains ions having energies up to 5 keV and less than 1% of the plasma, is consistent with a theoretical model based on the non-linear interaction of laser radiation with an inhomogeneous plasma proposed by Hora et al.[69] The behaviour of the inner core, which contains almost all of the plasma, is consistent with the fluid dynamic theory.

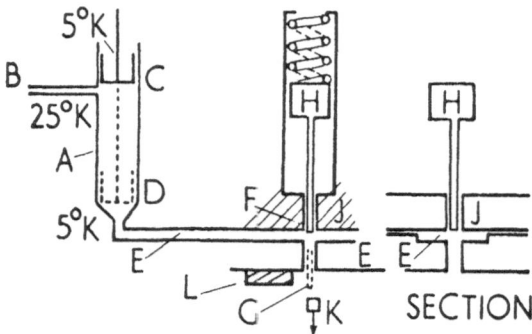

FIG. 4. Apparatus for the production of pellets of solid hydrogen in vacuo.[74]

A. Tube with temperature gradient; B. H$_2$ gas inlet; C. Cold piston (inlet position); D. Cold piston (extruding position); E. Rectangular extrusion channel; F. Initial position of punch; G. Final position of punch.

When piston (H) hits stop (J), the pellet (K) is ejected and a signal is taken from the microphone (L) to trigger the laser flash tubes.

Experiments on extruded filaments of hydrogen and deuterium have been carried out by Ascoli-Bartoli et al.,[70] Saunders et al.,[71] Colin et al.[72] and Sigel et al.[73] Ruby and Nd glass lasers have been used of power up to 2 GW. Observations have been made with streak cameras, ion collectors, microwaves and infra-red detectors. As in the experiments described earlier, the plasma is observed to expand with a velocity of $\sim 10^7$ cm/s corresponding to ion energies of ~ 1 keV and average thermal energy in the plasma of the order of 100 eV. From the infra-red emission Saunders et al. estimate that the electron density in the early stages of the expansion is at least 3×10^{20} cm^{-3}.

Preliminary studies of the plasma produced using small pellets of solid hydrogen and deuterium have been reported by Francis et al.[74] and Ascoli-Bartoli et al.[75] Francis et al. use a plunger (Fig. 4) to press out pellets 0·25 mm in diameter and 0·25 mm long from an extruded strip of solid hydrogen.

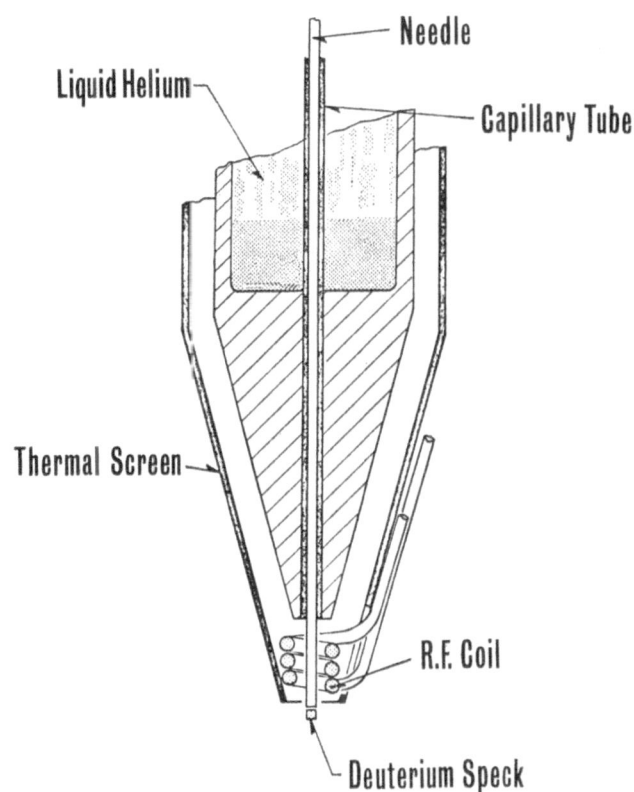

FIG. 5. Apparatus for detaching a small pellet of solid deuterium from the tip of an extruded filament by means of a r.f. heating coil.[75]

Ascoli-Bartoli *et al.* detach a small cylindrical pellet of solid deuterium 2.5×10^{-2} cm in diameter from the end of an extruded filament (Fig. 5) and have provision for studying the confinement of the plasma in a magnetic field of up to 450 kG. In both cases the laser is fired only when the pellet passes through the focal zone of the laser beam as determined by one or more auxiliary sensing beams.

Finally, mention should be made of recent measurements by Basov *et al.*[76] using a 10 joule Nd glass laser with a time duration of $\sim 10^{-11}$ s focused onto a LiD target prepared in an argon atmosphere to prevent oxidation of the surface. Out of fourteen shots neutron emission was detected on four occasions, coincident with the laser pulse to within 1 μs. No neutron emission was detected with longer pulses of 20 joules energy and 2 ns time duration (ruby laser), or 500 joules energy and 5 ns time duration (Nd glass laser).

REFERENCES

1. P. D. MAKER, R. W. TERHUNE and C. M. SAVAGE. 1963. *Third Conf. Quant. Elec., Paris.*
2. R. G. TOMLINSON. 1965. *Phys. Rev. Lett.,* **14**, 489.
3. R. G. MEYERAND, A. F. HAUGHT. 1963. *Phys. Rev. Lett.,* **11**, 401.
4. A. GOLD, and H. B. BEBB. 1965. *Phys. Rev. Lett.,* **14**, 60.
5. L. V. KELDYSH. 1965. *Soviet Physics JETP,* **20**, 1307.
6. G. S. VORONOV and N. B. DELONE. 1966. *Soviet Physics JETP,* **23**, 54.
7. D. H. GILL and A. A. DOUGAL. 1965. *Phys. Rev. Lett.,* **15**, 845.
8. H. T. BUSCHER, R. C. TOMLINSON and E. K. DAMON. 1965. *Phys. Rev. Lett.,* **15**, 847.
9. D. C. SMITH and A. F. HAUGHT. 1966. *Phys. Rev. Lett.,* **16**, 1085.
10. V. E. MITSUK, V. I. SAVOSKIN and V. A. CHERNIKOV. 1966. *JETP Letters,* **4**, 88.
11. A. V. PHELPS. 1966. *Proc. Phys. Quant. Elect. Conf.* McGraw-Hill.
12. D. C. SMITH and R. G. TOMLINSON. 1967. *Appl. Phys. Lett.,* **11**, 73.
13. F. V. BUNKIN and M. V. FEDOROV. 1966. *Soviet Physics JETP,* **22**, 844. See also F. V. BUNKIN and A. M. PROKHOROV. 1967. *Soviet Physics JETP,* **25**, 1072.
14. S. D. KAITMASOV, A. A. MEDVEDEV and A. M. PROKHOROV. 1968. *Doklady Acad. Nauk.,* **179**, 5.
15. A. J. ALCOCK and M. C. RICHARDSON. 1968. Paper presented at Fifth Intl. Conf. Quant. Elect., Miami, May 14–17.
16. S. A. RAMSDEN and W. E. R. DAVIES. 1964. *Phys. Rev. Lett.,* **13**, 227.
17. S. A. RAMSDEN and P. SAVIC. 1964. *Nature,* **203**, 1217.
18. YU. P. RAIZER. 1965. *Soviet Physics JETP,* **21**, 1009.
19. See, for example, J. W. DAIBER and H. M. THOMSON. 1967. *Physics Fluids,* **10**, 1162.
20. S. L. MANDELSHTAM, P. P. PASHININ, A. M. PROKHOROV, YU. P. RAIZER and N. K. SUKHODREV. 1966. *Soviet Physics JETP,* **22**, 91.
21. A. J. ALCOCK, P. P. PASHININ and S. A. RAMSDEN. 1966. *Phys. Rev, Lett.,* **17**, 528.
22. B. C. FAWCETT, A. H. GABRIEL, F. E. IRONS, N. J. PEACOCK and P. A. H. SAUNDERS. 1966. *Proc. Phys. Soc.,* **88**, 1051.
23. A. J. ALCOCK, E. PANARELLA and S. A. RAMSDEN. 1965. *Proc. VIIth Intl. Conf. Ion. Phen. in Gases.* Belgrade. See also A. J. ALCOCK and S. A. RAMSDEN. 1966. *Appl. Phys. Lett.,* **8**, 188.
24. A. J. ALCOCK, C. DE MICHELIS and K. HAMAL. 1968. *Appl. Phys. Lett.,* **12**, 148. See also A. J. ALCOCK, C. DE MICHELIS, K. HAMAL and B. A. TOZER. 1968. *Phys. Rev. Lett.,* **20**, 1095.
25. J. L. BOBIN, C. CANTO, F. FLOUX, D. GUYOT, J. REUSS and P. VEYRIE. 1968. Paper presented at Fifth Int. Conf. Quant. Elec., Miami, May 14–17.

26. V. V. KOROBKIN, S. L. MANDELSHTAM, P. P. PASHININ, A. V. PROKHINDEEV, A. M. PROKHOROV, N. K. SUKHODREV and M. YA. SCHELEV. 1968. *Soviet Physics JETP*, **26**, 79.
27. P. P. PASHININ. Private communication.
28. N. G. BASOV, V. A. BOIKO, O. N. KROKHIN and G. V. SKLIZKOV. 1967. *Doklady*, **173**, 538.
29. K. BUCHL, K. HOHLA, R. WIENECKE and S. WITKOWSKI. 1968. *Phys. Lett.*, **26A**, 248.
30. E. PANARELLA and P. SAVIC. 1968. *Can. J. Phys.*, **46**, 183.
31. T. P. EVTUSHENKO, G. M. MALYSHEV, G. V. OSTROVKSAYA, V. V. SEMENOV and T. YA. CHELIDZE. 1966. *Soviet Physics—Technical Physics*, **11**, 818.
32. A. N. ZAIDEL, G. V. OSTROVSKAYA, YU. I. OSTROVSKII and T. YA. CHELIDZE. 1967. *Soviet Physics—Technical Physics*, **11**, 1650.
33. R. W. MINCK. 1964. *J. Appl. Phys.*, **35**, 252.
34. Y. DURAND and P. VEYRIE. 1966. *Comptes Rendues*, **262**, 1283.
35. J. W. DAIBER and J. S. WINNANS. 1968. *J.O.S.A.*, **58**, 76.
36. M. M. LITVAK and D. F. EDWARDS. 1966. *J. Appl. Phys.*, **37**, 4462.
37. T. P. EVTUSHENKO, A. N. ZAIDEL, G. V. OSTROVSKAYA and T. YA. CHELIDZE. 1967. *Soviet Physics—Technical Physics*, **11**, 1126.
38. G. A. ASKARYAN, M. S. RABINOVITCH, M. M. SAVCHENKO and A. D. SMIRNOVA. 1965. *JETP Lett.*, **1**, 18, (trans. p. 162).
39. G. A. ASKARYAN, M. S. RABINOVITCH, M. M. SAVCHENKO and A. D. SMIRNOVA. 1965. *JETP Lett.*, **1**, 9 (trans. p. 5).
40. V. V. KOROBKIN and R. V. SEROV. 1966. *JETP Lett.*, **4**, 103 (transl. p. 70).
41. P. W. CHAN, C. DEMICHELIS and B. KRONAST. 1968. *Appl. Phys. Lett.*, **13**, 202.
42. L. E. VARDZIGULOVA, S. D. KAITMASOV and A. M. PROKHOROV. 1967. *Soviet Physics—JETP Lett.*, **6**, 253.
43. W. I. LINLOR. 1963. *Appl. Phys. Lett.*, **3**, 210.
44. W. I. LINLOR. 1964. *Phys. Rev. Lett.*, **12**, 383.
45. E. ARCHBOLD, D. W. HARPER and T. P. HUGHES. 1964. *Brit. J. Appl. Phys.*, **15**, 1321.
46. R. V. AMBARTSUMYAN, N. G. BASOV, V. A. BOIKO, V. S. ZUEV, O. N. KROKHIN, P. G. KRYUKOV, YU. V. SENAT-SKII and YU. YU. STOILOV. 1965. *Soviet Physics JETP*, **21**, 1061.
47. H. OPOWER and E. BURLEFINGER. 1965. *Phys. Lett.*, **16**, 37.
48. P. LANGER, G. TONON, F. FLOUX and A. DUCAUZE. 1966. *I.E.E.E.—J. Quant. Elect.*, **2**, 499.
49. E. FABRE, P. VASSEUR and G. BEVERNAGE. 1966. *Phys. Lett.*, **20**, 381.
50. A. W. EHLER. 1966. *J. Appl. Phys.*, **37**, 4962.
51. N. G. BASOV, V. A. BOIKO, V. A. DEMENT'EV, O. N. KROKHIN and G. V. SKLIZKOV. 1967. *Soviet Physics*, **24**, 659.
52. N. G. BASOV, O. N. KROKHIN and G. V. SKLIZKOV. 1967. *Applied Optics*, **6**, 1814.
53. J. BRUNETEAU, E. FABRE, H. LAMAIN and P. VASSEUR. 1967. *Phys. Lett.*, **26A**, 37.
54. D. W. GREGG and S. J. THOMAS. 1967. *J. Appl. Phys.*, **38**, 1729.
55. H. OPOWER, W. KAISER, H. PUELL and W. HEINICKE. 1967. *Zeits für Naturforschung*, **22a**, 1392.
56. D. W. KOOPMAN. 1967. *Physics Fluids*, **10**, 2091.
57. W. G. GRIFFIN and J. SCHLUTER. 1968. *Phys. Lett.*, **26A**, 241.
58. N. G. BASOV and O. N. KROKHIN. 1964. *Soviet Physics, JETP*, **19**, 123.
59. J. M. DAWSON. 1964. *Physics Fluids*, **7**, 981.
60. A. F. HAUGHT and D. H. POLK. 1966. *Physics Fluids*, **9**, 2047.
61. C. DEMICHELIS and S. A. RAMSDEN. 1967. *Phys. Lett.*, **25A**, 162.
62. P. E. FAUGERAS, M. MATTIOLI and R. PAPOULAR. 1968. Paper presented at A.I.A.A. Fluid and Plasma Dynamics Conference, Los Angeles, June 24–28.
63. A. F. HAUGHT, D. H. POLK and W. J. FADER. 1968. Paper presented at I.A.E.A. Conference on Plasma Physics and Controlled Thermonuclear Fusion Research, Novosibirsk, U.S.S.R., Aug. 1–7.
64. M. J. LUBIN, H. S. DUNN and W. FRIEDMAN. 1968. Paper presented at I.A.E.A. Conference on Plasma Physics and Controlled Thermonuclear Fusion Research, Novosibirsk, U.S.S.R., Aug. 1–7.

65. R. F. WUERKER, H. M. GOLDENBERG and R. V. LANGMUIR. 1959. *J. Appl. Phys.*, **30**, 441.
66. E. W. SUCOV, J. L. PACK, A. V. PHELPS and A. G. ENGELHARDT. 1967. *Physics Fluids*, **10**, 2035.
67. J. L. PACK, T. V. GEORGE and A. G. ENGELHARDT. 1968. *Rev. Sci. Insts.*, **39**, 1697.
68. A. G. ENGELHARDT, H. HORA, T. V. GEORGE and J. L. PACK. 1968. *Bull. Am. Phys. Soc.*, **13**, 887.
69. H. HORA, D. PFIRSCH and A. SCHLUTER. 1967. *Zeits für Naturforschung*, **20a**, 278.
70. U. ASCOLI-BARTOLI, C. DEMICHELIS and E. MAZZUCATO. 1966. Conference on Plasma Physics and Controlled Thermonuclear Fusion Research (International Atomic Energy Agency, Vienna).
71. P. A. H. SAUNDERS, P. AVIVI and W. MILLAR. 1967. *Phys. Lett.*, **24A**, 290.
72. C. COLIN, Y. DURAND, F. FLOUX, D. GUYOT, P. LANGER and P. VEYRIE. 1968. *J. Appl. Phys.*, **39**, 2991.
73. R. SIGEL, K. BUCHL, P. MULSER and S. WITKOWSKI. 1968. *Phys. Lett.*, **26A**, 498.
74. G. FRANCIS, D. W. ATKINSON, P. AVIVI, J. E. BRADLEY, C. D. KING, W. MILLAR, P. A. H. SAUNDERS and A. F. TAYLOR. 1967. *Phys. Lett.*, **25A**, 486.
75. U. ASCOLI-BARTOLI, B. BRUNELLI, A. CARUSO, A. DEANGELIS, G. GATTI, R. GRATTON, F. PARLANGE and H. SALZMAN. 1968. Paper presented at I.A.E.A. Conference on Plasma Physics and Controlled Thermonuclear Fusion Research, Novosibirsk, U.S.S.R., Aug. 1–7. See also, A. CECCHINI, A. DEANGELIS, R. GRATTON and F. PARLANGE. 1968. *J. Sci. Inst. (Journal of Physics E)*, Series 2, Vol. 1, 1040.
76. N. G. BASOV, S. D. ZAHAROV, P. G. KRYUKOV, YU. V. SENAT-SKII and S. V. CEKALIN. 1968. Paper presented at Fifth Int. Conf. Quant. Elect., Miami, May 14–17.

10

THE PRODUCTION AND CONTAINMENT OF
HIGH DENSITY PLASMAS

G. B. F. NIBLETT

U.K.A.E.A. Culham Laboratory

1

INTRODUCTION

RESEARCH on the production and containment of high density plasmas has had an important part to play in the burgeoning activity in plasma physics during the last fifteen years. This is to be expected, since much of the interest in laboratory plasmas has sprung from the attempt to generate controlled thermonuclear reactions among the light elements, and it is the spectacular uncontrolled release of energy at high density in the H-bomb that has given hope that controlled fusion power is possible. In this article we consider briefly those characteristics of high density plasmas which make them of particular interest, and outline some of the physical problems associated with their production and containment. A very useful collection of papers giving results of recent research on pulsed high-density plasmas is given in the proceedings of a conference held at Los Alamos, U.S.A. in September 1967.[1]

For the purpose of this article we define "high density plasma" to mean a fully ionized gas with density in excess of about 10^{16} particles/cm^3 and with temperature in excess of some 10^6 °K. Such a plasma is, generally speaking, collision-dominated and behaves in most respects as a fluid rather than as an assembly of independent particles. The instabilities to which the plasma is subject are consequently limited to the fluid or magnetohydrodynamic instabilities; we are not concerned with the so-called velocity-space instabilities or those which derive from anisotropies in the pressure distribution.

Since the yardstick of success for thermonuclear containment experiments is the product $n\tau$ of particle concentration and confinement time, and since the growth-rate of fluid instabilities is largely independent of density, it follows that the higher the plasma density the smaller the range of instabilities which can prevent the establishment of a favourable energy balance. If Lawson's criterion, i.e. the minimum value of $n\tau$ for a power-producing reactor, is met for a D–T system by confinement at 10^{17} ions/cm^3 for a millisecond, then instabilities whose growth-time is longer than this cease to be of crucial importance.

In general, we can say that magnetically-confined plasmas at high density have a high value of plasma β, i.e. the ratio of plasma pressure to external magnetic pressure. It follows that these plasmas are more economic in their use of magnetic field and they are less likely to suffer energy losses by cyclotron radiation.

One unfortunate consequence of working at high pressures is that these plasmas tend to be highly ephemeral, partly because it is difficult to contain for long periods the high pressure which the plasma exerts, and also because the loss of energy by radiation at high densities often sets a limit to the containment time. A corollary of this, of course, is that these pulsed plasmas are often valuable light sources giving copious emission of electromagnetic radiation at short wavelengths.

High density plasmas are produced by the rapid transfer of energy to the plasma from an external source. Various mechanisms for energy transfer can be used; they include the application of fast-rising magnetic fields as in the theta-pinch, the concentration of plasma from a plasma gun as in plasma focus, the impact of high powered laser beams, or systems driven by conventional high explosives. All these techniques have the same end in view —the rapid generation and concentration of ionized matter in a confined volume of space.

In order to examine some of the essential features of these systems it is helpful to consider in detail the mechanisms that take place in the theta-

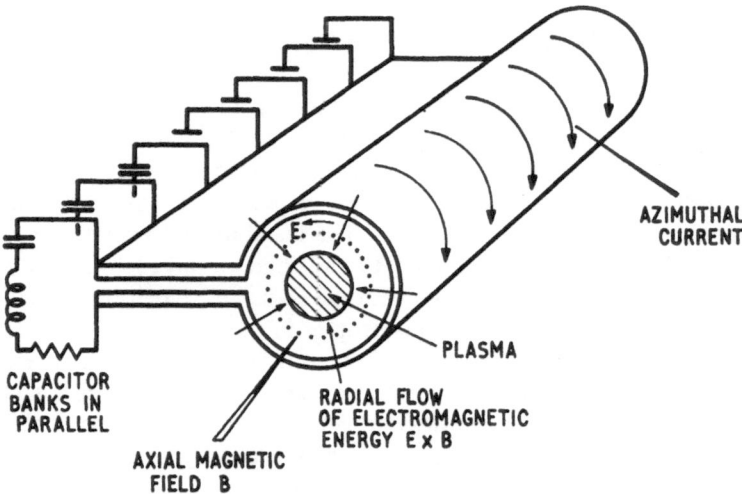

Fig. 1. Schematic diagram of theta-pinch experiment.

pinch for this is a device which has been thoroughly investigated in the last ten years and provides a convenient example of the principles involved in the rapid compression of plasmas using fast-rising magnetic fields.

2

THE THETA-PINCH

A schematic diagram of a theta-pinch configuration is shown in Fig. 1. In essence the system consists of a specially designed low-inductance capacitor bank connected to a single turn coil which encircles a discharge vessel containing pre-ionized hydrogen or deuterium gas. When the bank is triggered the rapidly increasing axial magnetic field which accompanies the current in the coil heats and compresses the plasma. Theta-pinch experiments are now being studied in many laboratories and for a variety of purposes. The stored energy in the capacitor bank and the dimensions of the coil vary over a wide range: one of the largest is at Culham where a megajoule bank is being used in conjunction with a coil almost eight metres long. Details of the engineering parameters of the larger theta-pinch experiments are shown in Table I.

A useful heuristic method of analysing the processes which follow on the application of the fast-rising axial magnetic field to the pre-ionized plasma is to consider the equation of power balance in the early stages of the theta-pinch. When the capacitor bank is switched on to the coil, there is an inward radial flow of energy $E \times B$ accompanying the azimuthal E_θ field and the axial B_z field. In the z-pinch the radial flow results from an E_z field crossed with B_θ but the consequence is the same: it is the radial flow of energy which acts as the driving force to heat the plasma.

An elementary expression for power balance in the theta-pinch is given in the following equation:

$$-2\pi r_0 \underbrace{[E \times B]}_{\substack{\text{Poynting} \\ \text{flux}}} \tfrac{1}{4}\pi = \frac{\partial}{\partial t} \int \underbrace{\left[\tfrac{3}{2}nk(T_e + T_i) + \tfrac{1}{2}\rho v^2\right]}_{\substack{\text{Thermal and directed energy} \\ \text{provided by } j^2\eta \text{ and } J \times B \cdot v \\ \text{terms}}} dx^3$$

$$+ \int (\underbrace{\varepsilon}_{\substack{\text{Radiation} \\ \text{flux}}} + \underbrace{\nabla \cdot q}_{\substack{\text{Conduction} \\ \text{loss}}}) dx^3 + \frac{\partial}{\partial t} \int \underbrace{\frac{B^2}{8\pi}}_{\substack{\text{Magnetic} \\ \text{energy}}} dx^3 \qquad \dots[2.1]$$

We consider unit length of an infinitely long theta-pinch so that only radial flow is considered. The left-hand side of the equation gives the power into the system, i.e. the rate of flow of energy radially into unit length of the coil given by the Poynting flux $E \times B$ integrated over the inner surface of the coil of radius r_0. The flow of energy through the *surface* of the coil is distributed throughout the *volume* of the plasma in the manner shown on the right-hand side of the equation.

TABLE I

Theta-pinch experimental parameters

		Los Alamos Scylla IV	N.R.L. Pharos	General Electric Schenectady	Munich Isar 1	Julich	Culham 2-metre coil	Culham 8-metre coil
Coil length	(cm)	100	180	36	150	128	200	772
Coil diameter	(cm)	10	10·5‡	20	10·6	10·5	10	11
Stored energy, fast	(MJ)	0·6	1·3	0·5	2·7	0·6	1·1	1·1
slow	(MJ)	3·8	—	—	—	—	—	—
Voltage	(kV)	50	18·7	50	40	20	40	40
Half cycle	(μs)	7·4	24	15	19	17	12	10·5
Duration with crowbar	(μs)	52	160	—	105	—	35	160
Peak magnetic field	(kg)	100	88‡	55	178	100	76	25
Peak electric field	(V/cm)	1100	315	600	700	300	550	250
Peak current	(MA)	10	12·7	3	21	10	12·1	15·4

‡ More recent experiments use a larger coil diameter and correspondingly reduced magnetic field (see Ref. 1: Papers A-5 & G-5).

The design of the capacitor assembly is aimed so as to maximize the Poynting flux; in practice this means establishing a high electric field E at the surface of the coil. Of course most of the power that enters the coil turns up on the extreme right-hand side of the equation as energy stored in the magnetic field and is in a sense wasted though in principle it can be recovered as the plasma expands.

3

SHOCK HEATING AND JOULE HEATING

In the implosion phase energy is supplied to the plasma as thermal and directed energy by Joule and shock heating. It is the term $[\frac{3}{2}nk(T_e+T_i)+\frac{1}{2}\rho v^2]$ that the process is designed to maximize. In the early days of theta-pinch experiments the clearly defined purpose was to heat the plasma by maximizing the $(j \times B) \cdot v$ term, that is the rate at which mechanical work is done on the plasma. This energy appears initially as directed energy of the massive ion component of the plasma and is accompanied by shock waves which convert it into randomized thermal energy. Fast compression systems in which the Poynting flux is particularly large, for example, the Scylla experiments at Los Alamos[2] and the experiments at G.E., Schenectady[3] and at

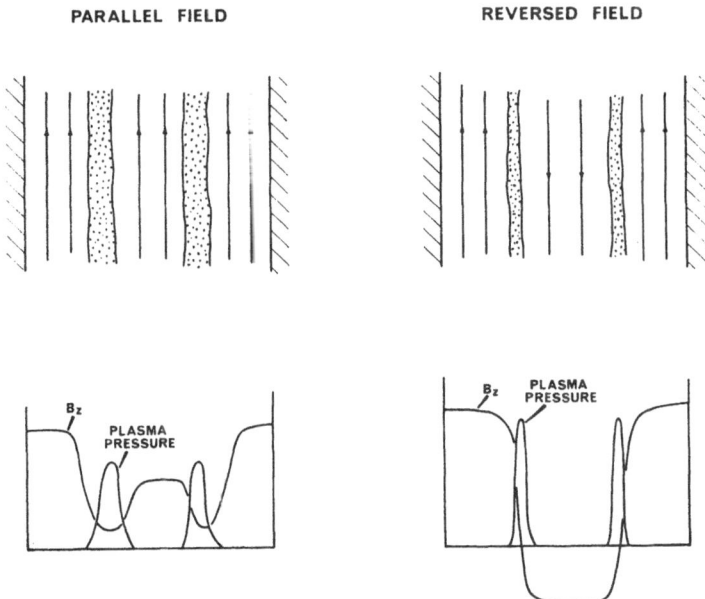

PARALLEL FIELD   REVERSED FIELD

FIG. 2. Radial distributions of plasma and magnetic field in parallel and reversed field theta-pinches.

Munich in Germany[4] have succeeded in shock heating deuterium ions to energies of several thousand electron volts. Thus there seems no doubt that, given a very fast capacitor bank, the shock heating process is an effective means of heating the ion component.

The Joule heating term $j^2\eta$ is also effective in heating the plasma—in this case the electron component. The resistive heating arises in two ways: either because the current density j is high, or because of an anomalously large resistivity η; either circumstance can result in Joule heating being the dominant heating process.

Large current densities arise in particular in reversed field configurations in which plasma containing trapped magnetic field of one sign is compressed by a field of the opposite sign. The resulting configuration exhibits a sharp gradient in the magnetic field, and the large current density in the neutral surface leads to Joule heating of the electrons. Such a process is shown schematically in Fig. 2 and is the technique used in the PHAROS experiment at the Naval Research Laboratories in the U.S.A. for producing elevated plasma temperatures.[5]

When large electric fields are applied to theta-pinch plasmas, particularly at initial filling pressures below 20 mtorr, the electrons are observed to be heated by Joule heating arising from an anomalously high resistivity. The

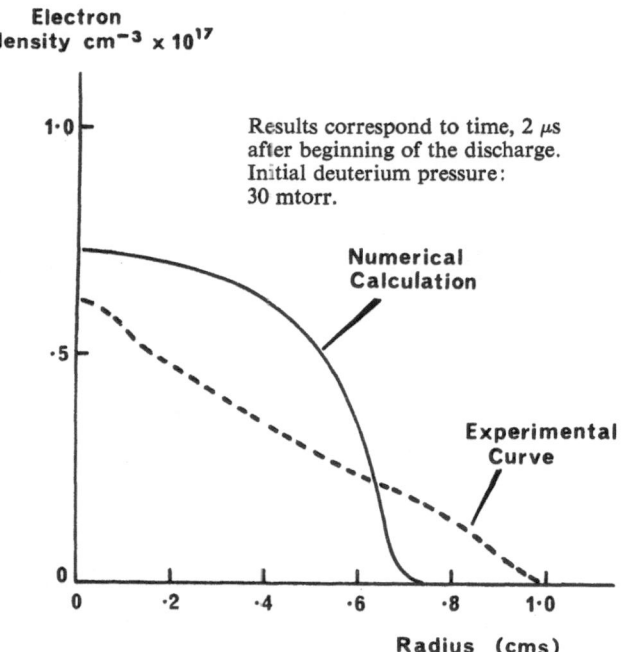

FIG. 3. Theoretical and experimental radial density distribution in the theta-pinch.

anomalous resistivity (whose presence is itself enhanced by reversed fields[6]) is a result of collisionless turbulence accompanying unstable drift motion of electrons relative to ions. This type of turbulent heating mechanism was proposed by Babykin et al.[7,8] to explain electron heating in rapidly alternating magnetic fields.

The anomalous resistivity in the early stages of the theta-pinch has two important consequences: it leads to a rapid increase in electron temperature; and it results in enhanced magnetic field diffusion. The field diffusion is illustrated in Fig. 3 which compares the experimentally-measured and theoretically predicted radial density distributions at the end of the implosion phase 2 μs after the beginning of a discharge formed in deuterium at an initial pressure of 30 mtorr. The discrepancy between the two curves is greater than can be explained by the errors in either, and the most likely explanation is that the enhanced diffusion observed experimentally is due to an anomalously high resistivity not included in the theoretical calculation. Recent work[9] has shown that if the classical resistivity is replaced by a semi-empirical non-classical expression suggested by Sagdeev[10] the experimental and theoretical curves can be brought into accord. The Sagdeev form of the resistivity exceeds the classical value whenever the electron-ion drift velocity is greater than the sound speed, but reverts to the classical value as diffusion proceeds and the current density decreases.

4

RADIATION AND CONDUCTION LOSSES

We now briefly consider the effect of the radiation and thermal conduction terms in equation [2.1]. Because this equation is concerned only with radial motion and assumes an infinitely long plasma, it takes no account of axial conduction; but we shall have to consider this. In general radial conduction can be ignored, because the effect of the strong axial magnetic field is to reduce by several orders of magnitude the coefficient of thermal conductivity in the radial direction. But the effect of axial conduction along the lines of force to the ends of the system plays an important role in the energy loss process.

The effect of radiation and thermal conduction on the power balance in the theta-pinch has been thoroughly examined in a paper by Green and his colleagues.[11] In this article we shall attempt to examine the rate of energy loss by elementary considerations. The characteristic cooling time τ_R for loss of energy by radiation is given by the plasma energy density divided by ε, the rate of loss of energy per cm^3, i.e.

$$\tau_R \simeq \frac{2nkT_e}{(\gamma - 1)} \cdot \frac{1}{\varepsilon},$$

where γ is the specific heat ratio and we have assumed equal ion and electron temperatures. If we assume that radiation losses follow the bremsstrahlung form then

$$\tau_R \simeq \frac{2nkT_e}{(\gamma-1)} \cdot \frac{1}{Bn^2 T_e^{\frac{1}{2}}} \simeq \frac{2kT_e^{\frac{1}{2}}}{(\gamma-1)Bn}$$

where B is a constant. We see that the cooling time increases as $T_e^{\frac{1}{2}}$, but is inversely proportional to the plasma density, so that radiation losses become less severe at high temperatures and correspondingly lower densities. The results of Ref. 11 show that for a deuterium plasma at a density of 5×10^{16} cm^{-3}, temperature of 300 eV and with 1% of oxygen impurity, the cooling time is about twenty to thirty microseconds: this could be increased by increase of temperature at constant pressure and by reduction of the impurity level.

The rate of loss of energy per cm^3 by thermal conduction from the centre of a coil of length $2l$ can be written as $\varepsilon \propto T_e^{\frac{7}{2}}/l^2$. The sensitive dependence of this loss rate on the electron temperature results from the five-halves power relationship between thermal conductivity and T_e. Thus the characteristic cooling time τ_c is given by

$$\tau_c \simeq \frac{2nkT_e}{(\gamma-1)} \cdot \frac{1}{\varepsilon} \simeq \frac{2}{(\gamma-1)} \cdot \frac{nkl^2}{cT_e^{\frac{5}{2}}}$$

which demonstrates the rapid decrease in cooling time with increase in temperature. This result led Green and his colleagues to predict a limiting temperature for open-ended theta-pinches of about 200–500 eV, depending on the length of the coil. Such a prediction is consistent with measurements reported from several different experiments with a wide spread of operating conditions.[11] It appears from this work that loss of energy by thermal conduction could be a major obstacle to the attainment of high electron temperatures in open-ended systems.

5

PLASMA FOCUS

Any discussion of the production of high density plasmas would be incomplete without mentioning the plasma focus experiment for this is a remarkable method of generating a laboratory plasma with extreme conditions of temperature and density.

A schematic diagram of the experimental configuration is shown in Fig. 4. It consists in essence of a coaxial plasma gun which propels a current sheath at high velocity in the annular region between the inner and outer coaxial conductors. When the plasma emerges from the end of the gun, it experiences a rapid compression towards the axis, resulting in the production of hot dense plasma foci as shown in the figure. Another way of looking at this is to note that as the current sheath reaches the end of the gun the change in mechanical configuration subjects the sheath to a sudden change in inductance, which is accompanied by an increase in electric field across the plasma. Thus the mechanical structure of the gun is an artificial method of generating an instantaneously high value of the Poynting flux.

This type of plasma experiment originated in Russia and was first reported by Filippov at the 1961 Salzburg Conference[12]; results from similar experiments have since been published by Mather[13] and Beckner[14] in the U.S.A., Long et al. at Culham[15] and Patou et al. in France.[16] These papers show that plasma foci are concentrated into isolated regions with a volume of no more than about a cubic millimetre, and last for about 0·1–0·2 microseconds. The feature which makes the plasma unique in the laboratory is its high temperature and density. The peak plasma temperature is of the order of 5 keV and the maximum particle concentration lies between 10^{19} and 10^{20} particles/cm^3. Despite the short life of the plasma, the density is sufficiently high for the ions and electrons to reach thermal equilibrium, and the temperature high enough for atoms of copper (which enter as impurities from the electrode) to be fully stripped of their electrons. Recent work[17] suggests that the compressed filamentary plasma is subject to hydromagnetic instabilities of the $m = 0$ type, i.e. the sausage-mode instability, and that the plasma breaks up and re-forms several times as it is carried downstream by the forward momentum given it by the coaxial gun.

Because of its high density and temperature, the plasma focus is an intense source of neutron and electromagnetic radiation. Mather, for example,[18] has measured a neutron output of 5×10^9 per pulse from the D–D reaction and has reached 4×10^{11} neutrons using a D–T mixture. Such a rate of emission of neutron radiation is greater than is produced by any other laboratory plasma.

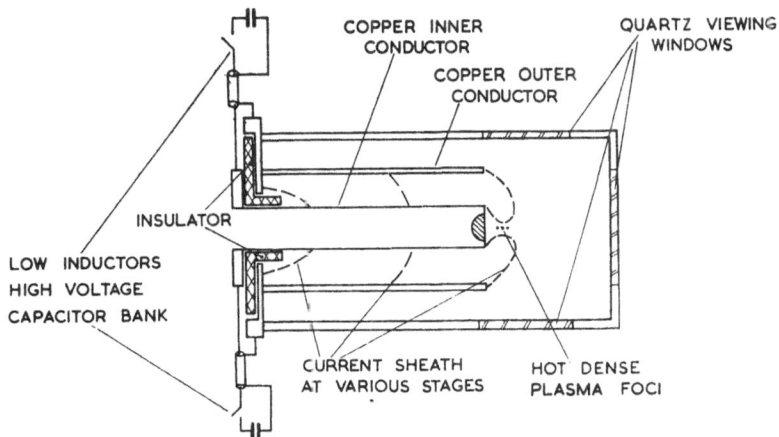

FIG. 4. Schematic diagram of plasma focus experiment.

The output of electromagnetic radiation is similarly intense and provides a copious emission of soft X-rays. Using a specially etched plane diffraction grating at a grazing angle of the order of ten minutes of arc Long et al.[15] have been able to record on photographic film emission spectra at wavelengths down to about 4 Å. In particular the Lyman-α line of hydrogen-like argon XVIII was recorded at 3·75 Å. Undoubtedly the intense X-radiation from these plasmas, coupled with novel techniques of dispersion at short wavelengths, is opening up a new domain of spectroscopy of highly ionized heavy atoms.

6

CONTAINMENT PROBLEMS

In the plasma focus experiments the containment of the plasma is largely achieved by inertial forces. Such confinement is of little help, however, for long term containment at densities of 10^{17} particles/cm^3 or less: magnetic confinement offers the only realistic means of laboratory confinement for the times necessary for useful power output. We shall now revert to a discussion of the theta-pinch plasma as a means of illustrating magnetic containment of high density plasmas.

In the straight, open-ended, theta-pinch we have to consider both the radial and axial losses, i.e. losses across and along the magnetic field lines. Few measurements have been reported on the radial particle losses for the simple reason that the end losses usually dominate. Recent measurements on an 8-metre theta-pinch at Culham, however, have led to estimates of the rate of radial diffusion.[19, 9] This coil is sufficiently long for measurements on undisturbed plasma to be made for times up to 40 microseconds. An estimate of the perpendicular diffusion coefficient has been made using calibrated image converter records to measure the variation with time of the radial plasma density distribution at the mid-plane of the coil. In the early stages of the discharge, when the plasma is subject to strong radial accelerations, the resistivity is anomalously high and the driving magnetic field diffuses rapidly into the plasma, reducing the mean values of β. At later stages, however, the diffusion rate is much slower and radial losses are negligible compared with losses by flow from the ends. Recent observations[9] made under conditions where the classical diffusion rate and the Bohm rate are in the ratio of 1:100, show that the measured diffusion is much closer to the classical rate.

7

AXIAL LOSSES

In the absence of reversed magnetic fields there must inevitably be a serious loss of energy in the straight theta-pinch because of convection of particles from the ends. For collision-dominated plasmas the loss of particles may be considered in terms of hydrodynamic flow through a flexible nozzle or aperture formed by the magnetic field lines. Fig. 5 is a schematic diagram of this type of nozzle in which B_e is the strength of the external magnetic field and B_i that of the internal or trapped magnetic field.

It is convenient to consider the problem in terms of the flow of perfectly conducting fluid along the magnetic field lines and out of the orifices at each end of the coil; it can readily be shown that the size of the orifices is determined largely by the strength of the trapped magnetic field, that is by the β value. The ratio of the area of the plasma at the centre of the coil to that in the mirror region is $R/(1-\beta)^{\frac{1}{2}}$ where R is the geometrical mirror ratio. Thus as the strength of trapped field is lowered the aperture of the escape nozzle is rapidly reduced until ultimately its dimensions are determined by the residual field which is trapped in the sheath.[20] This dependence of the loss rate on the "internal" mirror ratio $(1-\beta)^{-\frac{1}{2}}$ and on the "external" mirror ratio R has been confirmed experimentally.[21, 22, 23]

These elementary considerations of the loss process have been examined in detail[24] to provide an expression for the steady state efflux of plasma through a nozzle as a function of β and R. More recently, Wesson[25] has considered the problem of transient flow from a long column of plasma whose properties vary only in the axial direction and which is situated in an external B_z field constant in time. These assumptions are of course violated in practice, but Wesson's model nevertheless serves as a means of establishing the main characteristics of the flow and how they vary with the plasma temperature and the strength of trapped field.

Wesson's calculations show that at time zero centred rarefaction waves begin to propagate inwards from the ends of the coil towards its centre

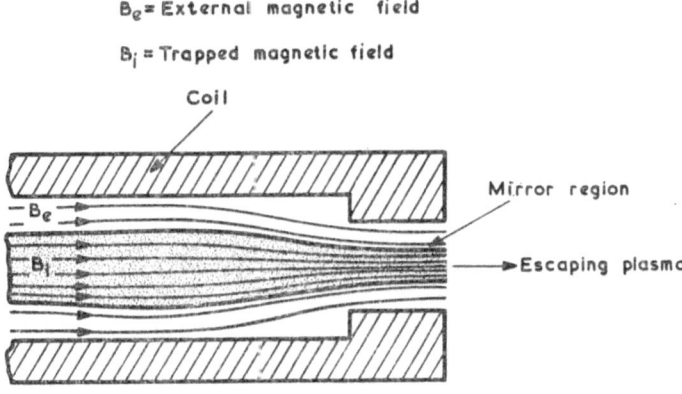

FIG. 5. Schematic diagram of plasma in the theta-pinch escaping through the "magnetic nozzle" at one end of the coil.

plane. The velocity v of these waves depends on β and on the ratio A_0/A of the plasma area to the coil area. The properties of these so-called area waves were first analysed by Taylor.[26] As β approaches zero, so the velocity v of the area waves becomes equal to the sound speed c_s in the plasma; whereas, as β approaches unity, the area wave velocity approaches zero. For A_0/A much less than 1 (which generally occurs in practice), and for the special case of specific heat ratio of 2, it can be shown that $v = c_s(1-\beta)^{\frac{1}{2}}$, so that the area wave velocity and the sound speed bear the same relationship to each other as does the area of the magnetic orifice to the cross-section of the undisturbed plasma.

The area wave velocity for various values of β is shown in Fig. 6 in which time is plotted in units of sound transit-time to the centre of the coil. The centred area waves travelling in at constant velocity from the ends of the coil are reflected at the centre and accelerated with respect to the laboratory frame as they travel back through the disturbed plasma. In Fig. 7 Wesson's results have been used to plot the line density as a function of time at various

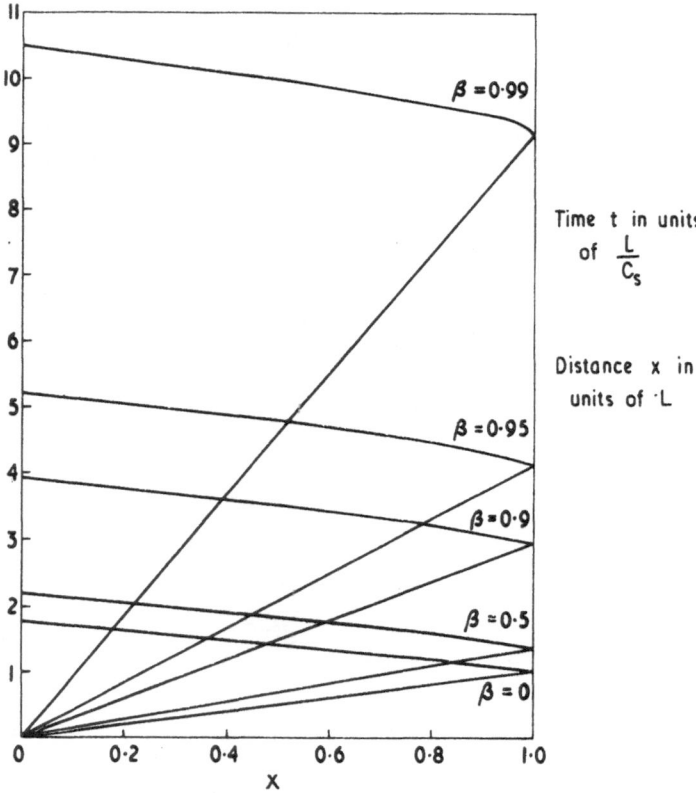

FIG. 6. Area wave diagram for various values of β.

positions along the coil. Time is here plotted in units of area wave speed v and the curves apply to a plasma with β of 0·9.

The predictions of this collisional model have been compared with experiments on a 2-metre theta-pinch at Culham[9] in which the mean plasma β was about 0·5. Interferometer measurements of the decay of plasma electrons were compared with the collisional theory and additional information on the loss mechanism was provided by piezo-electric pressure probes used to measure the dimensions of the axially escaping plasma.[27] Both measurements gave results which were consistent with the Taylor-Wesson model.

8

Consequences for Thermonuclear Systems

As we have seen, the finite length of the theta-pinch results in heavy losses of energy and particles by conduction and convection. Such losses make the straight theta-pinch an unsatisfactory means of long-term confinement on the time-scale demanded by a thermonuclear reactor.

An estimate of the minimum length of a power-producing thermonuclear reactor based on the straight theta-pinch has been given by Wesson[28] on the assumption that the plasma configuration is stable. For this calculation the area of the aperture at the ends of the coil is taken to be the size predicted by Taylor[20], i.e. $2\pi r r_m$ where r is the radius of the plasma column at its maximum value and r_m is the Larmor radius of the ions at the mirror region. The magnetic field strength is given by $B_m B^2 = 10^{16}$ gauss³, where B is the strength of the main field and B_m the mirror field. For a plasma lifetime

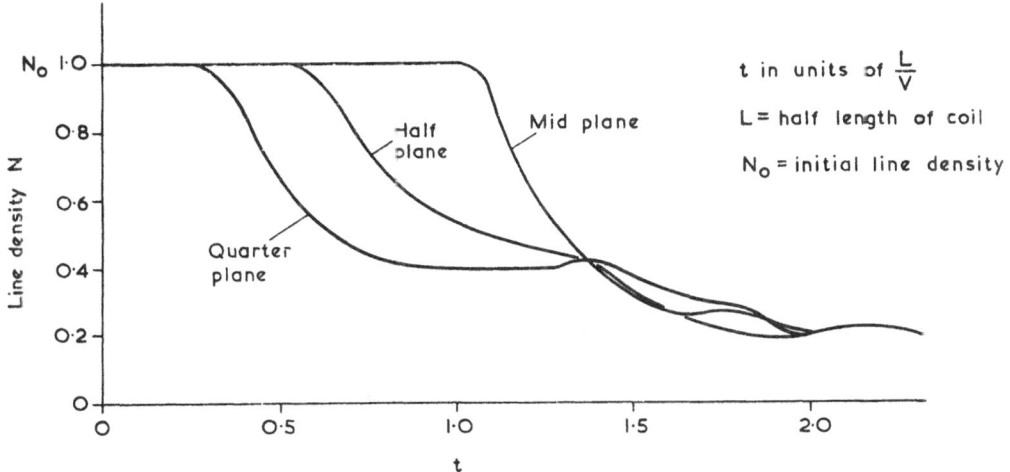

FIG. 7. Theoretical curves of the line density N as a function of time in various planes along the length of the coil.

$n\tau$ of 10^{15} s/cm^3, temperature of 10^8 °K and radius r of 10 cm, the minimum length L of the plasma is 200 metres. As a minimum length this is prohibitively long.

A novel proposal for reducing the flow of plasma from open ended magnetic confinement systems has been made by Tuck[29] who suggests that end losses can be mitigated by the use of magnetically rough walls. If the magnetic field at each end of a theta-pinch coil be spatially modulated by means of subsidiary solenoids, the plasma outflow will be impeded by the magnetic corrugations in a manner analogous to the impedance produced by rough tube walls in the Gaede molecular pump. Additional impedance may be provided by moving the corrugated field continuously towards the plasma. Preliminary calculations suggest that this magnetic roughness technique will reduce end-losses but hardly enough to make the high density open-ended systems attractive as thermonuclear containment vessels.

9

Toroidal Systems

The experimental evidence for heavy end-losses in the theta-pinch, points the need for experiments in which the plasma is confined in a closed magnetic system; for the only certain way of limiting the flow of energy and particles appears to be the interposition of a magnetic field. Thus future developments of the theta-pinch are likely to be in terms of toroidal geometry.

The problems posed by a toroidal system are twofold: how to produce an equilibrium configuration for the plasma; and how to ensure that this equilibrium is hydromagnetically stable. The equilibrium problem consists in overcoming the outward drift of the plasma to the outer wall of the torus; without additional applied fields, the plasma can find no equilibrium position. In a purely toroidal field the plasma drifts outwards with an acceleration $\ddot{r}$ given by $\ddot{r} = 2k(T_e + T_i)/(m_i R)$, where R is the mean radius of the magnetic field lines. The essential reason for this drift is that the field lines at the inside of the torus are shorter than those at the outside, and plasma equilibrium can be produced only by modifying the field so that each field line is the same length. This can be achieved by adding a transverse field or constructing a torus with a "bumpy" or corrugated inner wall. Such a solution, the so-called M & S configuration, was first proposed by Meyer and Schmidt[30, 31] and later in modified form by Pfirsch and Wobig.[32] Experiments on this type of high-β toroidal configuration have been proceeding for several years at the Max Planck Institute in Munich with encouraging results, in the sense that the existence of a plasma equilibrium position away from the walls has been demonstrated in a convincing manner.

A major experiment on toroidal confinement in theta-pinch geometry, the SCYLLAC experiment, has recently been proposed by the Los Alamos laboratory in the U.S.A. and preliminary studies are now far advanced.[33] The equilibrium configuration suggested is in essence similar to that of the Meyer-Schmidt experiment. In SCYLLAC two spatially-varying magnetic fields, one of which is a periodic mirror field and the other a periodic wavy field, are superimposed on the toroidal B_z field. Such a configuration undoubtedly provides a plasma equilibrium, but its stability remains to be tested and provision for dynamic stabilization is to be incorporated in the experiment by means of additional windings fed by high-Q capacitors.

The studies of straight theta-pinches at Culham is likewise being extended to toroidal geometry using a configuration which employs a carefully programmed combination of B_z and B_θ confining fields. In this way high magnetic shear is produced by currents flowing in the plasma. By programming the fields so that the axial electric and magnetic fields change sign in the outer regions of the plasma, the configuration can be made stable in terms of ideal magnetohydrodynamic theory, and the highly sheared fields provide theoretical stability against the growth of resistive instabilities and microinstabilities. The MHD stability of this type of configuration has been demonstrated by Ohkawa et al.[34] A larger-scale experiment forms a natural successor to the Culham 8-metre thetatron.

Undoubtedly, these toroidal experiments present formidable technical problems, since a large number of independent low inductance, high voltage capacitor banks capable of synchronous discharge are required to develop the programmed electric and magnetic fields. Not least of the problems is that of measuring the plasma properties in the awkward geometry that a closed field configuration presents. However, as we have seen, the development of toroidal systems is the only logical method of extending the time-scale of high-β confinement. Those who possess confidence in the high-β approach will be keenly awaiting the results of these experiments.

REFERENCES

1. *Proc. APS Topical Conf. on Pulsed High-Density Plasma, Los Alamos.* 1967. Published as *Los Alamos Scientific Laboratory Report LA-3770*; referred to below as *Report LA-3770.*
2. E. M. LITTLE et al. 1965. *Physics Fluids*, **8**, 1168.
3. L. M. GOLDMAN et al. 1965. *Physics Fluids*, **8**, 522.
4. A. ANDELFINGER et al. 1966. *Physics Lett.*, **20**, 491.
5. A. C. KOLB et al. 1966. Paper CN-21/98, *Proc. Culham Conf. on Plasma Physics and Controlled Nuclear Fusion* 1965. Published by I.A.E.A., Vienna, and referred to below as *Report CN-21*.
6. J. H. ADLAM and J. N. BURCHAM. 1967. *Physics Fluids*, **10**, 2458.
7. M. V. BABYKIN et al. 1963. *Soviet Phys. JETP*, **16**, 295.
8. M. V. BABYKIN et al. 1964. *Soviet Phys. JETP*, **19**, 349.
9. H. A. B. BODIN et al. 1968. (Culham Laboratory), paper presented at I.A.E.A. Conference on Plasma Physics and Controlled Nuclear Fusion Research, Novosibirsk.
10. R. Z. SAGDEEV. 1967. *Proc. Symp. on Applied Mathematics, XVIII, American Mathematical Society*, 281.

11. T. S. GREEN et al. 1967. *Physics Fluids*, **10**, 1663.
12. N V. FILIPPOV et al. 1962. *Nucl. Fusion Supplement*, Part 2, 577.
13. J. W. MATHER. 1965. *Physics Fluids*, **8**, 366.
14. E. H. BECKNER. 1966. *J. Appl. Phys.*, **37**, 4944.
15. J. W. LONG et al. 1967. *Report LA-3770*, Paper C5.
16. C. PATOU et al. 1967. *Report LA-3770*, Paper C2.
17. N. J. PEACOCK et al. 1968. (Culham Laboratory), paper presented at I.A.E.A. Conference on Plasma Physics and Controlled Nuclear Fusion Research, Novosibirsk.
18. J. W. MATHER. 1966. *Report CN-21*, Paper 80.
19. H. A. B. BODIN and A. A. NEWTON. 1967. *Report LA-3770*, Paper A-2.
20. J. B. TAYLOR. 1966. *Culham Report No. CLM-R58*.
21. T. S. GREEN. 1962. *Nucl. Fusion*, **2**, 92.
22. T. S. GREEN. 1963. *Physics Fluids*, **6**, 864.
23. A. HEISS et al. 1967. *Report LA-3770*, Paper D-7.
24. J. B. TAYLOR and J. A. WESSON. 1965. *Nucl. Fusion*, **5**, 159.
25. J. A. WESSON. 1966. *Report CN-21*, Paper 43.
26. J. B. TAYLOR. 1962. Appendix to paper by H. A. B. BODIN et al. *Nucl. Fusion Supplement II*, 511.
27. H. A. B. BODIN and C. A. BUNTING. 1967. *CLM-R70*. Available from HMSO.
28. J. WESSON. 1967. *Report LA-3770*, Paper B1.
29. J. L. TUCK. 1968. *Phys. Rev. Lett.*, **20**, 715.
30. F. MEYER and H. V. SCHMIDT. 1958. *Z. Naturf.*, **A13**, 1005.
31. W. LOTZ et al. 1964. *Nucl. Fusion*, **4**, 335.
32. D. PFIRSCH and H. WOBIG. 1966. *Report CN-21*, Paper 55.
33. E. KEMP et al. 1967. *Report LA-3770*. Paper 9-1.
34. T. OHKAWA, H. K. FORSEN, A. A. SCHUPP, JR., and D. W. KERST. 1963. *Physics Fluids*, **6**, 846.

11

LIGHT SCATTERING EXPERIMENTS

S. A. Ramsden

Department of Applied Physics, University of Hull

1

Introduction

With the advent of the laser it became possible to carry out experiments on the scattering of light from a plasma. Such experiments are of interest in relation to theories of the interaction of electromagnetic radiation with a plasma, developed initially in connection with work on the scattering of radar beams from the ionosphere, and also as a means of deriving fundamental information about laboratory plasmas. Scattering experiments provide information on one of the most basic properties of a plasma, namely the charged particle correlation function. They are likely to prove particularly useful in thermonuclear research where, because of the high temperature and energy density, other, more conventional, diagnostic techniques become difficult to use. Information may be derived about the density and velocity distribution of the ions and electrons, about collision processes, and about drifts and waves in the plasma. A wide range of experiments is possible on the interaction of two or more photons with a plasma and on non-linear effects. The range of plasma conditions which may be studied is extremely large, encompassing most plasmas normally produced in the laboratory, even though some of these may be more conveniently studied by other means. Scattering experiments have the considerable advantage that measurements may be made with true spatial resolution—that is, over a small element of volume at the intersection of the incident and scattered beams—without perturbing the plasma. During the last few years many of the basic features of the scattering process have been verified, and the technique developed to the point where it can shortly be expected to become a routine, although somewhat difficult, diagnostic tool.

In what follows, an account will first of all be presented of the scattering of light from a free electron gas. This not only corresponds to what is actually observed when the scale length of the scattering is less than the Debye length in the plasma, but also gives a useful insight into the physics of the problem.

It is followed by a brief account of the more generalized theory of the scattering of electromagnetic radiation from a thermal plasma, and of the effects of a magnetic field, collisions and drifts. An account is then given of the various factors determining the feasibility of a light scattering experiment, followed by a review of the experimental work carried out to date and a discussion of the techniques used.

2

SCATTERING FROM A FREE ELECTRON GAS

2.1. *Thomson scattering*

Under the action of the electric vector in the light wave the electrons are set into oscillation and form a set of radiating dipoles. If they are randomly distributed in space, the radiation scattered in directions other than that of the incident beam is incoherent. The scattered power is given by a summation of the powers scattered by the individual electrons, and is thus directly proportional to the electron density.

For a plane polarized wave incident on a free electron the differential scattering cross-section (the fraction scattered into solid angle $d\Omega$ at angle ψ to the electric vector) is

$$\frac{d\sigma}{d\Omega} = \frac{e^4}{m^2 c^4} \sin^2 \psi = r_0^2 \sin^2 \psi \qquad \qquad ...[2.1]$$

where $r_0 = e^2/(mc^2)$ is the classical radius of the electron.

The scattered flux has a dipole distribution, varying as $\sin \psi$, which is rotationally symmetric with respect to the electric vector. It is plane polarized in the same sense as the incident beam.

The total scattering cross-section, obtained by integrating over the entire solid angle, is given by

$$\sigma_T = \tfrac{8}{3}\pi r_0^2 = 6{\cdot}65 \times 10^{-25} \text{ cm}^2 \qquad \qquad ...[2.2]$$

which is the so-called Thomson cross-section. In terms of the Thomson cross-section the differential cross-section for the scattering of a plane polarized beam of light is thus

$$\frac{d\sigma}{d\Omega} = \frac{3}{8\pi} \sigma_T \sin^2 \psi. \qquad \qquad ...[2.3]$$

If the incident beam is unpolarized, the differential cross-section is

$$\frac{d\sigma}{d\Omega} = \frac{r_0^2}{2} (1+\cos^2 \theta) \qquad \qquad ...[2.4]$$

where θ is the angle between the incident and scattered directions. The scattered radiation is unpolarized in the forward and backward directions, linearly polarized in directions perpendicular to the incident beam ($\theta = 90°$), and partially polarized in all other directions.

2.2. *Effect of the motion of the electrons*

Equation [2.1] is wavelength independent, and thus, if the electron is at rest, the wavelength of the scattered radiation will be the same as that of the incident beam. If, on the other hand, the electron is in motion it will see a slightly different wavelength for the incident radiation, and the observer, or detector, will see a different wavelength for the scattered radiation, i.e. there will, in general, be a *double* Doppler shift of the wavelength of the incident beam.

The combined wavelength shift is given by the component of the electron velocity, v, perpendicular to the bisector of the angle θ between the incident and scattered directions, i.e.

$$d\lambda = \frac{2\lambda_i}{c} v \sin \frac{\theta}{2}, \qquad \ldots [2.5]$$

where λ_i is the wavelength of the incident radiation. This is important, in that it effectively introduces a spatial dependence of the results obtained upon the direction of observation.

For an isotropic Maxwellian distribution of velocities, the scattered radiation has a Gaussian intensity distribution of half-width

$$\Delta\lambda_{\frac{1}{2}} = 4 \sin \frac{\theta}{2} \left[\frac{2KT_e}{m_e c^2} \log_e 2 \right]^{\frac{1}{2}} \lambda_i \qquad \ldots [2.6]$$

where T_e is the electron temperature in the gas, m_e is the mass of the electron and K is Boltzmann's constant. From the half-width of the scattered spectrum we can thus determine a value for the electron temperature. The half-width is zero in the forward direction ($\theta = 0°$) and increases to a maximum of twice the radiative Doppler width at $\theta = \pi$ (backscatter).

3

SCATTERING FROM A PLASMA

3.1. *Phenomenological description*

In a plasma consisting of both ions and electrons the main scattering agents are still the electrons. Because of their greater mass, which appears in the denominator of equation [2.1], direct scattering by the ions is negligible by comparison. The above picture is, as we shall see, still valid in the limiting case when particle interactions can be neglected; but, in general, the spectrum of the scattered radiation is greatly modified by collective effects within the plasma.

As is well known, scattering only occurs as the result of deviations from uniformity in a medium and, in a plasma, this is due to variations in electron density. In a plasma there is an irregular, fluctuating, electrostatic field determined by the irregular distribution of charge density. Over distances greater than a Debye length λ_D, this electrostatic field causes fluctuations in electron density which almost exactly cancel out the spontaneous fluctuations due to the thermal motions of the electrons. The same electrostatic field also produces much slower fluctuations in electron density to neutralize the thermal motions of the ions. Thus, if the wavelength of the incident beam (or, more strictly, the scale length of the scattering $k^{-1} = \lambda_i/[4\pi \sin \frac{1}{2}\theta])$ is much greater than a Debye length, the spectrum of the scattered radiation no longer reflects the motion of the electrons, but, rather, reflects the motion of the ions, and consists of a narrow central peak whose width is of the order of the ion thermal velocity. In addition, a small amount of radiation is scattered into a pair of weak satellites, corresponding to scattering from longitudinal electrostatic plasma oscillations, which are separated from the wavelength of the incident beam by approximately the plasma frequency, ω_p.

The frequency spectrum of the scattered radiation is directly related to the frequency spectrum of the electron density fluctuations with a fixed wave vector $k = k_s - k_i$, where, as before, k_i and k_s are vectors in the direction of the incident and scattered beams respectively. That is, in a scattering experiment we effectively sample the electron density fluctuations in the plasma with a fixed wave vector k. The scattering of a plane wave through an angle θ may be regarded as a sampling of the amplitude of those electron density fluctuations of angular frequency ω and wave number $k = 4\pi \sin (\frac{1}{2}\theta)/\lambda_i$, which can give rise to constructive interference in the direction k_s in much the same way as X-rays are reflected from the planes of a crystal. The frequency shift of the scattered radiation may be viewed as the Doppler shift due to the propagation velocity, $v = \omega/k$, of the density fluctuations.

Alternatively, we may regard the particles as sorted out into groups such that the velocities of all particles belonging to the same group are almost identical and the particles within each group are randomly distributed in space. The power scattered by a single group is equal to the product of the scattering coefficient for a single particle and the number of particles in the group. The frequency of the scattered radiation differs from that of the incident beam by the Doppler shift due to the common velocity of the particles.

3.2. *Salpeter theory for scattering from a thermal plasma*

In mathematical terms, the differential scattering cross-section per unit angular frequency interval $d\omega$ per unit solid angle $d\Omega$ is given by

$$\frac{d^2\sigma}{d\omega d\Omega} = \sigma_T S(\omega, k) \qquad \qquad ...[3.1]$$

where σ_T is the Thomson cross-section and $S(\omega, k)$ is the frequency spectrum of electron density fluctuations with wave vector $k = k_s - k_i$, which is directly related to the time-space Fourier transform of the time dependent electron density pair correlation function.

There have been many calculations of the frequency spectrum of electron density fluctuations in both thermal[1] and non-thermal[4,5] plasmas using a variety of approaches. We follow here the treatment given by Salpeter for a plasma in thermal equilibrium.

For a Maxwellian distribution of velocities of both ions and electrons (but not necessarily $T_e = T_i$), neglecting collisions, assuming that the Coulomb interaction energy is small compared with the thermal kinetic energy (i.e. the fluctuating electrostatic field is only a small perturbation of the thermal motions of the ions and electrons), and that the observation time is long compared with the period of the fluctuation, the frequency spectrum is given by

$$S(\omega, k) = \frac{N}{k} \left\{ \left| \frac{1 - G_i}{1 - G_e - G_i} \right|^2 F_e \left[-\frac{\omega}{k} \right] + Z \left| \frac{G_e}{1 - G_e - G_i} \right|^2 F_i \left[-\frac{\omega}{k} \right] \right\},$$

$$\dots[3.2]$$

where:

$$\left. \begin{array}{l} \omega = \omega_i - \omega_s \\ k = |k_i - k_s| \end{array} \right\} \begin{array}{l} \text{are the angular frequency and wave number associated} \\ \text{with the electron density fluctuations in the plasma} \\ (\omega_i, k_i \text{ and } \omega_s, k_s \text{ being the corresponding quantities} \\ \text{for the incident and scattered radiation)} \end{array}$$

$F_e \left[-\dfrac{\omega}{k} \right]$ and $F_i \left[-\dfrac{\omega}{k} \right]$ are the Maxwell-Boltzmann velocity distribution

functions, $F[v] = \left[\left(\dfrac{m}{2\pi KT} \right)^{\frac{1}{2}} \exp \left(\dfrac{-mv^2}{2KT} \right) \right]$, for the electrons and ions respectively.

Z is the degree of ionisation.

$$G_e(\omega) = -\alpha^2 [1 - f(x) + i\pi^{\frac{1}{2}} x \exp(-x^2)]$$
$$G_i(\omega) = -Z(T_e/T_i) \alpha^2 [1 - f(y) + i\pi^{\frac{1}{2}} y \exp(-y^2)]$$
$$x = \frac{\omega}{\omega_e}, \ y = \frac{\omega}{\omega_i}, \ \omega_e = \left(\frac{2k^2 KT_e}{m_e} \right)^{\frac{1}{2}}, \ \omega_i = \left(\frac{2k^2 KT_i}{m_i} \right)^{\frac{1}{2}}$$
$$f(x) = 2x \exp(-x^2) \int_0^x \exp t^2 dt \text{ and } \alpha = 1/(k\lambda_D).$$

Examination of equation [3.2] shows that the first term represents a relatively wide spectral distribution, corresponding to Doppler spread frequencies characteristic of the electron thermal motion whereas the second term represents a much narrower spectral distribution, corresponding to the ion thermal motion. It is convenient to refer to the two terms as the electron and ion components respectively.

For $T_i \leqslant T_e$ a good approximation to equation [3.2] can be given in terms of a family of single parameter functions, Γ, of one variable:

$$S(\omega, k)d\omega = \frac{N}{k}\left[\Gamma_\alpha(x)\frac{d\omega}{\omega_e} + Z\left(\frac{\alpha^2}{1+\alpha^2}\right)^2\Gamma_\beta(y)\frac{d\omega}{\omega_i}\right], \qquad \ldots[3.3]$$

where:

$$\Gamma_\alpha(x) = \exp(-x^2)\{[1+\alpha^2-\alpha^2 f(x)]^2 + \pi\alpha^4 x^2 \exp(-2x^2)\}$$
$$\Gamma_\beta(y) = \exp(-y^2)\{[1+\beta^2-\beta^2 f(y)]^2 + \pi\beta^4 y^2 \exp(-2y^2)\}$$
$$\beta^2 = \frac{T_e}{T_i}\frac{Z\alpha^2}{1+\alpha^2}, \quad x = \frac{\omega}{\omega_e}, \quad y = \frac{\omega}{\omega_i}.$$

The total, integrated, intensities of the electron and ion components are $1/(1+\alpha^2)$ and $Z\alpha^4/[(1+\alpha^2)(1+\alpha^2+Z\alpha^2 T_e/T_i)]$, respectively. The sum varies only slowly with α. The electron component predominates for $\alpha \ll 1$ and the ion component for $\alpha \gg 1$. Both components have the same shape, determined by the function Γ, but cover different spectral regions. The ion component is only significant for small Doppler shifts of the order of ω/ω_i and the electron component for larger shifts of the order ω/ω_e. The shape of the electron component depends upon the parameter $\alpha = 1/(k\lambda_D)$, whereas the shape of the ion component depends upon the parameter $\beta = (T_e/T_i)^{\frac{1}{2}}[Z\alpha^2/(1+\alpha^2)]^{\frac{1}{2}}$. For $\alpha \gg 1$, when the ion component is the predominant feature, $\beta = Z(T_e/T_i)^{\frac{1}{2}}$ and the shape depends only on the ratio of the electron and ion temperatures.

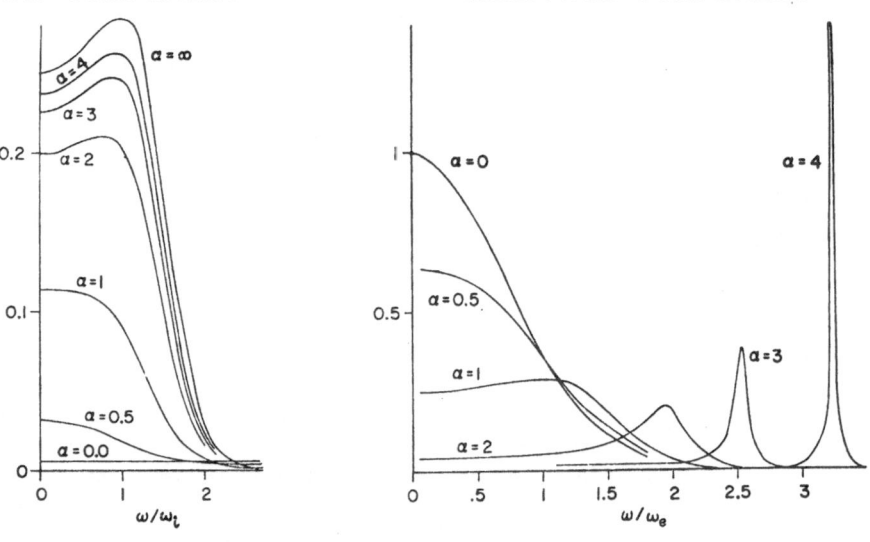

FIG. 1. Shape of the electron and ion components as functions of the parameter $\alpha = \lambda_i/(4\pi\lambda_D \sin \frac{1}{2}\theta)$ for the case $T_e = T_i$.

It is useful to consider further the shape of the scattered spectrum as a function of the parameter α (see Fig. 1).

(i) $\alpha \ll 1$

If $\alpha = 1/(k\lambda_D) \ll 1$, then $\lambda_i/[4\pi \sin (\frac{1}{2}\theta)] \ll \lambda_D$, and thus physically this corresponds to the case when the wavelength λ_i of the incident radiation (or, more strictly, $\lambda_i/[4\pi \sin (\frac{1}{2}\theta)]$) is much less than a Debye length λ_D. For small values of α only the first term in equation [3.3] is significant, and, as $\alpha \to 0$, $\Gamma_\alpha(x) \to \exp(-x^2)$ and thus, as expected, the scattered radiation has a Gaussian spectral distribution characteristic of the random thermal motion of the electrons.

(ii) $\alpha \simeq 1$

As α approaches unity, the electron spectrum becomes flat-topped and, for $\alpha > 1$, double humped. The ion component appears as a narrow feature

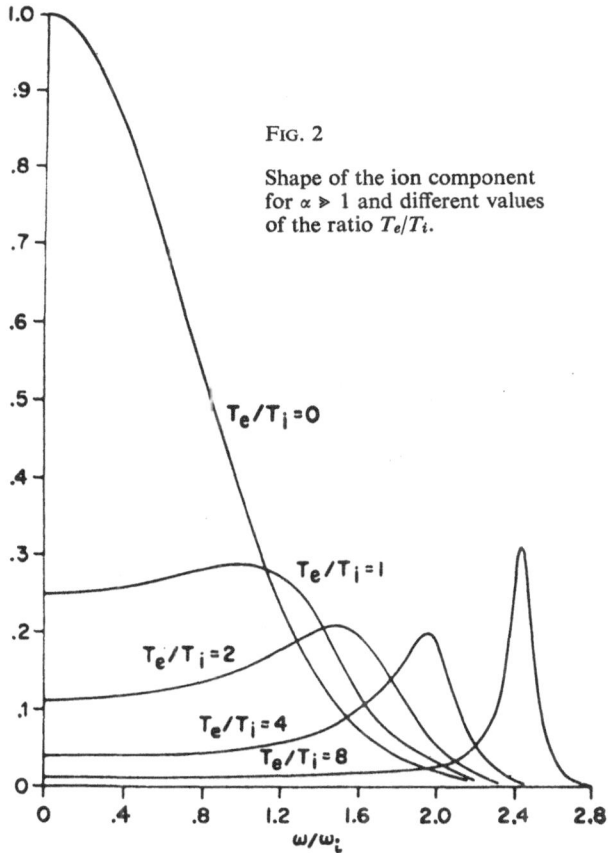

Fig. 2

Shape of the ion component for $\alpha \gg 1$ and different values of the ratio T_e/T_i.

in the centre of the distribution. For values of $\alpha \simeq 1$, it is possible to determine both the ion and electron temperatures from the one spectrum. The shape of the electron component is sensitive to the ratio of n_e to T_e allowing a value of the electron density to be determined also.

(iii) $\alpha \gg 1$

This corresponds to the case discussed earlier, when the wavelength of the incident radiation is much greater than the Debye length, and the spectrum is determined by collective interactions within the plasma.

For $\alpha \gg 1$, $\Gamma_\alpha(x)$ is small except near $x = \pm x_0$, where $x_0^2 = \frac{1}{2}(\alpha^2 + 3)$, i.e. $\omega_0^2 = \omega_p^2 + 3k^2 KT_e/m_e$. This expression for ω_0 is the well-known dispersion relation for longitudinal electrostatic oscillations. The electron component thus corresponds to a pair of weak (integrated intensity $\simeq 1/\alpha^2$) satellites at $\omega \simeq \pm\omega_p$, corresponding to scattering from electron plasma oscillations. In the absence of collisions, the damping of these oscillations is determined only by Landau damping and the width is very small.

Most of the scattered intensity resides in the ion component and thus occurs at small frequency shifts. As explained earlier, the shape of this central ion peak is determined by the same function Γ as the electron component with the parameter $\beta = Z(T_e/T_i)^{\frac{1}{2}}$ replacing the parameter α. Thus (Fig. 2) if $T_i \gg T_e$, ($\beta \ll 1$), the shape is a Gaussian; for $T_i \simeq T_e$, ($\beta \simeq 1$), the spectrum is flat-topped; and for $T_i \ll T_e$, ($\beta \gg 1$), the spectrum consists of a pair of lines corresponding to scattering by ion acoustic waves in the plasma, whose damping is small if $T_e \gg T_i$. From the ion component we may determine a value for the ion temperature in the plasma and, in favourable cases, as the shape depends upon the ratio T_e/T_i, a value for the electron temperature as well.

3.3. *Effect of a magnetic field*[2]

Magnetic fields influence the shape of the spectrum, and so give the possibility of measuring local magnetic fields in a plasma, but do not change the total scattering intensity. The spectrum function can usually be divided into two parts enabling separate discussion of the electron and ion components as before.

For $\alpha \ll 1$ and k perpendicular to the magnetic field B the spectrum consists of lines located at multiples of the electron gyrofrequency ω_{eH}. The envelope is approximately that of the Gaussian profile that would exist in the absence of a magnetic field. This fine structure is, however, rapidly smeared out if the angle between k and B is decreased, and thus the effect may be difficult to observe.

If k is parallel to B, a magnetic field has in general no effect on the spectrum. This statement is also approximately valid if

$$\cos \Phi \gg \frac{\omega_{eH}}{\omega_e} \quad \text{and} \quad \cos \Phi \gg \frac{\omega_{iH}}{\omega_i} \qquad \qquad \dots[3.4]$$

where Φ is the angle between k and B, and ω_e and ω_i are the frequency shifts, $(2k^2KT/m)^{\frac{1}{2}}$, characteristic of the electron and ion thermal velocities respectively.

For very strong magnetic fields, i.e.

$$\frac{\omega_{eH}}{\omega_e} \gg 1 \quad \text{and} \quad \frac{\omega_{iH}}{\omega_i} \gg 1 \qquad ..[3.5]$$

the spectrum is simply contracted by a factor $\cos \Phi$ by comparison with the spectrum without a magnetic field.

The case of weak magnetic fields is more complicated but, disregarding a fine structure, one can say that for weak magnetic fields and the additional criteria,

$$\cos \Phi \ll \frac{\omega_{eH}}{\omega_e} \ll 1; \quad \cos \Phi \ll \frac{\omega_{iH}}{\omega_i} \ll 1, \qquad ...[3.6]$$

the electron component shows maxima separated from each other by the electron gyrofrequency, and the ion component shows maxima separated by the ion gyrofrequency. These conditions can be satisfied for one component only; the other component then shows no maxima. For $\Phi = 90°$ these maxima become discrete lines and we have a line spectrum. The effect is, in general, appreciable only if the scattering vector and the magnetic field are nearly perpendicular. Collision effects can, however, be expected to smear out much of this structure.

3.4. Collisions[3]

Quite generally, collision effects become important if the collision frequency is of the order of the fine structure of the spectrum. They do not, however, affect the total intensity of the scattered radiation. In most cases of practical interest the collision frequency exceeds the electron gyrofrequency, and particularly the ion gyrofrequency, and it is probable that gyro-resonances can be observed only in low density plasmas, and even then only the electron resonances.

In the absence of a magnetic field, collisions may be expected to influence most greatly the electron and ion-acoustic plasma resonances. The width and shape of these are both related to the damping rate of the oscillations and hence the collision frequency. Thus, the ion acoustic resonances are considerably sharpened, since the damping of ion acoustic waves decreases with increasing collision frequency. On the other hand, the electron plasma resonance is broadened by collisions, as the damping of electron plasma oscillations increases with increasing collision frequency. In most cases of practical interest, however, the collision frequency is smaller than the bandwidth of the spectra, and the effect should be small.

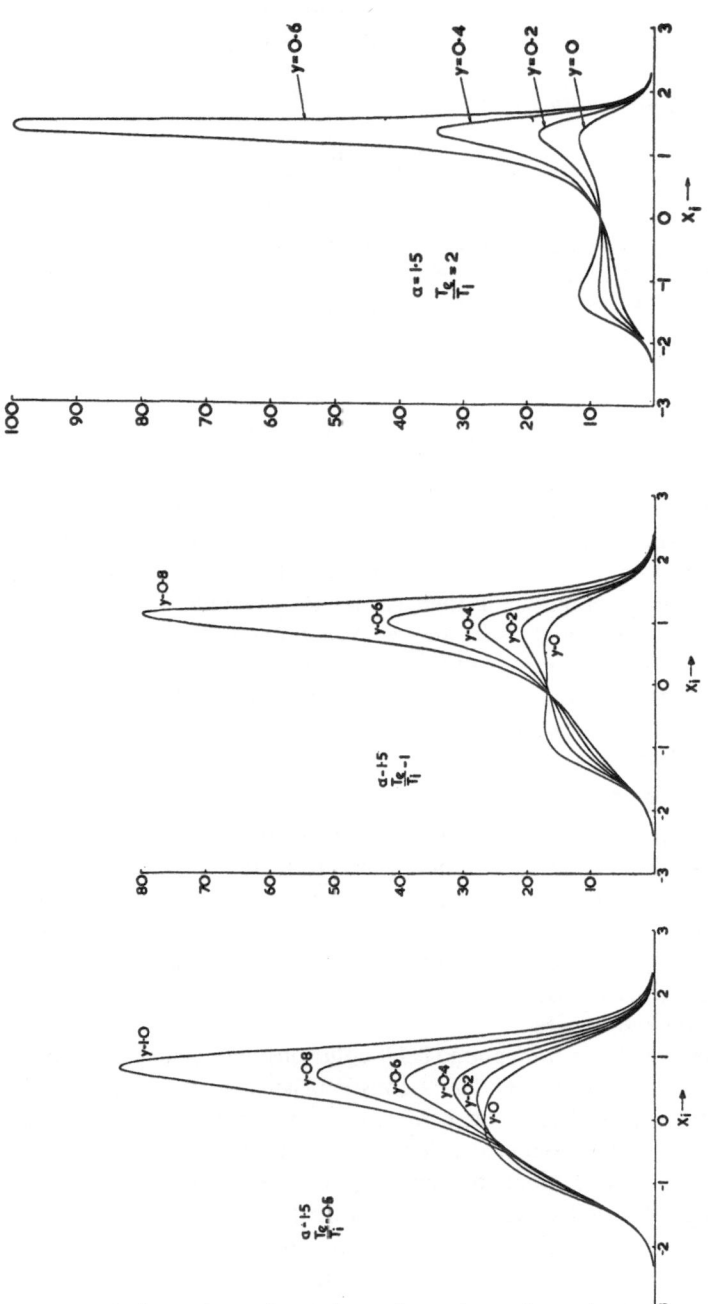

Fig. 3. Shape of the ion component for $\alpha = 1.5$ and different values of the ratio T_e/T_i, and of the ratio γ of drift velocity to electron thermal speed.

3.5. *Drifts*

Scattering from non-thermal plasmas has been considered by Rosenbluth and Rostoker[4] and by Lamb[5] using the dressed test particle approach. In particular, numerical calculations are given for plasmas with different electron and ion temperatures and for plasmas in which the electrons have a net drift velocity relative to the ions. As may perhaps be expected, the presence of drifts within the plasma results in an asymmetric ion spectrum (Fig. 3), the asymmetry increasing as the ratio of the electron drift velocity to the electron thermal velocity increases.

4

EXPERIMENTAL CONSIDERATIONS

4.1. *Light source*

Because the scattering cross-section is so very small a high incident photon flux is required if we are to obtain a scattered signal which is detectable and has good statistics. We can also be assured that stray light will be a problem.

The light source should also be preferably of high power and short duration, so that the scattered signal is detectable above the radiation from the plasma itself and the measurements may be made over a short period of time. Most plasmas are highly luminous, of short duration, and often unstable.

If we are to study the spectral distribution of the scattered radiation, we need a light source that is highly monochromatic. It is also desirable that the light be well collimated so that we can concentrate all the available energy within a small volume to obtain good spatial resolution. Further, the wavelength should be in a region where it is possible to obtain detectors of high gain and good quantum efficiency.

The Q-spoiled ruby laser satisfies all these requirements admirably and, as a result, has been the light source used for most scattering experiments to date.

4.2. *Scattered flux*

If F_i is the total flux in the incident beam and F_t the total flux scattered over length l of the beam into solid angle Ω in a direction perpendicular to the electric vector, then $F_t = 10^{-25} l\Omega F_i n_e$ cm^2/sr. Note that F_i is independent of the area of the beam, which can thus be made as small as is desirable to obtain good spatial resolution.

If we are to make measurements of the spectral distribution of the scattered radiation, we can, however, accept only a fraction of the total scattered flux. For simplicity, we restrict attention to the free electron case and take the flux $F_s \simeq F_t/10$, scattered over a bandwidth $\Delta/5$ where Δ is the half-width of the

Gaussian distribution. Using a reasonable value of $\Omega = 0.01$ sr, we obtain in this case that the ratio of scattered to incident flux is $F_s/F_i \simeq 10^{-28} l n_e$ cm^{-2}, where n_e is the electron number density in the plasma. For $n_e \simeq 10^{16}$ cm^{-3}, $l \simeq 1$ cm, $F_s/F_i \simeq 10^{-12}$ and, for a total energy in the beam of 1 Joule (equivalent to 3.5×10^{18} photons at 6943 Å) we obtain a total of about 10^6 scattered photons. We are thus assured of good statistics, even taking into account the relatively low quantum efficiency $\eta \simeq 5\%$ of the best detectors in the region (S–20 photomultiplier), but there are stringent requirements on the amount of stray light that can be tolerated. With a 100 J laser it might even be possible to perform scattering experiments at densities as low as 10^{11} cm^{-3}. A signal/noise ratio of 10:1 should still be possible, but the stray light ratio of 10^{-17} required will make measurements difficult near the laser wavelength. Measurements might still be carried out, however, in the wings of the distribution using a double monochrometer or a number of interference filters to discriminate against stray light.

4.3. *Plasma radiation*

As stated above the scattered radiation must be detected in the presence of intense radiation from the plasma itself. Fortunately, most laboratory plasmas show no intense line radiation at 6934 Å and this can usually be neglected.

For a fully ionized hydrogen plasma the continuous emission, which consists of free-free and free-bound radiation, decreases only slowly with increasing electron temperature T_e, whereas the half width of the scattered radiation $\Delta \propto T_e^{\frac{1}{2}}$. As a result, the plasma radiation collected over a bandwidth $\Delta/5$ varies only slowly with T_e and, for simplicity, is taken as approximately constant. Further, if we are to obtain spatial resolution, the volume of the plasma seen by the detection system will be greater than the volume illuminated by the beam, although this can be reduced to a minimum by the use of a suitable aperture imaging the scattering volume into the detection system. For the case of a laser beam of radius r passing down the axis of a cylindrical plasma of radius R, the ratio of continuous to scattered flux accepted by the detection system is, under the above assumption, independent of the electron temperature and given by $F_c/F_s \simeq 10^{-12}(n_e rR/\text{cm}^{-1})/(P/\text{kW})$, where P is the beam power. For an electron density of $n_e = 10^{16}$ cm^{-3}, a plasma radius $R = 1$ cm and a laser beam focused down to a radius $r = 0.1$ cm, we see that if the scattered flux seen by the detector is to exceed the continuous emission from the plasma itself, then we need a beam power $P > 1$ MW, which is readily obtained using a Q-spoiled ruby laser. The upper limit of electron density for which a scattering measurement can be made will be determined by the power and collimation of the beam. Using a 1000 MW ruby laser system is should be possible to carry out measurements on plasmas with an electron density as high as 10^{19} cm^{-3}, and, in some cases, perhaps higher than this, by the use of electrical filters or differential detection systems to discriminate against the radiation from the plasma itself. In this respect it should be noted that when $F_c > F_s$, since $F_c \propto n_e^2$ and $F_s \propto n_e$, the signal to noise ratio, $F_s/F_c{}^{\frac{1}{2}}$, is independent of n_e.

4.4. *Time resolution*

The use of the Q-spoiled ruby laser giving a pulse of time duration of 100 ns or less (5 ns is readily obtained, and pulse widths as short as 10^{-11} s can be obtained by cavity dumping techniques, albeit with much smaller energy) has the fortunate by-product that we can automatically obtain good time-resolution for measurements on transient or unstable plasmas. Measurements may, however, be made at only a single time during the discharge. If it is required to follow the time development of the plasma, this may be possible using longer pulse lasers in which steps are taken to reduce the intensity modulation to a minimum.

4.5. *Choice of scattering angle*

As we have seen, the spectral distribution of the scattered radiation and the relative intensities of the electron and ion components depend upon the parameter $\alpha = 1/(k\lambda_D) \propto (n_e/T_e)^{\frac{1}{2}}[1/\sin(\frac{1}{2}\theta)]$. For $\theta = 90°$, $\alpha < 1$ for all but low temperature, high density, plasmas. This means that by using a scattering angle of 90°—which is the most convenient to use in practice and most favourable for reducing stray light—we can readily observe the Gaussian feature due to scattering from free electrons and so determine the electron temperature in the plasma. Collective effects may also be observed at large scattering angles for high density, low temperature, plasmas. If, however, we wish to measure the ion temperature in a high temperature plasma, we must, in order to make $\alpha > 1$, usually make measurements at small forward scattering angles.

4.6. *Temperature range and wavelength resolution*

Apart from the restrictions imposed by α and the choice of scattering angle θ, the range of electron and ion temperatures over which measurements may be made covers most of the values likely to be encountered in laboratory plasmas. Ruby lasers are now available with line-widths less than 0·01 Å which is sufficient to measure the structure of the ion peak and, in favourable conditions, the profile of the electron plasma satellites when used in conjunction with high resolution, and high luminosity (e.g. Fabry-Perot) detection systems. It is possible that the fine structure due to magnetic field effects may also be observable, although the limitations imposed by the homogeneity of the magnetic field and of the plasma itself are likely to be severe.

Measurements of the electron feature appear to present no apparent difficulty, being most conveniently carried out using either a grating monochromator or interference filters.

4.7. *Rayleigh scattering*

So far no mention has been made of the possibility of scattering from atoms, ions or molecules in the case of a partially ionized plasma. The scattering cross-section for such particles is usually considerably less than the Thomson cross-section and thus unless the plasma is only very weakly ionized, the effect can usually be neglected. Even in extreme cases, however, useful information may still be derived from a scattering experiment if precautions are taken to discriminate against such scattered light, which occurs only at the wavelength of the incident beam.

4.8. *Perturbation of the plasma*

Finally, mention should be made of the possibility of perturbation of the plasma by the laser beam. Focused down to a small area, lasers can create a very high local electric field and it is not unreasonable to ask whether this will perturb the plasma. Although no definitive answer can be given at the present time, perturbation effects are, however, unlikely to be serious in the case of a fully ionized plasma. The velocity imparted to the electron is usually sufficiently small that corrections to the Thomson scattering cross section can be neglected. Further, although energy is imparted to the electrons, the frequency of the incident radiation is so much greater than the collision frequency that this is not coupled into the plasma. Perturbation effects may, however, occur in a partially ionized plasma due to photo-ionization of neutral atoms or ions which can cause a temporary, local, increase in electron density.

5

EXPERIMENTAL RESULTS

The first laboratory demonstration of the scattering of light by electrons, although not in a plasma, was carried out by Fiocco and Thompson[6] who, in 1963, observed the scattering of a normal mode ruby laser from an electron beam. They neatly side-stepped the problem of stray light by observing the scattering at 65° to the electron beam, thus Doppler shifting the scattered radiation by approximately 260 Å, and discriminating against stray light at the laser wavelength by means of a number of narrow band interference filters. The electron density in the beam was only $\sim 5 \times 10^9$ cm^{-3} and only $\sim 10^{-18}$ of the laser beam was scattered in the direction of observation. However, using a cooled S–20 photomultiplier they were nevertheless able to detect an average of 3 photoelectrons per laser pulse, which, although small, was sufficient to demonstrate the effect.

Observation of scattering of a ruby laser from a plasma was first reported by Schwarz[7] using a He hollow cathode discharge but the results obtained were not convincing and, although measurements were later[8] presented of the scattering signal as a function of time, in the free electron scattering regime down to electron densities of the order of 10^{13} cm^{-3}, the first unambiguous scattering signal was probably observed by Fünfer et al.[9] using a normal mode ruby laser and a θ-pinch plasma.

The first measurements of the spectral distribution of the scattered radiation in the free electron scattering regime ($\alpha < 1$) were made by Davies and Ramsden.[10] Using a double monochromator the scattered radiation was observed at right angles to the beam from a 10 MW Q-spoiled ruby laser focused in the centre of a small θ-pinch hydrogen plasma. Measurements were made in the afterglow to obtain a thermal plasma. The spectral distribution was, as predicted by theory, approximately a Gaussian corresponding to an electron temperature of 3·3 eV. The electron density in this experiment was 5×10^{15} cm^{-3}. Similar measurements, together with results showing the shoulders (see Fig. 4) predicted by theory for $\alpha \simeq 1$ were later published by Kunze et al.[11]

The electron Gaussian feature has also been observed by De Silva et al.[12] using a pulsed H$_2$ arc and a scattering angle of $\theta = 170°$. The electron density and temperature in the plasma were 10^{15} cm^{-3} and $\simeq 2$ eV respectively. Again a Q-spoiled ruby laser was used as the light source. Measurements were also made at a forward scattering angle of $\theta = 10°$, corresponding to $\alpha > 1$, which showed the existence of a narrow ion peak. In both cases an axicon

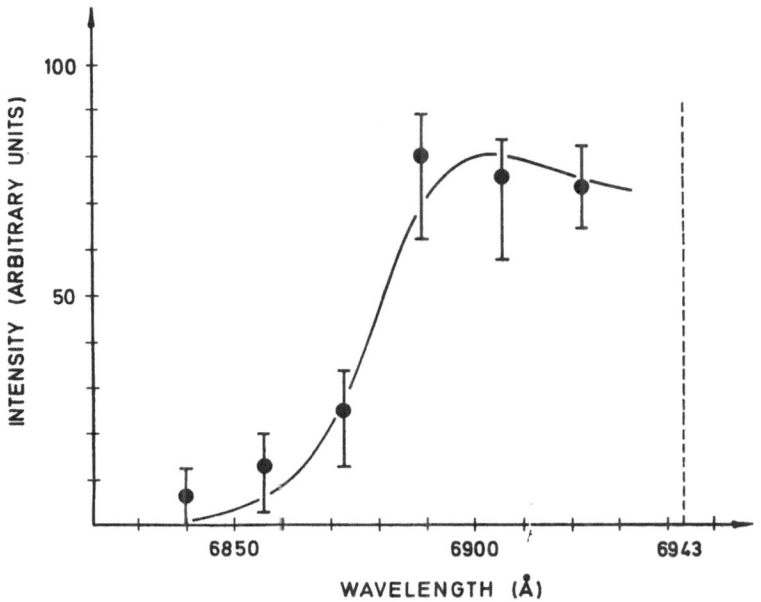

FIG. 4. Comparison of the measured spectrum[11,15] with theory for $\alpha = 0.97$, $T_e = 55,000°$K, $n_e = 4 \times 10^{16}$ cm^{-3}, $\theta = 90°$.

(conical lens) was used to collect all the light scattered at a given polar angle —a technique which is now standard for small-angle scattering. Rayleigh scattering from nitrogen at a pressure of $\frac{1}{4}$ atmosphere was used to check the absolute value of the scattered intensity.

An indication of the presence of an ion peak and satellites shifted by the plasma frequency was obtained by Ascoli-Bartoli et al.[13] in a forward-scattering experiment on a θ-pinch plasma. Some asymmetry was found which was attributed to drifts.

Electron density and temperature measurements in a θ-pinch plasma have been reported by Kunze et al.[14] and by Kunze.[15] Gerry and Rose[16] have made measurements of electron density and temperature on an argon arc plasma over the range 10^{13}–10^{14} cm^{-3} and 3–8 eV using a scattering angle of 45°, and have compared the results with Langmuir probe measurements. Patrick[17] has used 90° scattering to determine the density and temperature in the plasma produced by a magnetic annular shock tube. Excellent measurements (Fig. 5) have been reported by Anderson[18] for 90° scattering from a deuterium plasma in a sheet pinch; they show clearly not only the flat portion of the spectrum for $\alpha \simeq 1$ but also the central ion peak.

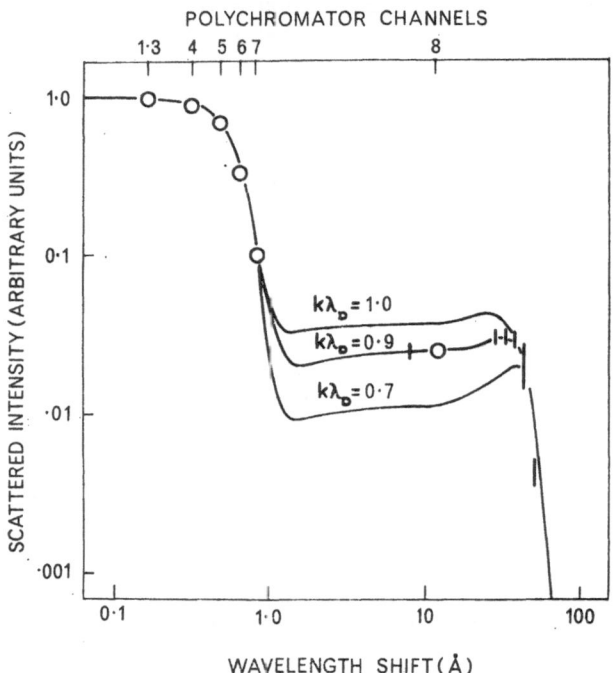

FIG. 5. Scattered spectrum[18] from a deuterium plasma ($\theta = 90°$, $n_e = 2\cdot2 \times 10^{16}$ cm^{-3}, $T_e = 2\cdot0$ eV) showing both the electron and ion components for the case $\alpha = 1$.

Observation of the electron plasma resonances has been reported by Chan and Nodwell[19] at a scattering angle of 45° on an argon plasma jet of density 10^{16}–10^{17} cm^{-3} and temperature 1–2 eV. The line width was greater than predicted on the collisionless theory, most probably due to variations of electron density within the scattering volume.

Observations of both the central ion feature and the satellite peaks have been made by Ramsden and Davies[20] at a scattering angle of 13° in a hydrogen θ-pinch plasma and corresponding to a value of $\alpha = 3 \cdot 0$. The Doppler broadened line for 90° scattering was also observed, and both sets of measurements (see Fig. 6) shown to be consistent with an electron density of $2 \cdot 4 \times 10^{15}$ cm^{-3} and an electron temperature of 1 eV.

Kronast et al.[21] have made measurements of the central ion peak at a

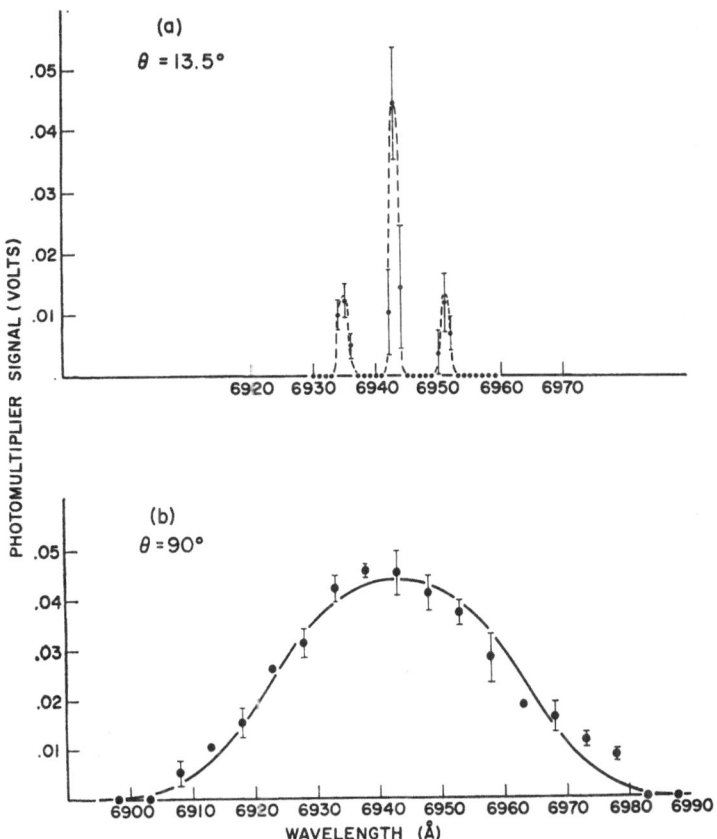

FIG. 6. Spectral distribution[20] of the radiation scattered at (a) $\theta = 13 \cdot 5°$ and (b) $\theta = 90°$ to the forward direction from a hydrogen plasma ($n_e = 2 \cdot 4 \times 10^{15}$ cm^{-3}, $T_e = 1 \cdot 0$ eV) corresponding to values of $\alpha = 3 \cdot 0$ (cooperative scattering) and $\alpha = 0 \cdot 5$ (free electron scattering) respectively.

forward scattering angle of 3° in a high temperature deuterium θ-pinch plasma. Spectrally resolved measurements were made, which indicated an ion temperature of 108 eV. When the stored energy in the capacitor bank was increased, they observed an asymmetric double humped distribution which was attributed either to unequal electron and ion temperatures or to an excited ion plasma wave. Somewhat similar measurements have also been reported by Ramsden et al.[22] This group have recently made further measurements[23] which give an ion temperature (Fig. 7) in good agreement with that determined from the neutron emission, thus providing evidence for a thermonuclear origin for the neutron emission observed.

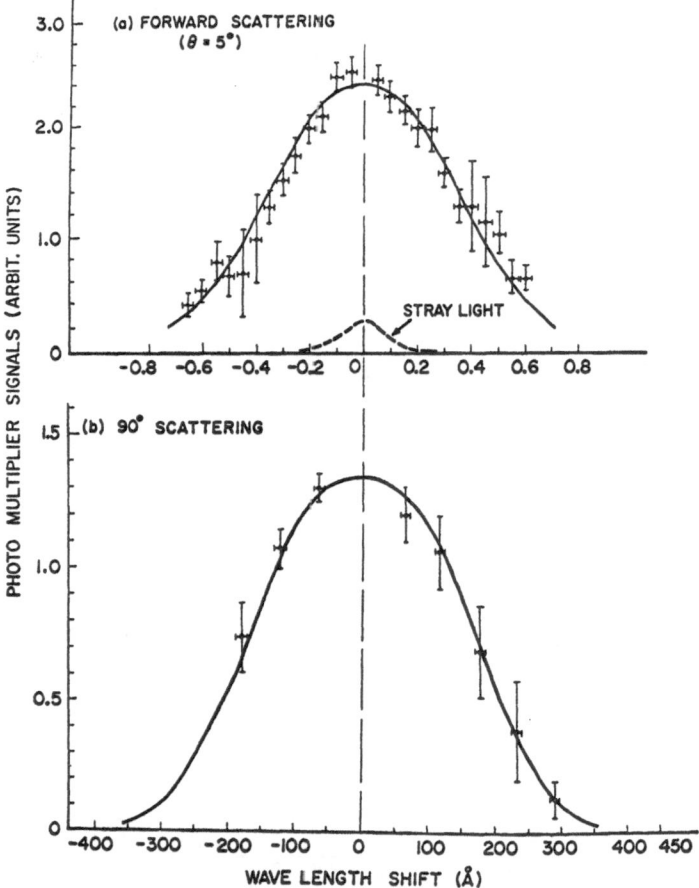

FIG. 7. Spectra[23] of light scattered at (a) $\theta = 5°$ and (b) $\theta = 90°$ to the forward direction from a high energy deuterium θ-pinch. The curves shown represent best fit analysis of the data and yield values of $T_e = 80^{+23}_{-16}$ eV, $n_e = (1·4 \pm 0·35) \times 10^{17}$ cm^{-3}, $\alpha = 0·44^{+0·1}_{-0·15}$ for $\theta = 90°$, and $T_i = 300 \pm 50$ eV for $\theta = 5°$.

Bottoms and Eisner[24] have used 90° scattering to determine the electron temperature in a plasma produced by a co-axial plasma accelerator and stagnated in a semi-cusp and have thereby been able to check the Spitzer formula for the electron-ion thermalization relaxation time.

Daehler and Ribe[25] have measured the shape of the ion peak observed at a scattering angle of $\sim 6°$ in the SCYLLA III θ-pinch device. They interpret the profile obtained in terms of scattering from suprathermal density fluctuations superimposed upon a broader thermal ion feature. Evans et al.[26] using a θ-pinch have observed at $\theta = 5°$ an asymmetric ion peak (Fig. 8) which they attribute to drifts and, in another paper,[27] they have described results which show the shoulders predicted by theory for the electron distribution at $\alpha \simeq 1$. Röhr[28] has reported spectrally resolved measurements of the ion peak (Fig. 9) and the electron plasma satellites (Fig. 10) in a 90° scattering experiment on a high density ($n_e = 5 \times 10^{17}$ cm^{-3}), relatively low temperature ($T_e = 32$ eV), θ-pinch.

Paul et al.[29] and De Silva and Stamper[30] have simultaneously published measurements of electron density and velocity distribution in a collisionless

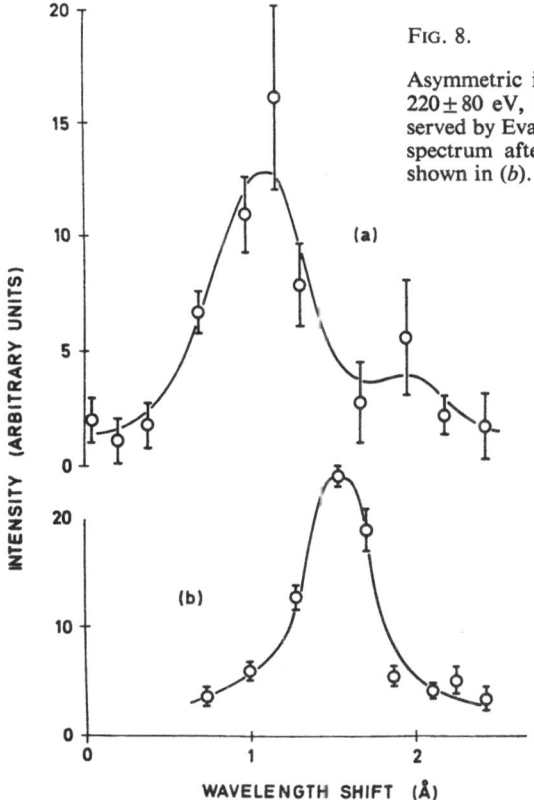

FIG. 8.

Asymmetric ion peak corresponding to $T_e = 220 \pm 80$ eV, $T_i = 220 \pm 105$ eV, $\gamma = 0\cdot55$ observed by Evans et al.[26] (a) shows the scattered spectrum after subtraction of the stray light shown in (b).

shock, Beach[31] has used the technique to map the electron density and temperature in a high energy θ-pinch, and Nodwell and Van der Kamp[32] have determined electron density profiles in a plasma jet. Dimock and Mazzucato[33] have carried out electron density and temperature measurements in

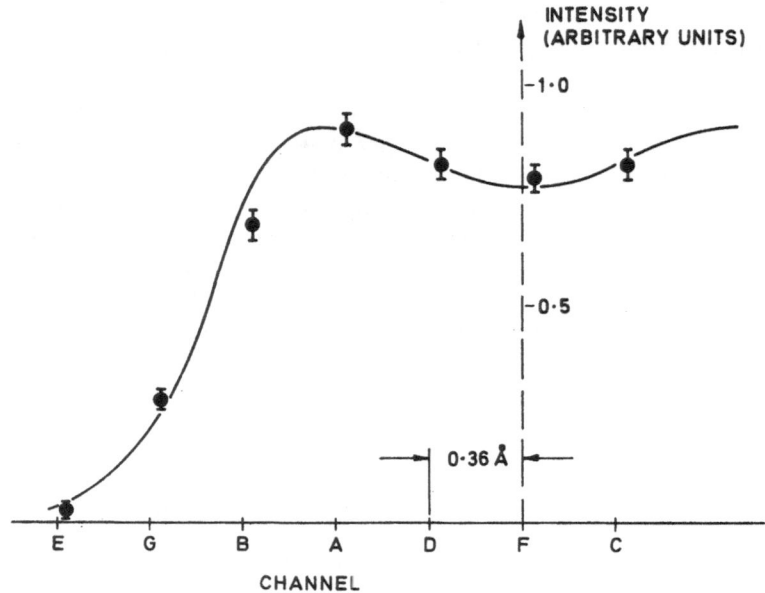

FIG. 9. Spectrum[28] of the ion component for $\theta = 90°$, corresponding to values of $T_e = T_i = 32$ eV, $n_e = 5 \times 10^{17}$ cm^{-3}.

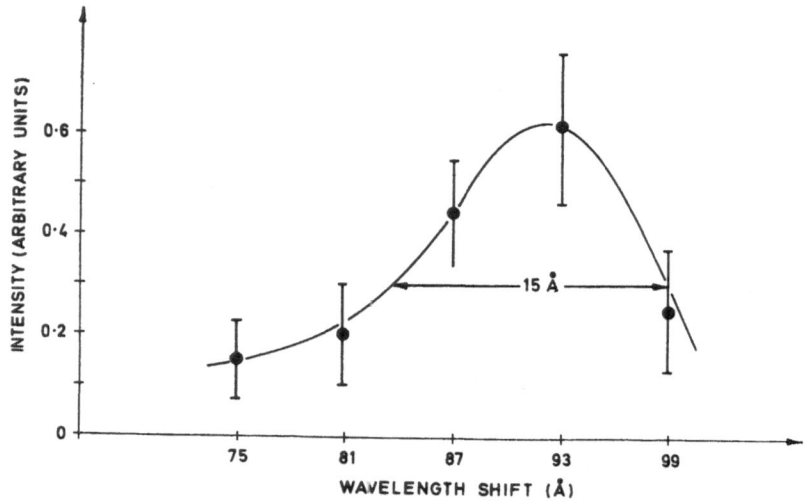

FIG. 10. Spectrum[28] of the electron satellite peak (same conditions as for Fig. 9).

the model-C Stellerator in the range 20–100 eV and 10^{12}–10^{13} cm^{-3} respetively, using a high energy (50–100 J) normal mode (~ 1 ms) ruby laser.

All the experiments mentioned so far have employed photomultiplier detection of the scattered radiation at one or more discrete wavelengths. A group[34] at the Institute of Nuclear Physics in Novosibirsk, U.S.S.R. has, however, succeeded in recording the complete spectrum for $\alpha < 1$, of the radiation scattered from a high density shock tube plasma using a three-stage image intensifier system. The signal/noise ratio was, however, somewhat poor.

6

TECHNIQUE

Although the actual experimental arrangements employed in the various scattering experiments carried out so far have varied to a certain extent, depending on the conditions of the experiment, many of the techniques involved are now becoming standard, and these will be illustrated with particular reference to a recent experiment[23] carried out by the author and his colleagues at the National Research Council of Canada, on a high energy θ-pinch, a schematic diagram of which is shown in Fig. 11.

6.1. *Ruby laser*

The ruby laser had an output of the order of 100 MW and a time duration of 30 ns. It was Q-switched by a rotating prism which was used to trigger the discharge of the θ-pinch capacitor bank in such a way that the laser pulse could be made to occur at any time during the discharge. Kerr or Pockels

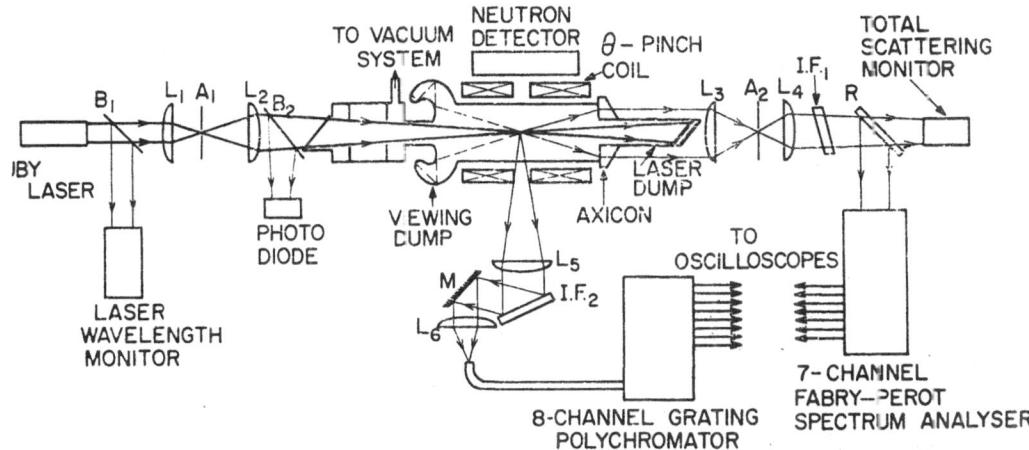

FIG. 11. Schematic diagram of a scattering experiment carried out on a high energy θ-pinch.[23]

cell Q-switches can also be used, although the Kerr cell has the disadvantage
of producing a rather large (~ 1 Å) half-width of the laser emission. The
laser was temperature controlled to $0.1°$ C to obtain wavelength reproducibility, and the output mode-selected by the use of a resonant front reflector,
consisting of a single quartz flat, to obtain a line width of better than 0.1 Å;
more elaborate mode selection techniques[35] can be used to reduce the line
width to less than 0.01 Å. The intensity and wavelength of the emission was
monitored using a photodiode and a Fabry-Perot interferometer, respectively.

6.2. *Stray light*

Although the ruby laser beam is intrinsically well collimated, there is a
small but significant amount of radiation emerging at large angles to the axis
which would give rise to considerable stray light within the scattering chamber. This was largely eliminated by focusing the beam using a plano-convex
lens (only simple lenses can be used as the beam damages cemented doublets)
on an aperture, made by drilling a fine hole in a blackened brass plate, which
also served to limit the beam divergence of the laser beam to ~ 3 mrad. A
further lens focused the laser beam to a spot 6 mm dia. at the centre of the
θ-pinch coil.

After passing through the scattering volume the laser beam was absorbed
in a piece of highly polished blue filter glass (Chance OB.10) placed at the
Brewster angle. This was located at the end of a glass tube ~ 20 cm long
which was blackened on the outside with an index matched black paint. A
Brewster angle entrance window, placed at the end of a long entrance tube
provided with a number of baffles, served to reduce reflections within the
scattering chamber. The window was far enough away from the scattering
volume to prevent the small amount of scattering that inevitably occurred at
the window itself from being seen by the axicon forward angle scattering
detection system. Further reduction of stray light was achieved by painting
the outside of the quartz scattering chamber and glass entrance tube with
black paint and providing a re-entrant "black" viewing dump for the axicon
detection system. In a situation where this was not possible Kronast[21] has
found some improvement in stray light by the use of a system of inverted
conical baffles in front of the axicon itself.

6.3. *Forward angle scattering detection system*

The use of an axicon or conical lens to detect small forward angle scattering is now standard. The axicon has the property that all rays scattered at a
given polar angle from the length of plasma within the field of view are
rendered parallel and can be focused on an aperture which serves to limit the
range of scattering angles observed. The length of plasma observed is determined by the aperture of the axicon and is usually much greater than the
diameter of the focused laser beam. Whilst this is of advantage in that a
greater scattering signal is obtained, the axial spatial resolution is poor. This

can, of course, be improved by masking down the aperture of the axicon but only at the expense of scattered signal.

The spectral distribution of the forward scattered radiation which, in this case, corresponded to the relatively narrow (~ 1 Å) ion peak, was measured using a combination of a narrow band interference filter and a multi-channel Fabry-Perot photomultiplier detection system of instrumental width 0·2 Å.

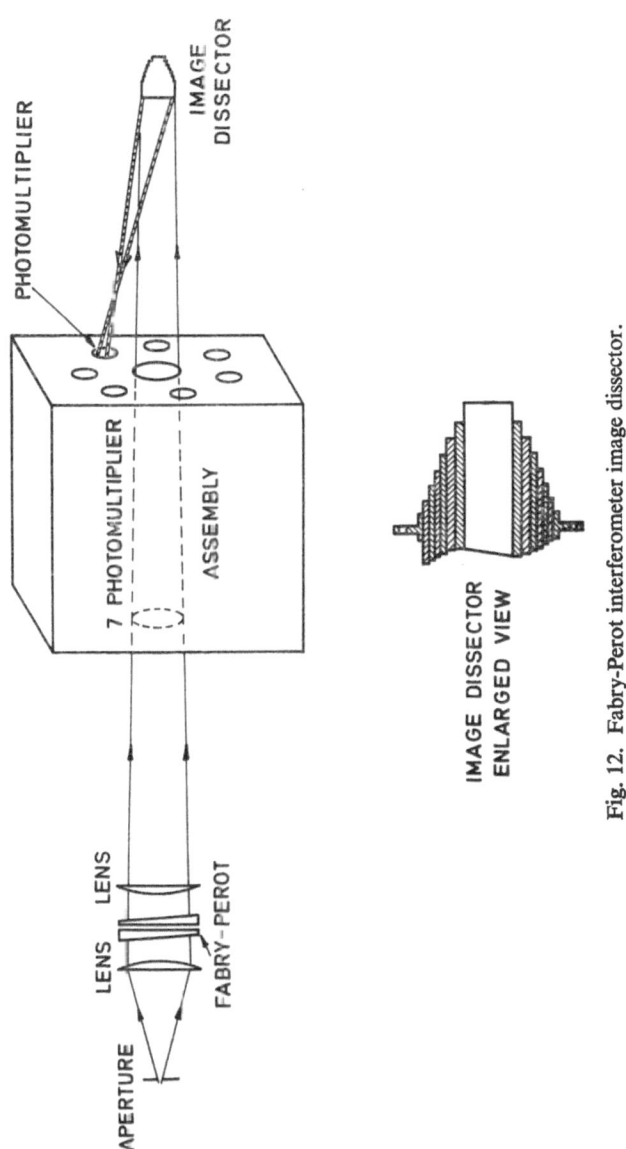

Fig. 12. Fabry-Perot interferometer image dissector.

Point-by-point scanning by varying the pressure in the Fabry-Perot chamber was tried but, because of irreproducibility of the plasma, the multi-channel system gave far better results and also reduced the data acquisition time. An annular nest of mirrors[36] was used to dissect the central fringe and, by inclining each at a different angle, direct different parts of the spectrum onto a set of seven S–20 photomultipliers, as shown in Fig. 12. Techniques are available, and have been used by other groups,[31] to display the outputs on a single oscilloscope, the outputs being delayed by multiples of an appropriate time, say 70 ns, using delay cables, so obtaining a direct display of the spectral distribution. This was difficult in the present case, because the scattering signal although clearly observable was not much greater than the continuous emission from the plasma. The use of electrical filters and differential amplifier systems to cut out this unwanted background presented difficulties, and so resort was made to displaying each signal individually using a number of oscilloscopes, the relative sensitivity of the different channels being determined from the uniform continuous emission from the plasma itself.

6.4. *90° scattering detection system*

At large scattering angles the spectrum is usually sufficiently broad that a monochromator or rotatable interference filter may be used to scan the distribution. Again, multichannel systems usually give better results and many have been developed. The system used in the present experiment was based on a design by Glock.[37] The scattering volume was imaged onto a fibre optic bundle, which acted as the entrance slit of a grating monochromator having eight fibre optic exit slits of average separation ~ 60 Å. As before, the outputs of the eight S–20 photomultiplier detectors were displayed on a number of oscilloscopes. Stray light at the laser wavelength—which was of little consequence in the measurement of the relatively broad (~ 4000 Å) electron feature and only contributed to stray light within the grating monochromator—was reduced by reflection at an interference filter which had a narrow pass-band at 6943 Å.

6.5. *Intensity calibration*

A convenient means of intensity calibration, and hence determination of the electron density in the plasma, is to compare the scattered signal with the scattering from dry N_2 at pressures of the order of one atmosphere, and this was used in the present experiment. The same technique may be used in the initial stages of the experiment to determine if a stray light level has been attained which will enable a scattering signal from the plasma to be detected. Precautions must be taken, however, to avoid scattering from dust particles —usually by careful filtering of the gas used and allowing a settling period of a few hours before a scattering measurement is made. It is also desirable to repeat the observations with two different gases (e.g. N_2 and CO_2) and check that the scattering signals obtained are in the ratio of the Rayleigh scattering cross-sections.

REFERENCES

1. See, for example, E. E. SALPETER. 1960. *Phys. Rev.*, **120**, 5. J. P. DOUGHERTY and E. P. FARLEY. 1960. *Proc. Roy. Soc. A*, **259**, 77. J. A. FEJER. 1961. *Can. J Phys.*, **39**, 716. W. B. THOMPSON. 1961. *Princeton University Plasma Physics Lab. Report* MATT–91. See also, W. B. THOMPSON: these Proceedings.
2. See, for example, E. E. SALPETER. 1961. *Phys. Rev.* **122**, 1663. J. A. FEJER. 1961. *Can. J. Phys.*, **39**, 716. D. FARLEY, J. DOUGHERTY and D. BARRON. 1961. *Proc. Rcy. Soe. A*, **263**, 238.
3. See, for example, D. F. DUBOIS and V. GILINSKI. 1964. *Phys. Rev.*, **133**, 1317. M. S. GREWAL. 1964. *Phys. Rev*, **134**, A86.
4. M. N. ROSENBLUTH and N. ROSTOKER. 1962. *Physics Fluids*, **5**, 776.
5. G. L. LAMB, *Los Alamos Scientific Laboratory Report* LA–2715.
6. G. FIOCCO and E. THOMPSON. 1963. *Phys. Rev. Lett.*, **10**, 89.
7. S. E. SCHWARZ. 1963. *Proc. IEEE*, **51**, 1362.
8. S. E. SCHWARZ. 1965. *J. Appl. Phys.* **36**, 1836.
9. E. FÜNFER, B. KRONAST and H. J. KUNZE. 1963. *Phys. Letters*, **5**, 125.
10. W. E. R. DAVIES and S. A. RAMSDEN. 1964. *Phys. Letters*, **8**, 179.
11. H. J. KUNZE, E. FÜNFER, B. KRONAST and W. H. KEGEL. 1964. *Phys. Letters*, **11**, 42.
12. A. W. DE SILVA, D. E. EVANS and M. J. FORREST. 1964. *Nature, Lond.*, **203**, 1321.
13. U. ASCOLI-BARTOLI, J. KATZENSTEIN and L. LOVISETTO. 1964. *Nature, Lond.*, **204**, 672.
14. H. J. KUNZE, A. EBERHAGEN and E. FÜNFER. 1964. *Phys. Letters*, **13**, 38.
15. H. J. KUNZE. 1965. *Z. Naturf.*, **20a**, 801.
16. E. T. GERRY and D. J. ROSE. 1966. *J. Appl. Phys.*, **37**, 2715.
17. R. M. PATRICK. 1965. *Physics Fluids*, **8**, 1985.
18. O. A. ANDERSON. 1966. *Phys. Rev. Lett.*, **16**, 978.
19. P. W. CHAN and R. A. NODWELL. 1966. *Phys. Rev. Lett.*, **16**, 122.
20. S. A. RAMSDEN and W. E. R. DAVIES. 1966. *Phys. Rev. Lett.*, **16**, 303.
21. B. KRONAST, H. ROHR, E. GLOCK, H. ZWICKER and E. FÜNFER. 1966. *Phys. Rev. Lett.*, **16**, 1082.
22. S. A. RAMSDEN, R. BENESCH, W. E. R. DAVIES and P. K. JOHN. 1966. *IEEE. J. Quant. Elect.* **QE–2**, 267.
23. S. A. RAMSDEN, P. K. JOHN, B. KRONAST and R. BENESCH. 1967. *Phys. Rev. Lett.*, **19**, 688.
24. P. J. BOTTOMS and M. EISNER. 1966. *Phys. Rev. Lett.*, **17**, 902.
25. M. DAEHLER and F. L. RIBE. 1967. *Phys. Letters*, **24A**, 745.
26. D. E. EVANS, M. J. FORREST and J. KATZENSTEIN. 1966. *Nature, Lond.*, **212**, 21.
27. D. E. EVANS, M. J. FORREST, J. KATZENSTEIN. 1966. *Nature, Lond.*, **211**, 23.
28. H. RÖHR. 1967. *Phys. Letters*, **25A**, 167.
29. J. W. M. PAUL, G. C. GOLDENEAUM, A. ILYOSHI, L. S. HOLMES and R. A. HARDCASTLE. 1967. *Nature, Lond.*, **216**, 363; see also J. W. M. PAUL, these *Proceedings*.
30. A. W. DE SILVA and J. A. STAMPER. 1967. *Phys. Rev. Lett.*, **19**, 1027.
31. A. D. BEACH. 1967. *J. Sci. Insts.*, **44**, 690; also A. D. BEACH, private communication.
32. R. A. NODWELL and G. S. J. P. VAN DER KAMP. 1968. *Can. J. Phys.*, **46**, 833.
33. D. DIMOCK and E. MAZZUCATO. 1968. *Phys. Rev. Lett.*, **20**, 713.
34. G. G. DOLGOV-SAVELEV. Private communication.
35. G. MAGYAR. 1967. *Rev. Sci. Insts.*, **38**, 517.
36. J. G. HIRSHBERG and P. PLATZ. 1965. *Appl. Optics*, **4**, 1375.
37. E. GLOCK. 1965. *Proc. VIIIth Int. Conf. Ion. Phenomena in Gases*, Belgrade.

12

PLASMA DIAGNOSTICS BASED ON REFRACTIVITY

U. Ascoli-Bartoli

Laboratori Gas Ionizzati (Associazione EURATOM-CNEN), Frascati

The idea of using a beam of light as a plasma probe occurred as soon as the need arose to diagnose a discharge without introducing large perturbations. The elementary mechanism underlying the propagation of radiation being the scattering of a photon by a free electron, the different aspects of refraction, diffusion and diffraction with which transmission manifests itself are characterized by the behaviour of the phase relationship of the scattered photons. In this discussion we limit ourselves to refraction, and this implies that the spatial density distribution of plasma—and hence the refractive index— does not vary greatly over distances of less than many light wavelengths.

1

Refractivity of Plasma

Quite a good deal of work has been done to date on this subject, but we mention here only those items which are strictly related to diagnostics.

The usual way is to deduce the wave equation from Maxwell's equations, where the characteristic plasma behaviour is introduced by relating the current density j to the velocities v_i and v_e of the plasma components under the effect of the electric field of the waves. In a first approximation the effect of the magnetic component of the wave can be disregarded, owing to the relationship

$$\frac{F_e}{F_m} = \frac{E}{H}\frac{1}{v\mu_0} = \frac{c}{v} \qquad \qquad ...[1.1]$$

between the electric and magnetic forces exerted on a charged particle having velocity v. In turn, v_i and v_e can be expressed by means of the momentum equations.

A discussion of the various phenomena taking place during the propagation is partially outside the purpose of these lectures. We limit our considerations to the few cases of practical importance, all referring to the case of a uniform plasma.

1.1. *Propagation in the absence of a static magnetic field*

A collision-free plasma propagates the e.m. wave isotropically with a refractive index n given by

$$n^2 - 1 = -\frac{\omega_{pe}^2}{\omega^2}\left(1 + Z\frac{N_i}{N_e}\frac{m_e}{m_i}\right) \simeq -\frac{\omega_{pe}^2}{\omega^2} \qquad ...[1.2]$$

where

$$\omega_{pe} = (4\pi N_e e^2/m_e)^{\frac{1}{2}} = 5\cdot64 \times 10^4 (N_e/\text{cm}^{-3})^{\frac{1}{2}}\text{ s}^{-1} \qquad ...[1.3]$$

is the plasma frequency. The e.m. wave propagates with a phase velocity greater than the speed of light *in vacuo*: the propagation is allowed for $\omega > \omega_{pe}$, whereas for $\omega < \omega_{pe}$ the wave is reflected by the plasma. For frequencies far from the cut-off $\omega = \omega_{pe}$ one obtains the equation $[1.2]$.

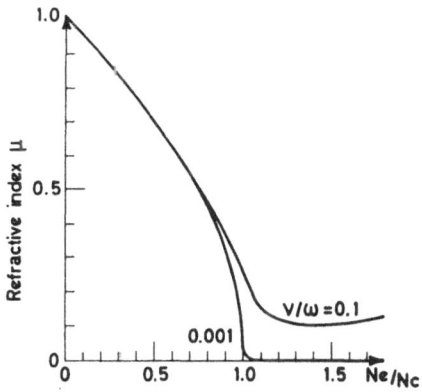

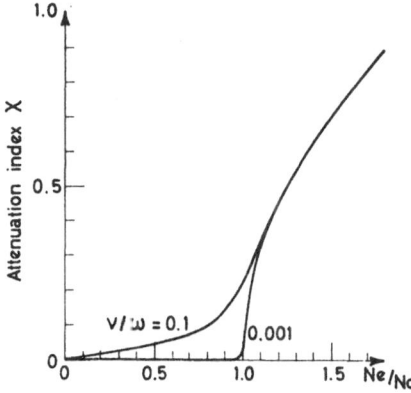

FIG. 1. Refractive index and attenuation for e.m. waves through a plasma.

When the collision frequency v is of the same order as the radiation frequency ω, its effect on the refractivity cannot be ignored: then

$$n^2 - 1 = \frac{-\omega_{pe}^2}{\omega^2 + v^2}\left(1 + i\,\frac{v}{\omega}\right). \qquad \qquad ...[1.4]$$

The presence of the imaginary part of the refractive index leads to the existence of a "skin depth" for the penetration of the e.m. wave into the plasma. This is the depth at which the wave is damped to about 1/2·7 of its original value by collision processes. The behaviour of the refractive index and of the attenuation versus $N_e/(N_e)_{\text{cut-off}}$ is shown in Fig. 1 for different values of v/ω. The question of the adequate formulation of v has been examined by Theimer and Taylor[1] and by Silin.[2] Theimer's formulation refers explicitly to the regime where the frequency of the applied electric field would be high enough for the subsequent amplitude of the electron oscillations to be smaller than the mean separation of electrons in the plasma. The suggested formula is

$$v = \frac{2\pi v N_e}{(16\pi)^{\frac{1}{3}}}\left(\frac{2\pi Z e^2}{m_e \omega}\right)^{\frac{4}{3}} \qquad \qquad ...[1.5]$$

which for laboratory plasmas at optical frequencies gives a $v/\omega \ll 1$.

In the case of very hot plasma a correction of [1.2] should be introduced:

$$n^2 - 1 = -\frac{\omega_{pe}^2}{\omega^2}\left(1 + \frac{kT_e}{m_e c^2}\right), \qquad \qquad ...[1.6]$$

whereas in the case of relativistic plasma $(kT \gg m_e c^2)$, $n^2 - 1 \simeq 0$.

1.2. Propagation in the presence of a static magnetic field B_z

In general, owing to the presence of a magnetic field, the electrons which are set in motion by the E-field of the wave experience the Lorentz force $F_L = v \times B_z$. Two cases are of interest:

(i) $E \equiv E_z \parallel B_z$. In this case $F_L = 0$ and the propagation is not influenced by the presence of the magnetic field;

(ii) $E_z \equiv 0$, and the propagation is in the direction of the static magnetic field.

In the absence of the wave the particles would be orbiting with random phases. In the presence of the wave the trajectories are altered in a manner which is dependent upon the relationship between ω and the cyclotron frequency ω_L. The plane e.m. wave can be considered composed of two waves, one right-circularly polarized and the other left-circularly polarized. Their interaction with the orbiting particles is different, and hence also are their propagation velocities (the Faraday effect). The refractive index is

$$n^2 - 1 = -\left(\frac{\omega_{pe}}{\omega}\right)^2 \frac{1 \mp \dfrac{\omega_L}{\omega} + \dfrac{iv}{\omega}}{\left(1 \mp \dfrac{\omega_{pe}}{\omega}\right)^2 + \left(\dfrac{v}{\omega}\right)^2}. \qquad ...[1.7]$$

The + and − signs refer to the right- and left-circularly polarized waves respectively.

Notice that, since $\omega_{L,e} = 1{\cdot}76 \times 10^7$ (B/gauss) s^{-1}, at optical frequencies $(\omega_{2000\,\text{Å}} = 9{\cdot}4 \times 10^{15}\,\text{s}^{-1}$, $\omega_{10\,000\,\text{Å}} = 1{\cdot}8 \times 10^{15}\,\text{s}^{-1})$, $\omega_L/\omega \ll 1$ for most plasma experiments.

1.3. Propagation in partially ionized plasma

Considering the plasma as a mixture of electrons, ions and residual gas atoms, the refractivity $(n-1)$ of the mixture can be expressed by

$$n - 1 = (n-1)_{\text{atoms}} + (n-1)_{\text{electrons}} = \sum_i K_i N_i, \qquad ...[1.8]$$

k_i being the specific refractivity and N_i the number density of the ith component (atoms in the various excited states, ions, electrons) of the mixture. The contribution from the atoms of the residual gas in various states of excitation can be of some importance in the case of low-energy discharges or, in any discharge, during the breakdown and the afterglow.

Both classical and quantum theories yield the well-known formula for refractivity:

$$n - 1 = \frac{2e^2}{m} \sum_l N_l \sum_k \frac{f_{lk}}{\omega_{lk}^2 - \omega^2}. \qquad ...[1.9]$$

which holds, provided that $\omega \neq \omega_{lk}$. Here ω_{lk} is the angular frequency of the line arising from the jump between the k and l states and f_{lk} is the corresponding oscillator strength. This can be calculated by means of quantum mechanics, degeneracy being taken into account. In applying the dispersion formula, the difficulty arises of including the transitions to the continuum, as well as those to the discrete energy states. This sometimes turns out to be of some importance.[3]

The population of the energy levels of the atoms depends upon the discharge: the refractivity of a discharge in thermal equilibrium was first studied by Kramers,[4] but no information can be given for other cases. Generally one could avoid the difficulty arising from the lack of information on the population of the excited levels by simply using a probe light beam of much lower frequency than that of the resonance lines. However, owing to the very large effects to be expected in a wavelength region where anomalous dispersion occurs, it seems reasonable to believe that the study of low energy discharges —the so-called "gaseous electronics"—could benefit from this rather neglected field of research.[5]

TABLE I

Refractivities of gases (0°C, 760 torr) referred to Cauchy's formula

	A	B/cm^2
He	$3·48 \times 10^{-5}$	$0·08 \times 10^{-14}$
Ne	$6·66 \times 10^{-5}$	$0·16 \times 10^{-14}$
A	$27·97 \times 10^{-5}$	$1·56 \times 10^{-14}$
Kr	$41·89 \times 10^{-5}$	$2·92 \times 10^{-14}$
Xe	$68·23 \times 10^{-5}$	$6·92 \times 10^{-14}$
H	$13·58 \times 10^{-5}$	$1·02 \times 10^{-14}$
N	$29·06 \times 10^{-5}$	$2·24 \times 10^{-14}$
Hg	$87·8 \ \times 10^{-5}$	$19·8 \ \times 10^{-14}$

If we consider only wavelengths far enough from the resonance lines, we can develop equation [1.9] in a series of powers of λ^{-1} which, stopped at λ^{-2}, gives Cauchy's formula

$$n-1 = A + \frac{B}{\lambda^2}. \qquad \qquad ...[1.10]$$

A and B are constants which are given for the most common gases in Table I and in terms of which refractivities (and polarizabilities) of most gases are tabulated.[6] In deriving [1.10] we have disregarded the contribution of excited states. This is still a rather questionable assumption; on the one hand, polarizability is proportional to the fourth power of the mean radius of the outermost electron,[7] but, on the other, the number density of excited atoms remains rather low because of the rather low temperature or because of the decrease of neutral atoms available in the case of hot plasmas.

A worse situation occurs in considering the ion refractivity. In this case, it seems hard to assume a low contribution from excited ions, at least in the vicinity of their resonance levels. In this case Cauchy's formula is meaningless. Methods are known for estimating the order of magnitude of ion polarizabilities.[8, 9] Generally speaking, it turns out that the refractivity of ions can be considered—apart from resonances—to be the same as that of the corresponding atoms, but of course a better knowledge of this would be helpful. These considerations naturally do not apply to proton and deuteron gases, whose refractivity can be obtained by using the formula for an electron gas. In the case of a partially ionized plasma the contribution to (n^2-1) of the free electrons is different from that of a neutral gas both in sign and in wavelength behaviour. The presence of neutral gas is not noticeable by means of refractivity if the condition

$$\frac{N_e}{N} \gg 0·83 \times 10^{-4} \frac{A}{(\lambda/cm)^2}, \qquad \qquad ...[1.11]$$

which can be obtained by comparing equations [1.10], [1.2] and [1.3], is fulfilled. Summarizing, the refractivity of a plasma is the result of contributions from the following: electron gas; ion gas; atoms, excited or not; molecules; collisions; and the magnetic field.

Thus, in planning an experiment based on refractivity, a good rule is to evaluate the relative importance of these contributions. Nothing can be said *a priori* for a general discharge, whereas in a high temperature plasma in hydrogen or deuterium the electron gas contribution predominates. Anyway, for the majority of cases of practical interest, the refractive index of a plasma can be considered to be linked to the plasma frequency and thus to the density, by means of [1.2] and [1.3], and, with easily evaluable good approximation,

$$n - 1 = -\frac{2\pi e^2}{m_e} \frac{N_e}{\omega^2} \qquad \qquad ...[1.12]$$

or, in practical terms

$$n - 1 = -4{\cdot}46 \times 10^{-14}(N_e/\text{cm}^{-3})(\lambda/\text{cm})^2. \qquad ...[1.13]$$

2

REVIEW OF METHODS AND TECHNIQUES IN REFRACTIVITY DIAGNOSIS

Let us consider a plasma contained in a vessel of suitable shape in order to allow electromagnetic waves, at least in the UV–IR wavelength interval, to cross the plasma without being deflected by irregularities in the transparent walls. The problem is that of observing what happens to the rays of the beam of light sent through the plasma, and predicting what can be inferred from these observations about the plasma itself. Here, as a first approach, the light illuminating the density inhomogeneities of the plasma is taken to consist of rays whose course is determined by Fermat's law. Let us now consider an individual light ray which in the absence of any disturbance would have reached the registering screen S at the point Q from the direction θ at the time t, but which actually reaches it at a point Q* from the direction θ* and at time t* (Fig. 2). The insertion of appropriate optical equipment in the light path will furnish on S a record of one of the following:

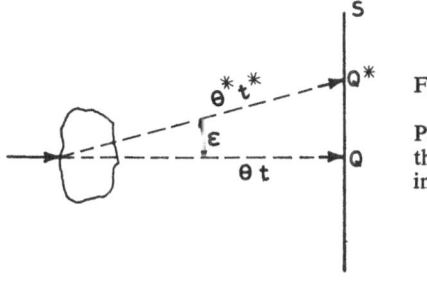

FIG. 2

Pattern of a light ray through a density inhomogeneity.

the phase lag $\tau = t^* - t$;
the deflection $\epsilon = \theta^* - \theta$;
the displacement $d = Q^*Q$;

or a function of two or all of them. The task of the experiment is:

  (a) to reconstruct from the record τ, ϵ and d;
  (b) to then deduce the values of n (x, y, z); and
  (c) to calculate local values of the density, from the law which relates the density and the refractivity.

An interferometer is a device which is able to record time lags, a schlieren system records deflections, and a shadowgraph records displacements. Since a change in travelling time means a change in refractive index and hence, by [1.10] and [1.13] a change in atom or electron density, an interferometric measurement gives the value of the mean density of particles along the light path. As may be anticipated, the schlieren method depends upon the first derivative, while the shadowgraph method depends upon the second derivative of the refractive index. Although these methods apply in principle to any density distribution—reviews of the method for evaluating the results being given for instance in Refs. 10, 11, 12, 13—we here refer only to a two-dimensional $n(x, y)$ and consider a light beam travelling in the z-direction.

The subject of how a beam of light is deflected in the presence of a spatial change of the refractive index has been extensively discussed in the literature. Two methods of representation can be distinguished. The first, which we can say to have been started by Mascart[14] in 1889 and was followed until a short time ago (see, for example, the article by Wolter[15]), regarded the problem in a strictly geometrical way, and is useful in those cases where the density of the medium follows slow and well-known variations along the trajectory of the light beam. The modern line of approach[63,64] treats the problem from the statistical point of view, which is suitable for consideration of turbulence in the medium and also in connection with the change in coherence of the light used.[66] This last aspect of the problem has become of practical interest thanks to the potential—still very little utilized—of the very high coherence of laser light.

Postponing a study of the modern approach till the last part of this review, the following considerations are based on the hypothesis that the total amount of deflection suffered by a light beam is small. We only need to remember that the trajectory of a thin beam of light, which at a given point of the propagating medium (plasma) makes an angle β with the direction of the density gradient, has a radius of curvature, R, given by

$$\frac{1}{R} = \frac{\sin \beta}{n} \nabla n. \qquad \qquad ...[2.1]$$

When a ray enters the discharge tube in a region where there is a constant gradient perpendicular to the ray, in a first approximation it suffers a devia-

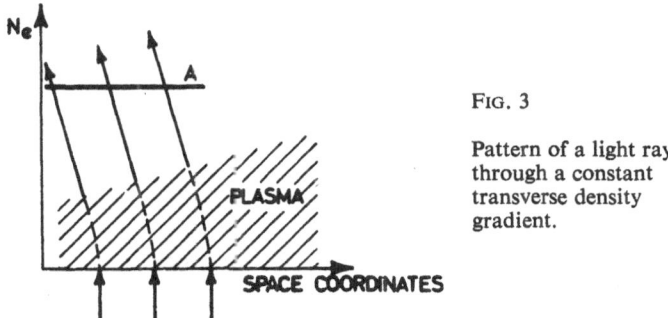

FIG. 3

Pattern of a light ray
through a constant
transverse density
gradient.

tion along an arc of a circle. Thus the trajectory remains roughly straight, since the bending is an effect of the second order caused by the density distribution. In the meantime the velocity of propagation of the light ray changes as it enters the vessel, and the same occurs for the related phase τ as compared with that of a ray travelling *in vacuo*. This gives an instructive explanation of the difference in behaviour between the interferometer and schlieren methods; in both cases the rays are bent but, whilst in a schlieren system this is a required feature, in the case of an interferometer this is something which is only a nuisance.

The shadowgraph shows the difference in displacement suffered by one ray with respect to that of an adjacent one, so that the net record will be zero if the displacements are equal, i.e. if the transverse refractive index gradient (and consequently the transverse density gradient) is constant (Fig. 3). A characteristic feature of the shadowgraph method is that a ripple in the space density distribution focuses a set of rays (Fig. 4), so that on a screen interposed for instance at A, the presence of a ripple is marked by a bright spot. With reference to the applications, we mention here the measurement of the thickness of shock waves and the mapping of turbulent regions. Generally it is clearly preferable to measure a quantity directly instead of obtaining it by integration; and thus schlieren and shadowgraph methods are to be supple-

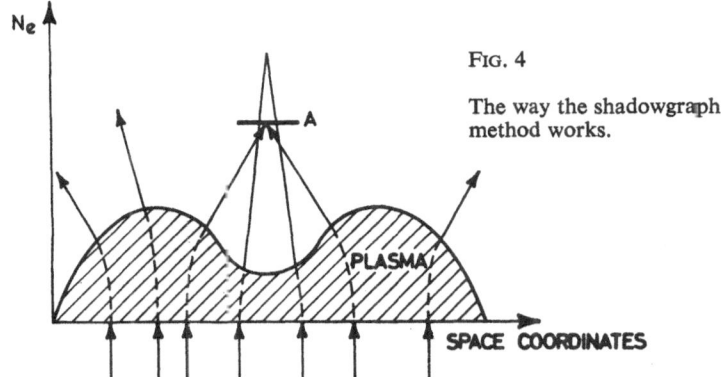

FIG. 4

The way the shadowgraph
method works.

mented by interferometry. But each of these methods has its own field of application where it is particularly suitable: interferometry in cases where the refractivity is slowly varying in space, the schlieren method when the value of gradients is needed, and the shadowgraph when rapid (even if small) changes in refractivity occur (as ripples) and are of interest.

3

SOME LIMITATIONS TO THESE METHODS

Each of the methods already mentioned requires an appropriate optical arrangement, which will be examined in the following paragraphs; these arrangements have, however, some features in common which can be described immediately. First of all, the light rays sent through the discharge chamber are generally parallel. This requires a light source (extended or not) at the focus of a lens or mirror. The receiver, for instance a photographic plate, is intended to take a picture of the cross-section of the tube. Therefore, on the side of the receiver there is an objective which focuses the mid-plane of the discharge or part of it, for example a diametrical slab, after the rays have passed the particular devices (beam splitters, knife edges, rulings and so on) which distinguish the method. Let us consider the following optical arrangement in order to show some characteristic limitations of the methods. It does not detect any particular effect; but it is common to all the devices we are going to study (Fig. 5).

A beam of parallel light is sent by the extended source through the discharge tube by means of the lens L_1. The objective L_2 images the mid-plane, diameter D, of the discharge on the photographic plate π. Owing to the finite size of the source, the rays entering the discharge tube form, at each point, a bundle of rays with an angle Γ given by

$$\Gamma \simeq \tan \Gamma = \delta/f_1. \qquad \ldots [3.1]$$

Clearly, by the geometrical optics approximation, all the information belonging to the cone *abcd* is recorded at the point P*, the image of P. Two image points P_1^* and P_2^* record completely different information, provided

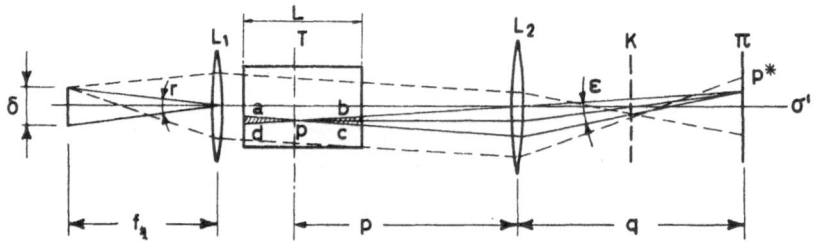

FIG. 5. Optical arrangement.

that they are images produced by two non-overlapping cones. As a consequence of this, along the diameter of the discharge tube only $D/(\Gamma L/2)$ regions can be regarded as being known independently of each other.

A slightly more careful evaluation leads to the conclusion that, as a first approximation, the above quantity might be increased to $D/(\Gamma L/6)$. But in principle we cannot rule out the possibility that some localized plasma disturbances inside the cone (e.g. a shock wave or turbulence) produce some diffracted light. Let us therefore suppose that an inhomogeneity of size d is present along the path of the rays; the order of magnitude of the diffraction angle is $\Gamma' = \lambda/d$. The cone containing "mixed" information is $\Gamma'' = \Gamma' + \Gamma$ and in the worst situation, when the disturbance is near the surface where the rays enter the tube, the size of the base of the cone of diffracted light on the other surface of the tube is $L\lambda/d$. If we want to have a record of the inhomogeneity d we need a Γ'' such that

$$\frac{\Gamma''L}{6} = \frac{L\lambda}{d} = d$$

and thus

$$\Gamma = 5(\lambda/L)^{\frac{1}{2}}. \qquad\qquad ...[3.2]$$

As a consequence of the above discussion, we find that one could make use of two different criteria for designing the optical equipment according to whether we want to collect information on very small regions where diffraction is the limitation, or whether we are satisfied with the geometrical optics approximation. In the first case, Γ is established by [3.2]. This has to be taken into consideration in deciding the parameters of the optical device, where one chooses that set of components which achieves the highest value of some "quality" of the instrument, for example the resolution or the "speed" (which is proportional to the light flux per unit area of the receiver). It is immediately seen that an increase of Γ diminishes the spatial resolution but increases the speed. The latter could also be improved independently by diminishing the size of the image, but at a certain point the finite resolving power of the receiver itself (e.g. about 20 lines/mm on a photographic plate) becomes of importance.

Generally one comes to some compromise between the conflicting requirements of the instrument. In the field of time-resolved plasma diagnostics a very strong requirement is that concerning the amount of light needed in order to obtain pictures with a very short exposure time. In order to obtain useful photographs with a photographic density of about one, the necessary illumination flux generally quoted[16] is 0·2 erg/cm^2 of blue light. It is rather easy to supply such a flux either under steady conditions or in a pulsed flash. In fact, even in the worst of these cases, when the required duration of the flash is of the order of 10^{-7} s and its dimensions are very small (e.g. a circular hole of 0·5 mm diameter) many methods have been successfully developed. But a very different problem occurs if the required amount of light has to be at longer wavelengths. Here one faces at the same time both the problem of the source and that of the receiver; however a set of methods, though not so useful as before, can be used[17-21]. The situation worsens, if a narrow wavelength interval $\Delta\lambda$ is required for interferometric purposes, owing to the condition

$$n\Delta\lambda = \lambda/2, \qquad \qquad ...[3.3]$$

which gives the highest number n of fringes which can be produced in an interval $\Delta\lambda$ around the central wavelength λ. In this case one is forced to make the most sparing use of the light at one's disposal. Let us evaluate the orders of magnitude of the parameters involved with the problem of the light source.

Let us call Σ the sensitivity of a photographic plate, defining $1/\Sigma$ as the luminous energy per unit area necessary to obtain a photographic density of one, and let E be the illumination of the plate, i.e. the light energy per unit time per unit area reaching the plate. If R takes into account the reciprocity failure of the plate and τ is the exposure time, then

$$E R\tau \geqslant 1/\Sigma \qquad \qquad ...[3.4]$$

is the condition required to get a photographic density not less than unity By definition $E = d\phi/d\sigma'$, where the flux ϕ is given by Lambert's law:

$$\phi = \tfrac{1}{4}\pi^2 B(\delta')^2 \sin^2 (\tfrac{1}{2}\Omega).$$

Here σ and σ' are corresponding surfaces in the image and object planes, i.e. at π and T respectively, B is the brightness of the source, and δ' is the size of the image of the source at K (Fig. 5). Since

$$\delta' = \delta(f_2/f_1) = \Gamma f_2; \; \sigma' = \sigma G^2 \text{ and } \sigma = \pi f_2^2 \tan^2 (\Omega/2),$$

calling G the magnification, we get for small values of Ω:

$$E = \tfrac{1}{4}\pi(B\Gamma^2/G^2). \qquad \qquad ...[3.5]$$

By means of [3.4] and [3.5] we get

$$\tfrac{1}{4}\pi(B\Gamma^2/G^2)R\tau \geqslant \frac{1}{\Sigma} \qquad \qquad ...[3.6]$$

where the ratio Γ/G has to be decided in order to suit the requirements of the experiment. Let n_r be the number of resolved lines per centimetre of the photographic plate and let us assume that for each piece of information we assign K lines on the plate, i.e. the point P is represented by K lines.

The resolving power of the optical arrangement is

$$N^{-1} = \tfrac{1}{3} \cdot \tfrac{1}{2}\Gamma L \qquad \qquad ...[3.7$$

at the discharge tube and therefore $GN^{-1} = \tfrac{1}{6}G\Gamma L$ at the plate.

Matching the two resolving powers gives

$$\frac{G}{N} = \frac{G\Gamma L}{6} = \frac{K}{n_r}.$$ ...[3.8]

Making use of this, equation [3.6] becomes

$$BR \, \Sigma \, n_r^2 \geqslant \frac{N^4 L^2 K^2}{\tau} \frac{1}{9\pi}.$$ ...[3.9]

If one wants to choose a value of Γ satisfying [3.2], since by [3.2] and [3.7] one has

$$N = \frac{6}{5} \frac{1}{(\lambda L)^{\frac{1}{2}}},$$ ...[3.10]

the following condition results:

$$BR \, \Sigma \, n_r^2 \geqslant \left(\frac{6}{5}\right)^4 \frac{1}{9\pi} \frac{K^2}{\tau} \frac{1}{\lambda^2}$$ .. [3.11]

which, being related by the above discussion to the resolution limit given by wave optics, is much harder to fulfil than [3.9]. Equations [3.9] and [3.11] are equations to be satisfied between the parameters of the light source and the photographic plate (B, n_r, Σ, R) in order to record, within an exposure time τ, a picture with a photographic density of unity. A similar relationship can be written for photocathode devices. The survey of light sources and receivers in terms of these constants is a matter for more specialized literature; some references will be briefly discussed here.

Disregarding stationary sources, photographic flashes and related commercially available sources, the following rough classification can be given for very fast, high brightness sources:

(a) free sparks in air or in xenon;
(b) guided discharges (Wood's magnesium slabs, capillary discharges);
(c) end-on viewed discharge channels; and
(d) exploding wires.

All of these make use of high voltage, fairly low capacitance condensers; the duration of the spark and its brightness depend on the inductance of the circuit and special care must be taken to reduce this. Probably the minimum time of duration of such light sources is due to the decay time of the excitation. The most advanced spark light sources are generally obtained by matching the impedance of the source to the spark gap, sometimes a low impedance transmission line being used. Also barium titanate transmission lines have been employed with considerable advantage.

When referring to data on the brightness of the sources, most authors use the photometric unit system, and since this is related to the emitted spectrum, it is often difficult to obtain a useful figure. Fortunately one of the most interesting light sources, which simply makes use of an extremely fast condenser and of a very low inductance circuit, due to Fisher,[22] is calibrated in M.K.S. units. He quotes a brightness of $B = 10^4$ W/(cm² sr) at $\lambda = 5500$ Å during a time $\tau = 2 \times 10^{-8}$ s.

Very different from the above mentioned sources, the laser constitutes the most singular light source ever known. Owing to its exceedingly small angular spread (less than 20″) the beam can be very precisely focused to give a point source. The brightness can be estimated by the formula

$$B = \frac{16}{\pi^2} \frac{f^2}{\delta^2 \phi^2} P, \qquad \qquad ...[3.12]$$

where the symbols are explained in Fig. 6 and P is the laser power. In the case of a coherent light source δ is of the order of the Airy disc. Using the usual figures, for a medium-powered monopulsed ruby laser the brightness given by [3.12] ranges from 10^7 to 10^{11} W/(cm² sr).

We wish now to make a comparison between Fisher's source and a medium-powered ruby laser light source from the point of view of equation [3.11], assuming the following typical values:

$R \simeq 1$
$\Sigma \simeq \frac{1}{3}$ erg^{-1} cm² (for instance Polaroid 3000 ASA)
$n_r \simeq 300$ lines/cm
$K \simeq 10$ lines
$B \simeq 1 \cdot 7 \times 10^4$ W/(cm² sr) (Fisher's source $\lambda \simeq 5500$ Å)
$B \simeq 10^9$ W/(cm² sr) (ruby laser $\lambda \simeq 6940$ Å)
$\tau \simeq 2 \times 10^{-8}$ s (Fisher's source)
$\tau \simeq 5 \times 10^{-8}$ s (ruby laser)

We get, for the right-hand side of [3.11]:

$$\left(\frac{6}{5}\right)^4 \frac{1}{9\pi} \frac{K^2}{\tau} \frac{1}{\lambda^2} = \begin{cases} 1 \cdot 2 \times 10^{17} \text{ s}^{-1} \text{ cm}^{-2} \text{ (Fisher's source)} \\ 3 \times 10^{16} \text{ s}^{-1} \text{ cm}^{-2} \text{ (ruby laser)} \end{cases}$$

and for the left side of [3.11]:

$$BR\Sigma n_r^2 = \begin{cases} 5 \times 10^{15} \text{ s}^{-1} \text{ cm}^{-2} \text{ (Fisher's source)} \\ 3 \times 10^{20} \text{ s}^{-1} \text{ cm}^{-2} \text{ (ruby laser)}. \end{cases}$$

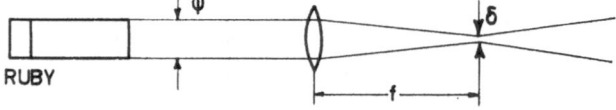

FIG. 6. Quantities appearing in the evaluation of the ruby laser performance.

We see that a ruby laser, even of modest performance, is able to give pictures of the best resolution, i.e. up to the limit set by wave optics, whereas one of the best known conventional light sources cannot reach this goal. As to the colour composition of the light source, taking into account equations [1.10] and [1.13] one realizes that the refractivity of atoms is larger in the violet region, whereas electron density can be better measured by using red or infrared light; also that, in order to be able to distinguish between electrons and atoms, both colours are necessary, so that a pair of equations of the kind [1.8] can be written one for each colour. For each colour, monochromaticity is not strictly needed. From this point of view, the ruby laser again turns out to be the more suitable, because of its well-known ability (due to the high power level) to excite the second harmonic (of wavelength $\lambda = 3471$ Å) when pulsing a suitable crystal (e.g. ADP).[23, 24, 25, 26]

In making use of the fundamental and the second harmonic at the same time, one gets both colours needed for analysing a plasma in the presence of neutral gas. On the receiver side one can simply make use of the fact that the two beams are polarized at 90° to one another because of the mechanism of second harmonic generation.

By using equations [1.10] and [1.13] in connection with the values given from Table I for the sensitivity, evaluated as

$$\sigma_{e,a} = \left| (n-1)/N_{e,a} \right|$$

one obtains the following values:

(i) for electrons, $\sigma_e = 2 \cdot 15 \times 10^{-22}$ cm^3 at $\lambda = 6943$ Å and four times less at $\lambda = 3471$ Å.

(ii) for atoms, $\sigma(H_2) \simeq \sigma(H) \simeq 4 \cdot 5 \times 10^{-24}$ cm^3, independent of the wavelength all along the visible spectrum.

Thus
$$\frac{\sigma_e}{\sigma_a} = \begin{cases} 48 & \text{(in the red)} \\ 12 & \text{(in the violet).} \end{cases}$$

This means that the refractivity is always sufficiently favourable to electron density measurements to make it possible for them to be made even in discharges where very low levels of ionization are present.

4

INTERFEROMETRIC MEASUREMENTS

The interferometric method is a very old established technique for measuring densities of any transparent material, and we discuss here only those items which are more strictly related to plasma.

The common idea underlying all interferometric work is that of producing fringes by interference of two coherent beams of light. The coherence is obtained by splitting the beam of light (e.g. by means of a semi-reflecting mirror) or by dividing the wave front issuing from a point source of light by using a mirror or a prism. Once these coherent beams of light are obtained, they can have quite different histories; for example, one can travel in the air, the other being inside the discharge tube. If not exactly parallel, these two beams cross somewhere, in a real or virtual position, and interference fringes appear there. Assuming for simplicity, that the indices of refraction n_1 and n_2 are constant over each path, and the geometrical lengths of the paths are l_1 and l_2 respectively, the places where the fringe maxima occur satisfy the condition

$$n_1 l_1 - n_2 l_2 = N\lambda, \qquad\qquad ...[4.1]$$

whereas for the fringe minima the above difference must be an odd multiple of $\lambda/2$: $(2N+1)\lambda/2$. A change in refractivity of the medium through which one beam passes causes a change in the place where [4.1] is satisfied, i.e. a shift in the fringe system in one direction or the other depending on whether the refractivity increases or decreases. The fringe shift, expressed in number of fringes, is proportional to the variation in the optical path

$$s = \frac{\Delta s}{\delta} = \frac{L\Delta(n-1)}{\lambda}, \qquad\qquad ...[4.2]$$

where L is the length in the plasma along the light path, δ the thickness of a fringe and $\Delta(n-1)$ is the variation of refractivity. In a discharge tube the change of refractivity due to the introduction of the plasma is

$$\Delta(n-1) = (n-1)_{\text{plasma}} - (n-1)_{\text{gas}}. \qquad\qquad ...[4.3]$$

The fringes are said to be localized at the place where the two beams intersect. The usual way of using the interferometer with a plasma is to let one beam travel along the discharge tube, the other one crossing the former virtually at the midplane of the tube itself. In this case there is a correspondence between the points of the midplane P of the discharge tube and the fringes. To each point on P there corresponds a particular point on one fringe. A displacement of the fringe following a change in refractivity changes the correspondence.

Permanent correspondence between the points of P and the plane at which the fringes are formed is easily established by forming the image of a wire grating on the plane P. Thus on the photographic plate which records the fringes the pattern of the wire grating is superimposed. The evaluation of densities (fringe shifts) is carried out for each point of the plane P simply by measuring the change in phase between two fringe patterns taken before and during the discharge at the relevant points. This implies that the photometric shape of the fringe must be accurately known, but this requirement is very difficult to meet for various practical reasons. One should perhaps be satisfied with a measurement of the fringe shift made at one point per fringe. The most sensitive points on a fringe are of course those where the change in photographic density is the fastest—in practice they are the only useful points of a fringe. Therefore we see that each fringe can give only the value of density regarding two points in the plane P. Thus the number of measured points in the mid plane of the discharge tube in the direction of the fringe shift is twice the number of fringes; in the direction perpendicular to this, the spatial resolution is that allowed by the optical device. Unfortunately this is not what is required for recording by means of a streak camera.

It is not the purpose of this article to review all the instruments. We limit ourselves here to some features of those which have been most used. The Jamin interferometer is relatively easy to use, but it suffers from two great drawbacks: the fringe localization is at infinity, and the two beams travel very close to one another. In the Michelson interferometer, because of the double transit, the localization occurs at one end of the discharge tube, and the sensitivity is twice that of a single transit interferometer. The Mach-Zehnder interferometer is in most use today owing to its very peculiar characteristic of being able to give fringes which can be localized in any plane along the discharge tube.

A very extensive literature has been written on the Mach-Zehnder, its history, properties and applications.[27-36]

As far as we are concerned here, the effect of the instrument can be described as that of producing on a plane M, considered as the locus of an infinite number of light sources, an image M' slightly tilted with respect to the former (see Fig. 7). Each pair of light sources, like points P and P', represent a coherent pair of sources. Q is a plane where interference fringes result. In the absence of the discharge tube, fringes of order K and wavelength λ are related to the ambient refractive index n_0 and the geometrical quantities ε and r by the relation:

$$K\lambda = n_0 \varepsilon r. \qquad \qquad ...[4.4]$$

Disregarding the possibility of having a three-dimensional distribution of refractivity as requiring too long a mathematical treatment to be presented

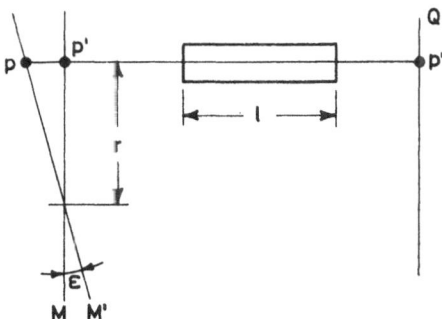

FIG. 7. Optical circuit equivalent to the Mach-Zehnder interferometer.

here, we confine our discussion to the problems arising from a two-dimensional density distribution. In this case the light is sent along the third axis.

A one-to-one correspondence between the geometrical space inside the tube and the plane of the fringes being assumed, variations in density in a given region of the discharge tube is noticed by displacement of the corresponding fringes. Considering the interferometric procedure, let us assume for a moment that the variation in density is constant across the discharge tube. The condition of interference, equation [4.4], in the case where a tube of length l containing a gas of refractive index n_1 is placed in the path P'P'', is

$$l(n_0 - n_1) - \varepsilon r n_0 = K\lambda.$$

Therefore

$$r = \frac{l(n_0 - n_1) - K\lambda}{\varepsilon n_0}. \qquad ...[4.5]$$

Calling n_1^* the refractive index of the subsequently formed plasma and r^* the corresponding distance according to [4.5], we find the fringe displacement due to the plasma

$$\Delta r = r^* - r = \frac{l}{\varepsilon n_0}(n_1 - n_1^*). \qquad ...[4.6]$$

Going from the Kth to the $(K+1)$th fringe, one travels the thickness of one fringe and from [4.5] this turns out to be

$$\delta = r_K - r_{K+1} = \frac{\lambda}{\varepsilon n_0}. \qquad ...[4.7]$$

From the last two equations we obtain the usual formula (cf. [4.2])

$$\frac{l(n_1 - n_1^*)}{\lambda} = \frac{\Delta r}{\delta}. \qquad ...[4.8]$$

But if we now suppose that the density (or refractive index) varies across the tube, and repeat the above calculation, supposing that

$$n_1^*(r^* + \Delta r) = n_1(r^*) + (dn_1^*/dr)_{r=r^*}\Delta r, \qquad ...[4.9]$$

but still with the hypothesis of negligible bending of the rays, we obtain for the fringe shift

$$\Delta r = \frac{l(n_1 - n_1^*)}{l(dn_1^*/dr)_{r^*} + \varepsilon n_0} \qquad ...[4.10]$$

and for the fringe thickness

$$\delta^* = \lambda/[\varepsilon n_0 + l(dn_1^*/dr)_{r^*}]. \qquad ...[4.11]$$

From [4.10] and [4.11] we thus obtain

$$s = \frac{\Delta r}{\delta^*} = \frac{l(n_1 - n_1^*)}{\lambda} \frac{[\varepsilon n_0 + l(dn_1^*/dr)_{r^*}]}{[\varepsilon n_0 + l(dn_1^*/dr)_{r^*}]} \qquad \ldots [4.12]$$

This enables us to use the same method of measuring $n_1^* - n_1$, even in the presence of constant gradients, provided that the fringe thickness is that of the fringes of the plasma interferogram, and that the density (and hence refractivity) gradient is unchanged in the space r^* to r.

From equations [4.7] and [4.11] one obtains a method for measuring gradients, which can be used as a check of the results:

$$l\left(\frac{dn_1^*}{dr}\right)_{r^*} = \lambda\left(\frac{1}{\delta^*} - \frac{1}{\delta}\right) \qquad \ldots [4.13]$$

But if the gradient produces a large variation across the field of one fringe, so that one can no longer grant the hypothesis underlying [4.12], the usefulness of interferometry is reduced. In order to discuss this point, let us distinguish between extended and point light sources. In the latter case the following occurs:

(a) a bent ray travels across regions of different refractivity, this giving rise to an additional phase change whose value can only be estimated on the basis of some assumption. Igenberg[36] has evaluated this phase shift $\Delta\phi$ on a first but largely valid approximation, using the hypothesis of a constant value of density (or refractivity) gradient. The result is

$$\Delta\phi = \frac{1}{12}\left(\frac{1}{n}\frac{dn}{dr}\right)^2 \frac{l^3}{\lambda} \qquad \ldots [4.14]$$

(b) still as a consequence of the bending, the magnification of the image of the cross section of the tube is no longer constant across the field itself as shown in Fig. 8.

Let us now turn to the extended source. Using a point source, the fringes of monochromatic light can be observed inside all the region where the beams

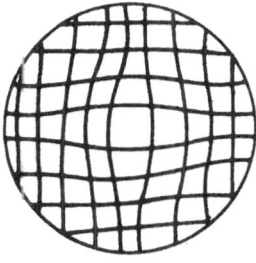

FIG. 8. Image of wire mask seen through a discharge.

intersect; so the focal depth of the fringe images is as long as this region. Displacement of the point source across the focal plane of the entering lens produces rotation of the planes of the fringes. Thus each point of an extended source of diameter d in the focal plane of the collimator (focal length f) gives a set of fringes whose plane is rotated with respect to that of the on-axis light point. This rotation has a maximum value of $\varepsilon' = d/f$. As a consequence of this, the depth of focus is reduced; it can be seen that now this is

$$T \simeq \frac{1 \cdot 22\lambda}{\varepsilon\varepsilon'} = \frac{1 \cdot 22\lambda f}{\varepsilon d}, \qquad \ldots[4.15]$$

ε being the usual angle between the two beams. The bending of the rays adds a rotation α to ε' and consequently T decreases to

$$T \simeq \frac{1 \cdot 22\lambda}{\varepsilon(\varepsilon'+\alpha)}. \qquad \ldots[4.16]$$

The evaluation of α can be made by integrating the equations of the trajectory of a light ray on the basis of some assumption about the density distribution (e.g. constant gradient).

But this is not the only drawback which occurs in the presence of gradients. A beam of light issuing from an extended light source produces, at a point P inside the discharge tube, bundles of rays which fill a cone having its apex at P (Fig. 5). Rays belonging to a given cone can experience different patterns and different phase shifts because of the different gradients possible along their trajectories. But, since all points inside T are focused on the recording plate, all rays belonging to the cone contribute to the same image point. Notwithstanding this, the superposition of different phases results in a blurring or diminution of contrast of the fringes and may even cause their disappearance.

Summarizing, we notice that the difficulties arising from the presence of gradients can be reduced by the use of a point source (e.g. from a laser), but of course one cannot go beyond the limit given by [4.14].

Turning to the practical uses of the interferometer, we now deal with its potential, which can be evaluated in terms of its sensitivity and the amount of information it is able to give. Let us first consider the measurement of uniform density. The amount of information which can be measured is given by the inverse of the least fringe shift which can be detected, or supposing that the fringe is not altered in shape during the shift, the number of parts into which a fringe can be meaningfully divided by the receiver (e.g. photographic plate plus microdensitometer). If the thickness δ^* of the fringe can be analysed by steps of size Δl the number of independent pieces of information contained in one fringe shift is $\delta^*/\Delta l$ and the sensitivity is

$$S_{\min} = \frac{\Delta l}{\delta^*}, \qquad \ldots[4.17]$$

whence, because of [*1.13*] and [*4.2*] which give

$$s = -4.46 \times 10^{-14}(N_e/cm^{-3})(\lambda/cm)(l/cm), \qquad ...[4.18]$$

we can evaluate $(N_e l)_{min}$.

Where the density is not uniform and $\mathcal{N}$ fringes are needed, the total information is $\mathcal{N}(\delta^*/\Delta l)$ or, by [*4.17*], $\mathcal{N}/S_{min}$. Now $\Delta l/\delta^*$ depends upon the manner in which the receiving system is used; two typical cases can be proposed according to whether the light for recording the fringes is energy- (or time-) limited or not.

If d is the dimension of the used part of the photographic plate,

$$d = \mathcal{N}\delta^*, \qquad ...[4.19]$$

thus using [*4.17*]

$$\frac{d}{\Delta l} = \frac{\mathcal{N}}{S_{min}}. \qquad ...[4.20]$$

In the first of the two cases mentioned above d could in principle be estimated on the basis of the condition in [*3.9*]; but let us assume, for the sake of brevity, that $d \simeq 5$ cm, $\Delta l \simeq 1/200$ cm as a typical case. By means of [*4.18*], [*4.20*] becomes

$$5 \times 200 = \mathcal{N}/\{4.46 \times 10^{-14}(\lambda/cm)[(N_e/cm^{-3})(l/cm)]_{min}\}, \qquad ...[4.21]$$

that is, at $\lambda = 7000$ Å the relationship for the minimum perceptible plasma density is

$$(N_e l)_{min} = 3.2\mathcal{N} \times 10^{14} \text{ cm}^{-2}. \qquad ..[4.22]$$

In doing this we have assumed, as is true in practice, that the limitation in the energy of the light source consequently limits the size of the plate.

In the other case, the size of the fringes can be made larger, and their evaluation is only limited by the microphotometer, the number of fringes being irrelevant. Therefore in [*4.17*] one could introduce for $\delta^*/\Delta l$ a value which, for a high-quality microphotometer, is of the order of 10^4, obtaining, at $\lambda = 7000$ Å,

$$(N_e l)_{min} = 3.2 \times 10^{13} \text{ cm}^{-2}.$$

Apart from the requirements for the optical devices, it seems to be very doubtful whether one could reach such a high sensitivity because, with conventional sources, the necessary amount of light can be delivered only in fairly long exposure times, and even in steady plasma the fluctuations in density could diminish the contrast of the fringes.

Practical values of $\Delta l/\delta^*$ quoted in the literature range between 1/10 and 1/100, and only a very accurate measurement obtained by Kennedy and cited in Ref. 37 reaches 1/1000. Satisfactory measurements of density distributions within low density wave fronts, equivalent to fringe movements of only 1/500 fringe, have been obtained using very sophisticated methods.[38, 39]

Another method of using the same instrumentation is to set up the interferometer in the so-called "teinte plate" condition, i.e. to spread the interference fringes so much that half of one fringe fills the whole field. In this way a local density variation, e.g. in a pinch, will be indicated by the appearance of closed fringes which form the contour lines of the density distribution. This representation is preferred when an intuitive description of the phenomenon under study by means of contour lines is desired rather than an exact measurement.

A few words and references on other kinds of interferometers that may be useful in plasma physics follow.

The diffraction grating interferometer based on the Ronchi method of optical testing is capable of giving fringes whose displacements are proportional either to the density[40] or to the density gradient,[41] depending upon the set-up used.

In plasma research the region surrounding the discharge tube is often occupied by coils, vacuum facilities, etc. and cannot therefore be used for the reference beam needed in interferometry. Clearly one possibility is that of using the so-called "series interferometer", the first application of which was in 1950 by Saunders and which is mentioned by Post[42, 43] in more general papers. It consists of three partial mirrors, arranged one behind the other with approximately equal optical separation. Apart from the advantage of mechanical simplicity, it is able to produce sharpened fringes (roughly δ-shaped) without any stringent restriction on mirror separation and monochromatic purity because of its multiple-beam character. This allows observations to be made in the 1/1000 fringe range.[44] The drawback lies in localization of the fringes. If the light source is a laser having a very high coherence length, one mirror could be removed.[45]

5

LASER INTERFEROMETRY

The interferometric techniques hitherto described can operate with both standard light sources and lasers. Recently a method was developed based on the use of a gas laser, which, notwithstanding some disadvantages with respect to the classical systems, offers the advantage of being less complicated optically. This method is based on the observation, made by King and Stewart,[46] that, if one returns part of the radiation into the cavity of, e.g. a gas laser, the intensity of the laser radiation is stronger or weaker than the intensity the laser would have without the radiation being reflected, according to the phase at which it is returned.

Ashby and Jephcott[47] took advantage of this principle, inserting the plasma between the reflection mirror and the laser and using a photomultiplier for detection (Fig. 9). They used a He–Ne laser which can emit simultaneously at $\lambda_1 = 6328$ Å and $\lambda_2 = 33\,900$ Å. These two radiations are produced by transitions from the same initial level, so that, for example, an increase of laser radiation at λ_2 reduces the population of the initial level and decreases the intensity at λ_1. A germanium filter allows only the radiation at λ_2 to pass into the part of the cavity occupied by the plasma. In this way the greater sensitivity of the interferometer at the higher wavelength λ_2 is exploited (cf. [1.13]) while the radiation at the lower wavelength λ_1 can be detected rather well with a photomultiplier. A schematic view of the arrangement is given in Fig. 9a. If the plasma in the region under study undergoes, in the course of time, a variation in its density, the photomultiplier will see a series of oscillations of light intensity which look like the reflection curves of a Fabry-Pérot, to which they are in fact closely linked.

Let us evaluate the sensitivity of the method. We assume that the external mirror EM is placed in such a position with respect to the plasma (Fig. 9a) that the laser radiation at λ_2, measured by the photomultiplier, is minimal: that is to say, the Fabry-Pérot formed by EM, LM and the plasma is in resonance. The propagation time of a photon going from LM to EM and back to EM is

$$t = (2/c)[D + l(n-1)]. \qquad \ldots[5.1]$$

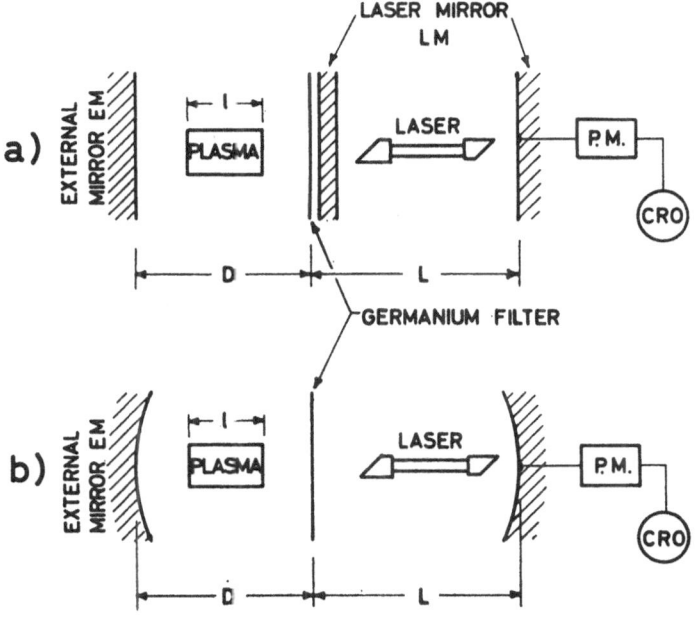

FIG. 9. Laser interferometer scheme.

We can express the resonance by saying that t is an integral multiple K of an oscillation period of the laser:

$$\frac{2}{c}[D+l(n-1)] = \frac{K}{v}. \qquad \qquad ...[5.2]$$

The $(K+1)$th order is the order of resonance closest to the Kth order: it will appear for a $(n+\Delta n)$ such that

$$\frac{2}{c}[D+l(n+\Delta n)-l] = \frac{K+1}{v} = \frac{K}{v}+\frac{1}{v}. \qquad ...[5.3]$$

Substituting in this expression the K/v of [5.2] we obtain

$$\Delta n = \frac{c}{2lv} = \tfrac{1}{2}\frac{\lambda}{l}.$$

To this Δn corresponds a minimum measurable density given by [1.13] which turns out to be

$$(N_e l)_{min} = 1 \cdot 14 \times 10^{13}(\lambda/\text{cm})^{-1}\ \text{cm}^{-2}. \qquad ...[5.4]$$

Through a comparison with [4.22] we see that this method is less sensitivy than the one based on classical interferometry. The most important reason is that, in this case, the possibility of measuring fractions of a fringe has been excluded. A considerable improvement is obtained if the light makes mane forward and backward passings across the plasma, sacrificing somewhat the spatial resolution.

A decisive improvement, however, has been obtained by Gerardo and Verdeyen[48,49] (Fig. 9b) using a spherical mirror in EM. In this case there can be various transverse modes for every longitudinal one, and the sensitivity is correspondingly increased. For plasma lengths of about 50 cm, densities of less than 10^{13} cm^{-3} can be measured in this way.

6

THE SCHLIEREN METHOD

The German word *schliere* is often used to indicate inhomogeneous regions in optical glasses which appear as streaks. Here, as in many other

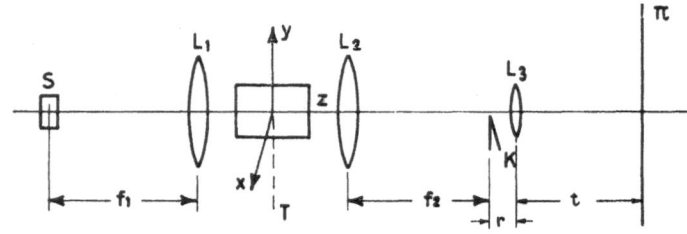

FIG. 10. Optical arrangement and quantities of Toepler's schlieren system.

cases of a similar nature, small changes of the refractive index of the transparent material are involved, so that direct visual observation of the refractive index is difficult. *Schlieren* is the name given to optical methods which are based on the refractive index change (or gradient) and are able to show such inhomogeneities. Extensive use is being made of this method in its various forms,[50, 15] but here we need recall only their general outlines, setting emphasis on their use for plasma problems.

Referring to the geometry of Fig. 10, it can be shown easily[15] that the deflection suffered by the beam of light travelling through a refractivity gradient $(\partial n/\partial x, \partial n/\partial y, 0)$ is, for small angles,

$$\epsilon_x = \frac{1}{n_0} \int_0^l \frac{\partial n}{\partial x} dz; \quad \epsilon_y = \frac{1}{n_0} \int_0^l \frac{\partial n}{\partial y} dz, \qquad ..[6.1]$$

n_0 being the ambient refractive index.

Schlieren techniques make possible detection of the small angles ϵ. To describe these methods, we choose Toepler's method, which translates this deviation into a variation of luminous intensity. The set-up is given in Fig. 10. The light source S is an extended one, but still of small size—its dimensions are estimated later in this paper. The discharge tube is crossed by a set of parallel beams, as in Fig. 5. The image of the source is obtained at K, L_1 and L_2 being placed at their focal distances from S and K respectively. A third lens forms the image of the test section T on the screen π. The same result can be obtained by using two concave mirrors in the places of the first two lenses. If now part of the image of the light source is intercepted by a partial screen (or "knife edge") placed at K, the illumination of the screen diminishes, each pencil of light being subjected to the action of the knife edge to the same extent; each point on the screen is darkened to the same degree. Let us now suppose that the knife edge is placed along the x direction. When an optical disturbance is present in the test section on a given pencil of light (e.g. at P in Fig. 5), the latter suffers a deflection given by ϵ_x, ϵ_y. At the knife edge the deviation of this bundle of rays is given by

$$\Delta b = \epsilon_x f; \quad \Delta a = \epsilon_y f \qquad ...[6.2]$$

(with $f = f_1 = f_2$). The deviation Δb displaces the image of the source along the knife edge and thus does not produce any change in the amount of light transmitted past the knife edge. The contrary is true for the Δa deviation, which produces, at the image point P* on the screen, a change in intensity which is proportional to $\epsilon_y f$. In this connection two parameters of the method are of interest—the sensitivity and the range.

In order to get the first, we must have an expression for the illumination E of the screen. In the same way as in Section 3, but making use of a rectangular-shaped light source, we have the following expression:

$$E = B \frac{(f+r)^2 ab \cos^2 (\Omega/2)}{t^2 f^2} \qquad ...[6.3]$$

where B is the brightness of the light source and a and b are the dimensions of the part of the image formed by rays which pass the plane of the knife. The other quantities are shown in Figs. 5 and 10. This expression is derived from the definition of illumination, $E = d\phi/d\sigma'$, Lambert's law $\phi = \pi Bab \sin^2 (\Omega/2)$, the magnification $\sigma'/\sigma = [t/(f+r)]^2$ and the geometrical relation $\sigma = \pi f^2 \tan^2 (\Omega/2)$. For small angles, and with $r \ll f$, we have:

$$E \simeq Bab/t^2. \qquad \qquad ...[6.4]$$

The optical disturbance at P* changes the illumination of its image by

$$\Delta E = B(b/t^2)\Delta a \qquad \qquad ...[6.5]$$

with Δa given by [6.2].

Thus, for the relative sensitivity $\Delta E/E$, the following relationship holds:

$$\frac{\Delta E}{E} = \frac{\epsilon_y f}{a} = \frac{\Delta a}{a}. \qquad \qquad ...[6.6]$$

One could also define a sensitivity per unit deflection:

$$S = \frac{d(\Delta E/E)}{d\epsilon_y} = \frac{f}{a}. \qquad \qquad ...[6.7]$$

With reference to the "range", let us suppose, as is usually the case, that the light source is imaged half on the knife and half above it. Then the possible range of displacement is $2a$, and the corresponding angle $\bar{\epsilon}$ is

$$\bar{\epsilon} = \frac{2a}{f}. \qquad \qquad ...[6.8]$$

As a consequence it appears that, once having established the range (that is, the size of the source), the sensitivity is independent of the properties of the optical system, and that the product of the sensitivity per unit deflection and range angle is a constant:

$$S\bar{\epsilon} = 2. \qquad \qquad ...[6.9]$$

This description of the Toepler schlieren arrangement, based as it is on geometrical optics, is seen to be not completely valid when the limitations due to wave optics are considered. This is of importance if a laser is used as the light source. The first difference is that the image of the source on the knife edge is altered by diffraction so that the edge of this image is not sharp. This may be regarded as being caused by a spurious deviation of magnitude

$$\epsilon_u = \frac{\lambda}{D}, \qquad \qquad ...[6.10]$$

which is present even in the absence of perturbations. The indeterminacy Δ^* in the position of the image edge is

$$\Delta^* = \epsilon_u f \qquad \qquad ...[6.11]$$

(of the order of $\lambda f/D$). So a blurring halo of size

$$a^* = \frac{\Delta^*}{2} \qquad \qquad ...[6.12]$$

is added to the contours of the image of the light source.

The same occurs if, instead of being limited by the finite diameter of the tube, D, the plane wave is limited to the width D^* of a pencil of light passing through some variation in refractivity. Consequently, if the highest sensitivity is to be reached, one cannot go to sizes of a smaller than a^*. Therefore the minimum observable deviation is, by equation [6.6],

$$\epsilon_{min}^* = \frac{\lambda}{2D^*}\left(\frac{\Delta E}{E}\right)_{min}, \qquad \qquad ...[6.13]$$

and depends on the spatial resolution (through D^*) and the minimum resolvable variation of the illumination, $(\Delta E/E)_{min}$. Indicating by $N = D/D^*$ the spatial resolution, we obtain from [6.10] and [6.13]

$$\epsilon_{min} = \tfrac{1}{2}\epsilon_u N\eta, \qquad \qquad ...[6.14]$$

where $\eta = (\Delta E/E)_{min}$ is a characteristic feature of the receiver (a photographic plate, or an image converter used as shutter or intensifier or both). It is a measure of the smallest fractional change in brightness which can be perceived by the receiving device, and is generally a function of the brightness itself and thus of the quality of the source.

Now ϵ_{min} can be written in terms of the minimum resolvable plasma gradient. Supposing that the plasma density is uniform in the z-direction we get from [6.1]

$$\epsilon_y = \frac{1}{n_0}\, l\nabla n, \qquad \qquad ...[6.15]$$

and from [1.13]

$$(\epsilon_y/cm) = -4{\cdot}46 \times 10^{-14}(\lambda/cm)^2(l/cm)(\nabla N_e/cm^{-4}). \qquad ...[6.16]$$

Thus the minimum resolvable gradient of electron or plasma density is, from [6.14] and [6.16],

$$\nabla N_e/cm^{-4} = -(\epsilon_u/cm)N\eta/[8{\cdot}92 \times 10^{-14}(\lambda/cm)^2(l/cm)]. \quad ..[6.17]$$

It might be worth while pointing out that this result is an aspect of the uncertainty principle for optics which is contained in [6.10].

A much better method for examining the theory of the schlieren effect follows from the use of Abbe's procedure for the theory of images. This has been done by Zernike and one of the results was the phase contrast method[15] which can be thought of as an improvement of the usual schlieren method. The starting-point lies in the statement that the distribution of the light just leaving the discharge tube and that at the plane of the knife edge are complex Fourier transforms of each other. The effects of the knife edge or whichever screening device is employed, or of a phase plate in the case of the phase contrast method, are described by dropping out or modifying the corresponding terms in the transform. Thus one reaches a unified representation of the various different forms of the schlieren method. By working out some particular case mathematically a set of relationships is obtained which must be satisfied in order to optimize the quality of the information received. For the Toepler system this has been done by H. J. Shafer.[51] We quote here some of his conclusions:

A large aperture gives a high aperture/disturbance-size ratio with consequent high contrast and density.

A large light source will give high density and low contrast.

The optimum-size light source is one whose geometrical image in the focal plane of the objective is equal to the width of the Airy disc of the objective.

The minimum-size light source is one whose half-width is equal to the distance from the central maximum of the diffraction pattern of the disturbance under study to the optical axis in the focal plane of the objective.

Hitherto we have dealt with the Toepler arrangement. It is the only one capable of giving information in a pictorial, intuitive form (Plate 1); but the results of various experiments made on plasma suggest that this arrangement cannot be used as a quantitative tool. Of the various practical reasons for this assertion it will be sufficient to remark that in most cases, wherever the ionization process is incomplete, it is necessary to have, for each instant of interest, two photographs taken at different wavelengths (violet and red) to be able to distinguish between atoms and electrons. Therefore a dichroic beam-splitter is required and a pair of interference filters. Generally for fast discharges requiring very short exposure times it is necessary to make use either of very fast photographic plates or of image-converter intensifiers.[52] In most cases these receivers do not reach a range of two decades of linear response, and the overall accuracy is very poor.

A slight improvement can be obtained, still using photographic densitometry, by including in the field of the test object a glass wedge or a lens of known characteristics (i.e. standard *Schlieren*) and thus making comparisons and interpolations between the observed darkenings of the plate. This approach assumes that the deflection produced by a thin wedge is equivalent to that of a thick layer, namely that

$$(1/n_0) \int_0^l (\partial n/\partial x)\, dz$$

is the same in the two cases. It can be shown that only within the limit of geometrical optics does this hold exactly.

In deriving [6.17] we assumed that the field of the discharge tube was divided into parts of extension D^*, supposing tacitly that inside this field the value of the density gradient was constant. This was consistent with the hypothesis that the plane of the knife edge was the focal plane of the lens L_2 (Fig. 10), whence [6.10] and the subsequent development. But this could not be so, for example, where the plasma has a bell-shaped radial density distribution. In this case the plasma behaves like a lens, and one is no longer allowed to take a constant value for f. Hence the knife edge is no longer at the right setting, and Fresnel diffraction fringes occur, as is demonstrated in Fig. 11. This effect is of course most easily seen in the case where, with the aim of increasing sensitivity, one makes use of very small a (see [6.7]) as in the case of ruby laser point sources. A theta-pinch schlieren photograph showing this effect is given in Plate 2. If the change in light intensity due to this de-focusing is to be a small fraction, c, of the change due to true deflection in the case of a point light source, the following relationship between the first and second derivatives of the density distribution must be obeyed.[45]

$$cD^* \left| \frac{d^2 N_e}{dx^2} \right| \leqslant \left| \frac{dN_e}{dx} \right|. \qquad \qquad ...[6.18]$$

This is generally a more stringent condition on $(\nabla N_e l)_{min}$ than [6.17].

It is therefore essential to look for other ways of recording the deflections experienced by a beam of light in the presence of density gradients. The most direct method is due to Lamm[53] and is mentioned in Ref. 54. The optical arrangement is that of Fig. 12, a point light source being necessary. Near the entrance window a transparent scale is added and is illuminated by a parallel beam of light from the source (not shown). The lens gives an image of the scale on the plate. In the absence of disturbance, the image of the scale is reproduced without any distortion. An optical disturbance bends the pattern of the ray inside the tube and the image of the scale becomes distorted. In

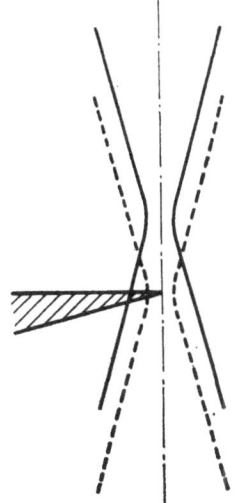

FIG. 11

Out-of-focus
Toepler schlieren system.

Fig. 12 the disturbance displaces the point image A′ of A to A″ which, in turn, in the absence of disturbance, is the image of A_1. The displacement $AA_1 = \Delta_1$ can be written in terms of ϵ:

$$\Delta_1 = a\epsilon, \qquad \qquad ...[6.19]$$

and the displacement $A′A″ = \Delta′$ can be evaluated in terms of Δ_1 and the magnification G:

$$\Delta′ = G\Delta_1. \qquad \qquad ...[6.20]$$

Since

$$\epsilon = \frac{1}{n_0} \int_0^l \frac{\partial n}{\partial x}\, dz$$
$$\simeq (1/n_0)l\nabla n \qquad \qquad ...[6.21]$$

i.e.
$$\epsilon \simeq l\nabla n, \qquad \qquad ...[6.22]$$

we have

$$\nabla n = \frac{\epsilon}{l} = \frac{\Delta_1}{al} = \frac{\Delta′}{aGl}. \qquad \qquad ...[6.23]$$

This displacement occurs at a value of x given by

$$x = z - \Delta_1 = z - (\Delta′/G) = (z′ - \Delta′)/G. \qquad ...[6.24]$$

It can be seen that the relationship holds exactly, provided that the trajectory is parabolic, because only a parabola among the elementary curves has the property that the tangent at C crosses parallel to the optical axis of the device at BP = BC/2 (Fig. 13); this requires that the ray travels through a region of constant density gradient. This method is very accurate, but requires a lengthy evaluation of the recording plate.

If the discharge has axial symmetry, measurements need be made along

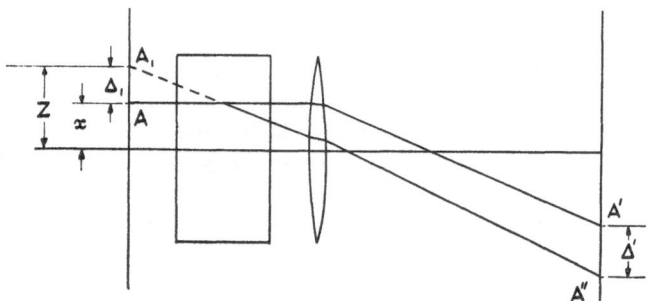

FIG. 12. Lamm's optical arrangement.

only one diameter. In this case the focal shift of the line image with respect to the straight diametrical line gives only the azimuthal component of the refractive index gradient; if the scale is dotted, one also obtains the radial component by measuring the spacing between the projections of the dots on the straight line. A different method[55] of recording has been studied by the author and co-workers. The optical arrangement is shown in Fig. 14. Here W is a point source of light (e.g. a ruby laser light beam focused by means of a lens whose focus is at W). The light is made parallel by means of a collimator, C. The image of the source, formed by L_S, is altered by means of a cylindrical lens L_C into a straight line (in the absence of disturbances) focused at P. The objective O focuses at P an image of the slit S which is placed in front of the discharge tube in order to restrict the observed region to a thin slab through the diameter of the tube. The lens L_C has no focusing effect on light in the x-direction, so that each coordinate corresponds to a given point of the tube diameter. In the presence of an optical disturbance the angular deviation is recorded on P with a displacement s such that

$$s = Gef_s, \qquad \qquad \dots [6.24]$$

f_s being the focal length of L_S, and G the magnification of the cylindrical lens. The effect of the objective O can be neglected. From equations [6.15] and [6.16] it follows that

$$\frac{\nabla n}{n_0} = \frac{s}{Glf_s}, \qquad \qquad \dots [6.25]$$

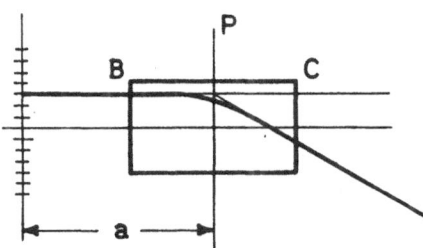

FIG. 13. The bending of the ray referred to in the text.

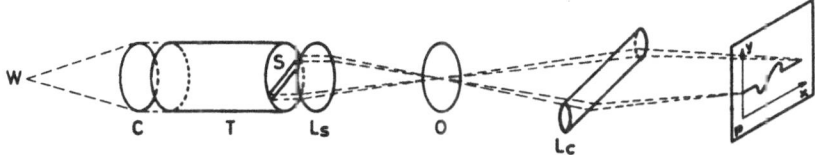

FIG. 14. Optical arrangement of a direct reading schlieren system.

and

$$\nabla N_e/\mathrm{cm}^{-4} = (s/\mathrm{cm})/[4{\cdot}46 \times 10^{-14} G(f_s/\mathrm{cm})(l/\mathrm{cm})(\lambda/\mathrm{cm})^2]. \quad ...[6.26]$$

Clearly ∇N_e is the *azimuthal* component of the refractivity gradient. To measure at the same time both radial and azimuthal components at each point of the slab of the discharge tube defined by the slit, a beam splitter and two cylindrical lenses are used. Their axes are perpendicular to each other and are inclined at 45° with respect to the slit S. In this arrangement the radial component causes displacements which are equal but in opposite directions in the two records, whereas the displacements resulting from the azimuthal component are equal and in the same direction. In the photographs the records are taken together with a zero line. For each value of r the azimuthal and radial components are thus obtained simply by summing or subtracting the corresponding displacements of the two records. A substantial improvement in the quality of the pictures can be obtained by using the *Kennzeichnung* method.[15]

All schlieren systems employing a parallel beam of light are unable to distinguish between the density gradients occurring at various positions along the light path. In many cases it would be desirable to obtain an image in which only density gradients in a given plane determine the image obtained. A schlieren system with multiple sources (Fig. 15) and corresponding knife edges has sharp-focusing properties. It is based on the principle that, for planes out of focus, superposition of the images blurs out the effects of the density gradients.

Shock waves in a nozzle have been successfully studied with such a device. In plasma research a great deal of work could be done with such an arrangement and the study of tearing modes and of end effects in discharge tubes could greatly benefit.

It is now worth putting on record some more methods which appear to be very relevant to plasma diagnostics. One is that called interferential strioscopy, based on the theory given by Françon.[15] By this method the entering window δ (Fig. 5) and the knife edge are replaced by two Savart plates or Wollaston prisms and a couple of polarizers. Fringes are obtained on π as a result of interference between two adjacent beams. These fringes are thus sensitive to density gradients. The advantage of this over the classical

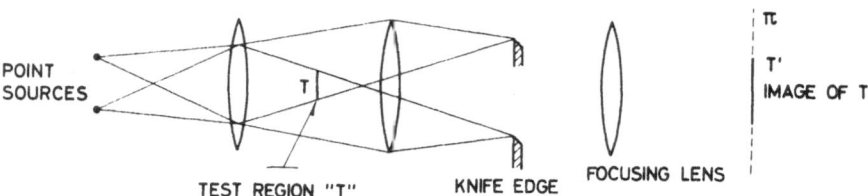

FIG. 15. The principle of the focus schlieren system.

Toepler method is that it is easy to get numerical data from the fringe shift; a disadvantage could be that of giving smaller spatial resolution, as in every method using fringes.

Another method in which some interest might be shown is the one based on an observation by Gayhart and Prescott and theoretically described by Temple.[26] According to these authors, interference fringes are observed in the schlieren system which make possible a quantitative evaluation of the starting constant needed when one wants to obtain a density profile by ntegrating the schlieren equations.

7

THE SHADOWGRAPH

Referring back to the presentation of Section 2, the shadowgraph is a method of observing the inhomogeneity of the plasma in which no optical equipment other than that producing the incident beam is required to obtain a record. The information is given by the mapping $Q \to Q^*$ and is of an implicit nature. A complex analysis would be required in order to go back from a shadowgraphic picture to the density distribution of the inhomogeneities.

Up to now, very little use has been made of the shadowgraph in plasma physics, the reason being the strong requirement for point light sources, as discussed in section 3. This method is quoted here, and deserves particular mention, because of the peculiar (potential) quality of the available information. The underlying idea having been already presented in Section 2, we need only mention here the most important features applicable to plasma diagnostics.

Referring to Fig. 16, let us introduce a rectangular system (x, y) $(\mathscr{S})$ in the plane π of the discharge tube. In the absence of disturbances, the beam of parallel light issuing from $\mathscr{S}$ will project another system of rectangular coordinates (x', y') $(\mathscr{S}')$ identical to the former on the photographic plate P,

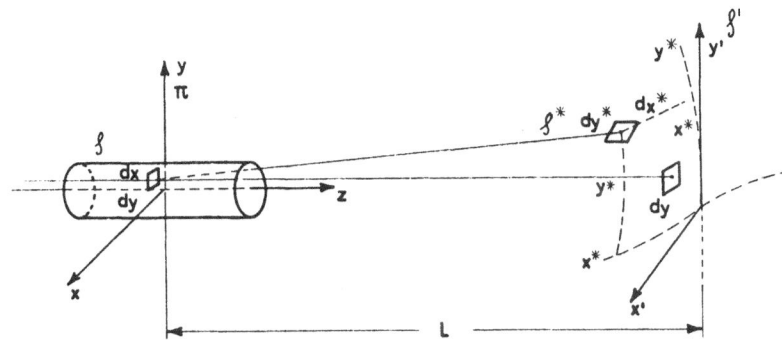

FIG. 16. Quantities related to the deflection of a beam.

whose distance from π is L. Disturbances in the plasma change the co-ordinate system $\mathscr{S}'$ in shape and position. Let us call (x^*, y^*) $(\mathscr{S}^*)$ the new coordinate system superimposed on $\mathscr{S}'$. The correspondence between $\mathscr{S}$ and $\mathscr{S}^*$ is given by a relationship of the form

$$x^* = \phi_1(x, y), \qquad \qquad ...[7.1]$$

$$y^* = \phi_2(x, y), \qquad \qquad ...[7.2]$$

where ϕ_1 and ϕ_2 depend on the disturbance.

Let $I(x, y)$ be the light intensity distribution at π in the absence of dis-turbances in the plasma, and $I^*(x^*, y^*)$ the corresponding one at P when disturbances are present. Then we have

$$I^* dx^* dy^* = \sum I dx dy, \qquad \qquad ...[7.3]$$

where the summation is extended to all area elements $dxdy$ which contribute to I^* by collapsing into $dx^* dy^*$. Since

$$dx^* dy^* = [\partial(\dot\phi_1, \phi_2)/\partial(x, y)] dx dy, \qquad \qquad ...[7.4]$$

equation [7.3] gives

$$I^* = \frac{\sum I}{\partial(\phi_1, \phi_2)/\partial(x, y)}. \qquad \qquad ...[7.5]$$

In general all these relationships are useless if one does not make the stipu-lation that the displacements must be infinitesimal both inside the tube and along the entire path from the tube as far as the plate. This is verified in practice by testing at different distances L, L', L'' and checking that a crossing of rays as shown in Fig. 17 does occur. Under this hypothesis

$$x^* = x + \Delta_x(x, y), \qquad \qquad ...[7.6]$$

and

$$y^* = y + \Delta_y(x, y), \qquad \qquad ...[7.7]$$

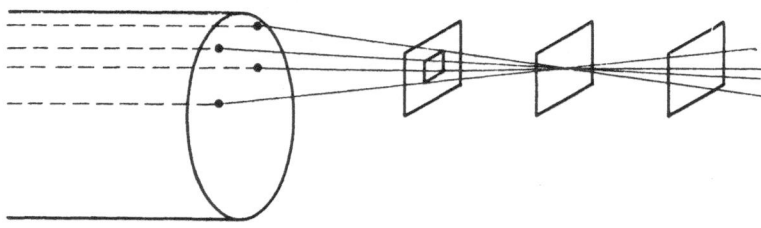

FIG. 17. Focusing effect of a shadowgraph.

where Δ_x, Δ_y are small quantities. Since

$$\mathcal{T} = \begin{vmatrix} \dfrac{\partial x^*}{\partial x} & \dfrac{\partial x^*}{\partial y} \\[2mm] \dfrac{\partial y^*}{\partial x} & \dfrac{\partial y^*}{\partial y} \end{vmatrix} = \begin{vmatrix} 1+\dfrac{\partial \Delta x}{\partial x} & \dfrac{\partial \Delta x}{\partial y} \\[2mm] \dfrac{\partial \Delta y}{\partial x} & 1+\dfrac{\partial \Delta y}{\partial y} \end{vmatrix} \simeq 1+\dfrac{\partial \Delta x}{\partial x}+\dfrac{\partial \Delta y}{\partial y} \qquad ...[7.8]$$

and

$$\left. \begin{array}{l} \Delta_x = L \tan \epsilon_x \simeq L\epsilon_x, \\ \Delta_y = L \tan \epsilon_y \simeq L\epsilon_y, \end{array} \right\} \qquad ...[7.9]$$

substitution into [7.5] gives the result

$$\frac{I(x, y)-I^*(x, y)}{I^*(x, y)} \simeq L\left(\frac{\partial \epsilon_x}{\partial x}+\frac{\partial \epsilon_y}{\partial y}\right). \qquad ...[7.10]$$

Now, since

$$\frac{I(x, y)-I^*(x, y)}{I^*(x, y)} \simeq \frac{I(x, y)-I^*(x, t)}{I(x, y)} = \frac{\Delta I}{I}, \qquad ...[7.11]$$

using equations [6.1] and [7.10] we get, finally,

$$\frac{\Delta I}{I} \simeq \frac{L}{n_0} \int_0^l \left(\frac{\partial^2 n}{\partial x^2}+\frac{\partial^2 n}{\partial y^2}\right) dz, \qquad ...[7.12]$$

or

$$\frac{\Delta I}{I} \simeq -4{\cdot}46 \times 10^{-14}(L/cm)(\lambda/cm)^2 \int_0^l \left[\left(\frac{\partial^2 N_e}{\partial x^2}+\frac{\partial^2 N_e}{\partial y^2}\right) \middle/ cm^{-5}\right] dz/cm.$$
$$...[7.13]$$

This is the working equation for the shadowgraph, in the same sense that [4.18] and [6.16] are working equations for interferometry and the schlieren method respectively. In this case the sensitivity—once the length of the discharge tube l is given—is proportional to L. Once the measurement of $\Delta I/I$ has been made point-by-point across the shadowgraph field, it is possible, in principle, to go back to the number density by making a double integration. Apart from the possible errors, the two integration constants involved must be obtained from other experiments, one using the schlieren method and the other an interferometric measurement. This is a disadvantage of the method, but its primary use is not quantitative measurement.

Before reviewing the possible application, let us first consider some limitations of the method. By its very nature a shadow picture involves uncertainty in the contour lines due to Fresnel diffraction. The size Δx of the minimum resolvable region can be evaluated by means of the uncertainty principle (see Ref. 15, p. 587):

$$\Delta x \geqslant (\lambda L)^{\frac{1}{2}},$$

so that, for example, for $\lambda = 6943$ Å (ruby laser) and $L = 1$ m, $\Delta x \simeq 0.83$ mm. This applies not only to the contours of the discharge tube, a picture of this phenomenon being given in Plate 3, but also to those images arising from the focusing effect of the plasma distribution, as seen in Plate 4, which shows a shadowgraph taken at the maximum compression of a θ-pinch at 0.1 torr.[56] In this latter case one can no longer speak about pure Fresnel diffraction, but one is still concerned with the uncertainty principle. Of course, one way of reducing this inconvenience is to reduce either λ or L as far as is compatible with the necessity of distinguishing between atoms and electrons (in the case of λ), and with the required sensitivity (in the case of L). Reducing L has the drawback of very much increasing the amount of (unwanted) plasma light on the recorder; this can be avoided by using (Fig. 18) the optical arrangement where T_1 is the image of T made by the lens L_2.

Another limitation has to be considered, especially when the shadowgraph is applied to gas discharges; the beam of rays is not only "focused" by the presence of small density ripples, but it is also deflected by the mean density distribution acting as a lens or wedge. This effect is capable of changing the topology of the representation, as indicated in Fig. 4.

Applications of the shadowgraph are all related to the above-mentioned property of being sensitive to sharp changes in the refractive index. In plasma physics these occur in the case of shock-waves, instabilities and other "peculiar events", like turbulence. But up to the present time this is only a list of potential applications, because, in spite of the relative ease of this method, especially when use is made of a monopulsed ruby laser arranged to give a point source, only a few exploratory experiments by the author and his co-worker are known to the author.

In aerodynamics extensive use has been made of the shadowgraph in the study of shockwaves (thickness, shape of the shock front and its velocity and Mach-number). All this could be translated into the field of plasma physics without any great effort.

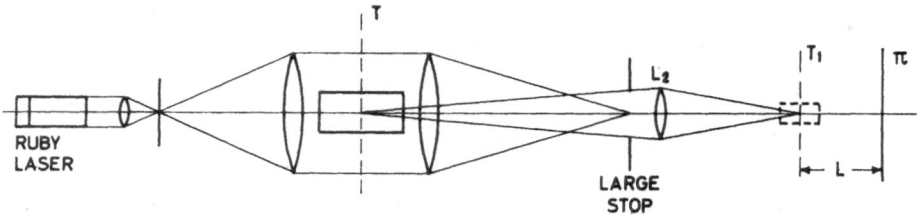

FIG. 18. A simple improvement in the shadowgraph technique.

8

APPLICATION OF HOLOGRAPHY IN OPTICAL PLASMA DIAGNOSTICS

After our review of the classical methods of optical plasma diagnostics, we will discuss the most recent technique, holography, which summarizes and complements the classical ones. This is not the occasion for a general explanation of the holographic technique,[57] and therefore, we shall limit ourselves to a description of the basic principles and those applications that are of interest in plasma physics.

To know the form of an object (for example its three-dimensional image) it is sufficient to know the front of the e.m. wave the object diffuses into the surrounding space when illuminated. Since the photographic plate is sensitive only to the square of the wave amplitude, one needs a way of registering the phase of this front to be able to record it; this is provided by interference. The light source S (Fig. 19) illuminates the object O: on the plate placed in π the wavefront coming from S and the one diffused by O overlap. If the wave coming from S is described by

$$U_0 = A_0 e^{i\psi_0} \qquad \qquad ...[8.1]$$

(A_0 and ψ_0 real) and the one coming from O by

$$U_1 = A_1 e^{i\psi_1}, \qquad \qquad ...[8.2]$$

the wave that arrives in π is

$$U = U_0 + U_1 = A_0 e^{i\psi_0} + A_1 e^{i\psi_1} = e^{i\psi_0}[A_0 + A_1 e^{i(\psi_1 - \psi_0)}]. \qquad ...[8.3]$$

The plate in π records, apart from a proportionality factor

$$I = A^2 = UU^* = A_0^2 + A_1^2 + 2A_0 A_1 \cos(\psi_1 - \psi_0), \qquad ...[8.4]$$

i.e. the interference between the primary and the diffracted beam is a measure of the phase $(\psi_1 - \psi_0)$. All this is possible if the radiation is coherent and monochromatic.

Once this photographic recording has been carried out, its reversal or

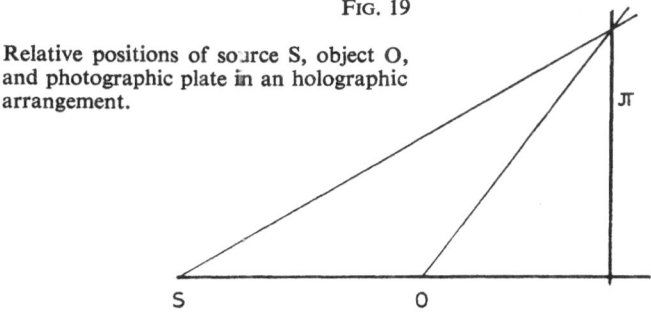

FIG. 19

Relative positions of source S, object O, and photographic plate in an holographic arrangement.

printing on transparent film, done with the right exposure, has a transparency t_p proportional to the intensity of the light that exposed the image:

$$t_p = kA^2. \qquad \qquad ...[8.5]$$

If we substitute this hologram for the plate π and remove the object, we will obtain diffracted radiation, when illuminating with S, given by

$$U_s = t_p U_0 = kA^2 U_0 = kA_0 e^{i\psi_0}[A_0^2 + A_1^2 + 2A_0 A_1 \cos(\psi_1 - \psi_0)]$$
$$= kA_0^2 e^{i\psi_0} \left[A_0 + \left(\frac{A_1}{A_0}\right)^2 + A_1 e^{i(\psi_1 - \psi_0)} + A_1 e^{-i(\psi_1 - \psi_0)} \right] \qquad ...[8.6]$$

having used the formula of Euler. The first and third term of [8.6] are the same as those appearing in [8.3], and therefore, after substituting in π the plate with the hologram, the radiation in π remains the same for as far as these two terms are concerned. The second term has the same phase as the radiation coming directly from S (intermodulation term) and causes no perturbation as generally $A_1 \ll A_0$.

The fourth term is equal to the third, but its phase has opposite sign. The function of the hologram hit by the spherical wave coming from S (Fig. 20), is evident. As would happen in the case of a transparent grating, the spherical wave coming from S, falling upon π, yields two diffracted spherical waves. If the hologram has had the right blackening, it behaves like a sinusoidal plate and yields therefore only two diffracted waves w_1, w_2, apart from the zero order w_0. Between them, w_1 and w_2 have opposite phases. Using the zone plate theory, we have that while w_1 diverges from O, w_2 converges on another point O', forming there an image of O; the two points O and O' are connected according to the law of thin lenses:

$$\frac{1}{O\pi} + \frac{1}{\pi O'} = \frac{1}{f} \qquad \qquad ...[8.7]$$

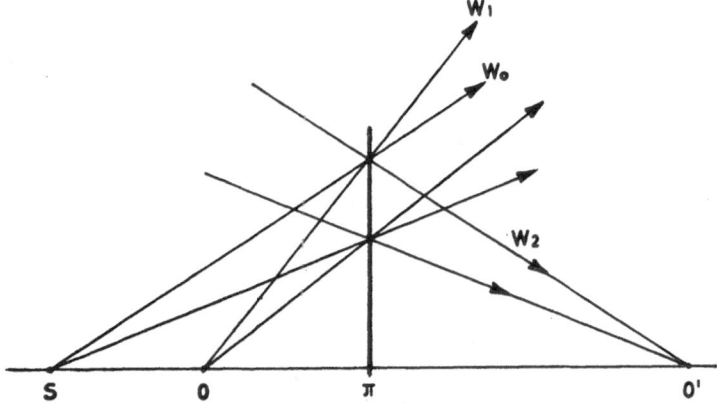

FIG. 20. Wavefront reconstruction.

where f is a function of the wavelength of the light used. This last property is used to change the enlargement of the reconstructed image.

Studying Fig. 20 it is obvious that the holographic method offers the possibility of obtaining a three dimensional picture of the scene in O. However, the simultaneous presence of the two images in the reconstruction process harms the unambiguousness of the information. This effect can easily be avoided by making sure that the beam of laser light which falls upon the object forms a rather large angle with the reference beam. In Fig. 21 a practical outline for a holograph device is shown.

An important aspect of holography is the resolving power of the system. A general characteristic of wave optics is the fact that the amount of data

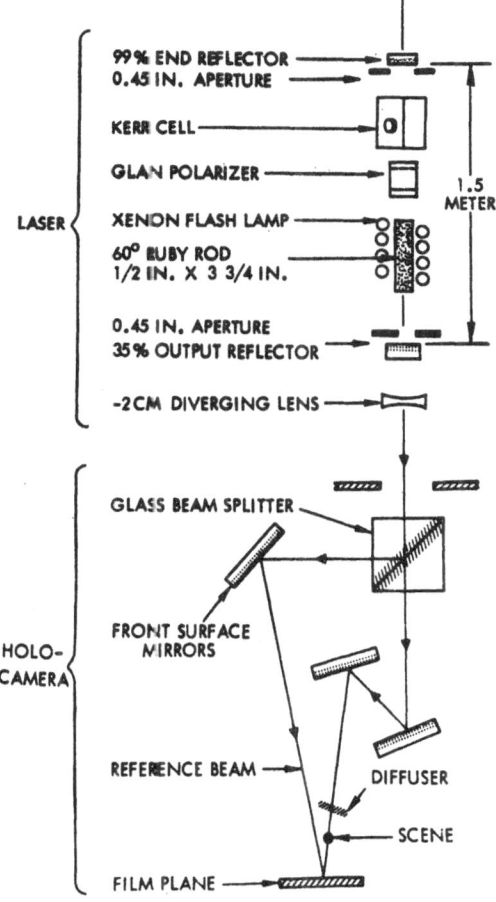

FIG. 21. Schema of apparatus used in the holographic photography of high speed phenomena (after R. E. Brooks *et al.*[68]).

given by a system of interference fringes depends directly on the number of fringes which effectively participate in the transfer of the information. As the optical system has to make use of a photographic plate, which at its best does not yet exceed 0·5 μ resolution, it is clear that the resolution of the system will be limited. In practice, this problem becomes important when the recording light is diffracted through large angles. The difficulty is the greater when making a hologram using the source and thus the object at small distances from the screen (giving a Fresnel transform hologram). However, when dealing with infinite distances (Fraunhofer diffraction) this difficulty is very much reduced, i.e. the number of fringes that can be registered is much larger. Think, for example, of the difference in aspect between the fringes of a zone plate and those of a Mach-Zehnder interferogram. This basic observation has led to the development of the high resolution Fourier-transform or Stroke-Fourier holography; in this case the images are at $\pm \infty$.

Apart from reconstruction of the image, which takes place by means of the first (± 1) orders, interferometry carried out, for example, with a Mach-Zehnder and a laser does not differ very much from this method. Holography supplements the interferometric techniques with the possibilities of reconstructing a three dimensional image and therefore of carrying out the schlieren and shadowgraph processes under ideal experimental conditions. However, all the activity in the field of holograms would be of little interest if holography had not also brought important by-products for plasma research. Here we want to mention holographic interferometry, Fourier transform spectroscopy with stationary interferometers, and, finally, the potential of holography in the study of turbulence.

8.1. Holographic interferometry

To obtain a hologram with strong spatial resolution, one needs a large number of fringes: practically, one comes close to the resolution limit of the plates. In the case of a plasma, it would be impossible to appreciate the fringe shifts caused by a ΔN_e. If we superimpose two holograms of the same object (e.g. a plasma tube), one taken in the presence and one in the absence of the plasma we obtain a "double exposure hologram" displaying the interference between the two fringe systems, which shows up only at those points of the image where the corresponding optical path has changed. This "moire" fringe system gives a "teinte plate" interference representation (see Section 4). Any imperfections in the optical system will remain unchanged and will therefore not appear in the resulting interferogram. This method of wiping out the imperfections of the optical system has already been used in interferometry.[67] It has the advantage over classical interferometry that it does not require optical windows of high quality, and that it is easily adjustable. It has the disadvantage, however, that it does require photographic plates with a high resolution, which are necessarily slower. The method begins to be used in plasma physics.

8.2. *Fourier-transform spectroscopy with stationary interferometers*

This application is a generalization of the holographic technique and combines its practical advantages with those of Fourier transform spectroscopy.[58]

The Fourier transform formulation of holographic imaging suggests an investigation of the extension of holography to spectroscopy.[59] We will give here only some information which can be understood intuitively, referring to Refs. 57 and 59. From [8.4], it follows that the superposition of two monochromatic coherent beams, forming between them a small angle θ, gives rise to interference fringes, which, if $A_1 = A_2$, are described as a function of the x axis of the photographic plate, by an expression of the type

$$I(x) = I_0[1 + \cos(2\pi x\theta/\lambda)]. \qquad \ldots[8.8]$$

The reconstruction process provides the image of the source. If the light of the source is not monochromatic, one will have, instead of [8.8],

$$I(x) = \int_0^\infty I_0(\sigma)(1 + \cos 2\pi\sigma x\theta)d\sigma$$

with $\sigma = 1/\lambda$, and the reconstruction process yields as many images of the light source as there are monochromatic components i.e. the spectrum of the source itself. This formulation is the same as that of Fourier transform spectroscopy[58] and the fact that it is realized with a single registration on the plate makes a direct reconstruction of the spectrum possible, without computing and with stationary interferometers. Of course the two methods vary in resolving power, sensitivity, etc.

8.3. *The statistical approach*

Optical methods based on refractivity discussed up to now make it possible to measure certain quantities (for example the plasma density) connected with refractivity, averaged along the trajectory of a light beam. However, the assumption that these quantities are constant along the trajectory, or have at least some form of symmetry, should be verified either directly or through the equation of Abel. This may be done for all manifestations of the plasma that show a certain regularity in the spatial distribution of the plasma density; e.g., the density distribution of a confined plasma or a shockwave.

As we have seen, the high resolving power of the optical instrumentation used does not in itself guarantee a perfect correspondence between a point of the photographic plate and a point of the plasma. This is essentially due to the bending of the trajectory when density gradients are present; in practice, it is almost impossible to distinguish the real phenomenon if we have a whole random distribution of density gradients along the trajectory.

This extreme case, which we will call "turbulence", is connected to phenomena which today enjoy great interest, and deserves a short discussion of its own, despite the fact that it is very complicated both from a theoretical and experimental point of view and has not yet received adequate attention from plasma researchers. In this case, we have to resort to a statistical description of the quantity we are interested in, for example N_e. Starting from the simplest case of a uniform plasma in a spatial region of volume V, and wanting to determine a hierarchy of information about N_e, we put in first order its average value $N_e = (1/V) \int_V N_e dV$, and in second order a quantity $G(\Delta P)$ (the autocorrelation) obtained by measuring N_e at a point P and at another point P_2 a distance ΔP from it. Taking the product and averaging over all points P_1 of the region of interest, we obtain: $G(\Delta P) = \langle N_e(P_1) N_e(P_2) \rangle$. Here ΔP can be defined with regard to space, time, or both. This quantity is a function of ΔP and not of P_1; it has obviously a maximum for $\Delta P = 0$. Our interest in this quantity is based on the fact that, if $N_e(P_1)$ and $N_e(P_2)$ are two stochastically completely independent variables, then $G(\Delta P) = 0$, while, where $G(\Delta P) \neq 0$ one has to conclude that $N_e(P_1)$ and $N_e(P_2)$ are subject to a common relationship, even if this is not strictly deterministic.

The distance ΔP^* within which we have still $G(\Delta P) \neq 0$ (the correlation length) is a characteristic of a given stochastical phenomenon, and is the first objective for a measurement in the domain of turbulent phenomena. The argument here can of course be generalized—we might consider measuring the average value of the quantity of interest in a group of n points (obtaining the nth-order correlation) and so describe the statistical behaviour of the medium in more and more detail. From the point of view of a description of a turbulent plasma, we shall have attained our goal when at least some elements of the hierarchy have been determined. Now we shall send a plane light wave into the medium sketched above. The electron density fluctuations, connected with the fluctuations of the refractive index of the medium, produce a diffraction of the incoming wave, and as a result, the wave function of the outgoing wave is relatively more or less modified in amplitude, phase and frequency. These modulations give a mapping of the random field corresponding to the turbulence. Therefore, examination of the statistical properties of these modulations is equivalent to determining the statistical characteristics of the turbulent medium. For this type of study light is again the ideal instrument.

We will now concern ourselves with how the statistical properties of the medium put their mark on the statistical characteristic of light, after the light has passed through the medium. However, we limit ourselves to modulations of the amplitude ($\langle \delta A \rangle$) and phase ($\langle \delta \phi \rangle$) of the wave, leaving out the variations in the frequency $\delta \nu$, which are of interest in scattering and which have been discussed here by Thompson and Ramsden. This is possible both conceptually and experimentally, as the geometrical conditions and the order of magnitude of the sensitivity required for the observations are well separated,

PLATE 1

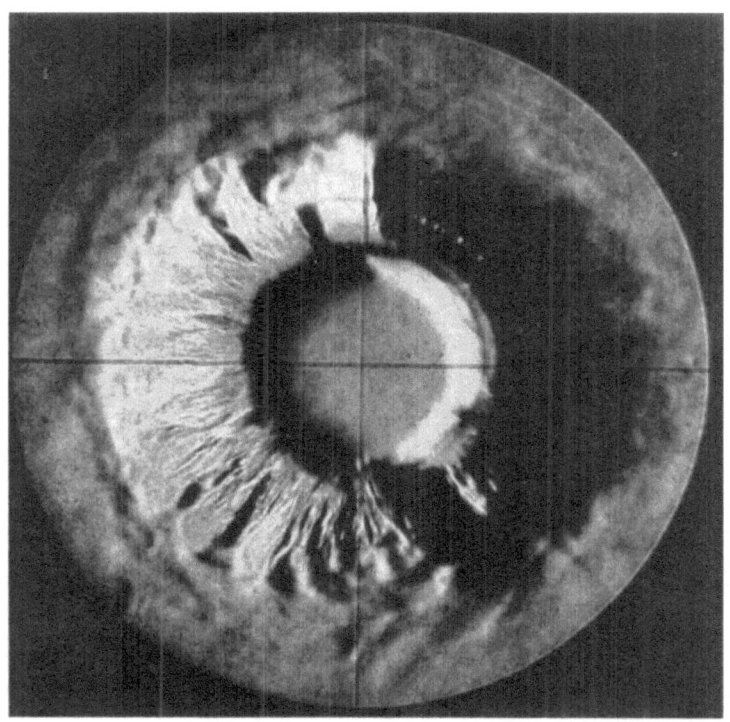

Toepler schlieren photograph of a theta-pinch discharge.

PLATE 2

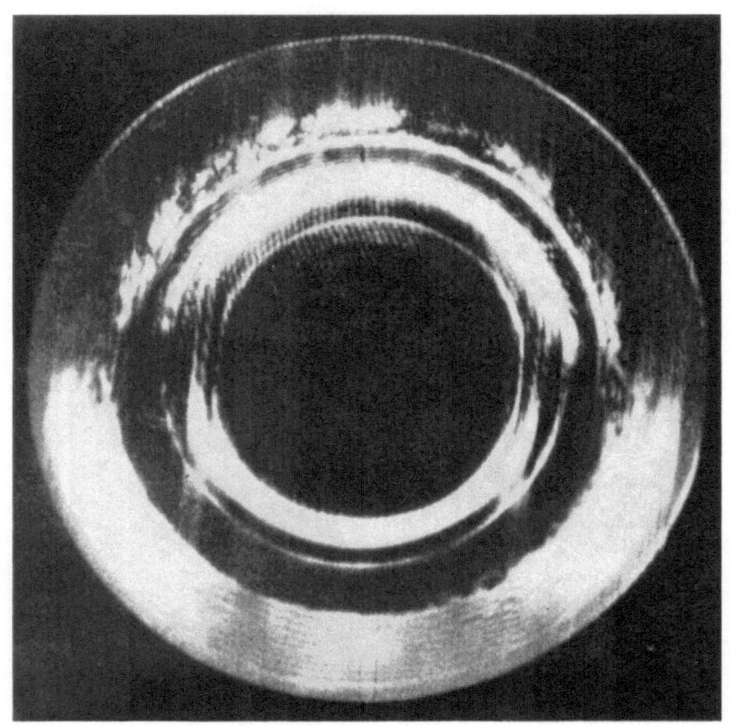

A Toepler schlieren picture taken with a ruby laser during a theta-pinch experiment showing the out-of-focus effect referred to in the text.

PLATE 3

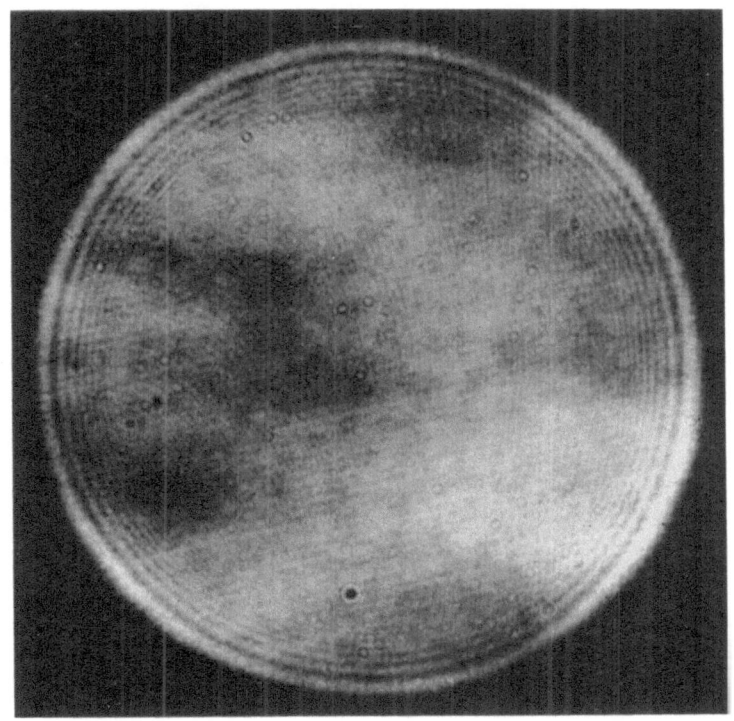

Diffraction effect of a laser beam.

PLATE 4

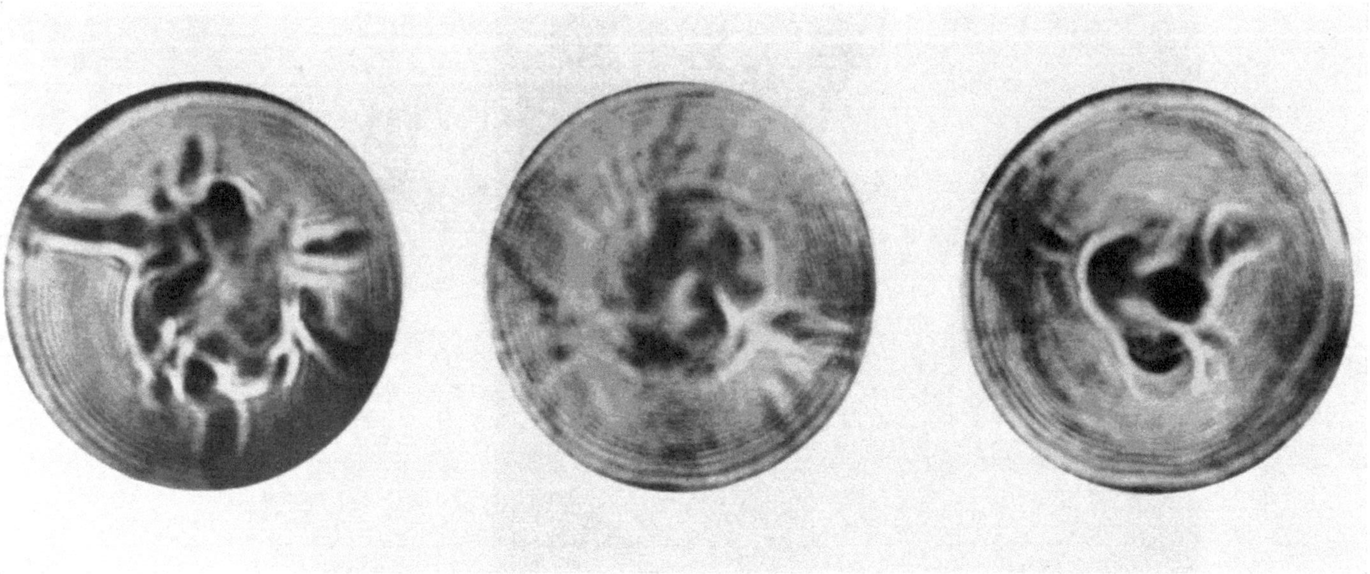

Shadowgraph of a pinch experiment at instants near the maximum of compression.

even though the phenomenon that causes them is unitary. We now have to find the algorithm that translates the statistics in δN (or δn) to those in δA, $\delta \phi$. The subject is rather involved, and an extensive series of studies has been devoted to it. They differ essentially in their assumptions about scaling (the dimensions of the region under study with respect to the correlation lengths of the plasma or the resolution time with respect to the fluctuation time), while special assumptions on the structure and nature of the random medium do not always appear in the descriptions. A large part of the literature treats these phenomena in view of the propagation of light in a turbulent atmosphere (for communications) and it is not yet clear how far this research can be applied directly to laboratory plasmas, where the presence of e.m. fields would automatically render the turbulent manifestations anisotropic. This, rather than being a disadvantage, could render the experimental study of such phenomena more fruitful, as it might provide, for once, useful indications for theoretical studies. It is however practicable to use a rather general approach, which connects a statistical representation of δn with those of $\delta \phi$ and δA, separating the two cases according to short or long observation time intervals taken with respect to the average evolution time of a density configuration.

Starting with the first, a very clear and general treatment is given by the work of H. Bremmer.[60] While the interested person will have to devote himself to a direct study of this remarkable train of thought, we must here limit ourselves to a recapitulation of it. To introduce the work of Bremmer in an intuitive way, we will consider a zone in space, limited by two planes $z = 0$ and $z = d$, filled with the medium under study (plasma) and assumed to be on average homogeneous. In this zone, the density is supposed to undergo fluctuations $\delta n(x, y, z)$ which have a random distribution, while we call l the average distance between the coordinates of the relative maxima of the density, which, due to our homogeneity assumption is independent of location. Consequently, the dimensions of the spatial zone in which one of these fluctuations takes place are smaller than, or equal to, l. Intuitively we conceive of "blobs" of particle density. Obviously we have $d \geqslant l$ and a similar inequality will be true for the transverse dimensions. We will now send a plane wave in the direction of the z-axis: this wave will be deformed. As a direct consequence of the crossing of the first blob, there will be phase-changes from point to point in the wavefront (with respect to an undisturbed passage), while the variations in amplitude will not be very large. The whole phenomenon can be described by means of Fresnel diffraction. As the wave continues its journey through the various blobs, the phase fluctuations between the various points of the front will diminish, while the trajectories of the elementary beams of light become more complicated and, due to the progressive growth of the beam divergence, the fluctuations of the wave amplitude increase. This long-distance behaviour is called Fraunhofer diffraction. It is obvious that the parameter determining the transition between the two zones is connected with the number of "blobs" passed. This number can be evaluated in the following way.

A light beam with wavelength λ, entering an inhomogeneous medium with lateral length l, will be diffracted with an angle of the order of λ/l and will form, at a distance h, a light cone with a base diameter of order $h\lambda/l$. This light cone completely passes through a blob in its path if the diameter of the blob is equal to the base diameter of the cone, i.e. if

$$\frac{h\lambda}{l} = l \quad \text{or} \quad h = \frac{l^2}{\lambda}.$$

Therefore

$$q = \frac{d}{h} = \frac{d\lambda}{l^2} \qquad \qquad ...[8.9]$$

is an estimate of the number of blobs of size l, passed through over a distance d by a radiation with wavelength λ. The medium, independent of time, is described by the refractive index

$$n(x, y, z) = 1 + \delta n(x, y, z),$$

and the propagation of the monochromatic wave $e^{i\kappa_0 z}$ is determined by

$$\nabla^2 U + \kappa_0^2 U = -2\kappa_0^2 \delta n(x, y, z) U, \qquad \qquad ...[8.10]$$

in which we have used Born's approximation, neglecting terms in δn beyond the first. This rather restrictive hypothesis is equivalent to saying that every light beam, along its trajectory, experiences at the most one elementary scattering action. It can be shown that as a consequence of this approximation no correlations that are of higher than second order can be thoroughly treated. However, a more refined treatment is possible.

The total field, the solution of [8.10] can be represented by

$$e^{i\kappa_0 d} + U_{sc} = e^{i\kappa_0 d}(1 + \delta A)e^{i\delta \psi}$$
$$\simeq e^{i\kappa_0 d}(1 + \delta A + i\delta \phi). \qquad \qquad ...[8.11]$$

The result obtained by Bremmer, in the small-distance approximation (assuming a small number of blobs), is

$$\delta \phi = \kappa_0 \int_0^d dz \delta n(0, 0, z) \qquad \qquad ...[8.12]$$

$$\delta A = -\tfrac{1}{2} \int_\delta^d dz(d-z) \left\{ \left(\frac{\partial^2}{\partial x^2} + \frac{\partial^2}{\partial y^2} \right) \delta n(x, y, z) \right\}_{x=y=0}. \qquad ...[8.13]$$

The equation [8.12] is of the well-known type on which interferometry is based, while in [8.13] we can recognize [7.13] (if we keep in mind that in [7.9] the origin of the deflection was placed at $z = 0$), and therefore it represents the lens effect caused by the blob of plasma. If we go to a distance such that several blobs can be included in the optical path, the equations we obtain have to contain an integration over the elementary effects [8.12] and [8.13] which are valid across each blob. In fact, one gets

$$\delta A + i\delta\phi = \frac{\kappa_0^2}{2\pi} \int_0^d \frac{dz}{d-z} \int \int_{-\infty}^{+\infty} dx\,dy\,\delta n(x, y, z) \exp\left[\tfrac{1}{2}i\kappa_0 \frac{(x^2+y^2)}{(d-z)}\right] \quad ...[8.14]$$

which can be shown to be the sum over the diffraction effects associated with the various elements of volume. In reality this reasoning is done, following Bremmer, in non-dimensional units, x/l, y/l, z/l, d/l, so as to be able to carry out the transition $q = 0$ to $q = \infty$.

Note that there is always a rather direct correspondence between δA, $\delta\dot\phi$ and δn. It is therefore possible to calculate directly from [8.12], [8.13] and [8.14] the corresponding moments $\langle \delta A(P_1)\, \delta A(P_2) \rangle$ or $\langle \delta\phi(P_1)\, \delta\phi(P_2) \rangle$. The formulae that then result contain on the right-hand side the expression $\langle \delta n(P_1)\, \delta n(P_2) \rangle$ thus establishing a direct link between the correlations of the optical quantities and those of the medium. Analytically we cannot go any further without explicitly knowing $\langle \delta n(P_1)\, \delta n(P_2) \rangle$. However, here some remarks have to be made. It is intuitively understood—and can be proved exactly—that if P_1 and P_2 are aligned in the direction of the z-axis, the resulting correlation distance is always d, independent of the properties of the medium. In fact, integration along z implies that the wavefront does not "remember" the structure of the fluctuations along this direction, giving in the end an average effect from which no statistical data can be deduced. It is not very reasonable to assume *a priori* that the fluctuations distribution in a plasma would be both homogeneous and isotropic, at least if we are talking about plasmas in which currents flow and which are subjected to magnetic fields. If, however, we assume that the anisotropy is in the z-direction, and renounce *a priori* information on this, we can make the assumption of isotropy in the x, y-plane. Without upsetting the general discussion, we can, for example, assume

$$\frac{\langle \delta n(P_1)\, \delta n(P_2) \rangle}{\langle \delta n^2 \rangle} = C\left(\frac{P_1 P_2}{l}\right) \qquad ...[8.15]$$

in which C denotes any function of the distance $P_1 P_2$, and the denominators represent certain suitable normalizations. We obtain thus

$$\left.\begin{array}{c} \langle \delta A(P_1)\,\delta A(P_2) \rangle \\[2mm] \langle \delta\Phi(P_1)\,\delta\Phi(P_2) \rangle \end{array}\right\} = \kappa_0^2 l d \langle \delta n^2 \rangle \left\{1 \mp \cos\left[\frac{q}{2\pi}\left(\frac{\partial^2}{\partial\alpha^2} + \frac{\partial^2}{\partial\beta^2}\right)\right]\right\}$$

$$\int_0^\infty dW\, C(\alpha^2 + \beta^2 + W^2)^{\frac{1}{2}} \qquad ...[8.16]$$

with

$$\alpha = \frac{x}{l}; \quad \beta = \frac{y}{l}; \quad W = \frac{z_1 - z_2}{l}.$$

From this we can see, apart from the different role played by the x, y coordinates with respect to those of z, that taking the sum of the two relations expressed in $[8.16]$ we obtain

$$\langle \delta A(\mathrm{P}_1)\, \delta A(\mathrm{P}_2)\rangle + \langle \delta\phi(\mathrm{P}_1)\, \delta\phi(\mathrm{P}_2)\rangle$$
$$= 2\kappa_0^2 ld\langle \delta n^2\rangle \int_0^\infty C dW = \kappa^2 \qquad\qquad ...[8.17]$$

which is a quantity independent of q. This result, which we had already foreseen as a consequence of the qualitative discussion above, is depicted in Fig. 22 for the case $\mathrm{P}_1 \equiv \mathrm{P}_2$ and is completely independent of what form the expression $C(\mathrm{P}_1\mathrm{P}_2)$ can assume. This also means that when q is very large, which is necessary to give a good statistical ensemble, both the measurements performed using the amplitude fluctuations of the electromagnetic field, and those using the phase fluctuations, have in principle the same sensitivity.

Having ascertained that there is a certain relationship between $\langle \delta A_1\, \delta A_2\rangle$ (or $\langle \delta\phi_1\, \delta\phi_2\rangle$) and $\langle \delta n_1\, \delta n_2\rangle$, for example the relations expressed in $[8.16]$ which are of the type $\langle \delta A_1\, \delta A_2\rangle = \mathscr{F}\,(\langle \delta n_1\, \delta n_2\rangle)$ (or $\langle \delta\phi_1\, \delta\phi_2\rangle = \mathscr{F}'\,(\langle \delta n_1\, \delta n_2\rangle)$), the only two problems that remain to be discussed to demonstrate the feasibility of a determination of $\langle \delta n_1\, \delta n_2\rangle$ are:

(i) to see how $\langle \delta A_1\, \delta A_2\rangle$ (or $\langle \delta\phi_1\, \delta\phi_2\rangle$) can be determined experimentally;

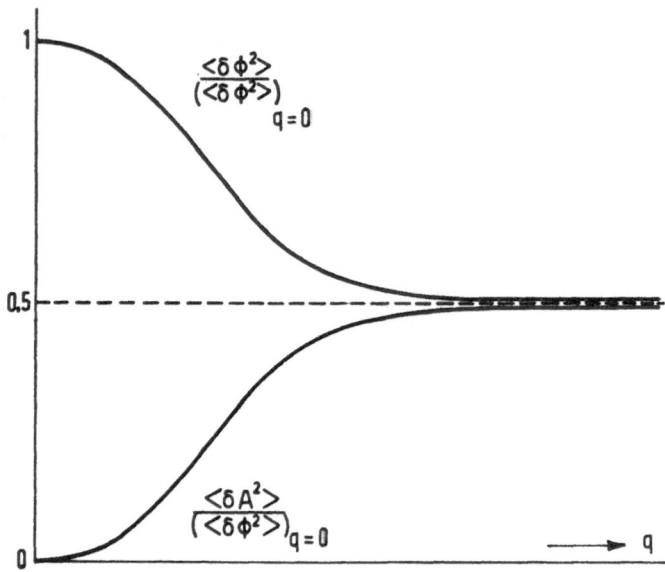

FIG. 22. Variation of the amplitude and phase fluctuations as functions of the transition parameter q.

(ii) to deduce $\langle \delta n_1 \, \delta n_2 \rangle$ from the theoretical formulations, (e.g. [8.16]) which can be written in the form

$$\langle \delta n_1 \, \delta n_2 \rangle = \mathscr{G}(\langle \delta A_1 \, \delta A_2 \rangle, \langle \delta \phi_1 \, \delta \phi_2 \rangle), \qquad \ldots [8.18]$$

where the experimental results from (a) are included via the functional $\mathscr{G}$. In practice the equations [8.16] must be solved for the unknown $\langle \delta n_1 \, \delta n_2 \rangle$, taking into account the definition [8.15].

The experimental determination of $\langle \delta A_1 \, \delta A_2 \rangle$ or $\langle \delta \phi_1 \, \delta \phi_2 \rangle$ is a modest example of optical data processing. There are several methods; but, as we do not want to stray too far from the subject, we will briefly examine only one connected with the shadowgraph and one connected with holography.

The simplest way to arrive at the first method is by examining the remarkable properties of the optical system shown in Fig. 23. In the focal plane of lens L_1 is placed a screen uniformly illuminated with an intensity I_0 (that means with an extended uniform light source); in the focal plane of L_2 the image, S_2, of S_1 is formed. T_1 and T_2 are two identical photographic plates, exactly aligned perpendicular to the optical axis, whose photographic transparency in the linear regime is given by $T(x, y)(\equiv T(x', y'))$. On S_2 the light, after having passed the two plates, has an intensity distribution $I(\xi'', \eta'')$ which we will demonstrate to be the autocorrelation function of $T(x, y)$. Due to the geometry of the optical system, those of the "∞^2" beams which pass the point (x', y') after having passed $T_1(x, y)$ will pass also through (ξ'', η'') if they satisfy the conditions

$$\frac{x' - x}{2t} = \frac{\xi''}{F}; \quad \frac{y' - y}{2t} = \frac{\eta''}{F}. \qquad \ldots [8.19]$$

As the transmittance of two elementary transparencies is equal to the product of the transmittance of each, the intensity $I(\xi'', \eta'')$ will be

$$I(\xi'', \eta'') = (I_0/A) \int\!\!\int_A T(x', y') T(x, y) \, dx \, dy, \qquad \ldots [8.20]$$

where A is the area of the collimated beam. Introducing [8.19] and putting $2t = F$ we have

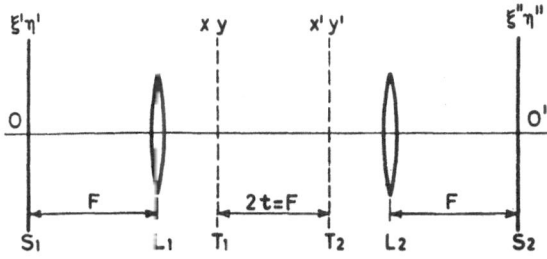

FIG. 23. Optical system to produce the correlations between the two transparencies T_1 and T_2.

$$\frac{I(\xi'', \eta'')}{I_0} = \frac{1}{A} \int\!\!\int_A T(x+\xi'', y+\eta'') T(x, y) dx dy$$
$$= (1/A)\langle T(x+\xi'', y+\eta'')T(x, y)\rangle, \qquad \ldots[8.21]$$

which is in agreement with our assumptions. The same result is obtained by taking advantage of the symmetry of the system and using a beam splitter. Instead of a smeared-out light source, we can use a point source and receive the signal $\mathscr{C}$ in O' by means of a photodiode. In this case, one has to let one of the transparencies (e.g. T_2) shift in its own plane (e.g. along x') to be able to obtain the correlation function $\mathscr{C}(x') [= \mathscr{C}(\xi'')]$. This system can be used if the eventual diffraction is not too great; otherwise one has to turn to somewhat more complicated systems in order to be able to use the zero-th order only.

Thus, if we now introduce in the system of Fig. 23 at T_1, T_2, two identical plates P_1, P_2, extracted from a shadowgraph, we shall have in S_2 an $I(\xi'', \eta'')$ which is the autocorrelation function of $T(x, y)$. We now want to see how a photographic registration carried out with an optical shadowgraph system can give us a mapping of $\langle \delta A_1 \delta A_2 \rangle$. The transfer function of the plate can be thought of as given by

$$t(P) \equiv t(x, y) = t_0 + \Delta t(x, y) \qquad \ldots[8.22]$$

(where t_0 refers to the background). The luminous intensity $I(P)$ on the plates is proportional to $t(P)$:

$$t(P) = \alpha I(P) = \alpha U(P) U^*(P), \qquad \ldots[8.23]$$

where $U(P)$ is the complex amplitude of the wavefront leaving the plasma and falling upon the photographic plates. The optical system of Fig. 23 gives

$$\langle t(P_1)t(P_2)\rangle = \alpha^2 \langle I(P_1)I(P_2)\rangle$$
$$= \alpha^2 \langle U(P_1)U^*(P_1)U(P_2)U^*(P_2)\rangle$$

and, since $U(P) \simeq e^{i\kappa_0 z}(1+\delta A+i\delta\phi)$,

$$\langle t(P_1)t(P_2)\rangle = \alpha\langle[(1+\delta A_1)^2+\delta\phi_1^2][(1+\delta A_2)^2+\delta\phi_2^2]\rangle$$
$$= \alpha(1+4\langle\delta A_1 \delta A_2\rangle+2\langle\delta A^2\rangle+2\langle\delta\phi)^2)$$
$$= \alpha \,.\, 4\langle\delta A_1 \delta A_2\rangle+\text{const.} \qquad \ldots[8.24]$$

Thus a shadowgraph combined with a suitable optical system is able in a first-order approximation to register the autocorrelation function of the amplitude of the wave leaving the plasma.

Let us now examine what information a hologram will give us. Because of its formal simplicity, we will refer to the holographic method in parallel light; and, using the symbols and figures whose meaning is by now clear, we will deduce the transfer function $t(P)$ of a transparency.

Referring to Fig. 24, the incident wave $e^{i\kappa_0 z}$ emerges from the plasma with a modulation in amplitude and phase according to $e^{i\phi(P)}A(P)e^{i\kappa_0 z}$. The reference beam which forms a very small angle θ with the latter, is described by $e^{i(\kappa_0 z + \alpha x)}$ where $\alpha = \kappa_0 \sin \theta \sim \kappa_0$. The total complex amplitude is therefore

$$U_{\text{tot}} = e^{i(\kappa_0 z + \alpha x)} + A(P) e^{i[\kappa_0 z + \phi(P)]}. \qquad \ldots [8.25]$$

Thus the photographic emulsion will be exposed to light of intensity

$$
\begin{aligned}
I(P) &= U_{\text{tot}} U_{\text{tot}}^* \\
&= [A(P) e^{i(\kappa_0 z + \phi)} + e^{i(\kappa_0 z + \alpha x)}][A(P) e^{-i(\kappa_0 z + \phi)} + e^{-i(\kappa_0 z + \alpha x)}] \\
&= A^2 + S e^{-i\alpha x} + S^* e^{i\alpha x} + 1, \qquad \ldots [8.26]
\end{aligned}
$$

where we have called

$$S = A_t e^{i\phi}. \qquad \ldots [8.27]$$

The transparency of the plates can be assumed to be $t = KI(P)$ with $I(P)$ given by [8.26]. In the evaluation or reconstruction process (Fig. 25), a plane wave $U_0 = e^{i\kappa_0 z}$ is incident on the plate. The emerging wave will be (using [8.23]):

$$
\begin{aligned}
U_x &= U_0 t = U_0 \kappa I(P) \\
&= \kappa e^{i\kappa_0 z}(A_1^2 + S e^{-i\alpha x} + S^* e^{i\alpha x} + 1), \qquad \ldots [8.28]
\end{aligned}
$$

which gives, as is known, the three diffracted beams:

(a) $\kappa S\, e^{i(\kappa_0 z + \alpha x)}$ of order $+1$;
(b) $\kappa(A_1^2 + 1)\, e^{i\kappa_0 z}$ (straight through) of zero-order;
(c) $\kappa S\, e^{i(\kappa_0 z - \alpha x)}$ of order -1.

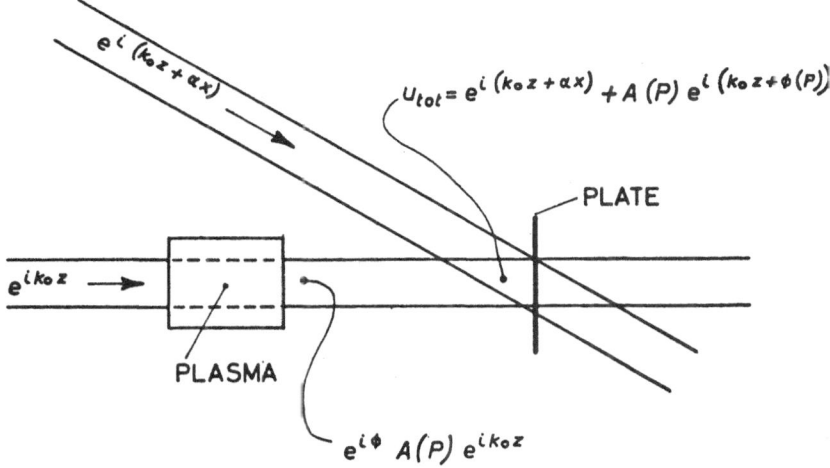

FIG. 24. Wavefronts of the hologram referred to in the text.

We now place across the beam (a) an identical transparency, which we will describe by $t = KI(P_2)$ and repeat the whole process. The incoming wave is now (a), and the outgoing wave will be

$$
\begin{aligned}
U_{sc} &= \kappa^2 S_1 \, e^{i(\kappa_0 z + ax)}[A_2^2 + 1 + S_2 e^{iax} + S_2^* e^{-iax}] \\
&= \kappa^2[(A_2^2 + 1)S_1 \, e^{i(\kappa_0 z + ax)} + S_1 S_2 \, e^{i(\kappa_0 z + 2ax)} + S_1 S_2^* \, e^{i\kappa_0 z}].
\end{aligned}
$$

$$...[8.29]$$

We will still have three diffracted beams

(aa) $\kappa^2 S_1 S_2 \, e^{i(\kappa_0 z + 2ax)}$ of order $+1$
(ab) $\kappa^2 S_1 (A_2^2 + 1) \, e^{i(\kappa_0 z + ax)}$ of zero order
(ac) $\kappa^2 S_1 S_2^* \, e^{i\kappa_0 z}$ of order -1

Analogously we can proceed for each of the orders given by the first transparency T_1. Each of these diffraction patterns: aa, ab, ac, ba, bb, bc, ca, cb, cc, can be subjected separately to a correlation analysis.

Among these the beams ac, bb, and ca deserve our attention. For example, for the correlation procedure on ac, keeping in mind that

$$
S = A \, e^{i\phi} \simeq 1 + \delta A + i\delta\phi,
$$

we have

$$
\begin{aligned}
U_{sc}(\langle ac \rangle) &= \langle [1 + \delta A_1 + i\delta\phi_1][1 + \delta A_2 - i\delta\phi_2] \rangle \\
&= 1 + \langle \delta A_1 \, \delta A_2 \rangle + \langle \delta\phi_1 \, \delta\phi_2 \rangle.
\end{aligned}
$$

$$...[8.30]$$

Taking into account that $\langle \delta A \rangle = \langle \delta\phi \rangle = 0$ and that $\langle \delta A_1 \, \delta\phi_2 \rangle = \langle \delta A_2 \delta\phi_1 \rangle$ the same is obtained for the pattern ca. Of course, what we obtain at the

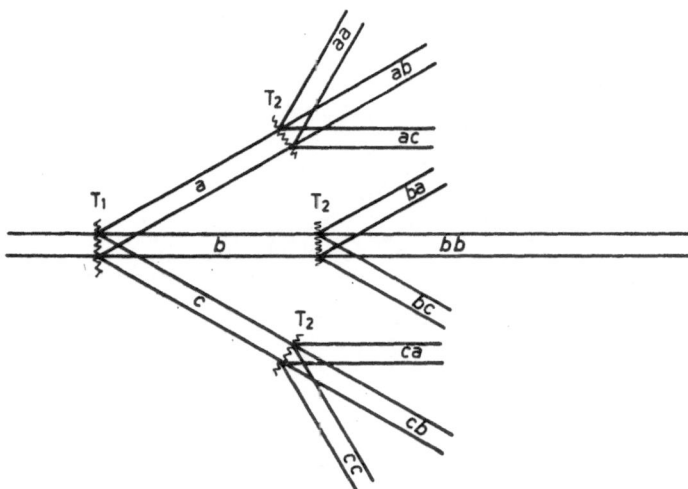

FIG. 25. Sketch of the correlation procedures referred to in the text.

correlation output is the square of $U_{sc}(\langle ac\rangle)$ but the important point is that $U_{sc}(\langle ac\rangle)$ is already a real quantity. However, as can be easily understood, the pattern bb reveals only autocorrelations $\langle \delta A_1 \, \delta A_2\rangle$, as is obvious. As we can see, comparing [8.30] with [8.17] and keeping in mind the observation made there, the holographic method is more complete and general than the shadowgraph in the study of statistical phenomena.

To wind up our investigation of the feasibility of this type of research, we should touch upon the experimental problems; two aspects can be distinguished here.

(1) The instrument readings are, in general, convolutions of the quantities we want to measure (because of instrumental effects, e.g. limited resolving power) and have the effect of smoothing out the observations.

(2) It often happens that what we measure is not directly the quantity desired but some functionally related magnitude, involving derivatives, integrals and the like (see, for example, [7.13] and the general problem posed by [8.18]).

The first problem, which is part of experimental method, is reduced by means of an integral equation (see, e.g., Refs. 61 and 62) to a knowledge of the instrumental effect. In the case of the optical systems that interest us, we can refer to an extensive series of studies devoted to the determination of what is called the transfer function.[63] In reality, if we consider as "the optical system" the whole of the instrumentation and the medium (which may, for example, be turbulent), the same technique used for determination of the transfer function can be applied to statistical study of the medium.[64] This line of approach is of interest only if we can rely on long observation times during which the macroscopic physical state of the medium does not change considerably.

The second problem, which is more mathematical in nature, has received a specific treatment with respect to the mapping and measurement of random fields by M. Uberoi and L. Kovasznay.[65] This is not a very up-to-date reference, and the availability of fast computers probably makes it possible to solve it in other ways. In fact, the autocorrelation process can also be worked out by a computer.

This section has been added, even though nothing—or very little—of what has been discussed has up to now been applied to plasma physics. The intention is only to indicate how many of the techniques that are today used in other fields of physics might be applied immediately to plasmas. We cannot end our discussion without noting how the most remarkable characteristic, viz. the coherence, of the most important ingredient of the experiments we have studied here—the laser—has not yet been fully exploited. Analogously, we have not yet spoken about higher than second order correlations in plasmas, or in transparent media in general. This is not without reason, as the coherence characteristics that e.m. radiation which is initially perfectly coherent assumes when passing through a medium are closely linked to the statistics of the medium itself. This fascinating study is now being fully developed in the field of atmospheric turbulence; but it also seems very promising in the field of plasma physics.[66] Unfortunately, however, this research requires a technological involvement which for the moment is far from the usual activity of plasma research laboratories.

REFERENCES

1. O. H. THEIMER and L. S. TAYLOR. 1960. *Ann. Phys. (N.Y.)*, **11**, 377.
2. V. P. SILIN. 1962. *Sov. Phys. JETP*, **14**, 617.
3 J. A. WHEELER. 1933. *Phys. Rev.*, **43**, 258.
4. H. A. KRAMERS. 1924. *Nature, Lond.*, **113**, 673 and **114**, 310.
5. R. LADENBURG. 1933. *Rev. Mod. Phys.*, **5**, 243.
6. W. C. ALLEN. 1955. *Astrophysical Quantities*, p. 86. Athlone Press.
7. R. B. BIRD, C. F. CURTISS and J. O. HIRSCHFELDER. 1954. *Molecular Theory of Gases and Liquids*, chapters 12–13. J. Wiley and Sons Inc., New York; Chapman and Hall Ltd., London.
8. R. A. ALPHER and D. R. WHITE. 1959. *Physics Fluids*, **2**, 153 and 162.
9. L. H. THOMAS and K. UMEDA. 1956. *J. Chem. Phys.*, **24**, 1113.
10. J. M. BARTOS and F. D. BENNET. 1957. *Ball. Res. Lab. Report No. 1027.*
11. F. D. BENNET, V. E. BERGDOLT and W. C. CARTER. 1952. *J. Appl. Phys.*, **23**, 453.
12. R. LADENBURG. J. WINCKLER and C. C. VAN VOORHIS. 1948. *Phys. Rev.*, **73**, 1359.
13. F. J. WEYL. 1945. *Nav. Ord. Rept.*, 211–45.
14. M. E. MASCART. 1889. *Traité d'Optique*. Gauthier Villards, Paris.
15. H. WOLTER. 1956. Schlieren-Phasecontrast and Lichtschnittverfahren. *Handbuch der Physik, Band XXIV*. Springer-Verlag, Berlin-Göttingen, Heidelberg.
16. M. BILTZ. 1933. *Phys. Zeit.*, **34**, 200.
17. J. W. BEAMS, A. R. KUHLTHAU, A. C. LAPSLEY, J. H. QUEEN, L. B. SNODDY and W. D. WHITEHEAD. 1947. *J. Opt. Soc. Amer.*, **37**, 868.
18. F. D. BENNET. 1951. *J. Appl. Phys.*, **22**, 184 and *J. Appl. Phys.*, **22**, 776.
19. J. A. FITZPATRICK, J. C. HUBBARD and W. J. THALL. 1950. *J. Appl. Phys.*, **21**, 1269.
20. L. S. G. KOVASZNAY. 1949. *Rev. Sci. Instrum.*, **20**, 696.
21. L. H. TANNER. 1956. *Proc. 3rd Int. Congr. on High-Speed Photography* (R. B. COLLINS ed.). Butterworths Scientific Publications, London.
22. H. FISHER. 1961. *J. Opt. Soc. Am.*, **51**, 5.
23. P. A. FRANKEN, A. E. HILL, C. W. PETERS and G. WEINREICH. 1961. *Phys. Rev. Lett.*, **7**, 118.
24. D. A. KLEINMAN. 1962. *Phys. Rev.*, **128**, 1761.
25. P. D. MAKER, R. W. TERHUNE, M. NISENOFF and C. M. SAVAGE. 1962. *Phys. Rev. Lett.*, **8**, 21.
26. E. B. TEMPLE. 1957. *J. Opt. Soc. Amer.*, **47**, 91.
27. F. D. BENNET and G. D. KAHL. 1953. *J. Opt. Soc. Am.*, **43**, 71.
28. V. E. BERGDOLT. 1949. *Ball. Res. Lab. Report 692.*
29. G. HANSEN. 1940. *Z. Instrumkde*, **60**, 325.
30. W. KINDER. 1946. *Optik*, **1**, 413.
31. R. W. LADENBURG. 1954. *Physical Measurements in Gas Dynamics and Combustion*, p. 39. Princeton University Press, New York.
32. L. MACH. 1892. *Z. Instrumkde*, **12**, 89.
33. H. SCHARDIN. 1933. *Z. Instrumkde*, **53**, 396.
34. J. WINCKLER. 1948. *Rev. Sci. Instr.*, **19**, 307.
35. L. ZEHNDER. 1891. *Z. Instrumkde*, **11**, 275.
36. P. P. IGENBERG. 1963. Interferometrische Messung der Elektronendichte beim Theta pinch. Diplomarbeit München Universität.

37. C. Candler. 1951. *Modern Interferometry*, pp. 126–7. Hilger & Watts.
38. J. Dyson. 1963. *Appl. Optics*, 487.
39. B. M. Leadon and F. D. Weener. 1953. *Rev. Sci. Instr.*, **24**, 121.
40. R. Kraushaar. 1950. *J. Opt. Soc. Amer.*, **40**, 480.
41. U. Ascoli-Bartoli and S. Martellucci. 1962. *Rapporto Interno CNEN RT/FI (62)*, 62.
42. D. Post. 1954. *J. Opt. Soc. Amer.*, **44**, 243.
43. D. Post. 1958. *J. Opt. Soc. Amer.*, **48**, 309.
44. W. Primak. 1958. *J. Opt. Soc. Amer.*, **48**, 375.
45. U. Ascoli-Bartoli, S. Martellucci and E. Mazzucato. 1964. *Nuovo Cimento*, **32**, 298.
46. P. G. R. King and G. J. Stewart. 1963. *New Scientist*, **17**, 180.
47. D. E. T. F. Ashby and D. F. Jephcott. 1963. *Appl. Phys. Lett.*, **3**, 13.
48. J. B. Gerardo and J. T. Verdeyen. 1963. *Appl. Phys. Lett.*, **3**, 13.
49. J. B. Gerardo, J. T. Verdeyen and M. A. Gusinow. 1965. *J. Appl. Phys.*, **36**, 2146.
50. H. Schardin. 1942. *Ergeb. Exakt. Naturwiss.*, **20**, 303.
51. H. J. Shafer. 1949. *J. Soc. Mot. Pict. Engnrs.*, **53**, 524.
52. U. Ascoli-Bartoli and S. Martellucci. 1963. *Nuovo Cimento*, **27**, 475.
53. O. Lamm. 1938. *Nova Acta Regiae Soc. Sci. Upsaliensis*, **10**, 1.
54. W. Thiele. 1968. *Veb. Carl Zeiss Jena Nachrichten*, 8 Folge, Heft 2.
55. U. Ascoli-Bartoli, S. Martellucci and E. Mazzucato. 1963. *VIème Conf. Inter. sur les Phén. d'Ionis. dans les Gas* (Paris), **4**, 41.
56. U. Ascoli-Bartoli, S. Martellucci and E. Mazzucato. 1963. *Ibid.*, p. 97.
57. G. W. Stroke. 1966. *An Introduction to Coherent Optics and Holography*. Academic Press, New York.
58. J. Connes. 1961. *Rev. Optique*, **40**, 45, 116, 171, 231.
59. G. W. Stroke and A. T. Funkhouser. 1965. *Phys. Letters*, **16**, 272.
60. H. Bremmer. 1964. Semi-geometric optical approaches to scattering phenomena. *Proc. of the Symposium on Quasi Optics*, p. 415. Polytechnic Press of the Polytechnic Inst. of Brooklyn, New York.
61. H. C. Van d. Hulst. 1941. *Bull. Astr. Netherlands*, **9**, 225.
62. H. C. Van d. Hulst. 1946. *Bull. Astr. Netherlands*, **10**, 75.
63. R. Barakat and R. F. van Ligten. 1965. *Appl. Optics*, **4**, 749.
64. R. E. Hufnagel and N. R. Stanley. 1964. *J. Opt. Soc. Am.*, **54**, 52.
65. M. S. Uberoi and L. S. G. Kovasznay. 1953. *Quarterly of Applied Mathematics*, **X**, 375.
66. F. De Marco and L. Pieroni. 1968. *Investigation of Turbulent Plasmas by Light Scattering*. L.G.I. 68/15.
67. H. I. Ashkenas and A. E. Bryson. 1951. *J. Aero. Sci.*, **18**(2), 82.
68. R. E. Brooks, L. O. Heflinger, R. F. Wuerker and R. A. Briones. 1965. *Appl. Phys. Lett.*, **7**, 92.